P9-CBG-971

Springer Tracts in Natural Philosophy

Volume 23

Hans J. Stetter

Analysis of Discretization Methods for Ordinary Differential Equations

With 12 Figures

Springer-Verlag Berlin Heidelberg New York 1973

Hans J. Stetter
Institut für Numerische Mathematik, Technische Hochschule Wien
A-1040 Wien/Austria

AMS Subject Classifications (1970): 34 A 50, 39 A 10, 65 J 05, 65 L 05, 65 Q 05

ISBN 3-540-06008-1 Springer-Verlag Berlin Heidelberg New York
ISBN 0-387-06008-1 Springer-Verlag New York Heidelberg Berlin

 Library of Congress Catalog Card Number 72-90188. Printed in Germany. Monophoto typesetting and offset printing: Zechnersche Buchdruckerei, Speyer. Bookbinding: Konrad Triltsch, Graphischer Betrieb, Würzburg.

Dem Andenken meines Vaters gewidmet

Preface and Introduction

Due to the fundamental role of differential equations in science and engineering it has long been a basic task of numerical analysts to generate numerical values of solutions to differential equations. Nearly all approaches to this task involve a "finitization" of the original differential equation problem, usually by a projection into a finite-dimensional space. By far the most popular of these finitization processes consists of a reduction to a difference equation problem for functions which take values only on a grid of argument points. Although some of these *finite-difference methods* have been known for a long time, their wide applicability and great efficiency came to light only with the spread of electronic computers. This in turn strongly stimulated research on the properties and practical use of finite-difference methods.

While the theory or partial differential equations and their discrete analogues is a very hard subject, and progress is consequently slow, the initial value problem for a system of first order ordinary differential equations lends itself so naturally to discretization that hundreds of numerical analysts have felt inspired to invent an ever-increasing number of finite-difference methods for its solution. For about 15 years, there has hardly been an issue of a numerical journal without new results of this kind; but clearly the vast majority of these methods have just been variations of a few basic themes. In this situation, the classical textbook by P. Henrici has served as a lighthouse; it has established a clear framework of concepts and many fundamental results. However, it appears that now—10 years later—a further analysis of those basic themes is due, considering the immense productivity during this period.

It is the aim of this monograph to give such an analysis. This text is *not* an introduction to the use of finite-difference methods; rather it assumes that the reader has a knowledge of the field, preferably including practical experience in the computational solution of differential equations. It has been my intention to make such a reader aware of the structure of the methods which he has used so often and to help him understand their properties. I really wanted—as the title of the book indicates—to *analyze* discretization methods in general, and particular classes of such methods, *as mathematical objects*. This point of view forced me to neglect to a considerable extent the practical aspects of solving differential equations numerically. (Fortunately, a few recent textbooks—noteably Gear [6] and Lambert [3]—help to fill that gap.)

Another restriction—which is not clear from the title—is the strong emphasis on *initial value problems* for ordinary differential equations. Only in the first chapter have I sketched a *general* theory of the mathematical objects which I have called "discretization methods". This part is applicable to a wide variety of procedures which replace an infinitesimal problem by a sequence of finite-dimensional problems. The remainder of the text refers to systems of first order differential equations only. Originally, I had intended to consider both initial and boundary value problems and to develop parallel theories as far as possible (using the concept of stability also in the boundary value problem context); but the manuscript grew too long and I had to restrict myself to initial value problems.

Within these severe restrictions I hope to have covered a good deal of material. The presentation attempts to make the book largely self-contained. More important, I have used a consistent notation and terminology throughout the entire volume, as far as feasible. At a few places, this has perhaps led to a "twist" in presentation; but I felt it worthwhile to adhere to a common frame of reference, since the exposition of such a frame has been one of the main motivations for writing this book. The numerous examples are never meant as suggestions for practical computation but as illustrations for the theoretical development.

As mentioned above, Chapter 1 deals with the *general structure of discretization methods*. Particular emphasis has been placed on the theory of asymptotic expansions and their application. It appears that they may also be used with various other approximation procedures for infinitesimal problems, procedures which are normally not considered as discretization methods but which may be fitted into our theory. (To facilitate such interpretations I have used a sequence of integers as discretization parameters in place of the stepsizes.) A section on "error analysis" attempts to outline several important aspects of error evaluations in a general manner.

Chapter 2 is devoted to special features of discretization methods for *initial value problems* for ordinary differential equations. The first two sections present mainly background material for an analysis which involves the well-known limit process $h \to 0$: a fixed finite interval of integration is subdivided by grids with finer and finer steps. The last section of Chapter 2, however, deals with a different limit process: the interval of integration is extended farther and farther while the stepsize remains fixed. It is to be expected that a theory of this limit process "$T \to \infty$"

may serve as a basis for an understanding of the behavior of discretization methods on long intervals, with relatively large steps, in the same way that the $h \to 0$ theory has proved to be a good model for the case of small steps on relatively short intervals.

The remainder of the treatise analyzes particular classes of methods. The two most distinctive features of such methods, the "multistage" and the "multistep" features, are treated separately at first. Thus, in Chapter 3, we consider *one-step methods* which "remember" only the value of the approximate solution at the previous gridpoint but use this value in a computational process which runs through a number of stages, with re-substitutions in each stage (this process may also be strongly implicit). Butcher's abstract algebraic theory of such processes permits a rather elegant approach to various structural investigations, such as questions of equivalence, symmetry, etc. The theory of asymptotic expansions provides a firm basis for the analysis of both the local and the global discretization error. The last section of Chapter 3 is devoted to the $T \to \infty$ limit process; here a number of results are derived and directions for further research are outlined. A number of intuitive conjectures are shown to be false without further assumptions.

In Chapter 4, we analyze discretization methods which use values of the approximate solution at several previous gridpoints but do not permit re-substitutions into the differential equations, i.e., *linear multistep methods.* After an exposition of the wellknown accuracy and stability theory for such methods, we outline a theory of "cyclic" multistep methods which employ different k-step procedures in a cyclic fashion as the computation moves along (as suggested by Donelson and Hansen). The investigations of Gragg on asymptotic expansions for linear multistep methods have been presented and supplemented by a general analysis of the symmetric case. Again the last section of the chapter has been reserved for the $T \to \infty$ theory.

After these preparations, we discuss *general multistage multistep methods* in Chapter 5. This class of methods is so wide that we have concentrated on the analysis of important subclasses, such as predictor-corrector methods, hybrid methods, and others. In agreement with the intentions of the book it appeared relevant to point out typical restrictions which permit farther-reaching assertions. New results concern the principal error function of general predictor-corrector methods and an extension of the concept of effective order to multistep methods. Cyclic methods are taken up again in a more general context.

Methods which explicitly use *derivatives of the right hand side* of the differential equation have been analyzed in the first section of Chapter 6; this class includes power series methods, Lie series methods, Runge-

Kutta-Fehlberg methods and the like. The *Nordsieck-Gear multivalue approach* did not fit into the pattern of Chapter 5; I have tried to give a consistent account of its theory. Last but not least I have included an analysis of *extrapolation methods* which have proved to be most powerful in practical computation; a particular effort was devoted to a clarification of the stability properties of such methods.

Naturally it has not been possible to achieve full coverage of the field. In particular, it seemed premature to force a rigid terminology upon developments which may not yet have found their final form. For this reason I have not included an account of the "variable coefficient" approach of Brunner [1], Lambert [1], and others, although it appears to be very promising. Also, the use of spline functions was omitted together with that of various non-polynomial local approximants (see, e. g., Gautschi [2], Lambert and Shaw [1], Nickel and Rieder [1]). Finally, I did not dare to propose a refined theory of "stiff methods" at this time, although I hope that my theory of strong exponential stability will provide one of the bases for such a theory.

A few remarks concerning the *bibliography* are necessary. I had originally planned to include a comprehensive bibliography of more than a thousand entries. However, the use of such a bibliography in the text would necessarily have been restricted to endless enumerations and thus not have been very helpful. Therefore, I decided to restrict my references essentially to those publications whose results I either quoted without proof or whose arguments I followed exceedingly closely. Thus the bibliography in this book is quite meager and contains—besides a number of "classics"—only a strangely biased selection from the relevant literature. I sincerely hope that the many colleagues whose important contributions I have not quoted will appreciate this reasoning and be reassured by the discovery of how many other important papers have not been quoted. Clearly, the ideas of innumerable papers have influenced the contents of this book though they are not explicitly mentioned.

Similarly, my thoughts about the subject could only mature through personal contact and discussion with many of the prominent numerical analysts all over the world. It would be unfair to enumerate a few of the many names; all colleagues whom I had a chance to ask about their views on discretization methods are silent co-authors of this text, and I wish to thank them accordingly.

The manuscript was prepared between the summers of 1969 and 1971. It could not have been written had I not been able to spend the year 1969/70—without much teaching obligation—at the University of California at San Diego under a National Science Foundation Senior Fellowship. More than two thirds of the manuscript originated during that pleasant year in Southern California for which I have to thank above all the chairman of the UCSD Mathematics Department, my friend Helmut Roehrl. The remainder of the manuscript I somehow managed to complete at weekends, evenings, etc. during the following year, despite my many duties in Vienna, and during a "working vacation" with my friends at the University of Dundee.

The typing of Chapters 1 through 4 was done by Lillian Johnson at UCSD, that of Chapters 5 and 6 by Christine Grill at the Technical University of Vienna. Miss Grill also re-typed and rere-typed many pages and did a lot of editorial work on the manuscript. Further editorial work and proofreading was done by a number of my young colleagues at Vienna, principally by W. Baron, R. Frank, and R. Mischak. To all these I wish to express my gratitude.

Another word of thanks is due to my friend and colleague Jack Lambert at Dundee. Although he was hard-pressed by the preparation of a manuscript himself, he kindly read the entire text and corrected many of my blunders with regard to the use of the English language. Further improvements in style were suggested by Kenneth Wickwire. (I ventured to write this book in English because it will be more easily read in poor English than in good German by 90% of my intended readers.)

The co-operation I have received from Springer-Verlag has been most pleasing. The type-setting and the production of the book are of the usual high Springer quality.

Finally, I wish to give praise to my dear wife Christine who suffered, without much complaint, severe restrictions of our home life during more than 2 years. At the same time, her loving care and reassurance helped me to get on with the work. If there is just one person in the world who will rejoice in the completion of this book it will be she.

Vienna, June 72

Hans J. Stetter

Table of Contents

Chapter 1

General Discretization Methods

In this introductory chapter we consider general aspects of discretization methods. Much of the theory is applicable not only to standard discretization methods for ordinary differential equations (both initial and boundary value problems) but also to a great variety of other numerical methods as indicated in the Preface (see also the end of Section 1.1.1). It should be emphasized that this is also true for the material on asymptotic expansions and their applications, although we have not elaborated on this. The chapter is concluded by a few remarks on the practical aspects of "solving" ordinary differential equations by discretization methods.

1.1 Basic Definitions

1.1.1 Discretization Methods

Def. 1.1.1. Throughout, the problem whose solution is to be approximated by a discretization method will be called the *original problem.* It is specified by a triple $\{E, E^0, F\}$ where E and E^0 are Banach-spaces and $F\colon E \to E^0$, with 0 in the range of F. A *true solution* of the original problem is an element $z \in E$ such that

$$Fz = 0. \tag{1.1.1}$$

We will always assume that a true solution of the original problem *exists* and is *unique* (the uniqueness may often be obtained by suitably restricting the domain of F); we will not consider existence and uniqueness questions for the original problem.

Since we are dealing with the numerical solution of (real) ordinary differential equations, E und E^0 will normally be spaces of continuous functions from an interval in $\mathbb{R}$ to $\mathbb{R}^s$ and the mapping F will be defined by a differential operator. Initial and/or boundary conditions are included in the definitions of E, F or the domain of F as seems appropriate.

Example.
$$E=C^{(1)}[0,1], \quad \text{with } \|y\|_E := \max_{t\in[0,1]} |y(t)|;$$

$$E^0=\mathbb{R}\times C[0,1], \quad \text{with } \left\|\begin{pmatrix} d_0 \\ d \end{pmatrix}\right\|_{E^0} := |d_0| + \max_{t\in[0,1]} |d(t)|;$$

(1.1.2)
$$Fy=\begin{pmatrix} y(0)-z_0 \\ y'(t)-f(y(t)) \end{pmatrix} \in E^0 \quad \text{for } y\in E,$$

where $z_0\in\mathbb{R}$, $f\in C(\mathbb{R}\to\mathbb{R})$ (f Lipschitz-continuous) are the data of the problem.

The true solution of the original problem thus specified (viz. the solution of the differential equation $y'=f(y)$, with initial condition $y(0)=z_0$) exists and is unique.

The basic idea of a discretization method is to replace the original problem by an infinite sequence of finite-dimensional problems each of which can be solved "constructively" in the sense of numerical mathematics. The replacement is to be such that the solutions of these finite-dimensional problems approximate, in a sense to be defined, the true solution z of the original problem better and better the further one proceeds in the sequence. Thus, one can obtain an arbitrarily good approximation to z by taking the solution of a suitably chosen problem in the sequence.

Of course, it will generally not be possible to obtain the solution of this finite-dimensional problem with arbitrary accuracy on a given computing tool and with a given computational effort (see Section 1.5). Nevertheless, the construction of discretization methods commonly follows the reasoning in the above paragraph.

Def. 1.1.2. *A discretization method* $\mathfrak{M}$ (*applicable to a given original problem* $\mathfrak{P}=\{E, E^0, F\}$) consists of an infinite sequence of quintuples $\{E_n, E_n^0, \Delta_n, \Delta_n^0, \varphi_n\}_{n\in\mathbb{N}'}$ where
E_n and E_n^0 are finite-dimensional Banach spaces;
$\Delta_n\colon E\to E_n$ and $\Delta_n^0\colon E^0\to E_n^0$ are linear mappings with

$$\lim_{n\to\infty} \|\Delta_n y\|_{E_n} = \|y\|_E \quad \text{for each fixed } y\in E,$$
$$\lim_{n\to\infty} \|\Delta_n^0 d\|_{E_n^0} = \|d\|_{E^0} \quad \text{for each fixed } d\in E^0;$$

and $\varphi_n\colon (E\to E^0)\to(E_n\to E_n^0)$, with F in the domain of all φ_n.

$\mathbb{N}'$ is an infinite subset of $\mathbb{N}$.

Def. 1.1.3. A *discretization* $\mathfrak{D}$ is an infinite sequence of triples $\{E_n, E_n^0, F_n\}_{n\in\mathbb{N}''}$ where E_n and E_n^0 are finite-dimensional Banach spaces; $F_n\colon E_n\to E_n^0$.

A *solution of the discretization* $\mathfrak{D}$ is a sequence $\{\zeta_n\}_{n\in\mathbb{N}''}$, $\zeta_n\in E_n$, such that

(1.1.3)
$$F_n\zeta_n=0, \quad n\in\mathbb{N}''.$$

$\mathbb{N}''$ is again an infinite subset of $\mathbb{N}$.

Def. 1.1.4. The discretization $\mathfrak{D}=\{\bar{E}_n, \bar{E}_n^0, \bar{F}_n\}_{n\in\mathbb{N}''}$ is called the *discretization of the original problem* $\mathfrak{P}=\{E, E^0, F\}$ *generated by the discretization method* $\mathfrak{M}=\{E_n, E_n^0, \Delta_n, \Delta_n^0, \varphi_n\}_{n\in\mathbb{N}'}$ if $\mathfrak{M}$ is applicable to $\mathfrak{P}$ and

$$\mathbb{N}''\subset\mathbb{N}',$$
$$\left.\begin{array}{ll}\bar{E}_n=E_n, & \bar{E}_n^0=E_n^0\\ \bar{F}_n=\varphi_n(F) & \end{array}\right\}\quad \text{for } n\in\mathbb{N}''.$$

In this case $\mathfrak{D}$ is denoted by $\mathfrak{M}(\mathfrak{P})$.

Remark. In the following, we will assume throughout (without loss of generality) that $\mathbb{N}''=\mathbb{N}'$ and that the sequences $\{E_n\}$ and $\{E_n^0\}$ of $\mathfrak{M}$ and $\mathfrak{M}(\mathfrak{P})$ resp. are identical.

Furthermore, we will always assume that the dimensions of E_n and E_n^0 are the same. This is a trivial necessary condition for the existence of a unique ζ_n which satisfies (1.1.3).

Example. Consider the original problem of the Example following Def. 1.1.1. The well-known Euler method (polygon method) may be characterized as a discretization method applicable to this problem as follows:

For $n\in\mathbb{N}'=\mathbb{N}$, let $\mathbb{G}_n:=\{\nu/n,\ \nu=0(1)n\}$ and

$$E_n=(\mathbb{G}_n\to\mathbb{R}),\quad \text{with } \|\eta\|_{E_n}:=\max_{\nu=0(1)n}\left|\eta\left(\frac{\nu}{n}\right)\right|;$$

$$E_n^0=(\mathbb{G}_n\to\mathbb{R}),\quad \text{with } \|\delta\|_{E_n^0}:=|\delta(0)|+\max_{\nu=1(1)n}\left|\delta\left(\frac{\nu}{n}\right)\right|;$$

$$(\Delta_n y)\left(\frac{\nu}{n}\right)=y\left(\frac{\nu}{n}\right)\quad \text{for } y\in E;$$

$$(\Delta_n^0 d)\left(\frac{\nu}{n}\right)=\begin{cases} d_0, & \nu=0,\\ d\left(\frac{\nu-1}{n}\right), & \nu=1(1)n,\end{cases}\quad \text{for } d=\begin{pmatrix} d_0\\ d(t)\end{pmatrix}\in E^0;$$

$$[\varphi_n(F)\eta]\left(\frac{\nu}{n}\right)=\begin{cases}\eta(0)-z_0, & \nu=0,\\ \dfrac{\eta\left(\frac{\nu}{n}\right)-\eta\left(\frac{\nu-1}{n}\right)}{1/n}-f\left(\eta\left(\frac{\nu-1}{n}\right)\right), & \nu=1(1)n.\end{cases}\tag{1.1.4}$$

The existence of a unique solution sequence $\{\zeta_n\}$ of the discretization is trivial since the values of the ζ_n are defined recursively.

A discretization method $\mathfrak{M}$ applicable to the original problem $\mathfrak{P}$ not only generates a discretization $\mathfrak{D}=\mathfrak{M}(\mathfrak{P})$ of $\mathfrak{P}$ but establishes at the same time, through its mappings Δ_n and Δ_n^0, relations between the spaces of $\mathfrak{P}$ and its discretization $\mathfrak{M}(\mathfrak{P})$ which permit an interpretation of $\mathfrak{M}(\mathfrak{P})$ and its solution $\{\zeta_n\}$ as an approximation of $\mathfrak{P}$ and its true solution z. This will be elaborated in the following sections.

The relations between the spaces and mappings involved in the original problem and its discretization can be visualized by the following diagram:

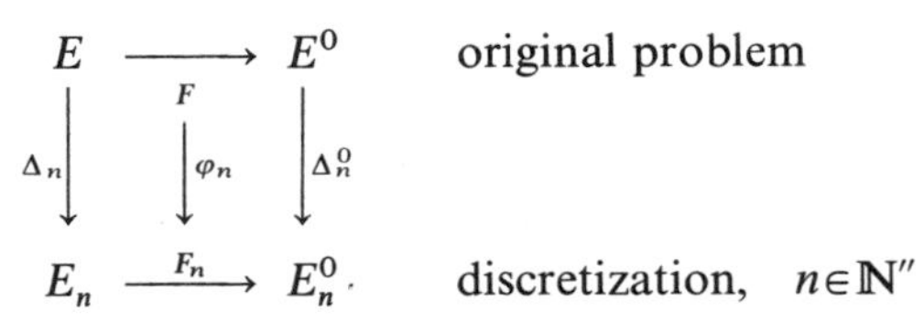

Fig. 1.1. Relation between the elements of a discretization method

The existence and uniqueness of a true solution z for the original problem does not automatically imply the existence and uniqueness of ζ_n which satisfy (1.1.3). Therefore, we will explicitly establish the unique solvability of $F_n \zeta_n = 0$ for many of the discretization methods under discussion. Furthermore, we will prove a result (Theorem 1.2.3) which guarantees the existence of a unique ζ_n for all sufficiently large n for reasonable discretization methods. For many classes of problems and discretization methods, the unique solvability of (1.1.3) is trivial.

As in the example following Def. 1.1.4, in our applications the spaces E_n and E_n^0 will normally be spaces of functions mapping discrete and finite subsets of intervals (so-called "grids") into $\mathbb{R}^s$. The mappings Δ_n and Δ_n^0 will "discretize" the continuous functions of E and E^0 into grid functions. The mappings $F_n = \varphi_n(F)$ will be defined by difference operators relative to the grids.

Def. 1.1.1—1.1.4 also admit essentially different interpretations. Consider, e.g., the original problem of the previous example, with all functions arbitrarily many times differentiable, and choose

$$E_n = \{\text{polynomials of degree } \leq n\};$$

$$\Delta_n y = \text{polynomial in } E_n \text{ obtained by the truncation of the formal power series of } y \in E;$$

$$E_n^0 = \mathbb{R} \times E_{n-1};$$

$$\Delta_n^0 \begin{pmatrix} d_0 \\ d(t) \end{pmatrix} = \begin{pmatrix} d_0 \\ \Delta_{n-1} d(t) \end{pmatrix};$$

$$F_n \eta = \begin{pmatrix} \eta(0) - z_0 \\ \eta'(t) - \Delta_{n-1} f(\eta(t)) \end{pmatrix}, \quad \text{for } \eta \in E_n.$$

For the generated discretization, (1.1.3) consists of a system of $n+1$ equations for the $n+1$ coefficients of ζ_n which is considered as an approximation to $\Delta_n z$.

A number of the general concepts and results also make sense for a variety of such unorthodox "discretizations" (e.g. Ritz and Galerkin's methods for boundary value problems). Also it is immediately clear that most results apply to standard discretizations of partial differential equations, at least when the side conditions are not too complicated. However, in the more specialized parts of this treatise we will restrict ourselves exclusively to genuine discretizations of ordinary differential equation problems.

1.1.2 Consistency

To be useful as a tool for obtaining an approximation to the solution z of (1.1.1), a discretization method should generate a discretization which approximates the original problem in the following sense:

Def. 1.1.5. A discretization method $\mathfrak{M}=\{E_n, E_n^0, \Delta_n, \Delta_n^0, \varphi_n\}_{n\in\mathbb{N}'}$ applicable to the original problem $\mathfrak{P}=\{E, E^0, F\}$ is called *consistent with* $\mathfrak{P}$ *at* $y\in E$ if y is in the domain of F and of $\varphi_n(F)\Delta_n$, $n\in\mathbb{N}'$, and

(1.1.5)[1]
$$\lim_{\substack{n\to\infty\\ n\in\mathbb{N}'}} \|\varphi_n(F)\Delta_n y-\Delta_n^0 F y\|_{E_n^0}=0\,.$$

$\mathfrak{M}$ is called *consistent with* $\mathfrak{P}$ if it is consistent with $\mathfrak{P}$ at each $y\in E$.

If $\mathfrak{M}$ is consistent with $\mathfrak{P}$ (at y) the discretization $\mathfrak{M}(\mathfrak{P})$ is also called consistent with $\mathfrak{P}$ (at y).

Def. 1.1.6. In the situation of Def. 1.1.5, $\mathfrak{M}$ and $\mathfrak{M}(\mathfrak{P})$ are called *consistent* with $\mathfrak{P}$ *of order* p at y if

(1.1.6)
$$\|\varphi_n(F)\Delta_n y-\Delta_n^0 F y\|_{E_n^0}=O(n^{-p}) \quad \text{as } n\to\infty\,.$$

In connection with the order of consistency, the reference "at z" may be omitted.

Example. As in Section 1.1.1. For any $y\in C^{(1)}[0,1]$, we have from (1.1.2) and (1.1.4).

$$[\varphi_n(F)\Delta_n y-\Delta_n^0 F y]\left(\frac{\nu}{n}\right)=\begin{cases}(y(0)-z_0)-(y(0)-z_0), & \nu=0,\\[2mm] \dfrac{y\left(\frac{\nu}{n}\right)-y\left(\frac{\nu-1}{n}\right)}{1/n}-f\left(y\left(\frac{\nu-1}{n}\right)\right) & \\ \qquad -\left[y'\left(\frac{\nu-1}{n}\right)-f\left(y\left(\frac{\nu-1}{n}\right)\right)\right], & \nu=1(1)n,\end{cases}$$

$$=\begin{cases}0, & \nu=0,\\[2mm] y'(\tilde{t}_\nu)-y'\left(\dfrac{\nu-1}{n}\right), & \nu=1(1)n, \quad \text{with } \tilde{t}_\nu\in\left(\dfrac{\nu-1}{n},\dfrac{\nu}{n}\right).\end{cases}$$

[1] *All* limit processes $n\to\infty$ carry the restriction $n\in\mathbb{N}'$ which we will omit from now on.

As

$$\max_{\nu=1(1)n} \left| y'(\tilde{t}_\nu) - y'\left(\frac{\nu-1}{n}\right) \right| \to 0 \quad \text{as} \quad n \to \infty$$

for any $y \in C^{(1)}$, the Euler methods is consistent with our original problem $\mathfrak{P}$. If z' is Lipschitz-continuous, our discretization method is consistent of order 1 with $\mathfrak{P}$.

Consistency at y requires the asymptotic (as $n \to \infty$) commutativity of the diagram

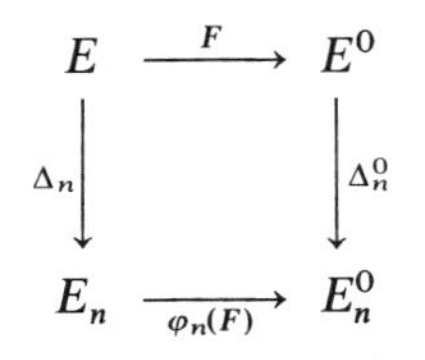

Fig. 1.2. Consistency diagram

for $y \in E$. Note that the element $(\varphi_n(F)\Delta_n - \Delta_n^0 F)y$ is in a *different* space E_n^0 for each n.

(1.1.5) also serves to make the relationship between Equation (1.1.1) and its discretization (1.1.3) unique, at least asymptotically as $n \to \infty$. Both (1.1.1) and (1.1.3) could otherwise be pre-multiplied by arbitrary regular linear operators which would not affect their solutions z and ζ_n respectively. In particular, (1.1.3) could be multiplied by some power of n, which is not possible if consistency is to be maintained; cf. (1.1.5).

Example (as above). If we had defined the mappings φ_n in the Euler method as

$$[\varphi_n(F)\eta]\left(\frac{\nu}{n}\right) = \begin{cases} \eta(0) - z_0, & \nu = 0, \\ \eta\left(\frac{\nu}{n}\right) - \eta\left(\frac{\nu-1}{n}\right) - \frac{1}{n} f\left(\eta\left(\frac{\nu-1}{n}\right)\right), & \nu = 1(1)n, \end{cases}$$

the method would not have been consistent with the original problem by our definition of consistency.

To obtain a more quantitative account of the approximative power of a discretization method we consider the *residual* generated by $\Delta_n z$ in (1.1.3):

Def. 1.1.7. Let the discretization method $\mathfrak{M} = \{E_n, E_n^0, \Delta_n, \Delta_n^0, \varphi_n\}_{n \in \mathbb{N}'}$ be consistent with the original problem $\mathfrak{P}$, with true solution z. The sequence $\{l_n\}_{n \in \mathbb{N}'}$, $l_n \in E_n^0$, with

$$l_n := \varphi_n(F)\Delta_n z, \qquad n \in \mathbb{N}', \tag{1.1.7}$$

is called the *local discretization error of* $\mathfrak{M}$ *and* $\mathfrak{M}(\mathfrak{P})$ *for* $\mathfrak{P}$.

By Def. 1.1.6 and 1.1.7, the local discretization error of $\mathfrak{M}$ for $\mathfrak{P}$ is $O(n^{-p})$ as $n \to \infty$ if $\mathfrak{M}$ is consistent of order p with $\mathfrak{P}$. With regard to the order of consistency, the normalization enforced by (1.1.5) becomes particularly important. The asymptotic behavior of $\{l_n\}$ would be meaningless if we could multiply φ_n with powers of n.

Example (as above). Since $z'(t)=f(z(t))$ and $z(0)=z_0$

$$l_n\left(\frac{v}{n}\right)=\begin{cases} z(0)-z_0 \\ \dfrac{z\left(\frac{v}{n}\right)-z\left(\frac{v-1}{n}\right)}{1/n}-f\left(z\left(\frac{v-1}{n}\right)\right)\end{cases}=\begin{cases} 0, & v=0, \\ \dfrac{1}{2n}z''(\tilde{t}_v), & v=1(1)n,\end{cases}$$

$\tilde{t}_v\in\left(\frac{v-1}{n},\frac{v}{n}\right)$, if $z\in C^{(2)}[0,1]$.

1.1.3 Convergence

We have stated that the motivation for discretizing a given problem lies in the fact that the solution sequence of the discretization is expected to approximate arbitrarily well the true solution of the original problem. We will now clarify how this approximation is to be understood.

Since the solution z of the original problem and the elements ζ_n of the solution $\{\zeta_n\}_{n\in\mathbb{N}'}$ of its discretization are elements of different spaces, they cannot be immediately related. There are two essentially different ways to effect a comparison:

1. To define with the discretization method a sequence of mappings $\nabla_n: E_n\to E$, with $\Delta_n\nabla_n=I$ (identity), and to consider the closeness of $\nabla_n\zeta_n$ and z in E. Intuitively, this amounts to interpolating the values of ζ_n defined on the grid and considering the approximation of z on the whole interval of integration.

2. To compare $\Delta_n z$ and ζ_n in E_n for each $n\in\mathbb{N}'$. Intuitively, this amounts to considering the approximation of z by ζ_n on the corresponding grid only.

Since the E_n are finite-dimensional (with the dimension normally growing with n) while E is infinite-dimensional, the choice of the "interpolation mappings" ∇_n in Approach 1 has to be arbitrary to a large extent; but it may strongly influence the size of the error $\nabla_n\zeta_n-z$. (This is the case even when the E_n are subspaces of E; there is no need to choose $\nabla_n=I$!) For this reason, the second approach has been used almost universally in the literature and we will also restrict ourselves to it.

From this point of view, the first approach may be formulated thus: Given the sequence of errors $\zeta_n-\Delta_n z\in E_n$ how should one choose the $\nabla_n: E_n\to E$ as to produce a minimal deviation $\nabla_n\zeta_n-z$ in E? The solution of this problem is independent of the way in which the ζ_n are generated and is the concern of the theory of interpolation or approximation.

Def. 1.1.8. Consider the discretization method $\mathfrak{M}=\{E_n, E_n^0, \Delta_n, \Delta_n^0, \varphi_n\}_{n\in\mathbb{N}'}$ applicable to the original problem $\mathfrak{P}$ with true solution z. Let the discretization $\mathfrak{M}(\mathfrak{P})$ possess a unique solution sequence $\{\zeta_n\}_{n\in\mathbb{N}'}$. The sequence $\{\varepsilon_n\}_{n\in\mathbb{N}'}$, with

$$(1.1.8) \qquad \varepsilon_n := \zeta_n - \Delta_n z \in E_n, \qquad n\in\mathbb{N}',$$

is called the *global discretization error of* $\mathfrak{M}$ (and $\mathfrak{M}(\mathfrak{P})$) for $\mathfrak{P}$.

Def. 1.1.9. In the situation of Def. 1.1.8, both $\mathfrak{M}$ and $\mathfrak{M}(\mathfrak{P})$ are called *convergent* for $\mathfrak{P}$ if

$$(1.1.9) \qquad \lim_{n\to\infty} \|\varepsilon_n\|_{E_n} = 0.$$

$\mathfrak{M}$ and $\mathfrak{M}(\mathfrak{P})$ are called *convergent of order p* for $\mathfrak{P}$ if

$$(1.1.10) \qquad \|\varepsilon_n\|_{E_n} = O(n^{-p}) \quad \text{as } n\to\infty.$$

Remark. According to our definition, convergence implies the existence and uniqueness of the solution ζ_n of (1.1.3). If unique ζ_n exist only for all sufficiently large $n\in\mathbb{N}'$, say for $n\geq n_0$, we will always replace $\mathbb{N}'$ by the set $\{n\in\mathbb{N}', n\geq n_0\}$.

Example. As in Section 1.1.1 and 1.1.2. In (1.1.2), let $f(y)=gy$, $g=\text{const}$. Then $z(\nu/n)=z_0\exp(g\nu/n)$ while the solution ζ_n of (1.1.4) is given by $\zeta_n(\nu/n)=z_0(1+g/n)^\nu$. Thus

$$\varepsilon_n\left(\frac{\nu}{n}\right) = z_0\left[\left(1+\frac{g}{n}\right)^\nu - \exp\left(g\frac{\nu}{n}\right)\right], \qquad \nu=0(1)n,$$

$$\|\varepsilon_n\|_{E_n} = |\varepsilon_n(1)| = |z_0|\left|\left(1+\frac{g}{n}\right)^n - e^g\right|$$

$$= \frac{g^2}{2n}|z_0|e^g + O(n^{-2}) \quad \text{as } n\to\infty.$$

Hence, the Euler method is convergent of order 1 for our problem.

Obviously, convergence asserts the asymptotic commutativity of the diagram

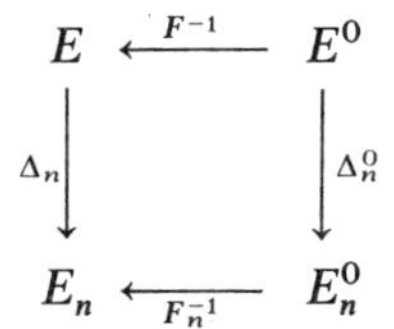

Fig. 1.3. Convergence diagram

for the zero element of E^0, assuming the existence of the inverse mappings. Or—as Watt [1] points out—convergence of $\{E_n, E_n^0, F_n\}_{n\in\mathbb{N}'}$ for $\{E, E^0, F\}$ is equivalent to consistency at 0 of $\{E_n^0, E_n, F_n^{-1}\}_{n\in\mathbb{N}'}$ for $\{E^0, E, F^{-1}\}$.

1.1.4 Stability

It may seem that a discretization method which is consistent with a given problem is automatically convergent for it. That this is not the case has been known for a long time (see Dahlquist [1], Richtmyer [1], and many others).

The examples given in these references show clearly that consistency at z is not sufficient for convergence.

The reason may be seen by comparing (1.1.7) and (1.1.8):

$$-\varepsilon_n = \Delta_n z - \zeta_n = F_n^{-1} F_n \Delta_n z - F_n^{-1} 0 = F_n^{-1} l_n - F_n^{-1} 0 .$$

The vanishing of l_n as $n \to \infty$ trivially implies the vanishing of ε_n if and only if the F_n^{-1} satisfy a Lipschitz condition *uniformly in n.*

Another view of the relation between consistency and convergence is the following: The global discretization error may be considered as the effect of solving $F_n \zeta_n = 0$ instead of $F_n \bar{\zeta}_n = l_n$ which—under the assumption of a unique solution—would yield $\bar{\zeta}_n = \Delta_n z$ by virtue of (1.1.7). In order that ε_n decrease with l_n we have to require that the relative effect of a perturbation of (1.1.3) can be bounded *independently of n.* Since perturbation insensitivity is denoted by "stability" almost universally in mathematics it has become customary in the treatment of discretization methods to use this term in the above situation.

Def. 1.1.10. Consider a discretization $\mathfrak{D} = \{E_n, E_n^0, F_n\}_{n \in \mathbb{N}'}$ and a sequence $\eta = \{\eta_n\}_{n \in \mathbb{N}'}$, $\eta_n \in E_n$. $\mathfrak{D}$ is called *stable at* η if there exist constants S and $r > 0$ such that, *uniformly* for all $n \in \mathbb{N}'$,

$$\|\eta_n^{(1)} - \eta_n^{(2)}\|_{E_n} \leq S \|F_n \eta_n^{(1)} - F_n \eta_n^{(2)}\|_{E_n^0} \tag{1.1.11}$$

for all $\eta_n^{(i)}$, $i = 1, 2$, such that

$$\|F_n \eta_n^{(i)} - F_n \eta_n\|_{E_n^0} < r . \tag{1.1.12}$$

S and r are called *stability bound* and *stability threshold*, resp.

Note that the definition of the stability of a discretization at some sequence η makes no reference to an original problem.

Def. 1.1.11. Consider a discretization method $\mathfrak{M}$ applicable to an original problem $\mathfrak{P}$ with true solution z. If the discretization $\mathfrak{M}(\mathfrak{P})$ is stable at $\{\Delta_n z\}$ both $\mathfrak{M}$ and $\mathfrak{M}(\mathfrak{P})$ are called *stable for* $\mathfrak{P}$.

Remarks. 1. Note that the important aspect of Def. 1.1.10 is not the existence of an estimate (1.1.11) for each n but its uniformity for $n \in \mathbb{N}'$, i.e., the fact that S and r may be chosen *independently of n.*

2. For discretizations $\{E_n, E_n^0, F_n\}$ with

$$F_n \eta_n = \hat{F}_n \eta_n + c_n, \qquad \textit{linear}, \quad c_n \in E_n^0, \tag{1.1.13}$$

stability requires simply the existence and uniform boundedness of the sequence of linear operators $\hat{F}_n^{-1}, n \in \mathbb{N}'$. There is *no dependence on* $\{\eta_n\}$ and $r = \infty$ always.

For a discretization $\{E_n, E_n^0, F_n\}_{n \in \mathbb{N}'}$ which is stable at $\{\eta_n\}$ the following argument is valid: Assume given an element $\tilde{\eta}_n \in E_n$; if it is known that $\|F_n \tilde{\eta}_n - F_n \eta_n\| \le \delta < r$, then $\|\tilde{\eta}_n - \eta_n\| \le S\delta$. The same estimate holds for arbitrarily large n.

This is the typical argument for which stability is used; see also the paragraph preceding Def. 1.1.10. It is therefore more natural to have (1.1.12) as a limit for the "domain" of stability than

(1.1.14) $$\|\eta_n^{(i)} - \eta_n\|_{E_n} < R,$$

which might first come to mind.

Actually, it will turn out (see Corollary 1.2.2) that stability under a condition (1.1.14)—with slight continuity assumptions—implies the existence of a stability threshold (1.1.12); the converse is not true, however. Thus, the stability requirement of Def. 1.1.10 is the weaker of the two possibilities, a fact which also makes it the preferable choice.

Example. As in previous sections. Assume that f in (1.1.2) possesses a uniform Lipschitz constant L:

(1.1.15) $$|f(x_1) - f(x_2)| \le L|x_1 - x_2| \quad \text{for arbitrary } x_i \in \mathbb{R}.$$

Choose $\eta_n^{(1)}, \eta_n^{(2)} \in E_n$ and let $\delta_n := F_n \eta_n^{(1)} - F_n \eta_n^{(2)}$, with F_n from (1.1.4). Then $\varepsilon_n := \eta_n^{(1)} - \eta_n^{(2)}$ satisfies

$$\varepsilon_n(0) = \delta_n(0),$$

and

$$\varepsilon_n\left(\frac{v}{n}\right) = \varepsilon_n\left(\frac{v-1}{n}\right) + \frac{1}{n}\left[f\left(\eta_n^{(1)}\left(\frac{v-1}{n}\right)\right) - f\left(\eta_n^{(2)}\left(\frac{v-1}{n}\right)\right)\right] + \frac{1}{n}\delta_n\left(\frac{v}{n}\right), \qquad v = 1(1)n.$$

Hence, by (1.1.15),

$$\left|\varepsilon_n\left(\frac{v}{n}\right)\right| \le \left|\varepsilon_n\left(\frac{v-1}{n}\right)\right|\left(1 + \frac{L}{n}\right) + \frac{1}{n}\left|\delta_n\left(\frac{v}{n}\right)\right|, \qquad v = 1(1)n,$$

which implies

$$\begin{aligned} \|\eta_n^{(1)} - \eta_n^{(2)}\|_{E_n} &= \max_{v=0(1)n}\left|\varepsilon_n\left(\frac{v}{n}\right)\right| \\ &\le |\delta_n(0)|\left(1 + \frac{L}{n}\right)^n + \frac{1}{n}\sum_{v=1}^{n}\left|\delta_n\left(\frac{v}{n}\right)\right|\left(1 + \frac{L}{n}\right)^{n-v} \\ &< e^L\left[|\delta_n(0)| + \max_{v=1(1)n}\left|\delta_n\left(\frac{v}{n}\right)\right|\right] = e^L \|\delta_n\|_{E_n^0}. \end{aligned}$$

Thus, when (1.1.15) holds, the discretization of our example is stable at any sequence $\{\eta_n\}$, with $S = e^L$ and $r = \infty$.

From the above, it is also evident how a finite stability threshold arises when f satisfies a Lipschitz condition on some restricted domain only. E.g., let (1.1.15) hold only for $x_i > 0$ and consider some $\{\eta_n\}_{n \in \mathbb{N}'}$ with $\eta_n(v/n) \geq 1$, $v = 0(1)n$. Any $\eta_n^{(i)}$ which satisfies $\|F_n \eta_n^{(i)} - F_n \eta_n\|_{E_n^0} < e^{-L}$ can have only positive values. Thus our discretization is stable at $\{\eta_n\}$ with $S = e^L$ as previously, but with stability threshold $r = e^{-L}$.

1.2 Results Concerning Stability

1.2.1 Existence of the Solution of the Discretization

While we generally assumed the existence and uniqueness of the true solution z of the original problem, we did not do so for the solution sequence $\{\zeta_n\}_{n \in \mathbb{N}'}$ of the discretization. We will now show that the existence and uniqueness of the ζ_n is implied, at least for sufficiently large n, by that of z and by the consistency and stability of the discretization. We first prove a fundamental lemma:

Lemma 1.2.1. *Let* $F_n : E_n \to E_n^0$ *be defined and continuous in*

$$B_R := \{\eta \in E_n : \|\eta - \overline{\eta}\| < R\}, \qquad R > 0,$$

with a fixed $\overline{\eta} \in E_n$. *Furthermore, for all* $\eta^{(i)} \in B_R$, $i = 1, 2$, *such that*

$$F_n \eta^{(i)} \in B_r^0 := \{\delta \in E_n^0 : \|\delta - F_n \overline{\eta}\| < r\}, \qquad r > 0,$$

let

$$\|\eta^{(1)} - \eta^{(2)}\| \leq S \|F_n \eta^{(1)} - F_n \eta^{(2)}\|, \qquad S > 0. \tag{1.2.1}$$

Then $F_n^{-1} : E_n^0 \to E_n$ *exists and is Lipschitz-continuous in* $B_{r_0}^0$ *with L-constant S, with*

$$r_0 := \min\left(r, \frac{R}{S}\right). \tag{1.2.2}$$

Proof. Since (1.2.1) guarantees the uniqueness of the inverse in $B_r^0 \cap F_n(B_R)$, it suffices to show that $B_{r_0}^0 \subset F_n(B_R)$.

Assume $\delta_0 \in B_{r_0}^0$, $\delta_0 \notin F_n(B_R)$, and let

$$\delta(\lambda) := (1 - \lambda) F_n \overline{\eta} + \lambda \delta_0, \qquad \lambda \geq 0.$$

Define

$$\overline{\lambda} := \begin{cases} \sup\{\lambda' > 0 : \delta(\lambda) \in F_n(B_R) \text{ for } \lambda \in [0, \lambda')\}, \\ 0 \text{ if the above set is empty}. \end{cases} \tag{1.2.3}$$

We show that the assumption $\overline{\lambda} \leq 1$ leads to a contradiction which makes the assumption $\delta_0 \notin F_n(B_r)$ absurd.

First we observe that $\bar{\delta} := \delta(\bar{\lambda}) \in F_n(B_R)$. For either $\bar{\lambda}=0$, or $0<\bar{\lambda}\leq 1$ and $\delta(\bar{\lambda}-\varepsilon)\in F_n(B_R)\cap B_r^0$ for $\varepsilon>0$ by (1.2.3). Hence, $F_n^{-1}(\delta(\bar{\lambda}-\varepsilon))$ exists and so, by (1.2.1), does $\lim_{\varepsilon\downarrow 0} F_n^{-1}(\delta(\bar{\lambda}-\varepsilon)) =: \bar{\zeta}$. Also, for all $\varepsilon>0$,

$$\begin{aligned}\|F_n^{-1}(\delta(\bar{\lambda}-\varepsilon))-\bar{\eta}\| &\leq S\|\delta(\bar{\lambda}-\varepsilon)-F_n\bar{\eta}\|\\ &= S(\bar{\lambda}-\varepsilon)\|\delta_0-F_n\bar{\eta}\| < S(\bar{\lambda}-\varepsilon)r_0 \leq (\bar{\lambda}-\varepsilon)R\end{aligned}$$

hence $\bar{\zeta}\in B_R$ and $\bar{\delta}=F_n(\bar{\zeta})$ by the continuity of F_n.

Now we can choose a closed ball $\bar{B}$ about $\bar{\zeta}$ such that $\bar{B}\subset B_R$ and $F_n(\bar{B})\subset B_r^0$; hence, F_n is one-to-one between $\bar{B}$ and $F_n(\bar{B})$. Due to the finite-dimensionality of $E_n\cong E_n^0$ (see Remark after Def. 1.1.4), this implies that $F_n(\bar{B})$ contains a neighborhood of $\bar{\delta}$ which is open in E_n^0. Thus $\bar{\lambda}\leq 1$ cannot satisfy (1.2.3). □

Corollary 1.2.2. *Consider a discretization* $\mathfrak{D}=\{E_n, E_n^0, F_n\}_{n\in\mathbb{N}'}$ *and a sequence* $\{\eta_n\}_{n\in\mathbb{N}'}$, $\eta_n\in E_n$. *Let the* F_n *be defined and continuous in* $B_R(\eta_n):=\{\eta\in E_n\colon \|\eta-\eta_n\|<R\}$, *and let them satisfy, for* $\eta_n^{(i)}\in B_R(\eta_n)$,

$$\|\eta_n^{(1)}-\eta_n^{(2)}\|\leq S\|F_n\eta_n^{(1)}-F_n\eta_n^{(2)}\|,$$

where both R and S are independent of n.

Then $\mathfrak{D}$ *is stable at* $\{\eta_n\}_{n\in\mathbb{N}'}$, *with stability bound S and stability threshold R/S.*

Proof. If in the proof of Lemma 1.2.1 we assume that (1.2.1) holds for *all* $\eta^{(i)}\in B_R(\eta_n)$, we find that F_n^{-1} is uniquely defined and Lipschitz-continuous with L-constant S in $\{\delta\in E_n^0\colon \|\delta-F_n\eta_n\|<R/S\}$. Since this is true for each $n\in\mathbb{N}'$, with uniform constants R and S, stability at $\{\eta_n\}$ follows (cf. Def. 1.1.10). □

This corollary establishes the fact that the uniform validity of a stability estimate (1.1.11) for all $\eta_n^{(i)}$ in balls with a fixed radius in E_n implies the existence of balls with radius independent of n in E_n^0 such that (1.1.11) holds for all $\eta_n^{(i)}$ whose images are in these balls (cf. the discussion in Section 1.1.4).

Theorem 1.2.3. *For the original problem* $\mathfrak{P}=\{E, E^0, F\}$ *with true solution z let the discretization method* $\mathfrak{M}=\{E_n, E_n^0, \Delta_n, \Delta_n^0, \varphi_n\}_{n\in\mathbb{N}'}$ *applicable to* $\mathfrak{P}$ *satisfy*:

(i) $F_n=\varphi_n(F)\colon E_n\to E_n^0$ *is defined and continuous in* $B_R(\Delta_n z)=\{\eta_n\in E_n\colon \|\eta_n-\Delta_n z\|<R\}$, $R>0$ *independent of n;*

(ii) $\mathfrak{M}$ *is consistent with* $\mathfrak{P}$ *at z;*

(iii) $\mathfrak{M}$ *is stable for* $\mathfrak{P}$.

Then the discretization $\mathfrak{M}(\mathfrak{P})$ *possesses unique solution elements* $\zeta_n\in E_n$ *for all sufficiently large* $n\in\mathbb{N}'$.

Proof. Assumptions (i) and (iii) form the hypotheses of Lemma 1.2.1 for $\bar{\eta}=\Delta_n z$. Hence there is an $r_0>0$ *independent of* n such that F_n^{-1} is uniquely defined in the ball $B_{r_0}^0(l_n):=\{\delta\in E_n^0: \|\delta-l_n\|<r_0\}$, where $l_n=F_n\Delta_n z$ is the local discretization error of $\mathfrak{M}$ for p (cf. Def. 1.1.7).

The consistency of $\mathfrak{M}$ implies $\lim_{n\to\infty}\|l_n\|=0$ (see Def. 1.1.5) so that

$$0\in B_{r_0}^0(l_n) \quad \text{for all sufficiently large} \quad n\in\mathbb{N}'. \quad \square$$

Example. Euler method. Assume that f is only defined and uniformly Lipschitz-continuous for positive arguments and that $z(t)\geq 1$ in $[0,1]$. Then the hypotheses of Lemma 1.2.1 are satisfied with $R=1$, $S=e^L$, $r=e^{-L}$ (see end of Section 1.1.4) so that $r_0=e^{-L}$. Thus we may be certain that the recursively defined solution of the discretization (without round-off errors) remains positive if

$$\|l_n\|_{E_n^0}=|l_n(0)|+\max_{\nu=1(1)n}\left|l_n\left(\frac{\nu}{n}\right)\right|<e^{-L}$$

which is equivalent to

$$n>\frac{e^L}{2}\max_{t\in[0,1]}|z''(t)|$$

as $l_n(0)=0$ and $l_n(\nu/n)=(1/2n)z''(\tilde{t}_\nu)$ for $\nu=1(1)n$, see end of Section 1.1.2.

Theorem 1.2.3 makes it clear why proofs for the existence of a (global) solution of a discretization in nontrivial cases are very similar to stability proofs and normally rest upon the same assumptions.

Theorem 1.2.3 is useful in conjunction with theorems which deduce the presence of stability in a complicated situation from its presence in a related simpler situation (see e.g. Section 1.2.3 and 1.2.4). Since the assumption (i) of Theorem 1.2.3 will appear frequently in the following we will abbreviate it to "F_n is continuous in $B_R(\Delta_n z)$". Analogously, we will use "F_n is continuous in $B_R(\eta_n)$" for the first hypothesis in Corollary 1.2.2.

1.2.2 The Basic Convergence Theorem

Theorem 1.2.4. *Under the assumptions of Theorem* 1.2.3, $\mathfrak{M}$ *is convergent for* $\mathfrak{P}$. *Furthermore, if* $\mathfrak{M}$ *is consistent with* $\mathfrak{P}$ *of order p then it is convergent for* $\mathfrak{P}$ *of order p.*

Proof. First, Theorem 1.2.3 guarantees the existence of unique solutions ζ_n of (1.1.3) for sufficiently large $n\in\mathbb{N}'$; cf. also the Remark after Def. 1.1.9. From (1.1.3), (1.1.7), and (1.1.8) we have

$$F_n\zeta_n-F_n\Delta_n z=F_n(\Delta_n z+\varepsilon_n)-F_n\Delta_n z=-l_n \tag{1.2.4}$$

where ε_n and l_n are the global and local discretization errors of $\mathfrak{M}$ for $\mathfrak{P}$. Due to the assumed stability, Eq. (1.2.4) implies

$$\|\varepsilon_n\|\leq S\|l_n\| \quad \text{if} \quad \|l_n\|<r \tag{1.2.5}$$

where S and r are the stability bound and threshold resp. Consistency with $\mathfrak{P}$ at z implies $\lim_{n\to\infty} \|l_n\| = 0$ so that $\|l_n\| < r$ for all sufficiently large $n \in \mathbb{N}'$ and (1.2.5) yields $\lim_{n\to\infty} \|\varepsilon_n\| = 0$. Analogously, $\|l_n\| = O(n^{-p})$ implies $\|\varepsilon_n\| = O(n^{-p})$. □

The assertion of Theorem 1.2.4 is often contracted into the slogan "Consistency + Stability = Convergence". The actual order of convergence may be higher than the order of consistency in some cases; see Example 1 in Section 2.2.4.

Example. In Section 1.1.3 we found that the Euler method was convergent of order 1 for a simple differential equation. Theorem 1.2.4 shows that the Euler method is convergent of order 1 for any differential equation $y' = f(y)$ with a solution from $C^{(2)}[0, 1]$; see the example at the end of Section 1.1.2.

In our abstract formulation both consistency and stability depend strongly on the choice of the spaces E_n^0 and their norms while the convergence can at most depend on the definition of the norm in E_n. Therefore, we cannot expect to prove the necessity of either stability or consistency for convergence as long as we do not settle for a particular sequence of spaces $\{E_n^0\}$ in conjunction with a particular class of discretization methods for a particular class of problems. The well-known theorems which state that consistency and stability are also necessary for convergence always refer to such special situations.

In a general context we can always make an otherwise sound method unstable or inconsistent by choosing a pathological sequence $\{E_n^0\}$. On the other hand, one may artificially construct discretizations which converge while being unstable or inconsistent in an intuitive sense.

Examples. 1. (Modelled after an example by M. N. Spijker, private communication): All spaces and mappings are as with the Euler method in the previous sections except that φ_n now maps F into $\tilde{F}_n$, defined by

$$(\tilde{F}_n \eta)\left(\frac{\nu}{n}\right) = \begin{cases} \eta(0) - z_0, & \nu = 0, \\ \dfrac{\eta\left(\frac{\nu}{n}\right) - \eta\left(\frac{\nu-1}{n}\right)}{1/n} - f\left(\eta\left(\frac{\nu-1}{n}\right)\right) - 2^{\nu-(n+1)}, & \nu = 1(1)n, \end{cases}$$

$$= (F_n \eta)\left(\frac{\nu}{n}\right) - \begin{cases} 0, \\ 2^{\nu-(n+1)}, \end{cases}$$

where F_n is from the Euler method.

Since we have shown that $\lim_{n\to\infty} \|F_n \Delta_n y - \Delta_n^0 F y\| = 0$ for any $y \in E$ we must have

$$\lim_{n\to\infty} \|\tilde{F}_n \Delta_n y - \Delta_n^0 F y\| = \tfrac{1}{2} \quad \text{for any } y \in E;$$

thus, the method is not consistent with our problem.

On the other hand, an additive term cannot harm the stability (cf. Def. 1.1.10). Furthermore, the method is still convergent as may be easily checked for $f(y)=gy$, where we get

$$\frac{1}{n}\sum_{\nu=1}^{n}\left(1+\frac{g}{n}\right)^{n-\nu} 2^{\nu-(n+1)} < \frac{1}{n}\left(1-\frac{g}{n}\right)^{-1}$$

as the maximal perturbation of the result from the Euler method (if $n>g$); this perturbation vanishes as $n\to\infty$. Actually, convergence is retained for an arbitrary Lipschitz-continuous f.

2. For the problem $y'=0$, $y(0)=0$ we can easily construct discretizations which are consistent but not stable and yet convergent, e.g. (we omit the details on the spaces, etc.)

$$(F_n\eta)\left(\frac{\nu}{n}\right)=\begin{cases}\eta\left(\frac{\nu}{n}\right), & \text{for } \nu=0,1\,,\\ \dfrac{\eta\left(\frac{\nu}{n}\right)-\eta\left(\frac{\nu-1}{n}\right)-\eta\left(\frac{\nu-2}{n}\right)}{1/n}, & \text{for } \nu=2(1)n\,.\end{cases}$$

We have (trivially) consistency at z since $F_n\Delta_n z=0$ for all n and convergence since $\zeta_n=\Delta_n z=0$ for all n. But

$$F_n(\Delta_n z+\varepsilon_n)-F_n\Delta_n z=\begin{cases}\delta, & \nu=0,1\,,\\ 0, & \nu=2(1)n\,,\end{cases}$$

produces $\varepsilon_n(\nu/n)=\beta_\nu\delta$ where β_ν is the ν-th Fibonacci number; thus $\|\varepsilon_n\|/|\delta|$ cannot be bounded uniformly in n and the "method" is unstable.

1.2.3 Linearization

According to Remark 2 following Def. 1.1.11, for mappings F_n of the form (1.1.13) – i.e., a linear mapping plus a constant – stability reduces to the existence and uniform boundedness of the sequence of inverses $\hat{F}_n^{-1}$ of the linear part. If the F_n are non-linear but Frechet-differentiable we may expect that a similar condition on the Frechet derivatives of the F_n implies stability.

Theorem 1.2.5. *For a given sequence* $\{\eta_n\}_{n\in\mathbb{N}'}$, $\eta_n\in E_n$, *let the mappings* F_n *of the discretization* $\mathfrak{D}=\{E_n, E_n^0, F_n\}_{n\in\mathbb{N}'}$ *be continuous in* $B_R(\eta_n)$. *Furthermore, for all* $\eta\in E_n$ *such that* $F_n\eta\in B_r^0(F_n\eta_n)$, $r>0$, *let* $[F_n'(\eta)]^{-1}$ *exist and*

$$\|F_n'(\eta)^{-1}\|\leq S \quad \textit{uniformly for } n\in\mathbb{N}'. \tag{1.2.6}$$

Then $\mathfrak{D}$ *is stable at* $\{\eta_n\}$ *with stability bound* S *and stability threshold* $r_0=\min(r, R/S)$.

Proof. The existence of unique inverses $F_n^{-1}: E_n^0\to E_n$ in the balls $B_{r_0}^0(F_n\eta_n)$, with $r_0=\min(r, R/S)$, for all $n\in\mathbb{N}'$ is shown as in the proof of

Lemma 1.2.1. In place of (1.2.1), (1.2.6) guarantees (by the inverse function theorem of functional analysis) the existence (and differentiability) of an inverse in an open (in E_n^0) neighborhood of each $F_n\eta$. Moreover, by the same theorem

$$(F_n^{-1})'(F_n\eta)=[F_n'(\eta)]^{-1}$$

so that (1.2.6) implies

$$\|(F_n^{-1})'(\delta)\|\leq S \quad \text{for } \delta\in B_{r_0}^0(F_n\eta_n), \qquad n\in\mathbb{N}'. \tag{1.2.7}$$

Since $B_{r_0}^0(F_n\eta_n)$ is convex, $(F_n^{-1})'$ exists and satisfies (1.2.7) at each element

$$F_n\eta^{(2)}+\lambda(F_n\eta^{(1)}-F_n\eta^{(2)}), \qquad 0\leq\lambda\leq 1,$$

if $F_n\eta^{(1)}$ and $F_n\eta^{(2)}$ are in $B_{r_0}^0(F_n\eta_n)$. For such $\eta^{(i)}$ we may thus apply the mean value theorem of functional analysis

$$\begin{aligned}\|\eta^{(1)}-\eta^{(2)}\|&=\|F_n^{-1}(F_n\eta^{(1)})-F_n^{-1}(F_n\eta^{(2)})\|\\ &\leq \max_{\delta\in B_{r_0}^0(F\eta_n)}\|(F_n^{-1})'(\delta)\|\,\|F_n\eta^{(1)}-F_n\eta^{(2)}\|\\ &\leq S\|F_n\eta^{(1)}-F_n\eta^{(2)}\| \quad \text{by (1.2.7).} \qquad \square\end{aligned}$$

Corollary 1.2.6. *Under the conditions of Theorem* 1.2.5 *if* (1.2.6) *holds for all* $\eta\in B_R(\eta_n)$, *then* $\mathfrak{D}$ *is stable at* $\{\eta_n\}$ *with stability bound S and stability threshold R/S.*

Proof (cf. Corollary 1.2.2). The F_n^{-1} are now uniquely defined and satisfy (1.2.7) in $B_{r_0}^0(F_n\eta_n)$ with $r_0=R/S$. The remainder of the proof of Theorem 1.2.5 is not affected. $\square$

Due to Theorem 1.2.5 and Corollary 1.2.6 stability analyses may in many cases be restricted to linear discretizations. Although this often makes little difference in the argument, the formalism of the proofs usually becomes simpler. The differentiability required for the application of the above theorems is normally assumed anyway since all "higher order methods" rely on a certain smoothness of the problem and its discretization.

Example. $-y''+f(t,y)=0$, $y(0)=z_0$, $y(1)=z_1$. We discretize this problem on the grids $\mathbb{G}_n:=\{(\nu/n),\ \nu=0(1)n\}$; the spaces $E_n=E_n^0=(\mathbb{G}_n\to\mathbb{R})$ are furnished with the max-norm, details are left to the reader. We assume that f is differentiable w.r.t. y.

The linearization of the mappings

$$(F_n\eta)\left(\frac{\nu}{n}\right)=\begin{cases}\eta(0)-z_0, & \nu=0,\\[1ex] -\dfrac{\eta\left(\frac{\nu+1}{n}\right)-2\eta\left(\frac{\nu}{n}\right)+\eta\left(\frac{\nu-1}{n}\right)}{\frac{1}{n^2}}+f\left(\frac{\nu}{n},\eta\left(\frac{\nu}{n}\right)\right), & \nu=1(1)n-1,\\[1ex] \eta(1)-z_1, & \nu=n,\end{cases} \tag{1.2.8}$$

at $\eta \in E_n$ is given by ($\varepsilon \in E_n$)

$$(F_n'(\eta)\varepsilon)\left(\frac{\nu}{n}\right) = \begin{cases} \varepsilon(0), & \nu=0, \\ -\dfrac{\varepsilon\left(\frac{\nu+1}{n}\right) - 2\varepsilon\left(\frac{\nu}{n}\right) + \varepsilon\left(\frac{\nu-1}{n}\right)}{\frac{1}{n^2}} + f_y\left(\frac{\nu}{n}, \eta\left(\frac{\nu}{n}\right)\right)\varepsilon\left(\frac{\nu}{n}\right), & \nu=1(1)n-1, \\ \varepsilon(1), & \nu=n. \end{cases}$$

It is easy to show that $F_n'(\eta)^{-1}$ exists and is uniformly bounded if $f_y\left(\frac{\nu}{n}, \eta\left(\frac{\nu}{n}\right)\right) \geq 0$, $\nu=1(1)n-1$, since the matrix representation of $F_n'(\eta)$ is diagonally dominant, see e.g. Henrici [1], Section 7.2–2.
Thus, if f_y is continuous in its second argument and

$$f_y(t, z(t)) > 0 \quad \text{in } [0,1]$$

there exists an $R>0$ such that $f_y((\nu/n), \eta(\nu/n)) \geq 0$ for all $\eta \in B_R(\Delta_n z)$. This implies the stability of (1.2.8) for our problem by Corollary 1.2.6; furthermore, since (1.2.8) is clearly consistent, it implies the existence of a solution to (1.2.8) and the convergence of the discretization (Theorems 1.2.3 and 1.2.4).

By an – admittedly artificial – counter-example we will now give a proof for the claim made in Section 1.1.4 that (1.1.11) with (1.1.12) is a weaker condition than (1.1.11) with (1.1.14):

Theorem 1.2.7. *Under the assumptions of Theorem 1.2.5 it is not in general true that there exist constants $R>0$ and S independent of n such that (1.1.11) holds for all $\eta_n^{(i)} \in B_R(\eta_n)$.*

Proof. By counter example. Let $E_n = E_n^0 = \mathbb{R}^2$, with Euclidean norm. Define $F_n: E_n \to E_n^0$ for $n \in \mathbb{N}$ by

$$F_n\begin{pmatrix} x_1 \\ x_2 \end{pmatrix} = \begin{pmatrix} (x_2+2)\cos n\pi x_1 \\ (x_2+2)\sin n\pi x_1 \end{pmatrix}$$

and let $\eta_n = \begin{pmatrix} 0 \\ 0 \end{pmatrix}$. In the unit disk $B_1(\eta_n)$ of E_n, the F_n are differentiable and

$$\left[F_n'\begin{pmatrix} x_1 \\ x_2 \end{pmatrix}\right]^{-1} = \begin{pmatrix} -\dfrac{\sin n\pi x_1}{(x_2+2)n\pi} & \dfrac{\cos n\pi x_1}{(x_2+2)n\pi} \\ \cos n\pi x_1 & \sin n\pi x_1 \end{pmatrix}$$

satisfies

$$\left\| F_n'\begin{pmatrix} x_1 \\ x_2 \end{pmatrix}^{-1} \right\| = 1 \quad \text{for all } n \in \mathbb{N}. \tag{1.2.9}$$

Moreover, since $F_n(B_1(\eta_n)) \supset B_1^0(F_n\eta_n)$, (1.2.9) also holds for all $\begin{pmatrix} x_1 \\ x_2 \end{pmatrix}$ such that $F_n\begin{pmatrix} x_1 \\ x_2 \end{pmatrix} \in B_1^0(F_n\eta_n)$ and the hypotheses of both Theorem 1.2.5 and Corollary 1.2.6 are satisfied.

But

$$F_n\begin{pmatrix} -\frac{1}{n} \\ 0 \end{pmatrix} = F_n\begin{pmatrix} \frac{1}{n} \\ 0 \end{pmatrix} = \begin{pmatrix} -2 \\ 0 \end{pmatrix} \quad \text{for all } n \in \mathbb{N};$$

hence there can be no $R>0$ independent of n such that

$$\|F_n\eta^{(1)} - F_n\eta^{(2)}\| \geq \frac{1}{S}\|\eta^{(1)} - \eta^{(2)}\| \tag{1.2.10}$$

would hold for all $\eta^{(i)} \in B_R(\eta_n)$. (Note that (1.2.10) certainly holds for all $\eta^{(i)}$ with $F_n\eta^{(i)} \in B_1^0(F_n\eta_n)$, with $S=1$.) □

1.2.4 Stability of Neighboring Discretizations

If the discretization $\{E_n, E_n^0, F_n\}_{n\in\mathbb{N}'}$ is stable at $\{\eta_n\}_{n\in\mathbb{N}'}$, it may be expected that a discretization $\{E_n, E_n^0, F_n + G_n\}_{n\in\mathbb{N}'}$ will also be stable at $\{\eta_n\}$ if the G_n are small in a suitable sense. (Actually, it will turn out that the rate of change of the G_n has to be sufficiently small.) Various results which assert the stability of a "neighboring" discretization may be formulated.

Theorem 1.2.8. *Consider the discretization* $\bar{\mathfrak{D}} = \{E_n, E_n^0, \bar{F}_n\}_{n\in\mathbb{N}'}$ *with* $\bar{F}_n = F_n + G_n$, *and a sequence* $\{\eta_n\}_{n\in\mathbb{N}'}$. *Assume that there exists an* $R>0$ *independent of n such that for all* $n \in \mathbb{N}'$

(i) *the* F_n *are continuous in* $B_R(\eta_n)$ *and satisfy, for all* $\eta^{(i)} \in B_R(\eta_n)$, $i=1,2$,

$$\|\eta^{(1)} - \eta^{(2)}\| \leq S\|F_n\eta^{(1)} - F_n\eta^{(2)}\|; \tag{1.2.11}$$

(ii) *the* G_n *satisfy, for all* $\eta^{(i)} \in B_R(\eta_n)$, $i=1,2$,

$$\|G_n\eta^{(1)} - G_n\eta^{(2)}\| \leq L\|\eta^{(1)} - \eta^{(2)}\| \quad \text{with } LS<1. \tag{1.2.12}$$

Then $\bar{\mathfrak{D}}$ *is stable at* $\{\eta_n\}_{n\in\mathbb{N}'}$, *with stability bound* $S/(1-LS)$ *and stability threshold* $(R/S)(1-LS)$.

Proof. For all $n \in \mathbb{N}'$ and all $\eta^{(i)} \in B_R(\eta_n)$ we have

$$\begin{aligned}\|\bar{F}_n\eta^{(1)} - \bar{F}_n\eta^{(2)}\| &\geq \|F_n\eta^{(1)} - F_n\eta^{(2)}\| - \|G_n\eta^{(1)} - G_n\eta^{(2)}\| \\ &\geq \left(\frac{1}{S} - L\right)\|\eta^{(1)} - \eta^{(2)}\|\end{aligned}$$

by (1.2.11) and (1.2.12). The assertion about the stability threshold follows from Corollary 1.2.2. □

Remark. Assumption (i) is stronger than the assumption of stability at $\{\eta_n\}$; cf. Theorem 1.2.7. If (1.2.11) is only satisfied for all $\eta^{(i)}$ such that $F_n\eta^{(i)} \in B_r^0(F_n\eta_n)$, with a *finite* $r>0$, the independence of a stability threshold for $\bar{\mathfrak{D}}$ on n may not be concluded without further assumptions. Of course, if F_n is *linear*, no such complications arise – see Remark 2 after Def. 1.1.11.

The following less immediate formulation has a greater range of applicability:

Theorem 1.2.9. *Consider the discretization* $\bar{\mathfrak{D}} = \{E_n, E_n^0, \bar{F}_n\}_{n\in\mathbb{N}'}$ *with* $\bar{F}_n = F_n + G_n$, *and a sequence* $\{\eta_n\}_{n\in\mathbb{N}'}$. *Assume that there exists* $R>0$ *independent of n such that for all* $n\in\mathbb{N}'$

(i) *the* F_n *are one-to-one between* $B_R(\eta_n)$ *and* $F_n(B_R(\eta_n))$, $F_n(B_R(\eta_n))$ *is convex, and*

$$\|(F_n^{-1})'(\delta)\| \le S \quad \text{for } \delta \in F_n(B_R(\eta_n)); \tag{1.2.13}$$

(ii) *for* $\delta \in F_n(B_R(\eta_n))$ *and* $\eta^{(i)} \in B_R(\eta_n)$, $\quad i=1,2,$

$$\|(F_n^{-1})'(\delta)[G_n\eta^{(1)} - G_n\eta^{(2)}]\| \le L'\|\eta^{(1)} - \eta^{(2)}\| \quad \text{with } L'<1. \tag{1.2.14}$$

Then $\bar{\mathfrak{D}}$ *is stable at* $\{\eta_n\}$ *with stability bound* $S/(1-L')$ *and stability threshold* $(R/S)(1-L')$.

Proof. Due to the convexity of $F_n(B_R(\eta_n))$ the linear operator $K_n\colon E_n^0 \to E_n$

$$K_n := \int_0^1 (F_n^{-1})'(F_n\eta^{(2)} + \lambda(F_n\eta^{(1)} - F_n\eta^{(2)}))\,d\lambda$$

is well-defined and $\|K_n\| \le S$ by (1.2.13). Furthermore,

$$\begin{aligned} K_n(\bar{F}_n\eta^{(1)} - \bar{F}_n\eta^{(2)}) &= K_n(F_n\eta^{(1)} - F_n\eta^{(2)}) + K_n(G_n\eta^{(1)} - G_n\eta^{(2)}) \\ &= \eta^{(1)} - \eta^{(2)} + K_n(G_n\eta^{(1)} - G_n\eta^{(2)}) \end{aligned}$$

so that

$$\begin{aligned} S\|\bar{F}_n\eta^{(1)} - \bar{F}_n\eta^{(2)}\| &\ge \|K_n(\bar{F}_n\eta^{(1)} - \bar{F}_n\eta^{(2)})\| \\ &\ge \|\eta^{(1)} - \eta^{(2)}\| - L'\|\eta^{(1)} - \eta^{(2)}\| \end{aligned}$$

by (1.2.14). Again, Corollary 1.2.2 produces the stability threshold. □

Remarks. 1. Trivially, (1.2.14) with (1.2.13) implies (1.2.12) with $L = L'/S$. Actually, condition (ii) of Theorem 1.2.9 is weaker than that of Theorem 1.2.8; it may sometimes be satisfiable even when there exists no uniform Lipschitz-constant L at all for (1.2.12).

2. The Remark after Theorem 1.2.8 applies analogously.

Theorem 1.2.9 becomes simpler when the F_n are of the form (1.1.13), i.e., a linear operator plus a constant term. Since this is by far the most important case for applications, we formulate this linear version separately for easy reference:

Corollary 1.2.10. *Consider the discretization* $\bar{\mathfrak{D}}=\{E_n, E_n^0, \bar{F}_n\}_{n\in\mathbb{N}'}$ *with* $\bar{F}_n=F_n+G_n$, *and a sequence* $\{\eta_n\}_{n\in\mathbb{N}'}$. *Assume that*

(i) *The* F_n *are of the form* (1.1.13) *and the discretization* $\mathfrak{D}=\{E_n, E_n^0, F_n\}_{n\in\mathbb{N}'}$ *is stable, with (see Remark 2 after Def.* 1.1.11)

$$\|\hat{F}_n^{-1}\|\le S \quad \text{for all } n\in\mathbb{N}'; \tag{1.2.15}$$

(ii) *There exists an* $R>0$ *such that, for all* $n\in\mathbb{N}'$ *and* $\eta^{(i)}\in B_R(\eta_n)$, $i=1,2$,

$$\|\hat{F}_n^{-1}(G_n\eta^{(1)}-G_n\eta^{(2)})\|\le L'\|\eta^{(1)}-\eta^{(2)}\| \quad \text{with } L'<1. \tag{1.2.16}$$

Then $\bar{\mathfrak{D}}$ *is stable at* $\{\eta_n\}$, *with stability bound* $S/(1-L')$ *and stability threshold* $(R/S)(1-L')$.

Examples. 1. As after Corollary 1.2.6: Denote (1.2.8) by $\bar{F}_n$ and let

$$(F_n\eta)\left(\frac{v}{n}\right)=\begin{cases}\eta(0)-z_0, & v=0,\\ -\dfrac{\eta\left(\frac{v+1}{n}\right)-2\eta\left(\frac{v}{n}\right)+\eta\left(\frac{v-1}{n}\right)}{\frac{1}{n^2}}, & v=1(1)n-1,\\ \eta(1)-z_1, & v=n,\end{cases}$$

$$(G_n\eta)\left(\frac{v}{n}\right)=\begin{cases}0, & v=0,n,\\ f\left(\frac{v}{n},\eta\left(\frac{v}{n}\right)\right), & v=1(1)n-1.\end{cases}$$

F_n is of the form (1.1.13); it is well-known (see e.g. Henrici [1], Section 7.2–2) that $\hat{F}_n^{-1}$ may be represented by the matrix $A=(a_{\mu v})$, $\mu, v=0(1)n$, with

$$a_{\mu 0}=\frac{n-\mu}{n},\qquad a_{\mu n}=\frac{\mu}{n},\qquad \mu=0(1)n,$$

$$a_{\mu v}=\begin{cases}\dfrac{\mu(n-v)}{n^3}, & \mu\le v,\\ \dfrac{v(n-\mu)}{n^3}, & \mu\ge v,\end{cases}\qquad v\neq 0,n.$$

An easy computation shows that

$$\|\hat{F}_n^{-1}\|=\|A\|_{\max}=\tfrac{9}{8} \quad \text{independently of } n.$$

Hence, Theorem 1.2.8 asserts the stability (at $\{\eta_n\}$) of the discretization $\{E_n, E_n^0, \bar{F}_n\}$ if f satisfies a Lipschitz condition w.r.t. its second argument (in $B_R(\eta_n)$) with $L<\frac{8}{9}$.

Applying Theorem 1.2.9, resp. Corollary 1.2.10 we obtain from

$$\|\hat{F}_n^{-1}(G_n\eta^{(1)}-G_n\eta^{(2)})\|\le\frac{L}{8}\|\eta^{(1)}-\eta^{(2)}\|$$

that $\{E_n, E_n^0, \bar{F}_n\}$ is stable if the Lipschitz constant L of f satisfies $L<8$.

Note that there is no assumption on the *sign* of f_y in this example.

2. Euler method, as previously: Let

$$(F_n\eta)\left(\frac{\nu}{n}\right)=\begin{cases}\eta(0)-z_0, & \nu=0,\\ \dfrac{\eta\left(\frac{\nu}{n}\right)-\eta\left(\frac{\nu-1}{n}\right)}{1/n}, & \nu=1(1)n,\end{cases}$$

$$(G_n\eta)\left(\frac{\nu}{n}\right)=\begin{cases}0, & \nu=0,\\ f\left(\eta\left(\frac{\nu-1}{n}\right)\right), & \nu=1(1)n.\end{cases}$$

It is easily verified that $(F_n^{-1}\delta)\left(\frac{\nu}{n}\right)=\delta(0)+\frac{1}{n}\sum_{\mu=1}^{\nu}\delta\left(\frac{\mu}{n}\right)$, $\nu=0(1)n$, so that, for our norm (see Section 1.1.1), $\|\hat{F}_n^{-1}\|=1$ for all $n\in\mathbb{N}$. Moreover, if f satisfies (1.1.15) we have

$$\|G_n\eta^{(1)}-G_n\eta^{(2)}\|\leq L\|\eta^{(1)}-\eta^{(2)}\|,$$

$$\|F_n^{-1}(G_n\eta^{(1)}-G_n\eta^{(2)})\|=\max_{\nu=0(1)n}\left|\frac{1}{n}\sum_{\mu=1}^{\nu}\left[f\left(\eta^{(1)}\left(\frac{\mu-1}{n}\right)\right)-f\left(\eta^{(2)}\left(\frac{\mu-1}{n}\right)\right)\right]\right|$$
$$\leq L\|\eta^{(1)}-\eta^{(2)}\|,$$

so that both Theorems 1.2.8 and 1.2.9 establish the stability of $\{E_n, E_n^0, \overline{F}_n\}$ for $L<1$ with a stability bound $1/(1-L)$. (But from Section 1.1.4 we know that this discretization is stable for an arbitrary L and has a stability bound e^L!)

The appearance of an artificial restriction – like $L<1$ – as a condition for stability is typical for applications of the theorems of this section to discretizations of *initial value problems*. The reason is quite obvious, at least for linear F_n: With initial value problems the matrix representations of both $\hat{F}_n$ and $\hat{F}_n^{-1}$ are lower triangular for any reasonable discretization. For such operators $\hat{F}_n$, much weaker conditions are sufficient for the assertions of our theorems.

For this reason, the analysis of the stability of neighboring discretizations will be taken up once more for "forward step methods" in Section 2.2.5 and relevant results will be obtained there.

1.3. Asymptotic Expansions of the Discretization Errors

1.3.1 Asymptotic Expansion of the Local Discretization Error

For the subsequent considerations it is important that the diagram of Fig. 1.2 (Section 1.1.2) can be made fully commutative for each $n\in\mathbb{N}'$, so that the image under F_n of the discretization of an element in E may be represented as the discretization of its image under an operation from E to E^0, see Fig. 1.4.

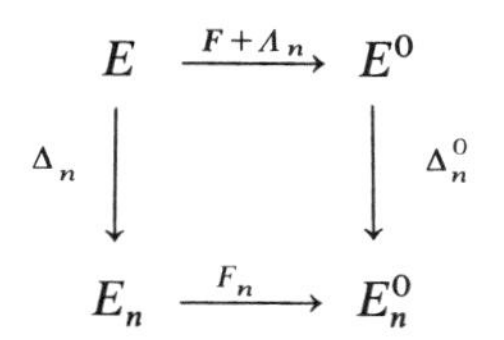

Fig. 1.4. Effect of a local error mapping

Def. 1.3.1. Consider the discretization method $\mathfrak{M}=\{E_n, E_n^0, \Delta_n, \Delta_n^0, \varphi_n\}_{n\in\mathbb{N}'}$ applicable to the original problem $\mathfrak{P}=\{E, E^0, F\}$. A sequence of mappings $\Lambda_n: E\to E^0$, $n\in\mathbb{N}'$, such that

$$\varphi_n(F)\Delta_n y := F_n\Delta_n y = \Delta_n^0[F+\Lambda_n]y \tag{1.3.1}$$

for all y in the domain of F is called a *local error mapping* of $\mathfrak{M}$ for $\mathfrak{P}$.

Remark. The *existence* of a local error mapping is not at all trivial for discretization methods where F_n is not simply a transcription of F into differences, e.g. for multistep methods for $y'=f(y)$ (see Section 4.4.1).

Naturally, since there are no unique inverses to the Δ_n^0, (1.3.1) does not define the Λ_n uniquely (if they exist). But the following further requirement on the sequence $\{\Lambda_n\}$ brings about this uniqueness in as far as it is needed.

Def. 1.3.2. A local error mapping $\{\Lambda_n\}_{n\in\mathbb{N}'}$ of a discretization method $\mathfrak{M}$ for an original problem $\mathfrak{P}$ *possesses an asymptotic expansion to order* J if there exist a non-empty subset $D_J\subset E$ and mappings $\lambda_j: D_J\to E^0$, $j=1(1)J$, independent of n, such that, for all $y\in D_J$,

$$\left\|\Delta_n^0\left[\Lambda_n y - \sum_{j=1}^{J}\frac{1}{n^j}\lambda_j y\right]\right\| = O(n^{-(J+1)}) \quad \text{as } n\to\infty. \tag{1.3.2}$$

The requirement that the λ_j be independent of n makes Λ_n unique except for terms of order $O(n^{-(J+1)})$ in all usual applications. If we have the situation of Def. 1.3.2 and $y\in D_L$ then

$$F_n\Delta_n y = \Delta_n^0\left[Fy + \sum_{j=1}^{J}\frac{1}{n^j}\lambda_j y\right] + O(n^{-(J+1)}). \tag{1.3.3}$$

In particular, if the true solution z of $\mathfrak{P}$ is in D_J, we have

$$l_n = F_n \Delta_n z = \sum_{j=1}^{J} \frac{1}{n^j} \Delta_n^0(\lambda_j z) + O(n^{-(J+1)}). \tag{1.3.4}$$

(1.3.4) is the *asymptotic expansion* (to order J) *of the local discretization error* $\{l_n\}$ of $\mathfrak{M}$ for $\mathfrak{P}$. Obviously (see Def. 1.1.6)

$$\lambda_j z = 0, \quad j = 1(1)p-1, \tag{1.3.5}$$

if $\mathfrak{M}$ is consistent with $\mathfrak{P}$ of order p.

Example. Euler method, as previously. Since

$$[(F_n \Delta_n - \Delta_n^0 F) y]\left(\frac{v}{n}\right) = \begin{cases} 0, & v = 0, \\ \dfrac{y\left(\frac{v}{n}\right) - y\left(\frac{v-1}{n}\right)}{1/n} - y'\left(\frac{v-1}{n}\right), & v = 1(1)n, \end{cases}$$

$$\Lambda_n : y \to \begin{pmatrix} 0 \\ \dfrac{y\left(t + \frac{1}{n}\right) - y(t)}{\frac{1}{n}} - y'(t) \end{pmatrix} \in E^0, \quad n \in \mathbb{N},$$

is a local error mapping (see the definition of Δ_n^0 in Section 1.1.1). Any term which vanishes at the grid points v/n may be added to Λ_n without invalidating (1.3.1).

For each $J \in \mathbb{N}$, the above $\{\Lambda_n\}_{n \in \mathbb{N}}$ possesses an asymptotic expansion to order J, with $D_J = C^{(J+2)}[0,1]$ and

$$\lambda_j : y \to \begin{pmatrix} 0 \\ \dfrac{1}{(j+1)!} y^{(j+1)}(t) \end{pmatrix}, \quad j = 1(1)J,$$

since

$$\frac{y\left(\frac{v}{n}\right) - y\left(\frac{v-1}{n}\right)}{1/n} = y'\left(\frac{v-1}{n}\right) + \sum_{j=1}^{J} \frac{1}{n^j} \frac{y^{(j+1)}\left(\frac{v-1}{n}\right)}{(j+1)!} + O(n^{-(J+1)})$$

for $y \in D_J$. Thus, if $z \in D_J$, the local discretization error $\{l_n\}$ has the asymptotic expansion

$$l_n\left(\frac{v}{n}\right) = \begin{cases} 0, & v = 0, \\ \displaystyle\sum_{j=1}^{J} \frac{1}{n^j} \frac{z^{(j+1)}\left(\frac{v-1}{n}\right)}{(j+1)!} + O(n^{-(J+1)}), & v = 1(1)n. \end{cases}$$

For $J = 1$, this reduces to (cf. end of Section 1.1.2)

$$l_n\left(\frac{v}{n}\right) = \frac{1}{2n} z''\left(\frac{v-1}{n}\right) + O(n^{-2}), \quad v = 1(1)n.$$

The choice of the mappings $\{\Delta_n^0\}$ strongly influences the form of the local error mapping and of a possible asymptotic expansion. E.g., if in the above example we replace the definition of Section 1.1.1 by

$$\left[\Delta_n^0\begin{pmatrix} d_0 \\ d(t)\end{pmatrix}\right]\left(\frac{\nu}{n}\right)=\begin{cases} d_0, & \nu=0, \\ d\left(\frac{\nu}{n}\right), & \nu=1(1)n, \end{cases}$$

we obtain

$$\frac{y(t)-y\left(t-\frac{1}{n}\right)}{1/n}-y'(t)+f(y(t))-f\left(y\left(t-\frac{1}{n}\right)\right)$$

for the second component of the Λ_n and the asymptotic expansion becomes more complicated.

An intelligent choice of $\{\Delta_n^0\}$ is particularly important for discretization methods whose local error mappings may possess an asymptotic expansion in *even* powers of $1/n$ for a proper choice of the Δ_n^0.

Example. The "implicit trapezoidal method" for $y'=f(y)$ is given by

$$(1.3.6)\quad [\varphi_n(F)\eta]\left(\frac{\nu}{n}\right)=\begin{cases} \eta(0)-z_0, & \nu=0, \\ \dfrac{\eta\left(\frac{\nu}{n}\right)-\eta\left(\frac{\nu-1}{n}\right)}{1/n}-\dfrac{1}{2}\left[f\left(\eta\left(\frac{\nu}{n}\right)\right)+f\left(\eta\left(\frac{\nu-1}{n}\right)\right)\right], & \nu=1(1)n. \end{cases}$$

If we define Δ_n^0 by

$$\left[\Delta_n^0\begin{pmatrix} d_0 \\ d(t)\end{pmatrix}\right]\left(\frac{\nu}{n}\right):=\begin{cases} d_0, & \nu=0, \\ d\left(\frac{2\nu-1}{2n}\right), & \nu=1(1)n, \end{cases}$$

and the second component of $\Lambda_n y$ by

$$\frac{y\left(t+\frac{1}{2n}\right)-y\left(t-\frac{1}{2n}\right)}{1/n}-y'(t)-\frac{1}{2}\left[f\left(y\left(t+\frac{1}{2n}\right)\right)+f\left(y\left(t-\frac{1}{2n}\right)\right)\right]+f(y(t)),$$

we have a local error mapping. Since $\Lambda_n y$ is obviously even in n we must have $\lambda_j=0$ for all odd j in the asymptotic expansion of $\Lambda_n y$.

If we had defined the Δ_n^0 differently, e.g. as with the Euler method, the asymptotic expansion of the local error mapping (which would have been different) would have contained all powers $n^{-j}, j\geq 2$. Thus, the important structural property of the discretization method of being even in n would have remained disguised.

1.3.2 Asymptotic Expansion of the Global Discretization Error

Def. 1.3.3. Consider the discretization method $\mathfrak{M}=\{E_n, E_n^0, \Delta_n, \Delta_n^0, \varphi_n\}_{n\in\mathbb{N}'}$ applicable to the original problem $\mathfrak{P}$. The global discretization error $\{\varepsilon_n\}_{n\in\mathbb{N}'}$ of $\mathfrak{M}$ for $\mathfrak{P}$ is said to *possess an asymptotic expansion to order J* if there exist elements $e_j \in E$, $j=1(1)J$, e_j independent of n, such that

$$\varepsilon_n=\Delta_n\left[\sum_{j=1}^{J}\frac{1}{n^j}e_j\right]+O(n^{-(J+1)}) \quad \text{as } n\to\infty. \tag{1.3.7}$$

(Naturally, if $\mathfrak{M}$ is convergent of order p for $\mathfrak{P}$, then $e_j=0, j=1(1)p-1$, $e_p \neq 0$.)

Remark. Even if the global discretization error of $\mathfrak{M}$ for $\mathfrak{P}$ possesses an asymptotic expansion to any arbitrary order, (1.3.7) need not converge as $J\to\infty$ for any fixed finite n. All that is implied by (1.3.7)—as well as by (1.3.2) through (1.3.4)—is a certain asymptotic behavior of the remainder term as $n\to\infty$.

Knowledge of the existence of an asymptotic expansion to some order $J\geq p$ of the global discretization error is of great practical importance; see Section 1.4. Actually, the asymptotic expansion of the local discretization error has been studied in the preceding section mainly as a tool for the analysis of the global discretization error.

In the following we will need some smoothness conditions on the asymptotic expansion of a local error mapping $\{\Lambda_n\}$.

Def. 1.3.4. In the situation of Def. 1.3.2, the asymptotic expansion (1.3.2) to order J of the local error mapping $\{\Lambda_n\}_{n\in\mathbb{N}'}$ is said to be *(J,p)-smooth* if the derivatives on the left-hand side of

$$\sum_{j=1}^{J}\frac{1}{n^j}\left[\lambda_j y+\sum_{m=1}^{\left[\frac{J-j}{p}\right]}\frac{1}{m!}\lambda_j^{(m)}(y)\left(\sum_{k=p}^{J}\frac{1}{n^k}e_k\right)^m\right] = \Lambda_n\left(y+\sum_{k=p}^{J}\frac{1}{n^k}e_k\right)+O(n^{-(J+1)}) \tag{1.3.8}$$

exist and (1.3.8) holds for arbitrary $y\in D_J$, $e_k\in D_J$, $k=p(1)J$.

If (1.3.2) is (J,p)-smooth we denote by $D_{J,k}\in E$, $k=p(1)J$ any domains such that (1.3.8) still holds for $y\in D_J$, $e_k\in D_{J,k}$, $k=p(1)J$.

Theorem 1.3.1. *Consider the discretization method* $\mathfrak{M}=\{E_n, E_n^0, \Delta_n, \Delta_n^0, \varphi_n\}_{n\in\mathbb{N}'}$ *applicable to the original problem* $\mathfrak{P}=\{E, E^0, F\}$ *with true solution z, and assume that* $\mathfrak{M}$ *and* $\mathfrak{P}$ *satisfy:*

(i) $\mathfrak{M}$ *is stable for* $\mathfrak{P}$;

(ii) $\mathfrak{M}$ *is consistent of order p with* $\mathfrak{P}$, *a local error mapping of* $\mathfrak{M}$ *for* $\mathfrak{P}$ *exists and possesses an asymptotic expansion to order J, and* $z \in D_J$;

(iii) *F possesses* $[J/p]$ *Lipschitz continuous Frechet derivatives in some ball* $B_R := \{y \in E: \|y-z\|_E < R\}$, $R>0$, *and the asymptotic expansion of assumption* (ii) *is* (J,p)*-smooth;*

(iv) $F'(z)^{-1}$ *exists.*

For $j=2p(1)J$ *define mappings* $g_j: D_{J,p} \times D_{J,p+1} \times \cdots \times D_{J,j-p} \to E^0$ *by equating the coefficients of* n^{-j} *in* (*cf. Assumption* (iii) *and Def.* 1.3.4)

$$\sum_{j=2p}^{J} \frac{1}{n^j} g_j(e_p, \cdots, e_{j-p}) \tag{1.3.9}$$
$$= \sum_{m=2}^{\left[\frac{J}{p}\right]} \frac{1}{m!} \left[F^{(m)}(z) + \sum_{j=1}^{J-mp} \frac{1}{n^j} \lambda_j^{(m)}(z) \right] \left(\sum_{k=p}^{J} \frac{1}{n^k} e_k \right)^m + O(n^{-(J+1)});$$

for $j=p(1)2p-1$ *we define* $g_j \equiv 0$. *If the recursive definition of elements* $e_j \in E$, $j=p(1)J$, *through* (*see Assumption* (iv))

$$F'(z)e_j = -\left[\lambda_j z + \sum_{k=1}^{j-p} \lambda_k'(z) e_{j-k} + g_j(e_p, \cdots, e_{j-p}) \right] \tag{1.3.10}$$

is possible, i.e. if successively for $j=p(1)J$

$$e_j \in D_{J,j}, \tag{1.3.11}$$

then the global discretization error $\{\varepsilon_n\}_{n \in \mathbb{N}'}$ *of* $\mathfrak{M}$ *for* $\mathfrak{P}$ *possesses the unique asymptotic expansion to the order J*

$$\varepsilon_n = \Delta_n \left(\sum_{j=p}^{J} \frac{1}{n^j} e_j \right) + O(n^{-(J+1)}).$$

Proof. For the e_j defined above consider (cf. Assumptions (ii) and (iii))

$$F\left(z + \sum_{k=p}^{J} \frac{1}{n^k} e_k \right) + \Lambda_n \left(z + \sum_{k=p}^{J} \frac{1}{n^k} e_k \right)$$
$$= F'(z) \sum_{k=p}^{J} \frac{1}{n^k} e_k + \sum_{j=p}^{J} \frac{1}{n^j} \lambda_j z + \sum_{j=1}^{J} \frac{1}{n^j} \lambda_j'(z) \left(\sum_{k=p}^{J-j} \frac{1}{n^k} e_k \right) \tag{1.3.12}$$
$$+ \sum_{m=2}^{\left[\frac{J}{p}\right]} \frac{1}{m!} \left[F^{(m)}(z) + \sum_{j=1}^{J-mp} \frac{1}{n^j} \lambda_j^{(m)}(z) \right] \left(\sum_{k=p}^{J} \frac{1}{n^k} e_k \right)^m$$
$$+ R_J(z, e_p, \cdots, e_J).$$

According to Assumption (iii) and (1.3.11) (see Def. 1.3.4), we have

$$\|R_J(z, e_p, \cdots, e_J)\| = O(n^{-(J+1)})$$

since $z + \sum_{k=p}^{J} (1/n^k) e_k \in B_R(z)$ for sufficiently large n. The other terms on the right-hand side of (1.3.12) vanish in virtue of (1.3.9) and (1.3.10) except for terms which are of order $O(n^{-(J+1)})$, as is easily checked. (Of course, (1.3.10) is constructed so as to achieve that.)

By Def. 1.3.1, the vanishing of the right-hand side of (1.3.12) except for $O(n^{-(J+1)})$ implies

$$F_n \Delta_n\left(z + \sum_{k=p}^{J} \frac{1}{n^k} e_k\right) = O(n^{-(J+1)}).$$

Since $F_n(\Delta_n z + \varepsilon_n) = 0$ by Def. 1.1.8 and (1.1.3), Assumption (i) implies (see Def. 1.1.11)

$$\varepsilon_n = \Delta_n\left(\sum_{j=p}^{J} \frac{1}{n^j} e_j\right) + O(n^{-(J+1)}).$$

The *uniqueness* follows from the property $\lim_{n\to\infty} \|\Delta_n y\|_{E_n} = \|y\|_E$ of the mappings Δ_n of a discretization method, cf. Def. 1.1.2:

Assume that there is another expansion

$$\varepsilon_n = \Delta_n\left(\sum_{j=p}^{J} \frac{1}{n^j} \hat{e}_j\right) + O(n^{-(J+1)}),$$

with coefficients $\hat{e}_j \in E$ independent of n. Subtract the two expansions and multiply by n^p to obtain

$$\Delta_n(\hat{e}_p - e_p) = O\left(\frac{1}{n}\right)$$

which implies $\hat{e}_p = e_p$. In the same manner, we obtain recursively $\hat{e}_j = e_j$, $j = p+1(1)J$. □

Remarks. 1. Note that the e_j are defined by original problems which are variational problems of $\mathfrak{P}$, see (1.3.10). The discretization method $\mathfrak{M}$ enters into (1.3.10) only via the coefficients $\lambda_j \in (E \to E^0)$ of the asymptotic expansion of a local error mapping of $\mathfrak{M}$ for $\mathfrak{P}$.

2. The only technicalities in the checking of the hypotheses of Theorem 1.3.1 for a given $\mathfrak{M}$ and $\mathfrak{P}$ are the verification of the (J,p)-smoothness of the asymptotic expansion of the local error mapping and of the condition (1.3.11), for a suitable choice of the $D_{J,j}$. In most applications this presents no difficulty (see the example below).

3. The original proof of an existence theorem for an asymptotic expansion of the global discretization error by Stetter [1] was considerably less transparent; it also involved some "silent" assumptions. It differed from the present proof in that it was carried out in E_n and E_n^0. The transfer to E and E^0 in the present proof became possible through the introduction of the local error mapping, Def. 1.3.1.

Example. Euler method for $y'=f(y)$, as previously. Assumptions (i) and (iv) are satisfied and we have consistency of order 1. The existence of a local error mapping with an asymptotic expansion to the order J has been shown in Section 1.3.1 for $D_J=C^{(J+2)}[0,1]^2$. The derivatives occuring in assumption (iii) are

$$F'(y)e=\begin{pmatrix} e(0) \\ e'(t)-f'(y(t))e(t)\end{pmatrix},\qquad F^{(m)}(y)e^m=\begin{pmatrix} 0 \\ -f^{(m)}(y(t))e(t)^m\end{pmatrix},\qquad m\geq 2;$$

for $j=1(1)J$:

$$\lambda_j'(y)e=\begin{pmatrix} 0 \\ \dfrac{1}{(j+1)!}e^{(j+1)}(t)\end{pmatrix}=\lambda_j e,\qquad \lambda_j^{(m)}(y)e^m=\begin{pmatrix}0\\0\end{pmatrix},\qquad m\geq 2.$$

Hence we have to require that f possess $J+1$ continuous derivatives in a neighborhood of $z(t)$, $t\in[0,1]$; this is in accordance with the requirement that $z\in C^{(J+2)}[0,1]=D_J$.

Due to the linearity of Λ_n and the λ_j we have (second components only)

$$\begin{aligned}\Lambda_n\left(y+\sum_{k=1}^{J}\frac{1}{n^k}e_k\right)&=\sum_{j=1}^{J}\frac{1}{n^j}\lambda_j\left(y+\sum_{k=1}^{J}\frac{1}{n^k}e_k\right)+O(n^{-(J+1)})\\ &=\sum_{j=1}^{J}\frac{1}{n^j}\left[\lambda_j y+\lambda_j\left(\sum_{k=1}^{J}\frac{1}{n^k}e_k\right)\right]+O(n^{-(J+1)}),\end{aligned}$$

for $y,e_1,\cdots,e_J\in D_J$; this relation remains valid for $e_k\in D_{j,k}:=D_{J-k}$, $k=1(1)J$, which verifies the $(J,1)$-smoothness.

For (1.3.9) we get

$$\begin{aligned}\sum_{j=2}^{J}\frac{1}{n^j}g_j(e_1,\cdots,e_{j-1})&=-\sum_{m=2}^{J}\frac{1}{m!}f^{(m)}(z(t))\left(\sum_{k=1}^{J}\frac{1}{n^k}e_k\right)^m+O(n^{-(J+1)})\\ &=\frac{1}{n^2}\left(-\frac{1}{2}f''(z)e_1^2\right)\\ &\quad+\frac{1}{n^3}\left(-f''(z)e_1e_2-\frac{1}{6}f'''(z)e_1^3\right)\\ &\quad+\frac{1}{n^4}\left(-\frac{1}{2}f''(z)(e_2^2+2e_1e_3)-\frac{1}{2}f'''(z)e_1^2e_2-\frac{1}{24}f^{\mathrm{IV}}(z)e_1^4\right)+\cdots;\end{aligned}$$

[2] Actually, the smoothness assumptions could be slightly relaxed in this example, cf. Theorem 3.4.5.

thus the $e_j, j=1(1)J$, are defined by (see (1.3.10))

$$e_j(0)=0,$$
$$e_j'(t)-f'(z(t))e_j(t)=b_j(t), \qquad t\in[0,1],$$

where

$$b_1 := -\tfrac{1}{2}z''$$
$$b_2 := -\tfrac{1}{6}z''' - \tfrac{1}{2}e_1'' + \tfrac{1}{2}f''(z)e_1^2,$$
$$b_3 := -\tfrac{1}{24}z^{\mathrm{IV}} - \tfrac{1}{6}e_1''' - \tfrac{1}{2}e_2'' + f''(z)e_1e_2 + \tfrac{1}{6}f'''(z)e_1^3,$$

etc.

From $z\in D_J$ we obtain recursively $e_j\in D_{J-j}$ as is easily seen. Thus we may conclude that these e_j are the coefficients of the asymptotic expansion of the global discretization error of the Euler method for (1.1.2).

1.3.3 Asymptotic Expansions in Even Powers of n

If $\mathfrak{M}$ possesses a local error mapping whose asymptotic expansion is in *even* powers of $1/n$, this property carries over to the expansion of the global discretization error.

Theorem 1.3.2 (Stetter [1]). *In the situation of Theorem* 1.3.1, *if the asymptotic expansion of the local error mapping of* $\mathfrak{M}$ *for* $\mathfrak{P}$ *is in even powers of* $1/n$, *i.e., if*

(1.3.13) $$\lambda_j y=0, \quad j \text{ odd, for all } y\in D,$$

then the asymptotic expansion of the global discretization error of $\mathfrak{M}$ *for* $\mathfrak{P}$ *is also in even powers of* $1/n$, *i.e.*

$$e_j=0, \quad j \text{ odd}.$$

Proof. Replace n by n^2 in the expansions in the proof of Theorem 1.3.1. □

Remark. It is *not* in general sufficient to have $\lambda_j z=0$, j odd, since this need not imply the vanishing of the $\lambda^{(m)}(z)$ for odd j.

Example. Theorem 1.3.2 shows that—with sufficient smoothness of $\mathfrak{P}$—the implicit trapezoidal method of the example at the end of Section 1.3.1 possesses an asymptotic expansion of the global discretization error in even powers of $1/n$ since the satisfaction of all assumptions of Theorem 1.3.1, with $p=2$, may be shown similarly as for the Euler method.

Some calculations show that the e_{2j} in $\varepsilon_n=\Delta_n \sum_{j=1}^{J}(1/n^{2j})e_{2j}$ satisfy

$$e_{2j}(0)=0, \qquad e_{2j}'-f'(z)e_{2j}=b_{2j},$$

with

$$b_2=\tfrac{1}{12}z''',$$
$$b_4=\tfrac{1}{480}z^{\mathrm{V}} - \tfrac{1}{24}e_2''' + \tfrac{1}{8}[(f'''(z)z'^2+f''(z)z'')e_2 + 2f''(z)z'e_2' + f'(z)e_2''] + \tfrac{1}{2}f''(z)e_2^2,$$

etc.

This example points out an interesting aspect of Theorem 1.3.1. While we observed in Section 1.3.1 that the local error mapping and its asymptotic expansion may depend strongly on the choice of the Δ_n^0, the e_j of the expansion (1.3.7) of the global discretization error and hence the values of the right-hand sides of (1.3.10) for $j=p(1)J$ cannot depend on the Δ_n^0. Thus, the construction mechanism of (1.3.9)/(1.3.10) must automatically reduce any unnatural complications which have been introduced through an improper choice of the Δ_n^0.

In particular, in the situation of Theorem 1.3.2, the right-hand sides of (1.3.10) must vanish for odd j even if the λ_j, j odd, themselves do not vanish due to an ill-advised choice of the Δ_n^0. Thus, while the Δ_n^0 are immaterial for the evenness of the asymptotic expansion of the global discretization error, it is clear that the verification of this evenness will be virtually impossible if it cannot be concluded via Theorem 1.3.2 from the evenness of the asymptotic expansion of a local error mapping.

Let us finally emphasize that Theorem 1.3.1 gives only *sufficient* conditions for the existence of an asymptotic expansion for the global discretization error.

1.3.4 The Principal Error Terms

The most immediate and most frequent application of asymptotic expansions in mathematical analysis consists of the use of the first term as an approximation for the quantity which admits the asymptotic expansion. As n increases this approximation becomes better and better. There are innumerable examples of this approach.

A similar situation exists in the analysis of the discretization error of a method which is convergent of order p. If the global discretization error $\{\varepsilon_n\}$ is known to possess an asymptotic expansion to order p, then the first and only term of this expansion may often be used as a reasonable approximation to the (unknown) discretization error. Even if n is not very large, the approximation will often be qualitatively satisfactory. This use of asymptotic expansions for discretization methods has been introduced and systematically exploited by Henrici in his treatises.

Def. 1.3.5. Assume that the discretization method $\mathfrak{M}$ is convergent of order p for the original problem $\mathfrak{P}$ and that both the local and the global discretization error of $\mathfrak{M}$ for $\mathfrak{P}$ possess asymptotic expansions to order p:

$$l_n = n^{-p}\Delta_n^0 \lambda_p z + O(n^{-(p+1)}), \tag{1.3.14}$$

$$\varepsilon_n = n^{-p}\Delta_n e_p + O(n^{-(p+1)}). \tag{1.3.15}$$

The (sequences of) elements $n^{-p}\Delta_n^0 \lambda_p z \in E_n^0$ and $n^{-p}\Delta_n e_p \in E_n$ are called the *principal terms* of the corresponding discretization errors.

If $\mathfrak{M}$ possesses a local error mapping $\{\Lambda_n\}$ for $\mathfrak{P}$ which has an asymptotic expansion to order p, then sufficient conditions for (1.3.15) to hold with e_p defined by

$$F'(z)e_p = -\lambda_p z \tag{1.3.16}$$

are easily obtained from Theorem 1.3.1. However, the assumption of the existence of a local error mapping may be replaced by a simpler, less stringent assumption if we are only interested in the principal term of the global discretization error.

Theorem 1.3.3. *Consider the discretization method* $\mathfrak{M}=\{E_n, E_n^0, \Delta_n, \Delta_n^0, \varphi_n\}_{n\in\mathbb{N}'}$ *applicable to the original problem* $\mathfrak{P}=\{E, E^0, F\}$ *with true solution* z, *and assume that* $\mathfrak{M}$ *and* $\mathfrak{P}$ *satisfy:*

(i) $\mathfrak{M}$ *is stable for* $\mathfrak{P}$;

(ii) *the local discretization error of* $\mathfrak{M}$ *for* $\mathfrak{P}$ *satisfies* (1.3.14);

(iii) *the* $F_n=\varphi_n(F)$ *are Frechet-differentiable at* $\Delta_n z$ *and*

$$\|F_n(\Delta_n z+\varepsilon_n)-F_n\Delta_n z-F_n'(\Delta_n z)\varepsilon_n\| = O(n^{-(p+1)})$$

for any sequence $\{\varepsilon_n\}_{n\in\mathbb{N}'}$, $\varepsilon_n\in E_n$, $\|\varepsilon_n\|=O(n^{-p})$;

(iv) $F'(z)^{-1}$ *exists;*

(v) *for the element* $e_p\in E$ *defined by* (1.3.16)

$$F_n'(\Delta_n z)\Delta_n e_p=\Delta_n^0 F'(z)e_p+O(n^{-1}). \tag{1.3.17}$$

Then $n^{-p}\Delta_n e_p$ *is the principal part of the global discretization error of* $\mathfrak{M}$ *for* $\mathfrak{P}$.

Proof. $F_n\left(\Delta_n z+\frac{1}{n^p}\Delta_n e_p\right)$

$$= F_n\Delta_n z+\frac{1}{n^p}F_n'(\Delta_n z)\Delta_n e_p+O(n^{-(p+1)}) \quad \text{by Assumption (iii)}$$

$$=\frac{1}{n^p}\Delta_n^0[\lambda_p z+F'(z)e_p]+O(n^{-(p+1)}) \quad \text{by Assumptions (ii) and (v)}$$

$$=O(n^{-(p+1)}) \quad \text{by (1.3.16).}$$

The stability of $\mathfrak{M}$ implies (1.3.15) since $F_n\zeta_n=0$. □

Remark. In place of the strong commutativity assumption inherent in the existence of a local error mapping (see beginning of Section 1.3.1) we have used the very weak commutativity assumption (1.3.17). It is possible to formulate a theorem for the general case of Theorem 1.3.1 which also replaces the assumption of a local error mapping by a set of more specialized commutativity assumptions. Its proof is a generalization of the above proof of Theorem 1.3.3 using expansions involving higher derivatives and the construction (1.3.9)/(1.3.10). However, the precise formulation of the necessary commutativity assumptions becomes quite awkward in the general case.

Example. Euler-method. The principal term of the local discretization error is determined by (see Section 1.3.1)

$$\lambda_1 z = -\tfrac{1}{2} z''$$

so that (1.3.16) becomes

$$e_1' - f'(z) e_1 = -\tfrac{1}{2} z'', \qquad e_1(0) = 0 .$$

Assumption (iii) of Theorem 1.3.3 requires

$$f\left(z\left(\frac{\nu-1}{n}\right) + \varepsilon_n\left(\frac{\nu-1}{n}\right)\right) = f\left(z\left(\frac{\nu-1}{n}\right)\right) + f'\left(z\left(\frac{\nu-1}{n}\right)\right)\varepsilon_n\left(\frac{\nu-1}{n}\right) + O(n^{-2})$$

for any sequence $\{\varepsilon_n\}$ with $\|\varepsilon_n\| = O(n^{-1})$, which is satisfied if f' is Lipschitz-continuous in a vicinity of the $z(t)$, $t \in [0,1]$. The commutativity condition (1.3.17) reduces to

$$\frac{e_1\left(\frac{\nu}{n}\right) - e_1\left(\frac{\nu-1}{n}\right)}{1/n} = e_1'\left(\frac{\nu-1}{n}\right) + O(n^{-1}),$$

i.e., to the Lipschitz-continuity of e_1'; but this is again implied by the Lipschitz-continuity of f' so that this is the only smoothness condition necessary.

Corollary 1.3.4. *Under the assumptions of Theorem* 1.3.3, *the element* $\lambda_p z \in E^0$ *in the principal term of the local discretization error* (cf. (1.3.14)) *is independent of the choice of* $\{\Delta_n^0\}$.

Proof. e_p is certainly independent of the Δ_n^0 and $\lambda_p z = -F'(z) e_p$ by (1.3.16). □

Remark. The independence of $\lambda_j z$ of $\{\Delta_n^0\}$ holds *only* for $\lambda_p z$.

Example. Implicit trapezoidal method; see end of Section 1.3.1. For $\nu = 1(1)n$,

$$l_n\left(\frac{\nu}{n}\right) = \frac{z\left(\frac{\nu}{n}\right) - z\left(\frac{\nu-1}{n}\right)}{\frac{1}{n}} - \frac{1}{2}\left(z'\left(\frac{\nu}{n}\right) - z'\left(\frac{\nu-1}{n}\right)\right)$$

$$= -\frac{1}{12n^2} z'''\left(\frac{2\nu-1}{2n}\right) - \frac{1}{480n^4} z^{\mathrm{V}}\left(\frac{2\nu-1}{2n}\right) - \cdots .$$

Thus, whenever (1.3.14) holds we have to have $\lambda_2 z = -\frac{1}{12} z'''$; however, the following terms in an asymptotic expansion (1.3.4) of l_n depend on the choice of Δ_n^0:

If we take Δ_n^0 as indicated at the end of Section 1.3.1, then $\lambda_3 z$ vanishes. If we choose

$$\Delta_n^0\begin{pmatrix} d_0 \\ d(t) \end{pmatrix}\left(\frac{v}{n}\right)=d\left(\frac{v}{n}\right) \quad \text{for } v=1(1)n$$

we have

$$l_n\left(\frac{v}{n}\right)=-\frac{1}{12n^2}z'''\left(\frac{v}{n}\right)+\frac{1}{24n^3}z^{\text{IV}}\left(\frac{v}{n}\right)$$

so that $\lambda_3 z=\frac{1}{24}z^{\text{IV}}$.

1.4. Applications of Asymptotic Expansions

1.4.1 Richardson Extrapolation

The basic idea of the so-called Richardson extrapolation technique is to consider the solution ζ_n of a convergent discretization as a *function of n*, to compute a few values ζ_{n_ρ} of this function and to interpolate these values by a suitable function χ of n. The limit process $n\to\infty$ which would lead to the true solution z of the original problem is then simulated by taking the value of the interpolation function χ at $n=\infty$

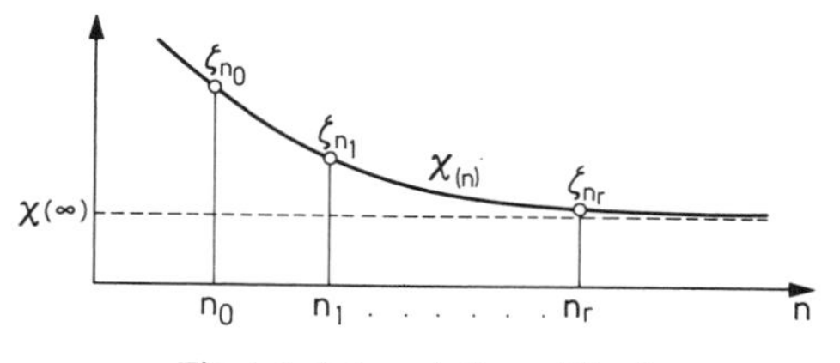

Fig. 1.5. Interpolation of the ζ_n

(see Fig. 1.5). Of course, Fig. 1.5 is only symbolic, since the ζ_{n_ρ} are elements of different spaces. For a formal treatment we have to "reduce" the ζ_{n_ρ} to a common space $E_{\bar{n}}$; in practical applications this means that we have to consider their values on the points of a reduced grid $\mathfrak{G}_{\bar{n}}$ which is contained in each of the grids $\mathfrak{G}_{n_\rho}$. The interpolation then takes place in $E_{\bar{n}}$ (i.e., for the values of the ζ_{n_ρ} on $\mathfrak{G}_{\bar{n}}$) and produces an approximation to $\Delta_{\bar{n}} z$.

The technique is named after L. C. Richardson ([1], [2]) who seems to have been the first to use it as a tool for improving the results gained from discretization methods.

Def. 1.4.1. Consider a discretization method $\mathfrak{M}=\{E_n, E_n^0, \Delta_n, \Delta_n^0, \varphi_n\}_{n\in\mathbb{N}'}$ applicable to $\mathfrak{P}=\{F, E^0, F\}$. For a given Banach space $\hat{E}$ and a linear mapping $\hat{\Delta}: E\to\hat{E}$ with $\|\hat{\Delta}\|=1$, if there exist an infinite set $\bar{\mathbb{N}}\subset\mathbb{N}'$

and a sequence of linear mappings $\pi_n: E_n \to \hat{E}$, $n \in \bar{\mathbb{N}}$, such that

$$\pi_n \Delta_n = \hat{\Delta} \quad \text{and} \quad \lim_{n\to\infty} \|\pi_n\| = 1, \tag{1.4.1}$$

then $\mathfrak{M}$ is called $\hat{\Delta}$-*reducible* and the π_n are the corresponding reduction mappings.

Example. With the Euler method, e.g., let $\mathbb{N}' = \mathbb{N}$, $\mathbb{G}_n := \{v/n, v = 0(1)n\}$, $E_n = (\mathbb{G}_n \to \mathbb{R})$ and $(\Delta_n y)(v/n) = y(v/n)$ for $y \in E = C[0, 1]$.

a) Consider an arbitrary fixed rational $\hat{t} = (p/q) \in [0, 1]$ and take $\hat{E} = \mathbb{R}$, $\hat{\Delta} y = y(\hat{t})$; then $\mathfrak{M}$ is $\hat{\Delta}$-reducible with $\bar{\mathbb{N}} = \{iq, i \in \mathbb{N}\}$ and $\pi_n \eta_n = \eta_n(p/q)$, $\eta_n \in E_n$, $n \in \bar{\mathbb{N}}$.

b) Take some fixed $\bar{n} \in \mathbb{N}$ and $\hat{E} = E_{\bar{n}}$, $\hat{\Delta} = \Delta_{\bar{n}}$; then $\mathfrak{M}$ is $\hat{\Delta}$-reducible with $\bar{\mathbb{N}} = \{i\bar{n}, i \in \mathbb{N}\}$ and the π_n are the restrictions of functions on $\mathbb{G}_n$ to $\mathbb{G}_{\bar{n}}$, $n \in \bar{\mathbb{N}}$.

If $\mathfrak{M}$ is $\hat{\Delta}$-reducible and if the global discretization error of $\mathfrak{M}$ for $\mathfrak{P}$ possesses an asymptotic expansion (1.3.7) to some order J, we have from (1.3.7) and (1.4.1)

$$\pi_n \zeta_n = \hat{\Delta} z + \hat{\Delta} \sum_{j=p}^{J} \frac{1}{n^j} e_j + O(n^{-(J+1)}), \qquad n \in \bar{\mathbb{N}}, \tag{1.4.2}$$

i.e., the sequence $\{\pi_n \zeta_n\}_{n\in\bar{\mathbb{N}}}$ of the reductions $\pi_n \zeta_n \in \hat{E}$ of the solutions ζ_n of $\mathfrak{M}(\mathfrak{P})$, $n \in \bar{\mathbb{N}}$, possesses an asymptotic expansion in the *fixed* space $\hat{E}$. ($\bar{\mathbb{N}}$ had to be an infinite set in order that the concept of an asymptotic expansion and the symbol $O(n^{-(J+1)})$ make sense.)

Since it is an expansion (1.4.2) rather than (1.3.7) which is needed for the application of Richardson extrapolation it is important to note that (1.3.7) is only a sufficient but *not a necessary condition* for the existence of an expansion (1.4.2) (see the example below).

Def. 1.4.2. If the discretization method $\mathfrak{M}$ applicable to $\mathfrak{P}$ is $\hat{\Delta}$-reducible and if the solution sequence $\{\zeta_n\}$ of $\mathfrak{M}(\mathfrak{P})$ satisfies

$$\pi_n \zeta_n = \hat{\Delta} z + \sum_{j=p}^{J} \frac{1}{n^j} \hat{e}_j + O(n^{-(J+1)}), \qquad n \in \bar{\mathbb{N}}, \tag{1.4.3}$$

where the $\hat{e}_j \in \hat{E}$, $j = p(1)J$, are independent of n, then the global discretization error of $\mathfrak{M}$ for $\mathfrak{P}$ is said to possess a $\hat{\Delta}$-*reduced asymptotic expansion to order* J.

Example. Cf. Sect. 4.4.2. For certain multistep methods for initial value problems the remainder term in (1.3.7) does not behave like $O(n^{-(J+1)})$ uniformly on the whole interval $[0, 1]$ of integration but only on each fixed subinterval $[\hat{t}, T]$, $\hat{t} > 0$. Thus, (1.3.7)—which implies that the *norm* of $\varepsilon_n - \Delta_n \sum (1/n^j) e_j$ is $O(n^{-(J+1)})$—does not hold. However, if we restrict attention to a fixed number of points in $[\hat{t}, T]$ which are common to infinitely many grids then a corresponding $\hat{\Delta}$-reduced asymptotic expansion (1.4.3) exists; see Theorem 4.4.2.

Def. 1.4.3. For some domain I and space X, let $C_r \subset (I \to X)$, $r \in \mathbb{N}$, be a family of functions from I to X such that

given disjoint $n_\rho \in I$ and $x_\rho \in X$, $\rho = 0(1)r$,

there is exactly one function $\chi \in C_r$ such that

$$\chi(n_\rho) = x_\rho, \qquad \rho = 0(1)r. \tag{1.4.4}$$

A sequence $\{C_r\}_{r \in \mathbb{N}}$ of such families is called a *class of interpolation functions from I to X*.

Example. It is well-known that the polynomials form a class of interpolation functions from $\mathbb{R}$ to $\mathbb{R}$; in this case the families C_r are the polynomials of degree not greater than r.

Remark. If X is a finite-dimensional linear space and hence isomorphic to some $\mathbb{R}^{\bar{m}}$, each class $\mathfrak{C}_0$ of interpolation functions from I to $\mathbb{R}$ generates a class of interpolation functions from I to X: It consists of the interpolations of the $\bar{m}$ individual *components* of the x_ρ by functions in $\mathfrak{C}_0$.

Def. 1.4.4. The class $\mathfrak{C}$ of interpolation functions is called *linear* if each C_r is a linear space (of dimension $r+1$); otherwise it is called *nonlinear*.

For a linear class $\mathfrak{C}$ of interpolation functions, the values of the function $\chi \in C_r$ satisfying (1.4.4) are linear combinations of the x_ρ:

$$\chi(n) = \sum_{\rho=0}^{r} \gamma_\rho(n) x_\rho.$$

Consider now a discretization method $\mathfrak{M}$ which is $\hat{\Delta}$-reducible and whose discretization error possesses, for some $\mathfrak{P}$, a $\hat{\Delta}$-reduced asymptotic expansion (1.4.3) to order J; $\hat{E}$, $\bar{\mathbb{N}}$ and π_n have the meaning given in Def. 1.4.1. Assume that we have a class $\mathfrak{C}$ of interpolation functions from $\mathbb{N}$ to $\hat{E}$ such that all functions in $\mathfrak{C}$ have a finite limit at infinity. If we have determined $r+1$ elements ζ_{n_ρ}, $n_\rho \in \bar{\mathbb{N}}$, $\rho = 0(1)r$, of the solution sequence of $\mathfrak{M}(\mathfrak{P})$, the elements $\pi_{n_\rho} \zeta_{n_\rho} \in \hat{E}$ determine a unique function $\chi \in C_r$ with

$$\chi(n_\rho) = \pi_{n_\rho} \zeta_{n_\rho}, \qquad \rho = 0(1)r, \tag{1.4.5}$$

according to Def. 1.4.3. Richardson extrapolation or "extrapolation to the limit" consists of determining the value

$$\chi(\infty) := \lim_{n \to \infty} \chi(n) \in \hat{E}$$

and using it as an approximation to $\hat{\Delta} z$. (This is the formal description of the procedure outlined at the beginning of this section.)

In order that this procedure be reasonable for values $\pi_n \zeta_n$ satisfying (1.4.2) we have to use a class $\mathfrak{C}$ of interpolation functions the asymptotic expansion of which in powers of n^{-1} has the same structure as (1.4.3), viz.

$$\chi(n) = \text{const.} + O(n^{-p}) \quad \text{for each } \chi \in \mathfrak{C}. \tag{1.4.6}$$

If (1.4.3) contains only even powers of n^{-1}, then the $\chi \in \mathfrak{C}$ should also be even functions of n.

Before we discuss details of the extrapolation process in the following sections, we give a rather simple-minded example:

Example. Euler-method for $y' = gy$, $y(0) = 1$. Choose $\bar{n} = 2$ in Example *b* after Def. 1.4.1, so that $\hat{E} = E_2 = (\{0, \frac{1}{2}, 1\} \to \mathbb{R}\}$, $\bar{N} = \{2i, i \in \mathbb{N}\}$,

$$\pi_n \zeta_n = \begin{pmatrix} \zeta_n(0) \\ \zeta_n(\frac{1}{2}) \\ \zeta_n(1) \end{pmatrix} \in \hat{E} \quad \text{for even } n.$$

Since the asymptotic expansion of the global discretization error of the Euler method contains all positive powers of $1/n$ (cf. end of Section 1.3.2) we may take for the class $\mathfrak{C}_0$ of interpolation functions from $\mathbb{N}$ to $\mathbb{R}$ to be used on the components of the $\pi_n \zeta_n$ (see the remark after Def. 1.4.3), e.g.,

a) the polynomials in $1/n$, with C_r the polynomials of degree not greater than r;
b) the rational functions with no poles in $[1, \infty)$; in C_r the degrees d_n and d_d of the numerator and denominator polynomials in $1/n$ have to satisfy certain restrictions.
Take $r = 1$ and assume $n_1 = 2n_0$, $n_0 \in \bar{\mathbb{N}}$. We obtain

$$\bar{\zeta}_0 := \pi_{n_0} \zeta_{n_0} = \begin{pmatrix} 1 \\ \left(1 + \dfrac{g}{n_0}\right)^{\frac{n_0}{2}} \\ \left(1 + \dfrac{g}{n_0}\right)^{n_0} \end{pmatrix}, \qquad \bar{\zeta}_1 := \pi_{n_1} \zeta_{n_1} = \begin{pmatrix} 1 \\ \left(1 + \dfrac{g}{2n_0}\right)^{n_0} \\ \left(1 + \dfrac{g}{2n_0}\right)^{2n_0} \end{pmatrix}.$$

The components of the interpolation function $\chi(n) \in C_1$ are formed from the components of $\bar{\zeta}_0$ and $\bar{\zeta}_1$ according to

a) $\chi(n) = \bar{\zeta}_1 + \left(1 - 2\dfrac{n_0}{n}\right)(\bar{\zeta}_1 - \bar{\zeta}_0)$, $\quad \chi(\infty) = \bar{\zeta}_1 + (\bar{\zeta}_1 - \bar{\zeta}_0)$; or

b) $\chi(n) = \dfrac{\bar{\zeta}_1}{1 + \left(1 - 2\dfrac{n_0}{n}\right)\dfrac{\bar{\zeta}_0 - \bar{\zeta}_1}{\bar{\zeta}_0}}$, $\quad \chi(\infty) = \bar{\zeta}_1 \left(1 - \dfrac{\bar{\zeta}_1 - \bar{\zeta}_0}{\bar{\zeta}_0}\right)^{-1}$.

With the above values of $\bar{\zeta}_0$ and $\bar{\zeta}_1$, this leads (after some computation) to

a)
$$\chi(\infty) = \begin{pmatrix} 1 \\ e^{\frac{g}{2}}\left[1 - \dfrac{g^3}{192}(3g + 16) n_0^{-2} + O(n_0^{-3})\right] \\ e^{g}\left[1 - \dfrac{g^3}{48}(3g + 8) n_0^{-2} + O(n_0^{-3})\right] \end{pmatrix},$$

b)

$$\chi(\infty)=\begin{pmatrix} 1 \\ e^{\frac{g}{2}}\left[1+\frac{g^3}{192}(3g-16)n_0^{-2}+O(n_0^{-3})\right] \\ e^g\left[1+\frac{g^3}{48}(3g-8)n_0^{-2}+O(n_0^{-3})\right] \end{pmatrix}.$$

It is evident that the values of $\chi(\infty)$ approach $\hat{\Delta}z=\begin{pmatrix}1\\ e^{g/2}\\ e^g\end{pmatrix}$ like n_0^{-2} as $n_0\to\infty$ and that they will be much better approximations to $\hat{\Delta}z$ than $\bar{\zeta}_0$ and $\bar{\zeta}_1$ even for small values of n_0.

1.4.2 Linear Extrapolation

If we perform the interpolation in $\hat{E}\cong\mathbb{R}^{\bar{m}}$ componentwise, i.e. separately for the individual components of each of the values of $\pi_n\zeta_n$, we have to perform a set of $\bar{m}$ independent one-dimensional interpolations (comp. the remark after Def. 1.4.3).

Consider the situation of Def. 1.4.2 and assume that the sequence $\{\pi_n\zeta_n\}_{n\in\bar{\mathbb{N}}}$ possesses the asymptotic expansion (1.4.3) to some order J. If we denote[3] the value of an arbitrary but fixed component of $\pi_{n_\rho}\zeta_{n_\rho}$ by T_0^ρ, $n_\rho\in\bar{\mathbb{N}}$, we have

$$T_0^\rho=\bar{z}+\sum_{j=p}^{J}\frac{1}{n_\rho^j}\bar{e}_j+R_J(n_\rho), \tag{1.4.7}$$

where $\bar{z}$ and $\bar{e}_j$ now denote the value of that particular component of $\hat{\Delta}z$ and $\hat{e}_j$ resp., and $R_J(n)=O(n^{-(J+1)})$ as $n\to\infty$.

For a given sequence $\{T_0^\rho\}$, $\rho=0,1,\ldots$, satisfying (1.4.7), with $n_{\rho+1}>n_\rho$, and a fixed class $\mathfrak{C}=\{C_r\}$ of interpolation functions from $[0,1]$ to $\mathbb{R}$ we denote[3] by $\chi_r^i\in C_r$ the function for which

$$\chi_r^i\left(\frac{1}{n_\rho}\right)=T_0^\rho,\qquad \rho=i(1)\,i+r, \tag{1.4.8}$$

and set

$$T_r^i:=\chi_r^i(0). \tag{1.4.9}$$

---

[3] With the denotation T_r^i etc. we follow the precedent set in a number of papers on the subject.

We thus obtain the well-known extrapolation table:

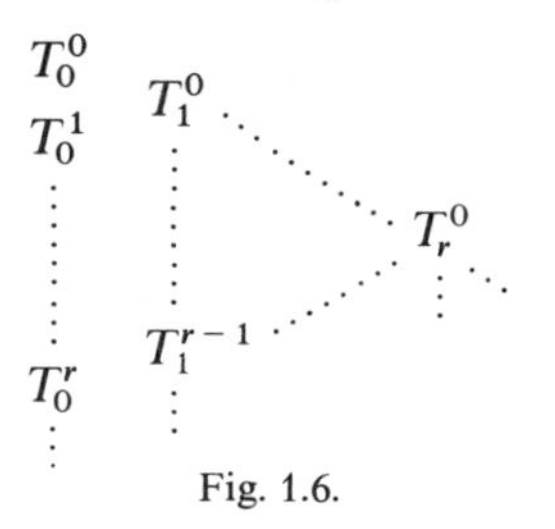

Fig. 1.6.

In the case of *linear* extrapolation (see Def. 1.4.4) the T_r^i are linear combinations of the T_0^ρ, $\rho = i(1)i+r$:

(1.4.10) $$T_r^i = \sum_{\rho=i}^{i+r} \gamma_{r\rho}^i T_0^\rho.$$

The coefficients $\gamma_{r\rho}^i$ depend on the class $\mathfrak{C}$ and on the sequence $\{n_\rho\}$.

Theorem 1.4.1. *Assume that* $\lim_{\rho\to\infty} T_0^\rho = \bar{z}$. *If the* $\gamma_{r\rho}^i$ *satisfy*

(1.4.11) $$\sum_{\rho=i}^{i+r} \gamma_{r\rho}^i = 1 \quad \text{for all } i \text{ and } r$$

then

(1.4.12) $$\lim_{i\to\infty} T_r^i = \bar{z} \quad \text{for each fixed } r>0.$$

If furthermore

(1.4.13) $$\sum_{\rho=i}^{i+r} |\gamma_{r\rho}^i| \le C < \infty \quad \text{uniformly for all } i \text{ and } r$$

and

(1.4.14) $$\lim_{r\to\infty} \gamma_{r\rho}^i = 0 \quad \text{for each fixed } i \text{ and fixed } \rho \ge i$$

then

(1.4.15) $$\lim_{r\to\infty} T_r^i = \bar{z} \quad \text{for each fixed } i \ge 0.$$

Proof. (1.4.12) follows immediately from (1.4.10) and (1.4.11). (1.4.15) follows from a well-known theorem of Toeplitz on transformations of sequences. □

Remark. (1.4.11) is satisfied whenever $\mathfrak{C}$ interpolates constant functions accurately. Assumptions (1.4.13) and (1.4.14) are not necessarily satisfied by an arbitrary class $\mathfrak{C}$ and sequence $\{n_\rho\}$.

Example. Let $\mathfrak{C}$ be the class of polynomials. By the well-known Lagrange representation of the interpolation polynomial, we have

$$\gamma_{r\rho}^i = \prod_{\substack{\rho'=i \\ \rho' \neq \rho}}^{i+r} \frac{0 - \frac{1}{n_{\rho'}}}{\frac{1}{n_\rho} - \frac{1}{n_{\rho'}}} = \prod_{\rho'=i}^{i+r}{}' \frac{1}{1 - \frac{n_{\rho'}}{n_\rho}}. \tag{1.4.16}$$

This implies (1.4.14) as the product (1.4.16) diverges to zero as $r \to \infty$. (1.4.13) holds if and only if there exists a $\delta > 0$ such that

$$\frac{n_{\rho+1}}{n_\rho} \geq 1 + \delta \quad \text{for all } \rho \in \mathbb{N}$$

(see e.g. Bulirsch [1]).

More interesting is the nature of the approximation to $\bar{z}$ provided by T_r^i for finite i and r.

Theorem 1.4.2. *Let the T_0^ρ possess an asymptotic expansion* (1.4.7) *to order $p+r$, with a known value of $p \geq 1$. If $\mathfrak{C}$ is taken to be the class of polynomials with*

$$\chi'(0) = \cdots = \chi^{(p-1)}(0) = 0 \tag{1.4.17}$$

then

$$T_r^i = \bar{z} + \left(\frac{1}{n_{i+r}^p} \prod_{\rho=i}^{i+r-1} \frac{1}{n_\rho} \right) K_{pr}(n_i, \ldots, n_{i+r}) \bar{e}_{p+r} + \sum_{\rho=i}^{i+r} \gamma_{r\rho}^i R_{p+r}(n_\rho) \tag{1.4.18}$$

where K_{pr} is homogeneous of order 0 *in its $r+1$ arguments (i.e., it depends only on their ratios).*

In particular, $K_{1r} \equiv (-1)^r$ independently of the n_ρ, so that for $p=1$

$$T_r^i = \bar{z} + (-1)^r \left(\prod_{\rho=i}^{i+r} \frac{1}{n_\rho} \right) \bar{e}_{r+1} + \sum_{\rho=i}^{i+r} \gamma_{r\rho}^i R_{r+1}(n_\rho). \tag{1.4.19}$$

Proof. By assumption, we have from (1.4.7)

$$T_0^\rho = \left[\bar{z} + \sum_{j=p}^{p+r-1} \frac{1}{n_\rho^j} \bar{e}_j \right] + \frac{1}{n_\rho^{p+r}} \bar{e}_{p+r} + R_{p+r}(n_\rho).$$

Due to the linearity, we may interpolate the three terms separately. The polynomials in C_r are of the form (see (1.4.17))

$$\chi(x) = c_0 + c_p x^p + \cdots + c_{p+r-1} x^{p+r-1}. \tag{1.4.20}$$

The values of the first bracket are values of a polynomial (1.4.20), with $c_0=\bar{z}$, $c_j=\bar{e}_j$, $j=p(1)p+r-1$; hence, the interpolation yields the exact value $\chi(0)=\bar{z}$.

For the interpolation of the term $(1/n_\rho^{p+r})\bar{e}_{p+r}$ we have to find a polynomial χ of the form (1.4.20) such that, with $x_\rho:=1/n_\rho$,

$$\chi(x_\rho)=x_\rho^{p+r}, \qquad \rho=i(1)i+r.$$

This leads to the linear system

$$\begin{pmatrix} x_i^p & \dots & x_i^{p+r-1} & 1 \\ \cdot & & \cdot & \cdot \\ \cdot & & \cdot & \cdot \\ \cdot & & \cdot & \cdot \\ x_{i+r}^p & \dots & x_{i+r}^{p+r-1} & 1 \end{pmatrix} \begin{pmatrix} c_p \\ \cdot \\ \cdot \\ c_{p+r-1} \\ c_0 \end{pmatrix} = \begin{pmatrix} x_i^{p+r} \\ \cdot \\ \cdot \\ \cdot \\ x_{i+r}^{p+r} \end{pmatrix}.$$

We use Cramer's rule to represent $\chi(0)=c_0$ as a quotient of two determinants. Expansion of the denominator determinant by its last column and the use of the formula for Vandermonde determinants gives

$$\begin{aligned} \frac{1}{c_0} &= \sum_{\rho=i}^{i+r} \frac{1}{x_\rho^p \prod_{\rho'=i}^{i+r}{}' (x_\rho - x_{\rho'})} \\ &= \left(\frac{1}{x_{i+r}^p} \prod_{\rho=i}^{i+r-1} \frac{1}{x_\rho}\right) \sum_{\rho=i}^{i+r} \left[\left(\frac{x_{i+r}}{x_\rho}\right)^{p-1} \prod_{\rho'=i}^{i+r}{}' \frac{x_{\rho'}}{x_\rho - x_{\rho'}}\right] \\ &=: \left(n_{i+r}^p \prod_{\rho=i}^{i+r-1} n_\rho\right) [K_{pr}(n_i, \dots, n_{i+r})]^{-1}, \end{aligned}$$

the homogeneity of order 0 of K_{pr} is evident. For $p=1$ we have

$$K_{1r} = \sum_{\rho=i}^{i+r} \prod_{\rho'=i}^{i+r}{}' \frac{\frac{1}{n_{\rho'}}}{\frac{1}{n_\rho} - \frac{1}{n_{\rho'}}} = (-1)^r \sum_{\rho=i}^{i+r} \gamma_{r\rho}^i = (-1)^r$$

by (1.4.16) and (1.4.11) which holds trivially for polynomials. □

Remark. Estimates of the remainder term in (1.4.18) or (1.4.19) are possible only when such estimates are available for (1.4.7). If the $\gamma_{r\rho}^i$ satisfy (1.4.13) then one at least knows

$$\left|\sum_{\rho=i}^{i+r} \gamma_{r\rho}^i R_{p+r}(n_\rho)\right| \le C \max_{\rho=i(1)n+r} |R_{p+r}(n_\rho)|. \tag{1.4.21}$$

Theorems which give more detailed estimates for special cases may be found, e.g., in Bulirsch-Stoer [1], and Gragg [1].

Corollary 1.4.3. *Under the conditions of Theorem* 1.4.2, *if the asymptotic expansion* (1.4.7) *is in even powers of* n^{-1}, $p=2$, *and* $\mathfrak{C}$ *is the class of even polynomials, then*

$$T_r^i=\bar{z}+(-1)^r\left(\prod_{\rho=i}^{i+r}\frac{1}{n_\rho^2}\right)\bar{e}_{2r+2}+\sum_{\rho=i}^{i+r}\bar{\gamma}_{r\rho}^i R_{2r+2}(n_\rho), \tag{1.4.22}$$

where the $\bar{\gamma}_{r\rho}^i$ *are the coefficients in* (1.4.10) *for interpolation with even polynomials.*

Proof. Take $p=1$ in Theorem 1.4.2 and replace n_ρ by n_ρ^2. □

Often the n_ρ are taken as fixed multiples of a basic parameter $\bar{n}$. In this case one may ask how the quality of the approximation furnished by the T_r^i improves as $\bar{n}$ increases:

Corollary 1.4.4. *In the situation of Theorem* 1.4.2, *if*

$$n_\rho=\beta_\rho\bar{n}, \qquad \rho=0(1)r, \qquad \beta_\rho \ \textit{fixed},$$

and if the corresponding $\gamma_{r\rho}^0$ *satisfy* (1.4.13) *uniformly for* $\bar{n}\in\mathbb{N}$, *then*

$$T_r^0=\bar{z}+O(\bar{n}^{-(p+r)}) \quad \textit{as } \bar{n}\to\infty .$$

In the situation of Corollary 1.4.3 *(expansion even in n), we have*

$$T_r^0=\bar{z}+O(\bar{n}^{-(2r+2)}) \quad \textit{as } \bar{n}\to\infty . \tag{1.4.23}$$

Proof. Follows immediately from (1.4.18) or (1.4.22), (1.4.21), and

$$R_{p+r}(n)=O(n^{-(p+r+1)}) . \qquad □$$

If the asymptotic expansion (1.4.7) may be extended to an arbitrary order, if the $\bar{e}_j$ are uniformly bounded as j grows, and if the K_{pr} are also uniformly bounded for the particular sequence $\{n_\rho\}$ to be used (for $p=1$, this is trivially satisfied, see (1.4.19)), then (1.4.18) shows that

$$\frac{T_{r+1}^i-\bar{z}}{T_r^i-\bar{z}}\approx\left(\frac{n_{i+r}}{n_{i+r+1}}\right)^{p-1}\frac{1}{n_{i+r+1}}\to 0 \quad \text{as } r \text{ increases} .$$

This is often referred to as "superlinear convergence" of the extrapolation along diagonals of the extrapolation table.

The algorithm for the construction of the extrapolation table in the case of polynomial extrapolation becomes very simple for the following two most frequent cases:

a) $p=1$, or $p=2$ and the expansion (1.4.7) is in terms of even powers of n [4]: In this case $\mathfrak{C}$ consists of all polynomials, or all even polynomials, and the wellknown *Neville algorithm* (see any text on numerical interpolation) permits the recursive construction of the T_r^i:

$$T_\rho^i = T_{\rho-1}^{i+1} + \frac{1}{\left(\frac{n_{i+\rho}}{n_i}\right)^p - 1}(T_{\rho-1}^{i+1} - T_{\rho-1}^i). \tag{1.4.24}$$

b) p arbitrary, but $n_\rho = b^\rho n_0$, $b>1$:

The theory of interpolation implies:

$$T_\rho^i = T_{\rho-1}^{i+1} + \frac{1}{b^{p+\rho-1}-1}(T_{\rho-1}^{i+1} - T_{\rho-1}^i). \tag{1.4.25}$$

Remark (see Bulirsch-Stoer [1]). Actually, (1.4.25) extends to the case where the asymptotic expansion (1.4.7) proceeds in arbitrary, not necessarily integer powers of $1/n$. Let $0<\beta_1<\beta_2<\cdots<\beta_r<B$ be known numbers and

$$T_0^\rho = \bar{z} + \sum_{j=1}^{r} \left(\frac{1}{n_\rho}\right)^{\beta_j} \bar{e}_j + O(n_\rho^{-B}), \qquad \rho = 0(1)r. \tag{1.4.26}$$

Then the recursive application of

$$T_\rho^i = T_{\rho-1}^{i+1} + \frac{1}{b^{\beta_\rho}-1}(T_{\rho-1}^{i+1} - T_{\rho-1}^i) \tag{1.4.27}$$

yields

$$T_r^0 = \bar{z} + O(n_0^{-B}).$$

1.4.3 Rational Extrapolation

The original approach to Richardson extrapolation which prevailed until the early 60's and which is still often used as an intuitive justification is the following: Consider (1.4.7) with $J \geq p+r-1$ for $\rho=0(1)r$, with a given sequence $\{n_\rho\}$. ($n_\rho = 2^\rho n_0$ was usually used in the early applications.) Find a linear combination of the T_0^ρ such that the $r-1$ lowest $\bar{e}_j$ are eliminated and $\bar{z}$ is reproduced. Use this linear combinations as an approximation to $\bar{z}$.

Since (1.4.7) starts as a polynomial in n^{-1}, this procedure is equivalent to the use of a suitable polynomial interpolation (cf. the proof of Theorem 1.4.2). But our approach also makes it clear that polynomial interpolation of the T_0^ρ is *not the only possibility.* As Bulirsch and Stoer [1] have pointed out one may just as well use suitable classes of rational functions for the purpose of Richardson extrapolation.

---

[4] or, more generally, the expansion (1.4.7) proceeds in powers of $(1/n_\rho)^p$.

Consider the following class $\mathfrak{C}$ of interpolation functions from $[0,1]$ to $\mathbb{R}$: Each C_r, $r \in \mathbb{N}$, contains the rational functions

$$\chi(x) = \frac{\varphi(x)}{\psi(x)}, \quad \text{with } \psi(x) \neq 0 \text{ in } [0,1],$$

where the maximal degrees d_φ and d_ψ of the polynomials φ and ψ are $d_\varphi = [r/2]$, $d_\psi = r - d_\varphi$. (I.e. the maximal degree of the denominator agrees with that of the numerator for even r and exceeds it by one for odd r.)

The asymptotic expansion of $\chi(1/n)$ begins with a polynomial in $1/n$, hence $\mathfrak{C}$ should be suitable for Richardson extrapolation under our assumptions. A simple example has been given at the end of Section 1.4.1. For the even case, we may use the $\chi(1/n^2)$.

Now the T_r^i are no longer linear combinations (1.4.10) of the T_0^ρ, $\rho = i(1)i+r$, but depend on these in a non-linear fashion; hence the analysis corresponding to Theorems 1.4.1—1.4.4 is much more involved for rational interpolation and beyond the purposes of this treatise. In a simple case it has been shown by Bulirsch-Stoer [1] that Corollary 1.4.4 continues to hold and that the principal term of $T_r^i - \bar{z}$ is proportional to $\prod_{\rho=i}^{i+r} (1/n_\rho)$, resp. $\prod_{\rho=i}^{i+r} (1/n_\rho^2)$ (cf. (1.4.19) and Corollary 1.4.3). For details, we refer to the above paper.

Practical experience has shown that rational Richardson extrapolation with the above class $\mathfrak{C}$ often gives better results than polynomial Richardson extrapolation. In the trivial example at the end of Section 1.4.1 we see that (at least for large n_0) polynomial extrapolation is better for $g<0$ while rational extrapolation is superior for $g>0$.

It is surprising that there exists a simple recursive algorithm for the construction of the extrapolation table with rational interpolation, at least in the case (a) discussed at the end of Section 1.4.2. This algorithm is a specialization of the *Stoer algorithm* for rational interpolation to the present case; see Bulirsch-Stoer [1]:

Assume $p=1$, or $p=2$ and an even expansion (1.4.7). Define $T_{-1}^i := 0$ for $i \geq 1$. Then the $T_\rho^i = \chi_\rho^i(0)$ are given recursively by

$$\text{(1.4.28)} \quad T_\rho^i := T_{\rho-1}^{i+1} + \left[\left(\frac{n_{i+\rho}}{n_i}\right)^p \left(1 - \frac{T_{\rho-1}^{i+1} - T_{\rho-1}^i}{T_{\rho-1}^{i+1} - T_{\rho-2}^{i+1}}\right) - 1\right]^{-1} (T_{\rho-1}^{i+1} - T_{\rho-1}^i).$$

The similarity between (1.4.28) and (1.4.24) is obvious and the effort of evaluating (1.4.28) is not much greater. It is clear that in the case of (1.4.28) there may arise a singular situation by the vanishing of the bracket. Practical experience seems to indicate, however, that this possibility causes no difficulties in the case of Richardson extrapolation.

To the author's knowledge, non-linear classes of interpolation functions other than those above have not thus far been used for Richardson extrapolation purposes.

1.4.4 Difference Correction

Another application of the asymptotic expansions of Section 1.3 consists of the recursive construction of improved discretizations to the original problem the solutions of which are better and better approximations of the true solution.

Assume that the local discretization error of a discretization method $\mathfrak{M}=\{E_n, E_n^0, \Delta_n, \Delta_n^0, \varphi_n\}_{n\in\mathbb{N}'}$ for the original problem $\mathfrak{P}=\{E, E^0, F\}$ possesses an asymptotic expansion to some order J. Obviously, if we knew the elements $\lambda_j z\in E^0, j=p(1)J$, the solution sequence $\{\zeta_n^J\}_{n\in\mathbb{N}'}, \zeta_n^J\in E_n$, of

$$F_n \zeta_n^J = \Delta_n^0 \sum_{j=p}^{J} \frac{1}{n^j} \lambda_j z \tag{1.4.29}$$

would satisfy

$$\zeta_n^J = \Delta_n z + O(n^{-(J+1)}) \quad \text{as } n\to\infty$$

if the discretization $\mathfrak{M}(\mathfrak{P})$ is stable.

The idea of the *difference correction procedure* is to obtain recursively a sufficiently accurate approximation to the right-hand side of (1.4.29) in the following manner:

Assume that there exist mappings $\varphi_{nj}^k: E_n\to E_n^0$ such that

$$\varphi_{nj}^k \Delta_n y = \Delta_n^0 \lambda_j y + O(n^{-k}), \qquad j=p(1)J, \qquad k=1,2,\ldots, \tag{1.4.30}$$

for y from a suitable subset of E. Let ζ_n be the solution of $\mathfrak{M}(\mathfrak{P})$ for a fixed $n\in\mathbb{N}'$ and define $\zeta_n^j\in E_n$, $j=p(1)J$, recursively by

$$\begin{cases} F_n \zeta_n^p = \dfrac{1}{n^p} \varphi_{np}^1 \zeta_n, \\[2ex] F_n \zeta_n^k = \displaystyle\sum_{j=p}^{k} \frac{1}{n^j} \varphi_{nj}^{k-j+1} \zeta_n^{k-1}, \qquad k=p+1(1)J. \end{cases} \tag{1.4.31}$$

Under suitable smoothness conditions and for stable $\mathfrak{M}$ it should then be true that

$$\zeta_n^k = \Delta_n z + O(n^{-(k+1)}), \qquad k=p(1)J, \tag{1.4.32}$$

since

$$F_n \Delta_n z = \Delta_n^0 \sum_{j=p}^{k} \frac{1}{n^j} \lambda_j z + O(n^{-(k+1)})$$

and

$$F_n \zeta_n^k = \sum_{j=p}^{k} \frac{1}{n^j} \varphi_{nj}^{k-j+1} \zeta_n^{k-1} = \Delta_n^0 \sum_{j=p}^{k} \frac{1}{n^j} \lambda_j(z+O(n^{-k})) + O(n^{-(k+1)}).$$

Of course, the last step made above is not justified by the assumption (1.4.30) since ζ_n^{k-1} is not the discretization of an element in E. However, if each of the ζ_n^k-sequences possesses an asymptotic expansion

$$\zeta_n^k = \Delta_n z + \Delta_n \left(\sum_{j=k+1}^{J_k} \frac{1}{n^j} e_j^k \right) + O(n^{-(J_k+1)})$$

with a sufficiently large J_k and if the mappings φ_{nj}^k possess Lipschitz constants of a sufficiently small order $O(n^{m_j})$ then the *assumption*

$$\frac{1}{n^j} \varphi_{nj}^{k-j+1} \zeta_n^{k-1} = \frac{1}{n^j} \Delta_n^0 \lambda_j z + O(n^{-(k+1)}) \tag{1.4.33}$$

may hold for $j=p(1)k$, $k=p+1(1)J$, which would justify the assertion (1.4.32).

In a general setting it is virtually impossible to specify simple assumptions on $\mathfrak{M}$ and $\mathfrak{P}$ which would guarantee the satisfication of (1.4.33); hence, we will not attempt to formulate a general theorem on the feasibility of the difference correction procedure except in the case of a simple correction. It has to be emphasized, however, that it is the verification of a condition of the type (1.4.33) (again a sort of commutativity condition) rather than that of (1.4.30) which normally is the difficult part in justifying the use of the difference correction procedure in a specific application.

Theorem 1.4.5. *Assume that the discretization method $\mathfrak{M}$ applicable to the original problem $\mathfrak{P}$ satisfies the assumptions of Theorem* 1.3.1, *with $J=\max(p,m_p)$ where m_p is from* (1.4.35) *below.*

Assume further that there exists a sequence of mappings $\varphi_{np}: E_n \to E_n^0$ such that

$$\varphi_{np} \Delta_n y = \Delta_n^0 \lambda_p y + O(n^{-1}) \tag{1.4.34}$$

for $y \in \tilde{D}_p \subset E$, and that φ_{np} satisfies a Lipschitz-condition (L a constant)

$$\|\varphi_{np} \eta^{(1)} - \varphi_{np} \eta^{(2)}\|_{E_n^0} \le L \cdot n^{m_p} \|\eta^{(1)} - \eta^{(2)}\|_{E_n} \tag{1.4.35}$$

for all $\eta^{(i)} \in E_n$ from a ball $B_R(\Delta_n z)$, $R>0$.

Let $\{\zeta_n\}_{n\in\mathbf{N}'}$ be the solution sequence of $\mathfrak{M}(\mathfrak{P})$ and $e_j \in E$, $j=p(1)J$, the coefficients of the asymptotic expansion (1.3.7). *If $z \in \tilde{D}_p$ and*

$$\varphi_{np} \Delta_n \left(z + \sum_{j=p}^{J} \frac{1}{n^j} e_j \right) = \varphi_{np} \Delta_n z + O(n^{-1}) \tag{1.4.36}$$

then the sequence $\{\zeta_n^p\}$, $\zeta_n^p \in E_n$, *determined by*

$$F_n \zeta_n^p = \frac{1}{n^p} \varphi_{np} \zeta_n \tag{1.4.37}$$

satisfies

$$\zeta_n^p = \Delta_n z + O(n^{-(p+1)}).$$

Similarly, if the F_n' *exist in* $B_R(\Delta_n z)$ *and satisfy a uniform Lipschitz condition, and if* $\{\varepsilon_n^p\}_{n\in\mathbb{N}'}$, $\varepsilon_n^p \in E_n$, *is defined by*

$$F_n'(\zeta_n)\,\varepsilon_n^p = \frac{1}{n^p} \varphi_{np} \zeta_n, \tag{1.4.38}$$

then

$$\zeta_n + \varepsilon_n^p = \Delta_n z + O(n^{-(p+1)}).$$

Proof. For sufficiently large $n \in \mathbb{N}'$, both ζ_n and $\Delta_n\left(z + \sum_{j=p}^{J} \frac{1}{n^j} e_j\right)$ are in $B_R(\Delta_n z)$. Hence,

$$\begin{aligned}
F_n \zeta_n^p &= \frac{1}{n^p} \varphi_{np} \zeta_n = \frac{1}{n^p} \varphi_{np}\left[\Delta_n\left(z + \sum_{j=p}^{J} \frac{1}{n^j} e_j\right) + O(n^{-(J+1)})\right] \\
&= \frac{1}{n^p} \varphi_{np}\left[\Delta_n\left(z + \sum_{j=p}^{J} \frac{1}{n^j} e_j\right)\right] + O(n^{-p+m_p-(J+1)}) \quad \text{by (1.4.35)} \\
&= \frac{1}{n^p} \Delta_n^0 \lambda_p z + O(n^{-(p+1)}) \quad \text{by (1.4.36), (1.4.34), and the choice of } J.
\end{aligned}$$

Since $F_n \Delta_n z = (1/n^p) \Delta_n^0 \lambda_p z + O(n^{-(p+1)})$ the stability of the discretization $\mathfrak{M}(\mathfrak{P})$ implies the first assertion.

Similarly, since $\|\varepsilon_n^p\| = O(n^{-p})$ under our assumption,

$$F_n(\zeta_n + \varepsilon_n^p) = F_n'(\zeta_n)\,\varepsilon_n^p + O(n^{-2p}) = \frac{1}{n^p} \Delta_n^0 \lambda_p z + O(n^{-(p+1)})$$

by the same argument as before. □

Remark. The computation of a correction ε_n^p to ζ_n by (1.4.38) was the original approach; of course, for linear F_n, (1.4.37) and (1.4.38) are equivalent. In the nonlinear case, the recursive procedure (1.4.31) may also be formulated for corrections to be obtained from a variational equation of type (1.4.38); the formulation becomes more complicated, however. See also Pereyra [1].

Example. Euler-method, cf. the example at the end of Section 1.3.2.

For $\lambda_1 y = \begin{pmatrix} 0 \\ \frac{1}{2} y''(t) \end{pmatrix}$ the mapping $\varphi_{n1}: E_n \to E_n^0$ defined by

$$(\varphi_{n1}\eta)\left(\frac{\nu}{n}\right) = \begin{cases} 0, & \nu = 0, \\ \dfrac{1}{2n^{-2}}\left[\eta\left(\dfrac{\nu+1}{n}\right) - 2\eta\left(\dfrac{\nu}{n}\right) + \eta\left(\dfrac{\nu-1}{n}\right)\right], & \nu = 1(1)n-1, \\ \dfrac{1}{2n^{-2}}\left[\eta(1) - 2\eta\left(\dfrac{n-1}{n}\right) + \eta\left(\dfrac{n-2}{n}\right)\right], & \nu = n, \end{cases}$$

satisfies
$$\varphi_{n1}\Delta_n y = \Delta_n^0 \lambda_1 y + 0(n^{-1}) \quad \text{for } y \in C^{(3)}[0,1] = \tilde{D}_1$$

and for arbitrary $\eta^{(i)} \in E_n$

$$\|\varphi_{n1}\eta^{(1)} - \varphi_{n1}\eta^{(2)}\|_{E_n^0} \le 2n^2 \|\eta^{(1)} - \eta^{(2)}\|_{E_n}$$

so that $m_1 = 2$. With $J = 2$ and the smoothness assumptions of Theorem 1.3.1 we have $e_1 \in C^{(3)}[0,1]$, $e_2 \in C^{(2)}[0,1]$, hence

$$\varphi_{n1}\Delta_n\left(z + \frac{1}{n}e_1 + \frac{1}{n^2}e_2\right) = \varphi_{n1}\Delta_n z + \Delta_n^0 \begin{pmatrix} 0 \\ \dfrac{1}{2n}e_1''(t) + \dfrac{1}{2n^2}e_2''(t) \end{pmatrix} + O(n^{-1})$$
$$= \varphi_{n1}\Delta_n z + O(n^{-1}).$$

Therefore, if we determine $\zeta_n^1 \in E_n$ from

$$\zeta_n^1(0) = z_0$$

$$\frac{\zeta_n^1\left(\frac{\nu}{n}\right) - \zeta_n^1\left(\frac{\nu-1}{n}\right)}{1/n} - f\left(\zeta_n^1\left(\frac{\nu-1}{n}\right)\right) = \frac{n}{2}\begin{cases} \zeta\left(\dfrac{\nu+1}{n}\right) - 2\zeta\left(\dfrac{\nu}{n}\right) + \zeta\left(\dfrac{\nu-1}{n}\right), & \nu = 1(1)n-1, \\ \zeta(1) - 2\zeta\left(\dfrac{n-1}{n}\right) + \zeta\left(\dfrac{n-2}{n}\right), & \nu = n, \end{cases} \tag{1.4.39}$$

we have

$$\zeta_n^1\left(\frac{\nu}{n}\right) = z\left(\frac{\nu}{n}\right) + O(n^{-2}).$$

Similarly, if

$$\varepsilon_n^1(0) = 0,$$

$$\frac{\varepsilon_n^1\left(\frac{\nu}{n}\right) - \varepsilon_n^1\left(\frac{\nu-1}{n}\right)}{1/n} - f'\left(\zeta\left(\frac{\nu-1}{n}\right)\right)\varepsilon_n^1\left(\frac{\nu-1}{n}\right) = \text{as in (1.4.39)},$$

then

$$\zeta_n\left(\frac{\nu}{n}\right) + \varepsilon_n^1\left(\frac{\nu}{n}\right) = z\left(\frac{\nu}{n}\right) + O(n^{-2}).$$

If we define φ_{n1}^2 by

$$(\varphi_{n1}^2\eta)\left(\frac{\nu}{n}\right) = \begin{cases} 0, & \nu = 0, \\ \dfrac{1}{2n^{-2}}\left[\eta\left(\dfrac{2}{n}\right) - 2\eta\left(\dfrac{1}{n}\right) + \eta(0)\right], & \nu = 1, \\ \dfrac{1}{2n^{-2}}\left[\eta\left(\dfrac{\nu}{n}\right) - 2\eta\left(\dfrac{\nu-1}{n}\right) + \eta\left(\dfrac{\nu-2}{n}\right)\right], & \nu = 2(1)n, \end{cases}$$

we have $\|\varphi_{n1}^2 \Delta_n y - \Delta_n^0 \lambda_1 y\|_{E_n^0} = O(n^{-2})$ for $y \in C^{(4)}[0,1]$, with a modified norm definition in E_n^0, see Section 2.2.1. (This norm leaves the Euler discretization stable.) Furthermore, the mappings $\varphi_{n2}^1: E_n \to E_n^0$ defined by

$$(\varphi_{n2}^1 \eta)\left(\frac{v}{n}\right) = \begin{cases} 0, & v = 0, \\ \dfrac{1}{6n^{-3}}\left[\eta\left(\dfrac{v+1}{n}\right) - 3\eta\left(\dfrac{v}{n}\right) + 3\eta\left(\dfrac{v-1}{n}\right) - \eta\left(\dfrac{v-2}{n}\right)\right], & v = 2(1)n-1, \end{cases}$$

with a shifted third difference for $v=1$ and n, satisfy

$$\|\varphi_{n2}^1 \Delta_n y - \Delta_n^0 \lambda_2 y\|_{E_n^0} = O(n^{-1}) \quad \text{for } y \in C^{(4)}[0,1];$$

the Lipschitz-constants of the φ_{n2} are of order $O(n^3)$. Thus if we are able to show that

$$\zeta_n^1 = \Delta_n\left(z + \sum_{j=2}^{3} \frac{1}{n^j} e_j^1\right) + O(n^{-4}) \tag{1.4.40}$$

and if we can verify (1.4.33) for $k=2, j=1,2,$ then we can apply the difference correction procedure once more. Actually, for $J=3$, (1.4.40) holds with $e_2^1 \in C^{(2)}[0,1]$, $e_3^1 \in C^{(1)}[0,1]$ which is sufficient for (1.4.33) to hold. Therefore, the solution of

$$F_n \zeta_n^2 = \frac{1}{n} \varphi_{n1}^2 \zeta_n^1 + \frac{1}{n^2} \varphi_{n2}^1 \zeta_n^1$$

satisfies $\zeta_n^2 = \Delta_n z + O(n^{-3})$.

In the case $f(y) = gy$, we have $\zeta_n(v/n) = (1+g/n)^v$ and we obtain for ζ_n^1

$$\zeta_n^1\left(\frac{v}{n}\right) = \begin{cases} \left(1+\dfrac{g}{n}\right)^v + v\left(1+\dfrac{g}{n}\right)^{v-1} \dfrac{g^2}{2n^2}, & v = 0(1)n-1, \\ \left(1+\dfrac{g}{n}\right)^n + \left[n\left(1+\dfrac{g}{n}\right)^{n-1} - \dfrac{g}{n}\left(1+\dfrac{g}{n}\right)^{n-2}\right]\dfrac{g^2}{2n^2}, & v = n. \end{cases}$$

After some computations, using the asymptotic expansion of $(1+g/n)^v$ we find

$$\zeta_n^1(1) = e^g\left[1 - \frac{g^3}{24}(3g+4)n^{-2} + O(n^{-3})\right]$$

which compares well with the results obtained at the end of Section 1.4.1 for one step of a polynomial or rational Richardson extrapolation.

Historically, the technique of difference correction was introduced by L. Fox [1] in connection with the numerical solution of linear two-point boundary-value problems by discretization methods. A general formulation of the difference correction procedure was given by Pereyra [1] and [2] who also has further references and examples.

The main advantage of the difference correction procedure over Richardson extrapolation is the fact that it uses one and the same value of n throughout. Thus, each of the discretization problems to be solved is of the same "size". This may be particularly important in the numerical solution of partial differential equations where the computational effort may increase with a power of n.

On the other hand, the construction of the mappings φ_{nj}^k may be quite complicated in practical problems and the precise smoothness requirements are hard to determine. Even if the problem is arbitrarily smooth the difference correction procedure is rarely used more than twice, except perhaps in linear problems. On the whole, both theoretical and practical aspects of this technique have received up to now less attention then the Richardson extrapolation technique.

1.5 Error Analysis

1.5.1 Computing Error

So far we have proceeded as though the solution sequence $\{\zeta_n\}$ of a discretization could be obtained exactly. In reality, however, values of a certain element ζ_n can normally be found only by numerical computation on a *digital computing tool.* (In this book, we will disregard the possibility of using analog devices to obtain approximate solutions of differential equations.) The results of such a computation do not generally agree fully with the correct values for two reasons:

1. *Round-off:* Any given computing tool can handle only finitely many rational numbers. In most cases, the exact result of an arithmetic operation with two machine-numbers as operands is not a machine number. The pseudo result which the computer furnishes is a machine number which is close to the exact result.

2. *Replacement of non-arithmetic operations:* Any non-arithmetic operation, in particular each evaluation of a non-arithmetic function, has to be approximated by a finite sequence of arithmetic operations. An *algorithm* for the computation of values of ζ_n has to specify these replacements in complete detail.

Def. 1.5.1. Assume that an algorithm has been given for the determination of the solution ζ_n of a discretization $\mathfrak{D}=\{E_n, E_n^0, F_n\}_{n\in\mathbb{N}'}$, for some or all $n\in\mathbb{N}'$. An element $\tilde{\zeta}_n\in E_n$ which is obtained by the execution of such an algorithm on a digital computing tool is called a *computational solution of* $\mathfrak{D}$.

Def. 1.5.2. The element $\rho_n := F_n\tilde{\zeta}_n\in E_n^0$ is called the *local computing error* of the computational solution $\tilde{\zeta}_n$ of $\mathfrak{D}$.

The term "local" in Def. 1.5.2 is used in the same sense as in Def. 1.1.7 of the local discretization error: ρ_n is the *residual* which arises from applying F_n to $\tilde{\zeta}_n$ instead of ζ_n. The mapping F_n in Def. 1.5.2 is the *exact* mapping F_n, not its execution on a computer.

Of course, it could happen that the computational solution $\tilde{\zeta}_n$ is no longer in the domain of F_n so that the local computing error ρ_n is not defined. We will not consider this abnormal case, which can usually be detected in the course of the computation (vanishing of a denominator or the like).

From the nature of the local computing error it is clear that it can generally not be expected to decrease with increasing n. Since the total number of arithmetic operations for the computation of $\tilde{\zeta}_n$ normally is proportional to n (or a positive power of n) we may rather expect that $\|\rho_n\|_{E_n^0}$ will *increase with n*.

Example. Euler-method, $s=1$.

Assume we have a computer in floating point arithmetic with rounding, with basis b and a fixed length l of the mantissa. Let $\oplus$ and $\otimes$ denote computer operations, $\tilde{f}(x)$ the computer evaluation of $f(x)$ for a machine-number x. If we assume that z_0 and $1/n$ are machine-numbers, then $\tilde{\zeta}_n$ is defined recursively by

$$\tilde{\zeta}_n(0) := z_0,$$

$$\tilde{\zeta}_n\left(\frac{v}{n}\right) := \tilde{\zeta}_n\left(\frac{v-1}{n}\right) \oplus \frac{1}{n} \otimes \tilde{f}\left(\tilde{\zeta}_n\left(\frac{v-1}{n}\right)\right), \qquad v = 1(1)n.$$

Let $\tilde{\zeta}_n((v-1)/n) \gg (1/n)\tilde{f}(\tilde{\zeta}_n((v-1)/n))$. Then the dominant contribution to the computing error arises from the rounding in the addition, the size of which can be bounded by $\frac{1}{2}|\tilde{\zeta}_n((v-1)/n)|\, b^{-(l-1)}$. Hence,

$$\left|(F_n\tilde{\zeta}_n)\left(\frac{v}{n}\right)\right| \le \frac{1}{2n^{-1}}\left|\tilde{\zeta}_n\left(\frac{v-1}{n}\right)\right| b^{-(l-1)},$$

$$\|F_n\tilde{\zeta}_n\|_{E_n^0} \le \frac{n}{2}\|\tilde{\zeta}\|_{E_n} b^{-(l-1)} = O(n^{+1}).$$

Although this is a matter of great practical importance we will not consider techniques for the estimation of the local computing error. General aspects of round-off error analysis are discussed, e.g., in Wilkinson [1]. Henrici [1], analyzes the local computing error occuring in various discretization algorithms. For realistic applications, the bounds which may be derived will generally strongly overestimate the actual error.

Naturally, the consideration of the residual $F_n\tilde{\zeta}_n$ is primarily intended to give some insight into the difference arising between the computational and the exact solution of the discretization:

Def. 1.5.3. The element $r_n := \tilde{\zeta}_n - \zeta_n \in E_n$ is called the *accumulated* (or *global*) *computing error* of the computational solution $\tilde{\zeta}_n$ of $\mathfrak{D}$.

Methods for an appraisal of the size of r_n from a knowledge of ρ_n are discussed in the following sections combined with the analogous problem for the discretization error.

1.5.2 Error Estimates

The term "error estimate" may be used in two different ways, either to denote a strict bound for the error or to denote a description of the qualitative and quantitative behavior of the error. A *bound* must take into account the worst possible coincidences and will thus often give much too pessimistic a picture. Naturally, bounding the error is the only way to make completely rigorous assertions about the error which do not depend on some quantities being "sufficiently" large or small in some given sense.

On the other hand, from the viewpoint of applications, a bound—even if it is the best possible one in the mathematical sense—which overestimates the error by orders of magnitude in, say, 9999 out of 10,000 applications is virtually useless. One may rather wish to take the risk of somewhat underestimating the error in exceptional cases (particulary if such cases can be recognized during the computation as critical ones so that precautions may be taken). Therefore the *estimation* of the error in this relaxed sense is also an important subject of investigation.

Another distinction concerns the basic approach in error analysis: We can attempt to follow the effects of each source of error until we arrive at an explicit bound or estimate for the deviation between the value of some quantity x defined by the original problem and the approximation ξ for x obtained computationally. This is commonly called *forward error analysis.*

Conversely, we can attempt to interpret the various local errors as effects of *perturbations of the data of the original problem.* The approximate value ξ for x now appears as the exact value of the true solution of an original problem the data of which differ from that of the given problem by some quantity d. We will hence wish to obtain a bound or an estimate for d. This is called *backward error analysis.*

The latter analysis, although it does not make any direct assertions about the deviation $\xi - x$, is quite adequate in many practical situations where the data are known with a limited accuracy only. If the backward error analysis reveals that the perturbation of the data which corresponds to the errors made in the computational solution of the problem is below the natural "noise-level" of the problem then ξ is as accurate a result as we may hope to get.

In our abstract formulation of discretization methods these approaches may be represented as follows:

Assume that the discretization method $\mathfrak{M}=\{E_n, E_n^0, \Delta_n, \Delta_n^0, \varphi_n\}_{n\in\mathbb{N}'}$ is stable for the original problem $\mathfrak{P}=\{E, E^0, F\}$ with true solution z, with a stability bound S and a stability threshold r, and that we have obtained a computational solution $\tilde{\zeta}_n \in E_n$ of $\mathfrak{D}=\mathfrak{M}(\mathfrak{P})$ for some $n\in\mathbb{N}'$.

If we can obtain bounds R and D such that

$$\|F_n\tilde{\zeta}_n\| = \|\rho_n\| \le R, \qquad \|F_n\Delta_n z\| = \|l_n\| \le D, \tag{1.5.1}$$

and if $R+D<r$, then the bound

$$\|\tilde{\zeta}_n - \Delta_n z\| \le S(R+D), \tag{1.5.2}$$

implied by the stability of $\mathfrak{M}$ for $\mathfrak{P}$, represents the result of a forward error analysis. (Of course, it may be possible to obtain much more specific bounds than (1.5.2) for special cases.)

If we find elements $\hat{\rho}_n$ and $\hat{l}_n$ in E_n^0 such that

$$\rho_n \approx \hat{\rho}_n, \qquad l_n \approx \hat{l}_n, \tag{1.5.3}$$

and determine $\hat{\varepsilon}_n \in E_n$ from

$$F_n'(\tilde{\zeta}_n)\hat{\varepsilon}_n = \hat{\rho}_n - \hat{l}_n, \tag{1.5.4}$$

then

$$\tilde{\zeta}_n - \Delta_n z \approx \hat{\varepsilon}_n$$

is an estimate for the total error by a forward error analysis. Traditionally, (1.5.3) is often written in the form

$$\rho_n = \hat{\rho}_n(1+O(n^{-1})), \qquad l_n = \hat{l}_n(1+O(n^{-1})); \tag{1.5.5}$$

but (1.5.5) tells us no more about ρ_n and l_n than (1.5.3) for one fixed finite value of n.

In the case of (1.5.1), we may alternatively choose a $d\in E^0$ such that $\|\Delta_n^0 d\| \ge \|\rho_n\| + \|l_n\|$ and consider $\tilde{z}\in E$ determined from

$$F\tilde{z} = d. \tag{1.5.6}$$

However, this does not permit us to conclude that

$$\|\tilde{\zeta}_n - \Delta_n z\|_{E_n} \le \|\tilde{z} - z\|_E. \tag{1.5.7}$$

Actually, even if the stability bound S of $\mathfrak{M}$ for $\mathfrak{P}$ is the smallest constant such that

$$\|y^{(1)} - y^{(2)}\|_E \le S\,\|Fy^{(1)} - Fy^{(2)}\|_{E^0},$$

equ. (1.5.7) could obvious still not be deduced. Instead we would need an estimate of the type

(1.5.8) $$\|\tilde{z}-z\|_E \geq s\,\|d\|_{E^0}$$

together with $\|\Delta_n^0 d\|_{E_n^0} \leq \|d\|_{E^0}$ to obtain

(1.5.9) $$\|\tilde{\zeta}_n - \Delta_n z\|_{E_n} \leq \frac{S}{s}\,\|\tilde{z}-z\|_E$$

as the result of our backward error analysis. Of course, the freedom in the choice of d for (1.5.6) may be used to obtain a reasonable estimate (1.5.8) (see the example below). In any case, a backward error analysis should not be considered complete with the determination of d.

This is also true if we are satisfied with an estimate of d. Assume (1.5.3) and choose $\hat{d} \in E^0$ such that

(1.5.10) $$\Delta_n^0 \hat{d} \approx \hat{\rho}_n - \hat{l}_n\,.$$

If and only if we know that the linear operators from E^0 to E_n $[F_n'(\tilde{\zeta}_n)]^{-1}\,\Delta_n^0$ and $\Delta_n[F'(z)]^{-1}$ are "close" to each other:

(1.5.11) $$[F_n'(\tilde{\zeta}_n)]^{-1}\,\Delta_n^0 \approx \Delta_n[F'(z)]^{-1},$$

we may conclude, with $F\hat{z} = \hat{d}$,

(1.5.12) $$\begin{aligned}\Delta_n \hat{z} - \Delta_n z &\approx \Delta_n[F'(z)]^{-1}\,\hat{d} \approx [F_n'(\tilde{\zeta}_n)]^{-1}\,\Delta_n^0 \hat{d}\\ &\approx [F_n'(\tilde{\zeta}_n)]^{-1}(\hat{\rho}_n - \hat{l}_n) \approx \tilde{\zeta}_n - \Delta_n z\,.\end{aligned}$$

(1.5.12) states that the deviation of $\tilde{\zeta}_n$ from $\Delta_n z$ is about the same as the deviation which would result from a perturbation $\hat{d}$ of the data of the original problem.

The condition (1.5.11) will be examined in more detail in Section 1.5.3; it is often called "strong stability".

Example. Euler method, as previously; we will only point out a few aspects of the various approaches.

1. With the norms introduced in Section 1.1.1, the Euler method has the stability bound e^L where L is a Lipschitz constant for f (see end of Section 1.1.4) so that (1.5.2) becomes

$$\max_{\nu=0(1)n}\left|\tilde{\zeta}_n\left(\frac{\nu}{n}\right) - z\left(\frac{\nu}{n}\right)\right| \leq e^L(R+D).$$

Better bounds for the individual $\nu < n$ may be obtained by using the norm $\|\eta\|_{E_n} = \max_{\nu=0(1)n}[e^{-(\nu/n)L}|\eta(\nu/n)|]$. It is easily seen from the example in Section 1.1.4 that now $S=1$, so that

$$\left|\tilde{\zeta}_n\left(\frac{\nu}{n}\right) - z\left(\frac{\nu}{n}\right)\right| \leq e^{\frac{\nu}{n}L}(R+D).$$

It is well-known that both bounds are very poor if $f' < 0$ (they depend only on $|f'|$).

2. From the example in Section 1.1.2 we know that

$$l_n\left(\frac{\nu}{n}\right) \approx \frac{1}{2n} z''\left(\frac{\nu}{n}\right) = \frac{1}{2n} f'\left(z\left(\frac{\nu}{n}\right)\right) f\left(z\left(\frac{\nu}{n}\right)\right) \approx \frac{1}{2n} f'\left(\tilde{\zeta}\left(\frac{\nu}{n}\right)\right) f\left(\tilde{\zeta}\left(\frac{\nu}{n}\right)\right) =: \tilde{l}_n\left(\frac{\nu}{n}\right).$$

Under the assumptions made in the example in Section 1.5.1 we may estimate the modulus of $|\rho_n(\nu/n)|$ say by

$$\left|\rho_n\left(\frac{\nu}{n}\right)\right| \approx \frac{n}{2}\left|\tilde{\zeta}\left(\frac{\nu-1}{n}\right)\right| b^{-l} =: \hat{\rho}_n\left(\frac{\nu}{n}\right).$$

We may then use (1.5.4) in the form

$$\frac{\hat{\varepsilon}_n\left(\frac{\nu}{n}\right) - \hat{\varepsilon}\left(\frac{\nu-1}{n}\right)}{\frac{1}{n}} - f'\left(\tilde{\zeta}_n\left(\frac{\nu-1}{n}\right)\right)\hat{\varepsilon}\left(\frac{\nu-1}{n}\right) = -\left[\tilde{l}_n\left(\frac{\nu}{n}\right) + \operatorname{sign}\left(\tilde{l}\left(\frac{\nu}{n}\right)\right)\hat{\rho}_n\left(\frac{\nu}{n}\right)\right]; \tag{1.5.13}$$

(1.5.13) can be solved computationally along with $F_n \zeta_n = 0$.

3. The fact that the constant in (1.5.8) may depend strongly on the choice of d can be seen from the following simple example:

Take $e' - g e = d(t)$, $e(0) = 0$, $g = \text{const.}$ For $d^1(t) \equiv 1$ and $d^2(t) = \cos 2n\pi t$ we have

$$\Delta_n^0\begin{pmatrix} 0 \\ d^1(t) \end{pmatrix} = \Delta_n^0\begin{pmatrix} 0 \\ d^2(t) \end{pmatrix} \quad \text{and} \quad \left\|\Delta_n^0\begin{pmatrix} 0 \\ d^i(t) \end{pmatrix}\right\|_{E_n^0} = 1;$$

but we obtain

$$e^1(t) = \frac{1}{g}\left[\exp(g t) - 1\right]$$

for d^1 while d^2 yields

$$e^2(t) = \frac{g}{4n^2\pi^2 + g^2}\left[\exp(g t) - \cos 2n\pi t\right] + \frac{2n\pi}{4n^2\pi^2 + g^2} \sin 2n\pi t.$$

Thus, if we use the norms of Section 1.1.1 in E and E^0, d^2 makes $s = O(n^{-2})$ in (1.5.8) which leads to a poor bound (1.5.9).

Finally let us remark that a reasonable *a priori* error analysis is virtually impossible in all except the very simplest applications of discretization methods but that one will normally have to rely strongly on information obtained during the computation of $\tilde{\zeta}$.

1.5.3 Strong Stability

In our definition of stability (Section 1.1.4), we considered the perturbation sensitivity of a discretization $\mathfrak{D}$ and required it to be uniformly bounded in the sense of (1.1.11). If $\mathfrak{D}$ has been generated by the application of a discretization method $\mathfrak{M}$ to an original problem $\mathfrak{P}$ and if $\mathfrak{M}$ is consistent with and stable for $\mathfrak{P}$, it should be expected that the perturbation sensitivity of $\mathfrak{M}(\mathfrak{P})$ is closely related to that of $\mathfrak{P}$ in some specified sense. This is, however, not a consequence of the properties considered in Section 1.1; it is referred to as *strong stability*.

The problem of comparing the perturbation sensitivities of $\mathfrak{M}(\mathfrak{P})$ and $\mathfrak{P}$ also arose naturally in the backward error analysis of a computational solution of $\mathfrak{M}(\mathfrak{P})$ (see (1.5.11)). There we attempted to interpret a perturbation δ_n of $\mathfrak{M}(\mathfrak{P})$ as the result of a perturbation d of $\mathfrak{P}$. But this makes sense only if the effect of δ_n on the solution ζ_n of $\mathfrak{M}(\mathfrak{P})$ may be expected to be of the same kind as the effect of d on the solution z of $\mathfrak{P}$.

In the following, we will only consider perturbations which are sufficiently small so that linear perturbation theory is satisfactory and will only be interested in the situation near the true solution z and its discretizations $\Delta_n z$. If the mappings F and F_n are Frechet-differentiable then the perturbation sensitivities of $\mathfrak{M}(\mathfrak{P})$ and $\mathfrak{P}$ may be characterized by the mappings $F_n'(\Delta_n z)^{-1}$ and $F'(z)^{-1}$ resp. These mappings are also called the *Green's functions* of the corresponding problems.

A natural formalization of the condition for strong stability would seem to be either

$$(1.5.14) \qquad \lim_{n\to\infty} \|[F_n'(\Delta_n z)^{-1}\Delta_n^0 - \Delta_n F'(z)^{-1}]d\|_{E_n} = 0 \quad \text{for each } d\in E^0 \quad \text{or}$$

$$(1.5.15) \qquad \lim_{n\to\infty} \|F_n'(\Delta_n z)^{-1}\Delta_n^0 - \Delta_n F'(z)^{-1}\|_{(E^0\to E_n)} = 0\,.$$

However both conditions turn out to be unsatisfactory:

(1.5.14) is too weak; it restricts the analysis to sequences of perturbations $\{\delta_n\}$ of the discretization which are obtained from *one and the same* $d\in E^0$ by discretization. But there are perturbations which are likely to occur with discretizations which are not of this kind. E.g., in our standard Euler method example, consider $\delta_n(\nu/n)=(-1)^\nu$, $\nu=0(1)n$, $n\in\mathbb{N}'$. There is no $d\in E^0$ such that $\delta_n=\Delta_n^0 d$ for all $n\in\mathbb{N}'$.

On the other hand, (1.5.15) is too strong except with a very careful definition of the norm in E^0. Otherwise, for each $n\in\mathbb{N}'$ we will be able to find d^1 and $d^2\in E^0$ such that

$$\|d^1\|_{E^0} = \|d^2\|_{E^0} = 1\,, \qquad \Delta_n^0 d^1 = \Delta_n^0 d^2$$

but

$$\Delta_n F'(z)^{-1} d^1 = O(1) \quad \text{while} \quad \Delta_n F'(z)^{-1} d^2 = O(n^{-r}), \qquad r>0\,.$$

This is clearly shown in example 3 at the end of Section 1.5.2 where $\|\Delta_n e^1\|_{E_n} = O(1)$ while $\|\Delta_n e^2\|_{E_n} = O(n^{-2})$.

Since other possible abstract definitions suffer from similar drawbacks, we will refrain from a formal definition of strong stability in the general setting of Chapter 1 but use "strong stability" only as a generic term to denote the "closeness" of the perturbation sensitivities of $\mathfrak{M}(\mathfrak{P})$ and $\mathfrak{P}$. For given classes of original problems and discretization methods it will be possible to give a precise meaning to the term. We will then also distinguish between asymptotic strong stability, i.e. *strong stability for sufficiently large n*, and *strong stability for given finite n* (see Sect. 2.3).

1.5.4 Richardson-Extrapolation and Error Estimation

Under the assumption that the computing error is negligible compared with the discretization error, we may also consider the Richardson extrapolation technique as an error estimation device (see Section 1.4.1—1.4.3 for concepts and notations): Since—under suitable smoothness assumptions—T_r^0 is a considerably better approximation for $\bar{z}$ than T_{r-1}^1 (see e.g. Theorem 1.4.2), we may use T_{r-1}^1 as our final approximation $\bar{\zeta}$ for $\bar{z}$ and $T_{r-1}^1 - T_r^0$ as an estimation for the error of $\bar{\zeta}$.

This approach leads to the well-known *paradox* of Richardson extrapolation: Either all the information from the T_0^ρ, $\rho=0(1)r$, is used to obtain the best possible approximation T_r^0 to $\bar{z}$, and then no estimation of the remaining error of T_r^0 is possible, or some of the information is used for an error estimation in the above fashion and then the accepted approximation T_{r-1}^1 is normally much poorer than the available approximation T_r^0.

Whether either interpretation is justified should be determined by an analysis in each application. A detailed study of the possibility of obtaining asymptotic error bounds for the T_ρ^i has been made by Bulirsch-Stoer [2] in a special case. The monotonicity of the values appearing in the extrapolation table may be used as a valuable (but, of course, not infallible) guide as to whether a truly asymptotic behavior has been reached.

The following consideration furnishes an additional justification for the use of (polynomial) Richardson-extrapolation and points to possible pitfalls at the same time:

The asymptotic expansion (1.4.7) to order J implies that there are two well-defined but unknown numbers $\underline{R}$ and $\bar{R}$, $\underline{R} \le \bar{R}$, such that

$$\underline{R}\, n_\rho^{-(J+1)} \le R_J(n_\rho) \le \bar{R}\, n_\rho^{-(J+1)} \quad \text{for } \rho=0(1)r\,.$$

Hence we know that $\bar{z}$ must be contained in each of the intervals

$$I_\rho := \left[T_0^\rho - \sum_{j=p}^{J} \frac{1}{n_\rho^j} \bar{e}_j - \bar{R}\, n_\rho^{-(J+1)},\ T_0^\rho - \sum_{j=p}^{J} \frac{1}{n_\rho^j} \bar{e}_j - \underline{R}\, n_\rho^{-(J+1)} \right], \qquad \rho=0(1)r. \tag{1.5.16}$$

If *no information concerning the relative signs* of the $\bar{e}_j$, $\bar{R}$, and $\underline{R}$ is known, the maximal possible error of a linear combination (1.4.10) of the T_0^ρ satisfying (1.4.11) is given by

$$\begin{aligned} M(\gamma_{r0},\dots,\gamma_{rr}) &:= \max_{\bar{z}\in\bigcap_\rho I_\rho} \left| \sum_{\rho=0}^{r} \gamma_{r\rho} T_0^\rho - \bar{z} \right| \\ &= \sum_{j=p}^{J} \left| \sum_{\rho=0}^{r} \gamma_{r\rho} \frac{1}{n_\rho^j} \right| |\bar{e}_j| + \left| \sum_{\rho=0}^{r} \gamma_{r\rho} \frac{1}{n_\rho^{J+1}} \right| \hat{R}, \end{aligned} \tag{1.5.17}$$

where

$$\hat{R} := \max(|R|, |\overline{R}|). \tag{1.5.18}$$

It seems intuitively clear that the minimal value of $M(\gamma_{r0},\ldots)$ will occur for that unique set of $\overline{\gamma}_{r\rho}$ which makes the first sum on the right-hand side of (1.5.17) vanish if this first sum represents the major contribution to M. This statement can be made more precise by the following analysis for the case $p=1$, $J=r$ (which includes, of course, the case of $p=2$ and an even asymptotic expansion through the substitution $n \to n^2$). This is a generalization of the argument for $J=p$ given in Stetter [2].

Let $\overline{\gamma}_{r\rho}$ be the coefficients $\gamma^0_{r\rho}$ of polynomial Richardson extrapolation (see (1.4.16)), then (1.5.17) yields

$$M(\overline{\gamma}_{r0}+\varepsilon_0,\ldots,\overline{\gamma}_{rr}+\varepsilon_r)-M(\overline{\gamma}_{r0},\ldots,\overline{\gamma}_{rr})$$

$$= \sum_{j=1}^{r} \left| \sum_{\rho=0}^{r} \varepsilon_\rho \frac{1}{n_\rho^j} \right| |\overline{e}_j| + \left[\left| \sum_{\rho=0}^{r} (\gamma_{r\rho}+\varepsilon_\rho) \frac{1}{n_\rho^{r+1}} \right| - \left| \sum_{\rho=0}^{r} \gamma_{r\rho} \frac{1}{n_\rho^{r+1}} \right| \right] \hat{R}$$

$$\geq \sum_{j=1}^{r} \left| \sum_{\rho=0}^{r} \varepsilon_\rho \frac{1}{n_\rho^j} \right| |\overline{e}_j| - \left| \sum_{\rho=0}^{r} \varepsilon_\rho \frac{1}{n_\rho^{r+1}} \right| \hat{R}. \tag{1.5.19}$$

Let

$$c_j = \sum_{\rho=0}^{r} \frac{1}{n_\rho^j} \varepsilon_\rho, \qquad j=0(1)r, \tag{1.5.20}$$

where $c_0=0$ due to the assumption (1.4.11) on the admissible $\gamma_{r\rho}$. (1.5.20) implies

$$C \sum_{j=1}^{r} |c_j| \geq \sum_{\rho=0}^{r} |\varepsilon_\rho|$$

where C is a bound on the sum norm of the inverse of the Vandermonde matrix in (1.5.20). Hence we may continue (1.5.19) by

$$\geq \min_{j=1(1)r} |\overline{e}_j| \sum_{j=1}^{r} |c_j| - \sum_{\rho=0}^{r} |\varepsilon_\rho| \frac{\hat{R}}{n_0^{r+1}} \geq \sum_{\rho=0}^{r} |\varepsilon_\rho| \left[C^{-1} \min_j |\overline{e}_j| - \frac{\hat{R}}{n_0^{r+1}} \right].$$

Hence the maximal possible error M takes a unique minimum at the coefficients $\overline{\gamma}_{r\rho}$ of polynomial Richardson-extrapolation if

$$\min_{j=1(1)r} |\overline{e}_j| > \frac{C}{n_0^{r+1}} \hat{R}. \tag{1.5.21}$$

From the well-known expression for the elements of the inverses of Vandermonde matrices (see e. g. Gautschi [1]), one obtains $C=O(n_0^r)$ if the n_ρ are assumed to be *fixed* multiples of n_0 so that

$$\frac{C}{n_0^{r+1}} \to 0 \quad \text{as } n_0 \to \infty$$

in this case. Thus, *if none of the* $\bar{e}_j$ *vanishes*, polynomial Richardson extrapolation is the best strategy for error minimization within the linear combination range and for sufficiently small n_ρ.

Our result again points to the importance of anticipating the vanishing of $\bar{e}_j$'s. It is well-known, for example, that no improvement can be expected in the R_1^ρ over the R_0^ρ when too low a value of p has been used in the extrapolation.

Let us finally consider the effect of the *computing errors* in the T_0^ρ on the values in the extrapolation table (without taking the round-off errors in forming the table entries into account). From (1.4.10) we have in the case of *linear extrapolation*

$$\tilde{T}_r^i := \sum_{\rho=i}^{i+r} \gamma_{r\rho}^i \tilde{T}_0^\rho = T_r^i + \sum_{\rho=i}^{i+r} \gamma_{r\rho}^i (\tilde{T}_0^\rho - T_0^\rho)$$

and

$$|\tilde{T}_r^i - T_r^i| \leq \sum_{\rho=i}^{i+r} |\gamma_{r\rho}^i| \, |\tilde{T}_0^\rho - T_0^\rho|. \tag{1.5.22}$$

If we have a uniform bound (1.4.13) on $\sum_\rho |\gamma_{r\rho}^i|$ which is reasonably small, we may conclude from (1.5.22) that the repeated extrapolation yields improved approximations (all other assumptions for this fact being satisfied) until the remaining discretization error in the T_ρ^i is of the order of the computing error in the T_0^ρ. This stage may normally be recognized from a non-monotonic behavior of the "high" extrapolates after a monotonic behavior of the previous values.

In the case of rational extrapolation the effect of the error in the T_0^ρ depends on the sizes and signs of the quantities involved in a complicated fashion.

1.5.5 Statistical Analysis of Round-off Errors

Although a computation in a digital computer is a strictly deterministic process, the individual round-off errors in the execution of a given algorithm on a given computer depend on the initial data and the sequence of operations in such a complicated fashion that they may often be considered as *random variables*. (Note that most of the so-called "chance events" also depend on initial data and extraneous influences in a deterministic but extremely complicated and globally intransparent way!) For a similar reason it is often natural to consider certain individual round-off errors as *independent* random variables.

In this treatise we will not enter into a discussion of the feasibility of a statistical analysis of round-off errors but we will attempt to indicate how one can proceed if one has decided to use the probabilistic approach. The following considerations are patterned after the well-known probabilistic treatment of round-off errors in Henrici's books [1] and [2].

Let the finite-dimensional real Banach space E_n^0 be isomorphic to $\mathbb{R}^N$ so that each element from E_n^0 may be considered to possess N "components". We will say that the local computing error $\rho_n = F_n \tilde{\zeta} \in E_n^0$ of a computational solution $\tilde{\zeta}_n$ of a discretization is a random variable if each of its N components is a (real-valued) random variable in the ordinary sense.

Then the *mean* $\mu(\rho_n)$ is the N-vector of the means of the components while the *variance* $\sigma^2(\rho_n)$ is the $N \times N$ covariance matrix of the components. We assume that $\mu(\rho_n)$ and $\sigma^2(\rho_n)$ can be approximately determined in some way (cf. Henrici).

As a function of ρ_n, the accumulated computing error $r_n \in E_n$ (see Def. 1.5.3) is a random variable in the same sense as ρ_n. (Note that for a stable discretization the dimensions of E_n and E_n^0 are the same.) Under the assumption that ρ_n is sufficiently small we may use first order perturbation theory to represent r_n as a linear transform of ρ_n:

(1.5.23) $$r_n = F_n'(\zeta_n)^{-1} \rho_n \approx F_n'(\Delta_n z)^{-1} \rho_n.$$

Let Γ_n be the matrix representation of the linear mapping $F_n'(\Delta_n z)^{-1}$ w.r.t. the components introduced above, then the definitions of mean and variance imply

(1.5.24) $$\mu(r_n) = \Gamma_n \mu(\rho_n),$$

(1.5.25) $$\sigma^2(r_n) = \Gamma_n \sigma^2(\rho_n) \Gamma_n^T,$$

for the mean and the variance of the random variable r_n.

We assume now that we can find an element $m_\rho \in E^0$ such that

(1.5.26) $$\mu(\rho_n) \approx n^{p_\mu} \Delta_n^0 m_\rho \quad \text{for } n \in \mathbb{N}';$$

the same arguments as in Section 1.5.2 (see (1.5.11)/(1.5.12)) then imply that *if the discretization method* $\mathfrak{M}$ *is strongly stable for* $\mathfrak{P}$, *then*

(1.5.27) $$\mu(r_n) \approx n^{p_\mu} \Delta_n [F'(z)]^{-1} m_\rho = n^{p_\mu} \Delta_n m_r,$$

where $m_r \in E$ is the solution of the "differential equation"

(1.5.28) $$F'(z) m_r = m_\rho.$$

In order to extend this treatment to the covariance matrices $\sigma^2(\rho_n)$ and $\sigma^2(r_n)$, we have to consider the structure of these quantities: If the elements of E_n and E_n^0 are mappings of $\mathbb{G}_n \to \mathbb{R}^s$ (where $\mathbb{G}_n$ is a "grid") then $\sigma^2(\rho_n)$ and $\sigma^2(r_n)$ are mappings of $\mathbb{G}_n \times \mathbb{G}_n \to \mathbb{R}^s$. We denote the corresponding spaces by ${}^2E_n^0$ and ${}^2E_n$. In order to interpret $\sigma^2(\rho_n)$ as the discretization of a "continuous" element we have to choose such an element v_ρ in the corresponding space ${}^2E_0$ (e.g. $C([0,1]\times[0,1])$ if $E^0=C[0,1]$) and must also construct a space ${}^2E$ in the analogous manner.

An operation on elements in E or E^0 can now be applied to either the first or the second "variable" of the elements in ${}^2E$ or ${}^2E^0$ resp. and we will distinguish these operations by superscripts in brackets. E.g., we will have discretization mappings

$${}^2\Delta_n=\Delta_n^{[1]}\Delta_n^{[2]} \quad \text{and} \quad {}^2\Delta_n^0=\Delta_n^{0[1]}\Delta_n^{0[2]}.$$

We now assume that there is an element $v_\rho\in{}^2E^0$ such that

$$\sigma^2(\rho_n)\approx n^{p\sigma}\,{}^2\Delta_n^0 v_\rho \quad \text{for } n\in\mathbb{N}'. \tag{1.5.29}$$

Then the element $\sigma^2(r_n)\in{}^2E_n$ defined in (1.5.25) satisfies—under the assumption of strong stability as above—

$$\sigma^2(r_n)\approx n^{p\sigma}\,{}^2\Delta_n[F'(z)^{[1]}]^{-1}[F'(z)^{[2]}]^{-1}v_\rho=n^{p\sigma}\,{}^2\Delta_n v_r, \tag{1.5.30}$$

where $v_r\in{}^2E$ is the solution of the "partial differential equation"

$$F'(z)^{[1]}F'(z)^{[2]}v_r=v_\rho. \tag{1.5.31}$$

In some cases where $\sigma^2(\rho_n)$ is a diagonal matrix, the partial differential equation (1.5.31) can be reduced to an ordinary differential equation (see Henrici [1], Section 3.4–2). In general, however, this is not possible.

The utilization of the results (1.5.26)—(1.5.31) is based on the following reasoning: As $n\to\infty$, the linear transformation $r_n=\Gamma_n\rho_n$ normally satisfies the hypotheses of the Central Limit Theorem on the distribution of weighted sums of large numbers of random variables (see Henrici, Section 1.5–4). Accordingly, we may assume that the components of r_n are random variables which are approximately *normally distributed* about their mean (1.5.24) with the variance (1.5.25), independently of the distribution functions of the components of ρ_n. From a knowledge of $\mu(r_n)$ and $\sigma^2(r_n)$ we can thus determine intervals in which the components of r_n lie with a prescribed probability.

These intervals provide *statistical estimates* for the size of the accumulated round-off error r_n. They are usually much smaller than the estimates obtainable from the methods of Section 1.5.2 and their dependence upon n is more favorable. For example, with the usual algorithms for initial value problems for ordinary differential equations of first order, the non-statistical estimates yield $\|r_n\|=O(n)$ while the statistical estimates yield $O(n^{1/2})$ for the lengths of the intervals in which the components of r_n lie with a pre-assigned probability (say 99%).

Careful numerical experiments by Henrici ([1] and [2]) have verified the results of these considerations not only qualitatively but even quantitatively to an amazing extent. Whether the use of these results is justified in a given situation will have to be established by a detailed analysis in each case.

1.6 Practical Aspects

An error analysis alone—even if it could be carried through to perfection—is not sufficient for practical purposes. In real life situations the *computational effort* plays a dominant role. Hence, what we would like to obtain is an approximate solution to one of the following two *optimization problems:*

a) Given an original problem $\mathfrak{P}$ and an accuracy specification, find an approximation of that accuracy to specified values of the true solution z *with the least total effort.*

b) Given an original problem $\mathfrak{P}$ and a certain amount of computing effort, find the most *accurate approximation* to the specified values of z with the given computing effort.

Depending on what we consider as fixed in the optimization and which context conditions we permit to vary, these optimization problems can become arbitrarily complex.

To start out in the simplest fashion, we may assume that we have a fixed discretization method, a fixed algorithm, and a fixed computing tool. Theoretically, here we need only scan the relation between the total error and the size of n until we reach the desired accuracy, resp. the permitted effort, which will normally grow with n. Of course, under these rigid side conditions, it may not be possible to achieve the desired accuracy at all or, in Version (b), the best possible accuracy may be achievable with less effort than specified. Note that normally there does exist a "best *possible* accuracy" since the computing error grows with n.

Furthermore, the answer will depend strongly on whether we want the accuracy mathematically guaranteed by a strict error bound, whether we are satisfied with a 99% probability of sufficient accuracy, or whether we would even accept an accuracy estimate which only specifies the order of magnitude. In any case, we will *need additional computing to establish the accuracy*—and this effort must be counted as part of the total effort!

This is where the definition of the optimization goal becomes blurred: The effort for achieving the optimization (no matter how well or badly) is actually part of the total effort, i.e., of the quantity to be minimized.

If we permit a variation of the method, first within a given class, then even between very different classes, or of the algorithm, or of the computer arithmetic, or of all of these, we soon loose sight of the outlines of our problem. If we assume that the decision between the various possibilities should be made by the computer according to an optimization algorithm and on the basis of the available data (both initially given and gained during the computation) we see that sticking to one general purpose method and choosing n by a crude rule of thumb may not be very far from optimal in the sense of *total* effort versus accuracy.

More realistically, a reasonable approach to the optimization problem seems to be along the following lines:

For each class of problems, a suitable class of discretization methods should be specified. A particular method from this class should be determined by a few parameters whose values should follow in a simple manner from the data of the problem and the accuracy and effort specifications. These parameters should also specify the value (or values) of n to be used (i.e., the stepsizes). Furthermore, each method should have some built-in simple optimization mechanism which makes use of the data that become available during the computation. The whole selection processes should be near-optimal in the sense of our optimization goals under the variations permitted for almost all members of the class of problems considered.

To obtain a basis for the construction of such algorithms, several research groups have carried out large-scale experiments to determine the practical efficiency of various discretization methods for certain classes of differential equations. One of the most recent efforts is that of Hull et alii [1]; their report also refers to previous work.

In the remainder of this treatise the many practical aspects of the numerical solution of ordinary differential equations by discretization methods will *not* be considered, as explained in the Preface.

Chapter 2

Forward Step Methods

In initial-value problems for ordinary differential equations the independent variable is commonly interpreted as *time* and the vector of unknown functions as a state vector varying in time. Given the state at one time instant t_0, the desired solution consists of the state vectors during some time interval $[t_0, t_0+T]$.

This "forward motion in time" is imitated by practically all discretization methods for initial-value problems. On a time interval $[t_0, t_0+T]$ which has been discretized into a discrete set of points, or grid, the process of solving the discretization (1.1.3) of the original differential equation (1.1.1) consists of recursively "stepping forward" along the grid and determining the values of the solution ζ of the discretization at one grid point after the other, starting at t_0. Such discretization methods will be called *forward step methods.*

In Chapter 2, we will introduce some common background and notation for the particular classes of forward step methods to be discussed in Chapters 3—6 and prove a few results which hold for forward step methods in general.

2.1 Preliminaries

2.1.1 Initial Value Problems for Ordinary Differential Equations

Throughout Chapters 2—6, we will be concerned with the following problem:

$$\left.\begin{cases} y_\sigma(t_0)=z_{0\sigma}, & \\ y'_\sigma=f_\sigma(t,y_1,\dots,y_s), & t\in[t_0,t_0+T], \end{cases}\right\} \sigma=1(1)s. \tag{2.1.1}$$

It is well-known (see any text on differential equations) that every initial-value problem for ordinary differential equations can be reduced to (2.1.1).

For the sake of simplicity we will always choose $t_0=0$ without loss of generality. Furthermore—except when we consider the effects of letting the length T of the integration intervall $[0, T]$ grow—we will replace (2.1.1) by the autonomous system

$$(2.1.2) \qquad \begin{cases} y_0(0)=0, & y_0'=1, \\ y_\sigma(0)=z_{0\sigma}, & y_\sigma'=f_\sigma(y_0, y_1, \dots, y_s), \qquad \sigma=1(1)s, \end{cases}$$

which is fully equivalent to (2.1.1). This results in a simplified notation; actually we will again number the components from 1 to s and will not make use of the special form of the first differential equation in (2.1.2). Finally, we will also choose $T=1$ without loss of generality whenever the interval of integration may be assumed fixed.

Def. 2.1.1. The following original problem $\{E, E^0, F\}$ (cf. Def. 1.1.1) will be denoted by *IVP* 1 *(initial value problem for first order systems):*

$$(2.1.3) \qquad E:=(C^{(1)}[0,1])^s \quad \text{with } \|y\|_E:=\max_{t\in[0,1]}\|y(t)\|_{\mathbb{R}^s};$$

$$(2.1.4) \qquad E^0:=\mathbb{R}^s\times(C[0,1])^s \quad \text{with } \left\|\begin{pmatrix} d_0 \\ d \end{pmatrix}\right\|_{E^0}:=\|d_0\|_{\mathbb{R}^s}+\int_0^1\|d(t)\|_{\mathbb{R}^s}dt.$$

The $\mathbb{R}^s$-norm used in (2.1.3) and (2.1.4) is arbitrary (but fixed) in general; s is a fixed natural number.

$$(2.1.5) \qquad F: y\in E \to \begin{pmatrix} y(0)-z_0 \\ y'(t)-f(y(t)) \end{pmatrix} \in E^0;$$

$z_0\in\mathbb{R}^s$ and $f\colon \mathbb{R}^s\to\mathbb{R}^s$ are assumed to be given, they constitute the *data* of a particular IVP 1 (besides the dimension s).

We will *always* assume that the mapping f of an IVP 1, with true solution z, satisfies the following:

There are constants $L>0$, $R>0$, such that

$$(2.1.6) \qquad \|f(\xi^1)-f(\xi^2)\|\le L\,\|\xi^1-\xi^2\|$$
$$\text{for all } \quad \xi^1,\xi^2\in B_R(z):=\bigcup_{t\in[0,1]}\{\xi\in\mathbb{R}^s\colon \|\xi-z(t)\|<R\}\subset\mathbb{R}^s.$$[1]

Remark. In requiring the Lipschitz condition (2.1.6) throughout, we have gone slightly beyond requiring existence and uniqueness of the true solution of the original problem (cf. Sect. 1.1.1). However, the assumption of (2.1.6) simplifies the presentation of many results and constitutes no serious restriction of generality.

---

[1] Note that this definition of $B_R(z)$ is not quite analogous to that of $B_R(\Delta_n z)$ in Chapter 1.

Often—in order to draw certain conclusions about particular forward step methods—we will need much stronger smoothness assumptions on the IVP 1, i.e., on f:

Def. 2.1.2. An $IVP^{(j)}$ *1*, $j \in \mathbb{N}$, is an IVP 1 with a mapping f whose $(j-1)st$ Frechet derivative exists and satisfies a uniform Lipschitz condition in $B_R(z)$ (see (2.1.6)):

$$\|f^{(j-1)}(\xi^1) - f^{(j-1)}(\xi^2)\| \leq L_{j-1} \|\xi^1 - \xi^2\|.$$

The true solution z of an $\text{IVP}^{(j)}$ 1 satisfies

(2.1.7) $\quad z \in D^{(j)} := \{y \in E : y^{(j)}$ exists and is Lipschitz-continuous in $[0,1]\}$.

Note that $\text{IVP } 1 = \text{IVP}^{(1)}\, 1$ due to our general assumption (2.1.6).

In order to handle formally discretizations of a system of differential equations on the *semi-infinite interval* $[0, \infty)$ we consider a sequence of initial-value problems

$$\text{(2.1.8)} \qquad \begin{cases} y(0) - z_0 = 0, \\ y' - f(t,y) = 0, \qquad t \in [0, T_i], \end{cases}$$

where $\mathbb{T} := \{T_i\}_{i \in \mathbb{N}}$ is a strictly increasing sequence of positive numbers with $\lim T_i = \infty$. The function f: $[0, \infty) \times \mathbb{R}^s \to \mathbb{R}^s$ is assumed to satisfy the Lipschitz-condition (2.1.6) (with the interpretation (2.1.2)) uniformly on the interval $[0, \infty)$; see (2.3.2).

Def. 2.1.3. An infinite sequence of initial value problems (2.1.8), with $\lim_{i \to \infty} T_i = \infty$ and the above restriction on f, is called a $\mathbb{T}$*-sequence of* *IVP* 1 or $\{\text{IVP } 1\}_{\mathbb{T}}$.

Note that each element of an $\{\text{IVP } 1\}_{\mathbb{T}}$ is an IVP 1 in the sense of Def. 2.1.1 after a reduction (2.1.2) to autonomous form and a normalization of the interval.

2.1.2 Grids

Def. 2.1.4. A *grid on* $[a,b]$ is any finite set of *grid-points* $t_\nu \in [a,b]$, $\nu = 0(1)n$, with

$$\text{(2.1.9)} \qquad t_0 = a, \qquad t_{\nu-1} < t_\nu, \qquad \nu = 1(1)n.$$

If the interval $[a,b]$ is not specified, the interval $[0,1]$ will be assumed.

The quantities $h_\nu := t_\nu - t_{\nu-1} > 0$, $\nu = 1(1)n$, are the *steps* of the grid $\mathbb{G} = \{t_\nu, \nu = 0(1)n\}$,

$$\text{(2.1.10)} \qquad \hat{h}_n := \max_{\nu = 1(1)n} h_\nu$$

is the *maximal step* of $\mathbb{G}$.

The sequence of grids $\{\mathbb{G}_n\}_{n\in\mathbb{N}'}$ on which the elements $\eta_n \in E_n$ and $\delta_n \in E_n^0$ of a discretization method $\mathfrak{M} = \{E_n, E_n^0, \Delta_n, \Delta_n^0, \varphi_n\}_{n\in\mathbb{N}'}$ are defined and on which the equations (1.1.3) generated by the φ_n are to be solved, plays a fundamental role in any forward step method. In practical applications, the grid $\mathbb{G}_n$ which belongs to a given discretization parameter $n\in\mathbb{N}'$ (which may be thought of as an accuracy requirement parameter) is often determined only during the solution process for (1.1.3); the rule for its recursive determination is actually part of the method $\mathfrak{M}$.

Since we are mainly interested in analyzing the fundamental structural properties of forward step methods we will assume during most of the following sections that $\mathbb{G}_n$ is *a priori* determined by n and will often take $n+1$ to be the number of points in $\mathbb{G}_n$, so that the grid determines the value of n rather than vice versa.

In the notations t_ν and h_ν for the gridpoints and steps of a grid we will normally not explicitly indicate the grid to which we refer by a separate subscript. It has to be kept in mind, however, that these quantities have a meaning only with reference to a *particular* grid $\mathbb{G}_n$, $n\in\mathbb{N}'$. It must be clear from the context which $\mathbb{G}_n$ is meant.

Def. 2.1.5. A grid sequence $\{\mathbb{G}_n\}_{n\in\mathbb{N}'}$ is called *linearly convergent* if there are two positive constants c_1 and c_2 such that all steps h_ν in each $\mathbb{G}_n$ satisfy

$$\frac{c_1}{n} \le h_\nu \le \frac{c_2}{n} = \hat{h}_n, \qquad n\in\mathbb{N}'. \tag{2.1.11}$$

If the contrary is not stated explicitly, we will generally assume that the grid sequences under consideration are linearly convergent. Then, for $n\to\infty$, $O(\hat{h}_n^p)$ and $O(h_\nu^p)$ are equivalent to $O(n^{-p})$.

Def. 2.1.6. A grid sequence $\{\mathbb{G}_n\}_{n\in\mathbb{N}'}$ is called *equidistant* if

$$\mathbb{G}_n := \left\{\frac{\nu}{n},\ \nu=0(1)n\right\}, \qquad n\in\mathbb{N}', \tag{2.1.12}$$

so that $h_\nu = 1/n$, $\nu=1(1)n$.

An equidistant grid sequence is trivially linearly convergent. The grids $\mathbb{G}_n$ of an equidistant grid sequence may be characterized by their unique step $h=1/n$ instead of n. h may even be used as the discretization parameter for the whole method.

In the literature on the subject the steps or "stepsizes" of the grids are almost always used in the role of our discretization parameter n. This step parameter h then varies in a set $H\subset(0,1]$, with 0 an accumulation point of H so that $\lim\limits_{h\to 0,\, h\in H}$ is well-defined. All of the concepts defined in Chapter 1 may naturally be defined analogously using this limit process.

2.1.3 Characterization of Forward Step Methods

Def. 2.1.7. A discretization method $\mathfrak{M}=\{E_n, E_n^0, \Delta_n, \Delta_n^0, \varphi_n\}_{n\in\mathbb{N}'}$ applicable to an IVP 1 is called a *forward step method (f.s.m.)* if

(i) both E_n and E_n^0 are spaces of functions from a grid $\mathbb{G}_n$ to some $\mathbb{R}^{\bar{s}}$ ($\bar{s}$ a fixed natural number),

(ii) for $\eta\in E_n$, $t_\nu\in\mathbb{G}_n$, $(\varphi_n(F)\eta)(t_\nu)$ depends only on the values $\eta(t_\mu)$, $\mu\le\nu$.

Of course, condition (ii) of Def. 2.1.7 does not yet guarantee that, given the values $\zeta_n(t_\mu)$, $\mu<\nu$, the relation

$$(F_n\zeta_n)(t_\nu)=0$$

can be solved for $\zeta_n(t_\nu)$ which is necessary for a recursive solution of (1.1.3). If this fact is not immediately evident for some special f.s.m. or if it does not follow from other results (e.g. via Theorem 1.2.3), then it has to be established. If it holds only for all $n\ge n_0$ we will naturally restrict $\mathbb{N}'$ to these n.

Example. The Euler method used in the examples of Chapter 1 is a f.s.m. Here $(\varphi_n(F)\eta)(t_\nu)$ depends only on $\eta(t_\nu)$ and $\eta(t_{\nu-1})$ for $\nu\ge 1$.

In most f.s.m., $(\varphi_n(F)\eta)(t_\nu)$ actually depends only on the values of η at a *fixed* number of previous gridpoints:

Def. 2.1.8. A f.s.m. is called a *k-step method* if $(\varphi_n(F)\eta)(t_\nu)$, $\nu\ge k$, depends only on the values of η at t_μ, $\mu=\nu-k(1)\nu$, and if k is the smallest such integer.

The relation $(\eta\in E_n)$

$$(\varphi_n(F)\eta)(t_\nu):=\Phi_{n\nu}(\eta(t_\nu),\dots,\eta(t_{\nu-k}))=0,\qquad \nu=k(1)n, \tag{2.1.13}$$

is called the *forward step procedure* of the k-step method.

Normally, the forward step procedure of a f.s.m. depends on ν and n only via the steps $h_{\nu-k+1},\dots,h_\nu$ (see the example below).

Obviously, the values at $t_0,\dots,t_{k-1}$ of the solution of the discretization generated by a k-step method cannot be obtained from the forward step procedure (2.1.13); they are usually specified directly in dependence upon the initial value z_0 of the IVP 1.

Def. 2.1.9. The specification $(\eta\in E_n)$

$$(\varphi_n(F)\eta)(t_\nu)=\eta(t_\nu)-s_{n\nu}(z_0)=0,\qquad \nu=0(1)k-1, \tag{2.1.14}$$

is called the *starting procedure* of a k-step method.

Example. In an equidistant grid sequence, the "2-step midpoint method" is characterized by

$$(\varphi_n(F)\eta)(t_\nu)=\begin{cases}\eta(t_\nu)-z_0-\nu h f(z_0), & \nu=0,1,\\[4pt] \dfrac{\eta(t_\nu)-\eta(t_{\nu-2})}{2h}-f(\eta(t_{\nu-1})), & \nu=2(1)n.\end{cases}$$

Here the functions $s_{n\nu}\colon \mathbb{R}^s\to\mathbb{R}^s$ of the starting procedure (2.1.14) are given by

$$s_{n\nu}(\xi_0)=\xi_0+\frac{\nu}{n}f(\xi_0), \qquad \nu=0,1,$$

the functions $\Phi_{n\nu}\colon(\mathbb{R}^s)^3\to\mathbb{R}^s$ of the forward step procedure (2.1.13) by

$$\Phi_{n\nu}(\xi_2,\xi_1,\xi_0)=\frac{\xi_2-\xi_0}{2/n}-f(\xi_1).$$

With many f.s.m., the computational process producing the value of the desired approximation to z at a given gridpoint t_ν runs through several (say m) *stages* in which intermediate values are produced. These may not only be used to determine the final value of the approximation at that gridpoint but they may also enter into the computation of the values at the following gridpoints. In these cases we therefore have to assume that the value at t_ν of the solution ζ_n of the discretization consists of all these m values. Thus, the elements of E_n must have values in $\mathbb{R}^{ms}$ rather than $\mathbb{R}^s$ (for an IVP 1 with s dependent variables).

Def. 2.1.10. A f.s.m $\mathfrak{M}$ is called an *m-stage method* if, for an s-dimensional IVP 1,

(i) $E_n=(\mathbb{G}_n\to\mathbb{R}^{ms})=(\mathbb{G}_n\to\mathbb{R}^s)^m$;

(ii) $\Delta_n\colon E\to E_n$ is defined for $y\in E$ by

$$(2.1.15)\qquad (\Delta_n y)(t_\nu)=\begin{pmatrix} y(t_\nu^1)\\ y(t_\nu^2)\\ \vdots\\ y(t_\nu^m)\end{pmatrix}\in\mathbb{R}^{ms}, \qquad t_\nu\in\mathbb{G}_n,$$

where the t_ν^μ, $\mu=1(1)m$, are points defined by $\mathfrak{M}$.

Example. Take $m=2$ and let $F_n\colon E_n\to E_n^0$ for $\boldsymbol{\eta}=\begin{pmatrix}\eta^1\\ \eta^2\end{pmatrix}\in E_n$ be defined by

$$(F_n\boldsymbol{\eta})(t_\nu)=\begin{cases}\begin{pmatrix}\eta^1(0)-z_0\\ \eta^2(0)-z_0\end{pmatrix}, & \nu=0,\\[8pt] \begin{pmatrix}\dfrac{\eta^1(t_\nu)-\eta^2(t_{\nu-1})}{h/2}-f(\eta^2(t_{\nu-1}))\\[8pt] \dfrac{\eta^2(t_\nu)-\eta^2(t_{\nu-1})}{h}-f(\eta^1(t_\nu))\end{pmatrix}, & \nu=1(1)n;\end{cases}$$

an equidistant grid sequence has been assumed. This is a well-known 2-stage Runge-Kutta method, the associated Δ_n has $t_\nu^1 = t_\nu - (h/2)$, $t_\nu^2 = t_\nu$, $\nu = 1(1)n$.

Remark. In dealing with m-stage methods, $m > 1$, one has to be aware of the fact that the various stages of the global discretization error $\zeta_n - \Delta_n z$ normally have a different asymptotic behavior as $n \to \infty$. Except when the norm in E_n has been chosen very specially to remedy this situation, the concept of the "order of convergence" as defined in Def. 1.1.9 is no longer meaningful in these cases. Thus, one has either to consider the individual orders of convergence of the various stages or to restrict attention to the order of that stage (usually the m-th one) which is regarded as *the* approximation to z.

In the remainder of Chapter 2, we will always assume that we are dealing with an *m-stage k-step method*, the letters m und k being used only in this sense.

2.1.4 Restricting the Interval

It is typical for initial value problems like IVP 1 that a change in the defining relations for $t > t_0 > 0$ does not affect the solution in $[0, t_0]$. A discretization generated by a f.s.m. should show the same behavior.

Let E_n, E_n^0 be from a f.s.m.; for $\varepsilon \in E_n$, $\delta \in E_n^0$, let

$$\varepsilon^{[\nu]}(t_\mu) := \begin{cases} \varepsilon(t_\mu), & \mu = 0(1)\nu, \\ 0, & \mu = \nu+1(1)n, \end{cases} \qquad \delta^{[\nu]}(t_\mu) \quad \text{analogously.}$$

Then define

$$\|\varepsilon\|_{E_n}^{[\nu]} := \|\varepsilon^{[\nu]}\|_{E_n}, \qquad \|\delta\|_{E_n^0}^{[\nu]} := \|\delta^{[\nu]}\|_{E_n^0}. \tag{2.1.16}$$

Of course, (2.1.16) do not define norms in E_n and E_n^0, resp.

Lemma 2.1.1. *Let the discretization* $\mathfrak{D} = \{E_n, E_n^0, F_n\}_{n \in \mathbb{N}'}$ *be generated by a f.s.m. and let* F_n *possess a unique inverse in* $\{\delta \in E_n^0 \colon \|\delta - F_n \eta_n\|_{E_n^0} < r\}$, $r > 0$, $\eta_n \in E_n$. *If for any* ν, $0 \le \nu \le n$, $\eta^{(1)}, \eta^{(2)} \in E_n$ *satisfy*

$$(F_n \eta^{(1)})(t_\mu) = (F_n \eta^{(2)})(t_\mu), \qquad \mu = 0(1)\nu, \tag{2.1.17}$$

$$\|F_n \eta^{(i)} - F_n \eta_n\|_{E_n^0}^{[\nu]} < r, \qquad i = 1, 2, \tag{2.1.18}$$

then $\eta^{(1)}(t_\mu) = \eta^{(2)}(t_\mu)$ *for* $\mu = 0(1)\nu$.

Proof. Intuitively, the assertion is a natural consequence of the definition of a f.s.m. Formally, we define for $i = 1, 2$,

$$\tilde{\eta}^{(i)}(t_\mu) := \eta^{(i)}(t_\mu), \qquad \mu = 0(1)\nu, \tag{2.1.19}$$

$$(F_n \tilde{\eta}^{(i)})(t_\mu) := (F_n \eta_n)(t_\mu), \qquad \mu = \nu+1(1)n. \tag{2.1.20}$$

By Def. 2.1.7, this defines $F_n\tilde{\eta}^{(i)}$ for all $t_\mu \in \mathbb{G}_n$ and

$$\|F_n\tilde{\eta}^{(i)} - F_n\eta_n\|_{E_n^0} = \|(F_n\tilde{\eta}^{(i)} - F_n\eta_n)^{[\nu]}\|_{E_n^0} < r \tag{2.1.21}$$

by (2.1.18) so that the $\tilde{\eta}^{(i)}$ exist and are unique by our hypothesis. Since $F_n\tilde{\eta}^{(1)} = F_n\tilde{\eta}^{(2)}$ by (2.1.17) and (2.1.19)/(2.1.20) this implies $\tilde{\eta}^{(1)} = \tilde{\eta}^{(2)}$ and our assertion. □

Remark. The existence and uniqueness of the inverses F_n^{-1} may be obtained, e. g., from Lemma 1.2.1.

We will now show that—under a natural assumption on the norm in E_n—a discretization is stable on each sub-interval $[0, t]$ if it is stable in the sense of Def. 1.1.10. Consider the two assertions

(i) $\mathfrak{D}$ is stable at $\{\eta_n\}_{n\in\mathbb{N}'}$, with stability bound S and stability threshold r.

(ii)

$$\|\eta^{(1)} - \eta^{(2)}\|_{E_n}^{[\nu]} \leq S\|F_n\eta^{(1)} - F_n\eta^{(2)}\|_{E_n^0}^{[\nu]} \tag{2.1.22}$$

holds for all $\eta^{(1)}, \eta^{(2)} \in E_n$ satisfying (2.1.18) and for each ν, $0 \leq \nu \leq n$.

Theorem 2.1.2. *In the situation of Lemma* 2.1.1, *if*

$$\|\varepsilon\|_{E_n}^{[\nu]} \leq \|\varepsilon\|_{E_n} \quad \textit{for each } \varepsilon \in E_n \textit{ and } \nu \leq n \tag{2.1.23}$$

then assertion (i) *implies assertion* (ii). ((ii) *trivially implies* (i).)

Proof. For arbitrary fixed ν, $0 \leq \nu \leq n$, let $\eta^{(1)}, \eta^{(2)}$ satisfy (2.1.18). The associated $\tilde{\eta}^{(i)}$ defined by (2.1.19)/(2.1.20) exist and satisfy (2.1.21) = (1.1.12), hence the stability of $\mathfrak{D}$ implies

$$\|\tilde{\eta}^{(1)} - \tilde{\eta}^{(2)}\|_{E_n} \leq S\|F_n\tilde{\eta}^{(1)} - F_n\tilde{\eta}^{(2)}\|_{E_n^0}. \tag{2.1.24}$$

Now, by (2.1.19) and (2.1.23)

$$\|\eta^{(1)} - \eta^{(2)}\|_{E_n}^{[\nu]} = \|\tilde{\eta}^{(1)} - \tilde{\eta}^{(2)}\|_{E_n}^{[\nu]} \leq \|\tilde{\eta}^{(1)} - \tilde{\eta}^{(2)}\|_{E_n}$$

and by (2.1.20) and Def. 2.1.7

$$\|F_n\tilde{\eta}^{(1)} - F_n\tilde{\eta}^{(2)}\|_{E_n^0} = \|F_n\tilde{\eta}^{(1)} - F_n\tilde{\eta}^{(2)}\|_{E_n^0}^{[\nu]} = \|F_n\eta^{(1)} - F_n\eta^{(2)}\|_{E_n^0}^{[\nu]}$$

so that (2.1.24) implies (2.1.22). □

Remark. All norms which are reasonably considered for f.s.m. satisfy (2.1.23) as well as the corresponding assumption

$$\|\delta\|_{E_n^0}^{[\nu]} \leq \|\delta\|_{E_n^0} \quad \text{for each } \delta \in E_n^0 \text{ and } \nu \leq n. \tag{2.1.25}$$

We conclude this section by proving two results which are convenient tools in later proofs:

Lemma 2.1.3. *For some* $n \in \mathbb{N}$, *consider two sets of non-negative reals* $\{\varepsilon_\nu, \nu=0(1)n\}$ *and* $\{\delta_\nu, \nu=0(1)n\}$ *which satisfy*

$$\varepsilon_\nu \le \delta_0 + \frac{S_1}{n}\sum_{\mu=1}^{\nu}\delta_\mu + \frac{S_2}{n}\sum_{\mu=0}^{\nu}\varepsilon_\mu, \quad \text{for } \nu=0(1)n, \tag{2.1.26}$$

where S_1 *and* S_2 *are positive constants. If* $n > S_2$, *the* ε_ν *also satisfy*

$$\varepsilon_\nu \le \left(1+\frac{\bar{S}_2}{n}\right)^{\nu}\bar{\delta}_0 + \frac{\bar{S}_1}{n}\sum_{\mu=1}^{\nu}\left(1+\frac{\bar{S}_2}{n}\right)^{\nu-\mu}\delta_\mu \quad \text{for } \nu=0(1)n \tag{2.1.27}$$

where

$$\bar{\delta}_0 := \frac{\delta_0}{1-\dfrac{S_2}{n}}, \qquad \bar{S}_i := \frac{S_i}{1-\dfrac{S_2}{n}}, \qquad i=1,2.$$

If the rightmost sum in (2.1.26) *extends only to* $\nu-1$ *then the* ε_ν *even satisfy*

$$\varepsilon_\nu \le \left(1+\frac{S_2}{n}\right)^{\nu}\delta_0 + \frac{S_1}{n}\sum_{\mu=1}^{\nu}\left(1+\frac{S_2}{n}\right)^{\nu-\mu}\delta_\mu, \qquad \nu=0(1)n. \tag{2.1.28}$$

Proof. By induction: (2.1.27) is implied by (2.1.26) for $\nu=0$. Assume that (2.1.27) has been verified for $\nu=0(1)\nu_0-1$, then it is also valid for $\nu=\nu_0$:

We state (2.1.26) for $\nu=\nu_0$, take ε_{ν_0} from the right-hand side to the left and divide by $1-S_2/n$. Then we increase the right-hand side of the thus modified inequality by introducing the estimates (2.1.27) for $\nu=0(1)\nu_0-1$. This yields

$$\begin{aligned}\varepsilon_{\nu_0} &\le \bar{\delta}_0 + \frac{\bar{S}_1}{n}\sum_{\mu=1}^{\nu_0}\delta_\mu + \frac{\bar{S}_2}{n}\sum_{\mu=0}^{\nu_0-1}\left[\left(1+\frac{\bar{S}_2}{n}\right)^{\mu}\bar{\delta}_0 + \frac{\bar{S}_1}{n}\sum_{\mu'=1}^{\mu}\left(1+\frac{\bar{S}_2}{n}\right)^{\mu-\mu'}\delta_{\mu'}\right] \\ &= \left(1+\frac{\bar{S}_2}{n}\right)^{\nu_0}\bar{\delta}_0 + \frac{\bar{S}_1}{n}\sum_{\mu=1}^{\nu_0}\left(1+\frac{\bar{S}_2}{n}\right)^{\nu_0-\mu}\delta_\mu\end{aligned}$$

by summation of the geometric progression. Starting with (2.1.26) and introducing (2.1.28) we may prove the second assertion in an analogous manner. □

Corollary 2.1.4. *In the situation of Lemma* 2.1.3, *if* (2.1.26) *is replaced by*

$$\varepsilon_\nu \le \delta_0 + \frac{S_1}{n}\sum_{\mu=1}^{\nu}\delta_\mu + \frac{S_2}{n}\sum_{\mu=0}^{\nu}\max_{\mu'=0(1)\mu}\varepsilon_{\mu'}, \qquad \nu=0(1)n, \tag{2.1.29}$$

the estimates (2.1.27) *and* (2.1.28) *still hold.*

Proof. Again we proceed by induction (the assertion is trivial for $\nu=0$) and assume the validity of (2.1.27) for $\nu=0(1)\nu_0-1$. At $\nu=\nu_0$, we increase the right-hand side of (2.1.29) by introducing the estimates (2.1.27) for $\nu = 0(1)\nu_0 - 1$ which leaves the "max" only with $\mu = \nu_0$. Now either $\varepsilon_{\nu_0} >$ estimate (2.1.27) for $\nu = \nu_0 - 1$, so that $\max_{\mu' = 0(1)\nu_0} \varepsilon_{\mu'} = \varepsilon_{\nu_0}$ and we proceed as in the proof of Lemma 2.1.3, or ε_{ν_0} satisfies (2.1.27) trivially. □

2.1.5 Notation

The following notational conventions will be adhered to in as much as is feasible throughout Chapters 2–6:

The elements of E and E^0 will be denoted by lower-case Latin letters, y and d being the generic elements of E and E^0 resp. while z will denote the true solution of the IVP 1.

The elements of E_n and E_n^0 will be denoted by lower-case Greek letters; η and δ will usually refer to an arbitrary element of these spaces while ζ will refer to the solution of the discretization. We will often not indicate the space E_n or E_n^0 to which an element belongs by an extra subscript n but will rather use the subscript position to indicate a *value* of an element at a particular gridpoint—as is common practice in the literature dealing with discretization methods:

$$\eta_\nu := \eta(t_\nu).$$

To avoid confusion with our previous subscript notation (η_n for an element of E_n) we will normally use ν (or another Greek letter) as the subscript varying along the grid while n will refer to the spaces in question.

In the case of an m-stage method, $m > 1$, the collection of the m "simple" values of $\eta \in E_n$ at any gridpoint t_ν will be denoted by (cf. Section 2.1.3)

$$\text{(2.1.30)} \qquad \boldsymbol{\eta}_\nu = \begin{pmatrix} \eta_\nu^1 \\ \vdots \\ \eta_\nu^m \end{pmatrix} \in \mathbb{R}^{ms}, \qquad \eta_\nu^\mu \in \mathbb{R}^s, \qquad \mu = 1(1)m.$$

Similarly, for $f: \mathbb{R}^s \to \mathbb{R}^s$,

$$\text{(2.1.31)} \qquad \mathbf{f}(\boldsymbol{\eta}_\nu) := \begin{pmatrix} f(\eta_\nu^1) \\ \vdots \\ f(\eta_\nu^m) \end{pmatrix} \in \mathbb{R}^{ms}.$$

The norms used in $\mathbb{R}^{ms}$ will vary with the application but we will assume that they are generated from a given norm in $\mathbb{R}^m$ and one in $\mathbb{R}^s$ by the following mechanism:

For $\xi = (\xi^1, \ldots, \xi^m) \in \mathbb{R}^m$ let $\|\xi\|_{\mathbb{R}^m} = \mathfrak{N}_m(|\xi^1|, \ldots, |\xi^m|)$, then

$$\text{(2.1.32)} \qquad \|\boldsymbol{\eta}_\nu\|_{\mathbb{R}^{ms}} := \mathfrak{N}_m(\|\eta_\nu^1\|_{\mathbb{R}^s}, \ldots, \|\eta_\nu^m\|_{\mathbb{R}^s}).$$

Example. $\mathfrak{N}_m(|\xi^1|, \ldots, |\xi^m|) = \max_{\mu=1(1)m} |\xi^\mu|$ generates

$$\|\boldsymbol{\eta}_\nu\|_{\mathbb{R}^{ms}} = \max_{\mu=1(1)m} \|\eta_\nu^\mu\|_{\mathbb{R}^s}.$$

The sets $\{t_\nu^\mu, \mu=1(1)m\}$ associated with each $t_\nu \in \mathbb{G}_n$ by an m-stage method, $m>1$, (see Def. 2.1.10) will be denoted by

$$\mathbf{t}_\nu := \begin{pmatrix} t_\nu^1 \\ \vdots \\ t_\nu^m \end{pmatrix} \in \mathbb{R}^m. \tag{2.1.33}$$

The values of the elements $\Delta_n y \in E_n$, $y \in E$, in the same situation will be denoted by

$$\mathbf{y}(\mathbf{t}_\nu) := \begin{pmatrix} y(t_\nu^1) \\ \vdots \\ y(t_\nu^m) \end{pmatrix} \in \mathbb{R}^{ms}, \quad \mathbf{t}_\nu \text{ from (2.1.33)}, \tag{2.1.34}$$

while the element of $\mathbb{R}^{ms}$ consisting of m copies of $y(t_\nu)$ will be denoted by

$$\mathbf{y}(t_\nu) := \begin{pmatrix} y(t_\nu) \\ \vdots \\ y(t_\nu) \end{pmatrix} \in \mathbb{R}^{ms}. \tag{2.1.35}$$

An analogous notation will be used for elements of E^0 and E_n^0.

If—as usual—the effect of Δ_n^0 on the second component of an element of E^0 (see Def. 2.1.1) is a restriction to the grid $\mathbb{G}_n$, we may define Δ_n^0 for an arbitrary function on $[0,1]$ as this restriction to $\mathbb{G}_n$. This permits a representation of the steps in the grids of certain *non-equidistant grid sequences* (cf. Section 2.1.2):

Def. 2.1.11. A linearly convergent grid sequence (cf. Def. 2.1.5) is called *coherent* if there exists a piecewise constant function $\theta: [0,1] \to [c_1, c_2]$, $0 < c_1 \leq c_2$, such that

$$h_\nu = \frac{1}{n} (\Delta_n^0 \theta)(t_\nu), \qquad \nu = 1(1)n, \tag{2.1.36}$$

holds for the steps of *each* $\mathbb{G}_n, n \in \mathbb{N}'$. θ is the *stepsize function* of $\{\mathbb{G}_n\}_{n\in\mathbb{N}'}$.

Example (see Fig. 2.1). Let the $\mathbb{G}_n$ of a grid sequence $\{\mathbb{G}_n\}_{n\in\mathbb{N}'}$, $\mathbb{N}' = \{3n, n \in \mathbb{N}\}$, consist of the gridpoints $\{\frac{3}{4}(\nu/n),\ \nu = 0(1)\frac{2}{3}n;\ \frac{3}{2}(\nu/n) - \frac{1}{2},\ \nu = \frac{2}{3}n + 1(1)n\}$; then $\{\mathbb{G}_n\}$ is a coherent grid sequence and has the stepsize function (for $(\Delta_n^0 d)(t_\nu) := d(t_{\nu-1})$)

$$\theta(t) = \begin{cases} \frac{3}{4}, & \text{for } t \in [0, \frac{1}{2}), \\ \frac{3}{2}, & \text{for } t \in [\frac{1}{2}, 1]. \end{cases}$$

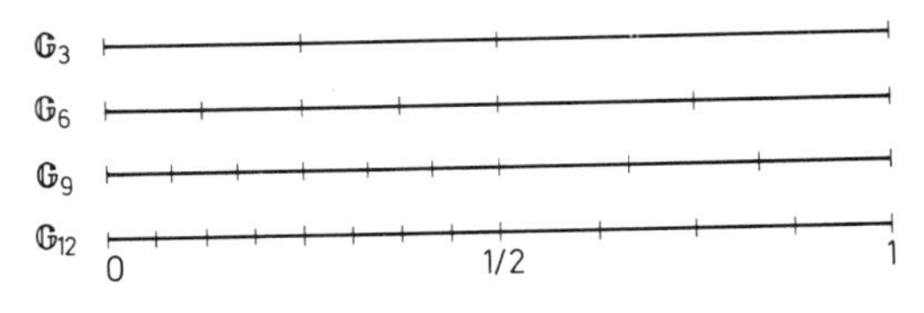

Fig. 2.1. Grids from a coherent grid sequence

2.2 The Meaning of Consistency, Convergence, and Stability with Forward Step Methods

Following our general analysis of discretization methods in Chapter 1, the precise meaning of the concepts of consistency, convergence, stability, etc. with f.s.m. is fully determined by the choice of the spaces E_n and E_n^0, particularly by the choice of the norms in these spaces. However, the same concepts have been defined in the literature in various other ways, mostly in connection with specific classes of f.s.m. In order to avoid confusion we will attempt to point out the analogies and differences between the various concepts which have become known under the same name. Finally, we will derive a result concerning stability which is peculiar to f.s.m.

2.2.1 Our Choice of Norms in E_n and E_n^0

Although the spaces E_n and E_n^0 are finite-dimensional, the stability of a discretization $\{E_n, E_n^0, F_n\}_{n\in\mathbb{N}'}$ in the sense of Def. 1.1.10 may depend on the choice of norms in $\{E_n\}$ and $\{E_n^0\}$. This is due to the fact that the various norms in $\mathbb{R}^n$ are *not* equivalent as $n\to\infty$. (Trivial example: Take $e_n\in\mathbb{R}^n$, with all components equal to 1. Then $\|e_n\|_{\max}=1$ for all n while $\|e_n\|_{\text{Eucl.}} =\sqrt{n}$ and $\|e_n\|_1=n$.)

Of course, the solution $\{\zeta_n\}$ of the discretization generated by a f.s.m. for a particular IVP 1 does not depend on the definition of the norms in $\{E_n\}$ and $\{E_n^0\}$. Therefore, the choice of these norms has to be oriented toward its sole purpose: To permit the formulation of meaningful *general* assertions about f.s.m.

The most important property of a discretization method is its *convergence*. Since convergence (see Def. 1.1.9) rests upon the definition of the norms in the E_n, this definition should be such that the formal concept of convergence agrees with the intuitive expectation that the values of the ζ_n tend to those of z uniformly[2] in $[0,1]$ as $n\to\infty$. Therefore, in

[2] Only rarely will another type of convergence be meaningful.

connection with m-stage f.s.m. (cf. the remark at the end of Section 2.1.3) the norm in E_n will usually be of the form $(g_{n\nu}>0)$

$$(2.2.1) \qquad \|\eta\|_{E_n} := \max_{\nu=0(1)n} g_{n\nu} \|\boldsymbol{\eta}_\nu\|, \qquad \eta \in E_n,$$

where $\|\boldsymbol{\eta}_\nu\|$ is of the form (2.1.32).

In a particular application, a sophisticated choice of the weights $g_{n\nu}$ may increase the significance of $\|\eta\|_{E_n}$ (see e.g. Example 1 in Section 1.5.2). For our general considerations we will take $g_{n\nu}\equiv 1$ so that

$$(2.2.2) \qquad \|\eta\|_{E_n} := \max_{\nu=0(1)n} \|\boldsymbol{\eta}_\nu\|, \qquad \eta \in E_n.$$

All general results which hold with (2.2.2) may easily be extended to (2.2.1).

Remark. For an m-stage method, $m>1$, it may be more meaningful to consider the convergence orders of each stage separately (see Def. 5.1.6). Many of the following results may be extended to the case of stagewise orders of convergence simply by using, in place of (2.1.32), a norm which weights the various stages with appropriate powers of n.

The properties of *consistency and stability* have been introduced mainly as a means of proving convergence. They both rest heavily upon the definition of the norms in the E_n^0. Thus, these norms should be defined in a manner that is "reasonable"; i.e. convergent f.s.m. should be both consistent and stable. Note that a norm definition in the E_n^0 which makes it easy for a discretization to be consistent normally poses a strong stability requirement and vice versa (cf. Def. 1.1.5 and 1.1.10, and also Section 2.2.4).

We also want to take into account the interpretation of the elements of E_n^0 as *perturbations*. It is for this reason that a max-norm like the one in (2.2.2) does not seem appropriate for the E_n^0: A perturbation $\delta \in E_n^0$ which is of unit modulus along the entire grid $\mathfrak{G}_n$ should have a larger norm than a $\delta \in E_n^0$ with one unit value and vanishing values elsewhere on $\mathfrak{G}_n$. Also a perturbation of the starting values should be weighed differently from a perturbation of the forward step procedure. Thus, for a k-step method (cf. the remark at the end of Section 2.1.3) we will generally use the following norm with E_n^0:

$$(2.2.3) \qquad \|\delta\|_{E_n^0} := \max_{\nu=0(1)k-1} \|\boldsymbol{\delta}_\nu\| + \frac{1}{n} \sum_{\nu=k}^{n} \|\boldsymbol{\delta}_\nu\|, \qquad \delta \in E_n^0.$$

Of course, weight factors could be introduced into the sum, and the maximum over the norms of the initial "perturbations" could be replaced by a weighted sum; but again this does not change any of the significant general results.

We will find that the norms (2.2.2) and (2.2.3) conform with the considerations stated above. Also, the concrete definitions of consistency and stability which arise from Def. 1.1.5 and Def. 1.1.10 for particular classes of f.s.m. under this choice of norms coincide with the definitions which are most widely used for these methods in the literature. However, it will turn out in Section 2.2.4 that there exists another norm definition of considerable importance for the E_n^0 which is not equivalent to (2.2.3).

An immediate consequence of (2.1.14) is the following:

Lemma 2.2.1. *The stability of a k-step method depends only on its forward step procedure.*

Proof. From (2.1.14) we have for $\eta^{(1)}, \eta^{(2)} \in E_n$

$$(\varphi_n(F)\eta^{(1)})(t_\nu) - (\varphi_n(F)\eta^{(2)})(t_\nu) = [\boldsymbol{\eta}_\nu^{(1)} - s_{n\nu}(z_0)] - [\boldsymbol{\eta}_\nu^{(2)} - s_{n\nu}(z_0)] = \boldsymbol{\eta}_\nu^{(1)} - \boldsymbol{\eta}_\nu^{(2)}, \quad \nu = 0(1)k-1. \tag{2.2.4}$$

Thus

$$\|F_n\eta^{(1)} - F_n\eta^{(2)}\|_{E_n^0} = \max_{\nu=0(1)k-1} \|\boldsymbol{\eta}_\nu^{(1)} - \boldsymbol{\eta}_\nu^{(2)}\| + \frac{1}{n}\sum_{\nu=k}^{n} \|\Phi_{n\nu}(\boldsymbol{\eta}_\nu^{(1)}, \ldots) - \Phi_{n\nu}(\boldsymbol{\eta}_\nu^{(2)}, \ldots)\|$$

is independent of the $s_{n\nu}$. □

Of course, Lemma 2.2.1 does not imply that a perturbation of the starting values would have no effect on the solution values generated. On the contrary, the analysis of the effect of starting errors as $n \to \infty$ will be crucial in the stability analysis of k-step methods. However, this effect depends only on the sizes and signs of the starting errors and not on the particular starting procedure that may have produced them.

2.2.2 Other Definitions of Consistency and Convergence

Virtually all consistency concepts deal with the asymptotic closeness of the difference equation generated by the f.s.m. and the differential equation of the IVP 1. In most cases, only consistency at z, the solution of the IVP 1, is considered, i.e., consistency is taken to be equivalent to the vanishing of the local discretization error as $n \to \infty$ (cf. Def. 1.1.5 and Def. 1.1.7).

As we have pointed out in Section 1.1.2, this leaves open the normalization of the mappings F_n. There are two main possibilities:

1. To take the difference equation in a form which is an immediate transcript of the differential equation, i.e., a difference quotient (over

at most $k+1$ gridpoints in a k-step method) plus a linear combination of values of f.

2. To take the difference equation in a form in which it is "solved" for η_ν, the next value to be computed:

$$\boldsymbol{\eta}_\nu - H_n(\boldsymbol{\eta}_\nu, \boldsymbol{\eta}_{\nu-1}, \ldots, \boldsymbol{\eta}_{\nu-k}) = 0,$$

where H_n is a contraction w.r.t. $\boldsymbol{\eta}_\nu$, at least for sufficiently large n.

Our Definition 1.1.5 clearly corresponds to the first approach. Since we are dealing with systems of differential equations of *first* order, the second approach yields local discretization errors which are obtained from ours by multiplication with n^{-1} (or h) and possibly a numerical factor. Therefore, authors who use the second approach require

$$\|l_n\| = o(n^{-1}) \quad \text{or } O(n^{-2})$$

for consistency. If Def. 1.1.6 is retained, the orders of consistency are one unit larger than in our approach (and also one unit larger than the order of convergence for the same method). This has to be kept in mind when one is comparing the results of different publications.

Example. Euler method (cf. Section 1.1.2). The second approach would consist in taking F_n in the form

$$\eta_\nu - \left[\eta_{\nu-1} + \frac{1}{n} f(\eta_{\nu-1})\right] = 0$$

which yields

$$\hat{l}_n(t_\nu) := z(t_\nu) - z(t_{\nu-1}) - \frac{1}{n} f(z(t_{\nu-1})) = \frac{1}{2n^2} z''(t_{\nu-1}) + O(n^{-3})$$

and $\|\hat{l}_n\| = O(n^{-2})$ for the local discretization error $\hat{l}_n$.

The *consistency of the starting procedure* of a k-step method is often referred to separately and almost universally defined in the following way which fits well into our pattern and which we will also use (cf. Def. 2.1.9):

Def. 2.2.1. The *starting procedure* of an m-stage k-step method is called *consistent* with a given IVP 1 if

$$\lim_{n\to\infty} \max_{\nu=0(1)k-1} \|\mathbf{z}(\mathbf{t}_\nu) - \mathbf{s}_{n\nu}(z_0)\| = 0. \tag{2.2.5}$$

It is called consistent *of order p* (with the IVP 1) if

$$\max_{\nu=0(1)k-1} \|\mathbf{z}(\mathbf{t}_\nu) - \mathbf{s}_{n\nu}(z_0)\| = O(n^{-p}) \quad \text{as } n \to \infty. \tag{2.2.6}$$

Note that consistency of the starting procedure requires simply

$$\lim_{n\to\infty} \mathbf{s}_{n\nu}(z_0) = \mathbf{z}_0, \qquad \nu = 0(1)k-1. \tag{2.2.7}$$

A concept of *semiconsistency* has been introduced by Butcher [1] for certain classes of f.s.m. This concept is of a rather formal nature and we will not use it in this treatise.

Let us finally consider the influence of using

$$\|\delta\|_{E_n^0} = \max_{\nu=0(1)n} \|\boldsymbol{\delta}_\nu\|$$

in place of (2.2.3): Under normal situations, when $[F_n \Delta_n y - \Delta_n^0 F y](t_\nu)$ is of the same order in n for $\nu = k(1)n$, the two choices are clearly equivalent. However, with (2.2.3) we find that a local error which is of order $O(n^{-(p-1)})$ may be admitted a *fixed* number of times without invalidating the consistency of order p; this will be important in judging the effects of a step change. Results of this type cannot be obtained immediately with a maximum norm in E_n^0.

With respect to *convergence*, there is universal agreement in the literature concerning the general requirement of uniform approximation of the true solution at the points of the grids $\mathbb{G}_n$ (see Section 2.2.1), merely the side conditions vary.

Many authors consider a forward step procedure together with an *arbitrary* consistent starting procedure and require that each such f.s.m. be convergent (see e.g. Henrici [1], Section 5.2-3). Except in pathological situations, this requirement is no stronger than ours (cf. also Lemma 2.2.1):

Lemma 2.2.2. *Consider a k-step method* $\mathfrak{M}$ *which is convergent of order p for an IVP 1. If* $\mathfrak{M}$ *is stable for the IVP 1, then it remains convergent of order p if its starting procedure is replaced by an arbitrary starting procedure which is consistent of order p.*

Proof. Let $\{\zeta_n^{(1)}\}$ be the solution of $\mathfrak{M}$ (IVP 1) and $\{\zeta_n^{(2)}\}$ the solution with the new starting procedure. Then

$$\begin{aligned}\|\zeta_n^{(1)}(t_\nu) - \zeta_n^{(2)}(t_\nu)\| &\le \|\zeta_n^{(1)}(t_\nu) - \mathbf{z}(\mathbf{t}_\nu)\| + \|\zeta_n^{(2)}(t_\nu) - \mathbf{z}(\mathbf{t}_\nu)\| \\ &= O(n^{-p}) \quad \text{for } \nu = 0(1)k-1\end{aligned}$$

since the first term is $O(n^{-p})$ due to the assumed p-th order convergence of $\mathfrak{M}$ and the second due to the assumed p-th order consistency of the new starting procedure. Since

$$(F_n \zeta_n^{(1)})(t_\nu) = (F_n \zeta_n^{(2)})(t_\nu) = 0, \qquad \nu = k(1)n,$$

by the definition of $\zeta_n^{(1)}$ and $\zeta_n^{(2)}$, the stability of $\mathfrak{M}$ for the IVP 1 implies (see (2.2.3))

$$\|\zeta_n^{(1)} - \zeta_n^{(2)}\|_{E_n} \le S\Big[\max_{\nu=0(1)k-1} \|\zeta_n^{(1)}(t_\nu) - \zeta_n^{(2)}(t_\nu)\|\Big] = O(n^{-p}). \qquad \square$$

Dahlquist's concept of *stable convergence* will be considered in Section 2.2.3.

2.2.3 Other Definitions of Stability

Most stability definitions in the literature refer only to a specific class of f.s.m.; in the following, we will interpret them as applications of a general stability definition to the particular class. There are two major causes for discrepancies:

a) Different basic definitions are used (in place of Def. 1.1.10);

b) Different norms are taken in E_n^0.

Let us first look at a). The most common deviation from Def. 1.1.10 concerns the absence of a stability threshold which in turn makes it meaningless to speak of stability "at". While the resulting concept of "*uniform stability*" is natural for linear discretizations (see Remark 2 after Def. 1.1.11), it is clear that hardly any discretization of a non-linear problem will be stable in this sense. Therefore, severe and unrealistic restrictions on the data of the IVP 1 are normally assumed (e. g. that f satisfies a uniform Lipschitz condition for arbitrary values of its arguments).

Other authors introduce at least a qualitative stability threshold by requiring that (1.1.11) should hold for $\eta^{(i)}$ satisfying (1.1.12) with a sufficiently small value of r (and $\eta_n = \Delta_n z$ normally). Our concept of a well-defined stability threshold is no more restrictive but facilitates the concise formulation of many results. Note also that theorems like Theorem 1.2.5, which relate the stability of a non-linear discretization to that of its linearization, produce a definite value for the stability threshold of the non-linear discretization in a natural manner.

For the derivation of many results concerning the discretization error it would suffice to require the validity of (1.1.11) for arbitrarily small perturbations:

(2.2.8) $$\overline{\lim_{\delta\to 0}}\left[\frac{1}{\delta}\sup_{\|\delta_n\|_{E_n^0}\le\delta}\|\tilde{\eta}_n-\eta_n\|_{E_n}\right]\le S \quad \text{for all } n\in\mathbb{N}'$$

with $\delta_n := F_n\tilde{\eta}_n - F_n\eta_n$. Here $\{\eta_n\}$ is the fixed sequence at which the stability occurs while the supremum is taken over all sequences $\{\tilde{\eta}_n\}$ which satisfy $\|F_n\tilde{\eta}_n - F_n\eta_n\| \le \delta$. Such an "*asymptotic stability*" is suggested (as one of several alternatives) by Babuška et al [1]. This stability concept does not permit the consideration of perturbations (like computing errors) which do not become arbitrarily small as $n\to\infty$.

The same objection can be raised against an "*m-restricted stability*" (e. g. Stetter [3]) which requires a stability threshold proportional to n^{-m}.

Some authors (e. g. Hull-Luxemburg [1], Babuška et al [1]) relax the equi-Lipschitz-continuity of the sequence $\{F_n^{-1}\}$ required in (1.1.11) to a mere *equi-continuity*. With this weaker stability concept many important quantitative results (on orders of convergence, asymptotic expansions, etc.) cannot be obtained. Furthermore, with discretizations of ordinary differential equations there are hardly any (non-pathological) applications where this stability concept carries further than ours.

Dahlquist, in his classical papers [1] and [2], combined stability and convergence into one concept:

Def. 2.2.2 (Dahlquist). A k-step method $\mathfrak{M}$ with the starting procedure $s_{n\nu}(z_0)=\mathbf{z}(\mathbf{t}_\nu)$, $\nu=0(1)k-1$, where z is the true solution of $\mathfrak{P}$, is called *stably convergent for* $\mathfrak{P}$ if there are numbers S and $r>0$ such that

$$\overline{\lim_{n\to\infty}} \|\tilde{\zeta}_n-\Delta_n z\|_{E_n}\leq S\delta \tag{2.2.9}$$

for each sequence $\{\tilde{\zeta}_n\}_{n\in\mathbb{N}'}$ such that

$$\|F_n\tilde{\zeta}_n\|_{E_n^0}\leq\delta<r \quad \text{for all } n\in\mathbb{N}'. \tag{2.2.10}$$

(In [2], Dahlquist considers only arbitrarily small perturbations using a formulation analogous to (2.2.8).)

It is trivial that this stable convergence of a f.s.m. $\mathfrak{M}$ for $\mathfrak{P}$ implies the convergence of $\mathfrak{M}$ for $\mathfrak{P}$ in the sense of Def. 1.1.9: We may take $\delta=0$ in (2.2.10) for $\{\zeta_n\}$, the solution sequence of $\mathfrak{M}(\mathfrak{P})$. On the other hand, the stability of $\mathfrak{M}(\mathfrak{P})$ at $\{\zeta_n\}$ and its convergence for $\mathfrak{P}$ imply the stable convergence of $\mathfrak{M}$ for $\mathfrak{P}$: Let S be the stability bound and r the stability threshold; for any sequence $\{\tilde{\zeta}_n\}$ satisfying (2.2.10) we have

$$\begin{aligned}\|\tilde{\zeta}_n-\Delta_n z\| &\leq \|\tilde{\zeta}_n-\zeta_n\|+\|\zeta_n-\Delta_n z\|\\ &\leq S\|F_n\tilde{\zeta}_n-F_n\zeta_n\|+\|\zeta_n-\Delta_n z\|\leq S\delta+\|\varepsilon_n\|\end{aligned}$$

by (2.2.9). But $\|\varepsilon_n\|\to 0$ by the assumed convergence.

While stable convergence of $\mathfrak{M}$ for $\mathfrak{P}$ does not formally imply stability in the sense of Def. 1.1.10 at $\{\zeta_n\}$ or $\{\Delta_n z\}$ (since (2.2.9) does not permit estimates for finite n) it is equivalent to stability in all practical situations; it is quite clear that convergence without stability is of only academic interest.

Let us now consider b), the choice of norms for E_n^0. The stability definitions of many authors imply the use of a max-norm in E_n^0. Theoretically, this leads to a weaker stability requirement than the one obtained from (2.2.3) since

$$\max_{\nu=0(1)k-1}\|\boldsymbol{\delta}_\nu\|+\frac{1}{n}\sum_{\nu=k}^{n}\|\boldsymbol{\delta}_\nu\|\leq 2\max_{\nu=0(1)n}\|\boldsymbol{\delta}_\nu\| \tag{2.2.11}$$

while there is no constant c independent of n such that

$$\max_{\nu=0(1)n}\|\boldsymbol{\delta}_\nu\|\leq c\left[\max_{\nu=0(1)k-1}\|\boldsymbol{\delta}_\nu\|+\frac{1}{n}\sum_{\nu=k}^{n}\|\boldsymbol{\delta}_\nu\|\right] \quad \text{for all } n. \tag{2.2.12}$$

However, this weakening of the stability concept does not make any (non-pathological) classes of f.s.m. stable which would not be stable with (2.2.3).

In the following section, however, we will discuss a reasonable alternative to the norm (2.2.3) in E_n^0 which does change the meaning of consistency and stability in a significant manner.

2.2.4 Spijker's Norm for E_n^0

With the sequence $\{E_n^0\}_{n\in\mathbb{N}'}$ consider two sequences of norms, $\|\cdots\|_{E_n^0}^{(1)}$ and $\|\cdots\|_{E_n^0}^{(2)}$. Since they refer to the same finite-dimensional Banach space for each given n there is always a smallest number $c_{12}(n)$ such that

(2.2.13) $$\|\delta_n\|_{E_n^0}^{(1)} \le c_{12}(n)\,\|\delta_n\|_{E_n^0}^{(2)} \quad \text{for all } \delta_n \in E_n^0.$$

To permit the concise statement of some of the following results we call

$$\|\cdots\|_{E_n^0}^{(2)} \begin{cases}\text{not weaker}\\ \text{weaker}\end{cases} \text{than } \|\cdots\|_{E_n^0}^{(1)} \text{ if } \begin{cases} c_{12}(n) \le C < \infty & \text{for all } n,\\ \lim_{n\to\infty} c_{12}(n) = \infty. \end{cases}$$

Thus, by (2.2.11)/(2.2.12), the norm (2.2.3) is weaker than the max-norm in E_n^0.

The following is a special case of a more general investigation in Spijker's thesis [1] which we have adapted to fit into our context.

Def. 2.2.3. For the E_n^0 of a k-step method,

(2.2.14) $$\|\delta_n\|_{E_n^0}^{*} := \max_{\nu=0(1)k-1} \|\boldsymbol{\delta}_\nu\| + \frac{1}{n} \max_{\nu=k(1)n} \left\| \sum_{\mu=k}^{\nu} \boldsymbol{\delta}_\mu \right\|$$

is called the *Spijker norm.* (Actually, Spijker uses $\sum_{\nu=0}^{k-1} \|\boldsymbol{\delta}_\nu\|$ in place of $\max_{\nu=0(1)k-1} \|\boldsymbol{\delta}_\nu\|$ but this is of no importance.)

Theorem 2.2.3. (2.2.14) *is a norm in* E_n^0. *It is weaker than the norm* (2.2.3), *the best possible estimate* (2.2.13) *being*

$$\|\delta_n\|_{E_n^0} \le (2n-2k+1)\,\|\delta_n\|_{E_n^0}^{*}.$$

Proof. We initially disregard the part referring to the initial values and consider for $x \in \mathbb{R}^N$, with components ξ_ν, $\nu = 1(1)N$,

$$\|x\|_1 := \sum_{\nu=1}^{N} |\xi_\nu| \quad \text{and} \quad \|x\|^* := \max_{\nu=1(1)N} \left| \sum_{\mu=1}^{\nu} \xi_\mu \right|.$$

Trivially, $\|x\|^* \le \|x\|_1$ for each $x \in \mathbb{R}^N$ since

$$\max_{\nu=1(1)N} \left| \sum_{\mu=1}^{\nu} \xi_\mu \right| \le \max_{\nu=1(1)N} \sum_{\mu=1}^{\nu} |\xi_\mu| = \sum_{\mu=1}^{N} |\xi_\mu|.$$

We now show by induction on N that

$$\|x\|_1 \le (2N-1)\|x\|^* \tag{2.2.15}$$

is the best possible estimate in the reverse direction: (2.2.15) holds for $N=1$; assume it holds for $N=1(1)N_0$. Then

$$\begin{aligned}\sum_{\nu=1}^{N_0+1} |\xi_\nu| &\le (2N_0-1) \max_{\nu=1(1)N_0} \left|\sum_{\mu=1}^{\nu} \xi_\mu\right| + |\xi_{N_0+1}| \\ &\le 2N_0 \max_{\nu=1(1)N_0} \left|\sum_{\mu=1}^{\nu} \xi_\mu\right| + \left[|\xi_{N_0+1}| - \left|\sum_{\nu=1}^{N_0} \xi_\mu\right|\right] \\ &\le 2N_0 \max_{\nu=1(1)N_0} \left|\sum_{\mu=1}^{\nu} \xi_\mu\right| + \left|\sum_{\nu=1}^{N_0+1} \xi_\nu\right| \le (2N_0+1) \max_{\nu=1(1)N_0+1} \left|\sum_{\mu=1}^{\nu} \xi_\mu\right|.\end{aligned}$$

The fact that (2.2.15) is best possible is seen from the example $\xi_1=1$, $\xi_\nu=(-1)^{\nu-1}2$, $\nu \ge 2$, for which $\|x\|^*=1$, $\|x\|_1=2N-1$. The definiteness of $\|x\|^*$ now follows from (2.2.15), the homogeneity and the triangle inequality are trivial, thus $\|\cdot\cdot\|^*$ is a norm.

These results trivially imply the assertions of Theorem 2.2.3 as inspection of (2.2.3) and (2.2.14) shows. □

Since the Spijker norm is weaker than the norm (2.2.3), if it is used for E_n^0 in place of (2.2.3) it must make it easier for a f.s.m. to be consistent of some order p but harder to be stable. While the same relationship between the norm (2.2.3) and the max-norm has no practical effects there are meaningful applications where the use of the Spijker norm in place of (2.2.3) does give different results.

Examples. 1. Let a 1-step method for IVP 1 be defined by

$$(\varphi_n(F)\eta)\left(\frac{\nu}{n}\right) = \begin{cases} \eta_0 - z_0, & \nu=0, \\ \dfrac{\eta_\nu - \eta_{\nu-1}}{1/n} - f(\eta_{\nu-1}), & \nu \text{ odd}, \quad 1 \le \nu \le n, \\ \dfrac{\eta_\nu - \eta_{\nu-1}}{1/n} - f(\eta_\nu), & \nu \text{ even}, \quad 2 \le \nu \le n. \end{cases} \tag{2.2.16}$$

If we apply (2.2.16) to an IVP$^{(2)}$ 1 (see Def. 2.1.2) we obtain

$$l_n\left(\frac{\nu}{n}\right) = (\varphi(F)\Delta_n z)\left(\frac{\nu}{n}\right) = \begin{cases} 0, & \nu=0, \\ \dfrac{1}{2n} z''\left(\dfrac{\nu}{n}\right) + O(n^{-2}), & \nu \text{ odd}, \\ -\dfrac{1}{2n} z''\left(\dfrac{\nu-1}{n}\right) + O(n^{-2}), & \nu \text{ even}. \end{cases}$$

While $\|l_n\|_{E_n^0} = O(n^{-1})$ if $z'' \not\equiv 0$ the odd and even terms cancel in the formation of $\|l_n\|_{E_n^0}^*$:

$$\|l_n\|_{E_n^0}^* = \max\left[\frac{1}{2n^2}\left\|z''\left(\frac{1}{n}\right)\right\|, 0, \frac{1}{2n^2}\left\|z''\left(\frac{3}{n}\right)\right\|, 0, \dots\right] = O(n^{-2}).$$

Thus (2.2.16) is consistent of order 2 with the Spijker norm while it is only consistent of order 1 with (2.2.3).

On the other hand, the linear part of (2.2.16) is stable w.r.t. both norms: Let

$$(\hat{F}_n\eta)\left(\frac{v}{n}\right) = \begin{cases} \eta_0, & v=0, \\ \dfrac{\eta_v - \eta_{v-1}}{1/n}, & v=1(1)n; \end{cases}$$

then

$$\eta_v = (\hat{F}_n\eta)(0) + \frac{1}{n}\sum_{\mu=1}^{v}(\hat{F}_n\eta)\left(\frac{\mu}{n}\right) \tag{2.2.17}$$

which implies

$$\|\eta\|_{E_n} = \max_{v=0(1)n}\|\eta_v\| \le \|\eta_0\| + \frac{1}{n}\max_{v=1(1)n}\left\|\sum_{\mu=1}^{v}(\hat{F}_n\eta)\left(\frac{\mu}{n}\right)\right\| = \|\hat{F}_n\eta\|_{E_n^0}^* \le \|\hat{F}_n\eta\|_{E_n^0}.$$

By Corollary 2.2.7 which we will prove in Section 2.2.5 this implies the stability of (2.2.16).

By Theorem 1.2.4 we obtain thus the convergence of order 2 of (2.2.16) if we use the Spijker norm. The derivation of the same result with the norm (2.2.3) is less straightforward.

2. The 2-step midpoint rule (see the example after Def. 2.1.9) has the linear part

$$(\hat{F}_n\eta)\left(\frac{v}{n}\right) = \begin{cases} \eta_v, & v=0,1, \\ \dfrac{\eta_v - \eta_{v-2}}{2/n}, & v=2(1)n. \end{cases} \tag{2.2.18}$$

This implies

$$\eta_v = \begin{cases} \eta_0 + \dfrac{2}{n}\displaystyle\sum_{\mu=1}^{\frac{v}{2}}(\hat{F}_n\eta)\left(\frac{2\mu}{n}\right), & v \text{ even}, \\ \eta_1 + \dfrac{2}{n}\displaystyle\sum_{\mu=1}^{\frac{(v-1)}{2}}(\hat{F}_n\eta)\left(\frac{2\mu+1}{n}\right), & v \text{ odd}, \end{cases}$$

so that

$$\|\eta\|_{E_n} \le 2\left[\max_{v=0,1}\|\eta_v\| + \frac{1}{n}\sum_{\mu=2}^{n}\left\|(\hat{F}_n\eta)\left(\frac{\mu}{n}\right)\right\|\right] = 2\|\hat{F}_n\eta\|_{E_n^0}$$

which implies the stability of (2.2.18) w.r.t. the norm (2.2.3).

Now take $\eta_v = 1 + (-1)^v v\delta$, $v=0(1)n$, with some $\delta \in (0,2]$; then, for $v=2(1)n$,

$$\frac{1}{n}\sum_{\mu=2}^{v}(\hat{F}_n\eta)\left(\frac{1}{n}\right) = \frac{1}{2}[\eta_v + \eta_{v-1} - \eta_1 - \eta_0] = \frac{1+(-1)^v}{2}\delta.$$

Hence $\|\hat{F}_n\eta\|_{E_n^0}^* = 1+\delta$ while $\|\eta\|_{E_n} = 1 + n\delta$ (for even n) and (2.2.18) cannot be stable w.r.t. the Spijker norm.

Actually, Spijker's norm (2.2.14) poses the strongest stability requirement which is still meaningful with f.s.m. applicable to IVP 1. A norm which would, for example, render the Euler method unstable for $y'=0$ should not be seriously considered.

Theorem 2.2.4. *Any norm $\|\cdot\cdot\|_{E_n^0}^{\wedge}$ w.r.t. which the Euler method is stable for $y'=0$ cannot be weaker than the Spijker norm.*

Proof. The F_n of Euler's method for $y'=0$ are identical to the $\hat{F}_n$ of Example 1 above so that (2.2.17) holds. But (2.2.17) implies for arbitrary $n\in\mathbb{N}$:

$$\begin{aligned}\|\eta\|_{E_n} &= \max_{\nu=0(1)n}\left\|\eta_0+\frac{1}{n}\sum_{\mu=1}^{\nu}(F_n\eta)\left(\frac{\mu}{n}\right)\right\| \\ &\geq \frac{1}{3}\left[\|\eta_0\|+\left(\max_{\nu=0(1)n}\left\|\eta_0+\frac{1}{n}\sum_{\mu=1}^{\nu}(F_n\eta)\left(\frac{\mu}{n}\right)\right\|+\|\eta_0\|\right)\right] \\ &\geq \frac{1}{3}\left[\|\eta_0\|+\frac{1}{n}\max_{\nu=1(1)n}\left\|\sum_{\mu=1}^{\nu}(F_n\eta)\left(\frac{\mu}{n}\right)\right\|\right]=\frac{1}{3}\|F_n\eta\|_{E_n^0}^{*}.\end{aligned}$$

As stability w.r.t. $\|..\|_{E_n^0}^{\wedge}$ is equivalent to

$$\|\eta\|_{E_n^0}\leq S\,\|F_n\eta\|_{E_n^0}^{\wedge} \quad \text{with } S<\infty \quad \text{independent of } n$$

we have $\|F_n\eta\|_{E_n^0}^{*}\leq 3S\,\|F_n\eta\|_{E_n^0}^{\wedge}$ uniformly in n which proves our assertion. □

Remark. It can be shown similarly that any norm which leaves the k-step Adams-methods stable for $y'=0$ cannot be weaker than the Spijker norm for k-step methods.

It is the unique feature of Spijker's norm (2.2.14) that it permits a *two-sided* estimate

$$s\,\|F_n\eta^{(1)}-F_n\eta^{(2)}\|_{E_n^0}^{*}\leq\|\eta^{(1)}-\eta^{(2)}\|_{E_n}\leq S\,\|F_n\eta^{(1)}-F_n\eta^{(2)}\|_{E_n^0}^{*}$$

for mappings F_n which occur in (non-pathological) stable discretizations.

2.2.5 Stability of Neighboring Discretizations

Our general results (Theorems 1.2.8—1.2.10) on the stability of neighboring discretizations required that the "variation" G_n of the mappings F_n be uniformly Lipschitz-bounded, with a *sufficiently small Lipschitz constant.* This restriction was found to be unnatural in the case of initial value problems (see Example 2, Section 1.2.4).

For f.s.m., Spijker [1] has proved a theorem which is a generalization of similar results known for specific classes of f.s.m. (see e.g. Stetter [4], Törnig [1]). The following theorem is an adaptation of Spijker's theorem to our context and terminology. We have formulated it so that it applies both to the norm (2.2.3) and to Spijker's norm (2.2.14) in E_n^0.

Theorem 2.2.5. *Let the discretizations* $\mathfrak{D}=\{E_n, E_n^0, F_n\}_{n\in\mathbb{N}'}$, *with continuous* F_n, *and* $\bar{\mathfrak{D}}=\{E_n, E_n^0, F_n+G_n\}_{n\in\mathbb{N}'}$ *each be generated by a f.s.m., with norm (2.2.2) in* E_n *and a norm satisfying (2.1.25) in* E_n^0. *For some sequence* $\{\eta_n\}_{n\in\mathbb{N}'}$, $\eta_n\in E_n$, *assume that for all* $\eta^{(1)}$, $\eta^{(2)}\in B_R(\eta_n)$, $R>0$, *and for* $v=0(1)n$

(i) *the* F_n *satisfy the stability estimate (cf. Theorem 2.1.2)*

$$\|\eta^{(1)}-\eta^{(2)}\|_{E_n}^{[v]} \le S\|F_n\eta^{(1)}-F_n\eta^{(2)}\|_{E_n^0}^{[v]} \tag{2.2.19}$$

uniformly in n,

(ii) *the* G_n *satisfy the Lipschitz condition*

$$\|G_n\eta^{(1)}-G_n\eta^{(2)}\|_{E_n^0}^{[v]} \le \frac{L}{n}\sum_{\mu=0}^{v}\|\eta^{(1)}(t_\mu)-\eta^{(2)}(t_\mu)\| \tag{2.2.20}$$

uniformly in n, and let $\bar{n}$ *be such that* $1-LS/\bar{n}=:\rho^{-1}>0$.

Then, upon restriction of $\mathbb{N}'$ *to* $\{n\in\mathbb{N}': n\ge\bar{n}\}$, $\bar{\mathfrak{D}}$ *is stable with stability bound and threshold*

$$\bar{S}=\rho S e^{\rho LS}, \qquad \bar{r}=\frac{R}{\bar{S}}. \tag{2.2.21}$$

Proof. Let $F_n+G_n=:\bar{F}_n$; (2.2.19) implies

$$\begin{aligned}
&\|\eta^{(1)}(t_v)-\eta^{(2)}(t_v)\| \le \max_{\mu=0(1)v}\|\eta^{(1)}(t_\mu)-\eta^{(2)}(t_\mu)\| \\
&= \|\eta^{(1)}-\eta^{(2)}\|_{E_n}^{[v]} \le S\|(\bar{F}_n\eta^{(1)}-\bar{F}_n\eta^{(2)})-(G_n\eta^{(1)}-G_n\eta^{(2)})\|_{E_n^0}^{[v]} \\
&\le S\|\bar{F}_n\eta^{(1)}-\bar{F}_n\eta^{(2)}\|_{E_n^0}^{[v]}+\frac{LS}{n}\sum_{\mu=0}^{v}\|\eta^{(1)}(t_\mu)-\eta^{(2)}(t_\mu)\| \quad \text{by (2.2.20)} \\
&\le S\|\bar{F}_n\eta^{(1)}-\bar{F}_n\eta^{(2)}\|_{E_n^0}+\frac{LS}{n}\sum_{\mu=0}^{v}\|\eta^{(1)}(t_\mu)-\eta^{(2)}(t_\mu)\| \quad \text{by (2.1.25)}.
\end{aligned} \tag{2.2.22}$$

An application of Lemma 2.1.3, with $\varepsilon_v=\|\eta^{(1)}(t_v)-\eta^{(2)}(t_v)\|$, $\delta_0=S\|\bar{F}_n\eta^{(1)}-\bar{F}_n\eta^{(2)}\|_{E_n^0}$, $\delta_\mu=0$, $\mu>0$, yields (see (2.1.27))

$$\|\eta^{(1)}(t_v)-\eta^{(2)}(t_v)\| \le \rho S\left(1+\rho\frac{LS}{n}\right)^v\|\bar{F}_n\eta^{(1)}-\bar{F}_n\eta^{(2)}\|_{E_n^0}$$

for each $v=0(1)n$ so that

$$\|\eta^{(1)}-\eta^{(2)}\|_{E_n} \le \rho S e^{\rho LS}\|\bar{F}_n\eta^{(1)}-\bar{F}_n\eta^{(2)}\|_{E_n^0}$$

holds for all $\eta^{(1)},\eta^{(2)}\in B_R(\eta_n)$. By Corollary 1.2.2, this implies the stability of $\bar{\mathfrak{D}}$ with stability bound $\bar{S}$ and stability threshold $R/\bar{S}$. □

Corollary 2.2.6. *Under the assumptions of Theorem* 2.2.5, *if* $(G_n\eta)(t_\nu)$ *depends only on the values of* η *at* $t_\mu < t_\nu$ *then* (2.2.21) *may be replaced by*

$$\bar{S} = S e^{LS}, \quad \bar{r} = R/\bar{S}. \tag{2.2.23}$$

Proof. Under the present assumption, the sums on the right hand side of (2.2.20) and (2.2.22) can run only to $\nu-1$ so that Lemma 2.1.3 implies the assertion (see (2.1.28)). □

Remarks. 1. The requirement (i) of a stability estimate for all elements from a ball in E_n is against the spirit of our approach in Section 1.1.4. It cannot be avoided, however, if we want to relate a stability estimate for the $\bar{F}_n$ to one for the F_n. Otherwise we would have to know a priori that a bound on $\|\bar{F}_n\eta^{(i)} - \bar{F}_n\eta_n\|_{E_n^0}^{[\nu]}$ implies a certain bound on $\|F_n\eta^{(i)} - F_n\eta_n\|_{E_n^0}^{[\nu]}$, for $\nu = 0(1)n$.

2. The estimate (2.2.19) for $\nu = 0(1)n$ for all $\eta^{(1)}, \eta^{(2)} \in B_R(\eta_n)$ is not generally implied by an assumption $\|\eta^{(1)} - \eta^{(2)}\|_{E_n} \le S\|F_n\eta^{(1)} - F_n\eta^{(2)}\|_{E_n^0}$ for $\eta^{(1)}, \eta^{(2)} \in B_R(\eta_n)$, contrary to the situation in Theorem 2.1.2. No matter how we construct $\tilde{\eta}^{(i)}$ (cf. (2.1.19)/(2.1.20)) such that

$$\|F_n\tilde{\eta}^{(1)} - F_n\tilde{\eta}^{(2)}\|_{E_n^0} = \|F_n\tilde{\eta}^{(1)} - F_n\tilde{\eta}^{(2)}\|_{E_n^0}^{[\nu]}$$

we cannot deduce $\tilde{\eta}^{(i)} \in B_R(\eta_n)$ without further assumptions.

If the F_n are essentially *linear*, i.e., of the form (1.1.13), the two complications just referred to are not present. Since this is by far the most frequent application we formulate the corresponding result separately:

Corollary 2.2.7. *Let the discretization* $\bar{\mathfrak{D}} = \{E_n, E_n^0, F_n + G_n\}_{n\in\mathbb{N}'}$ *be generated by a f.s.m. and let the* F_n *be of the form* (1.1.13). *If*

(i) *the discretization* $\{E_n, E_n^0, F_n\}_{n\in\mathbb{N}'}$ *is stable, with stability bound S,*

(ii) *the* G_n *satisfy the Lipschitz condition* (2.2.20) *for all* $\eta^{(1)}, \eta^{(2)} \in B_R(\eta_n)$, $R > 0$, *and* $\nu = 0(1)n$,

then $\bar{\mathfrak{D}}$ *is stable at* $\{\eta_n\}$ (for $n \ge \bar{n}$) *with stability bound and threshold given by* (2.2.21), *or by* (2.2.23) *in the case referred to in Corollary* 2.2.6.

Proof. For linear F_n there is no stability threshold (see Remark 2 after Def. 1.1.11) so that Theorem 2.1.2 implies (2.2.19) for arbitrary $\eta^{(1)}, \eta^{(2)}$. □

Example. Euler method, cf. Section 1.2.4, Example 2: Take

$$(F_n\eta)(t_\nu) = \begin{cases} \eta_0, & \nu = 0, \\ \dfrac{\eta_\nu - \eta_{\nu-1}}{h}, & \nu = 1(1)n, \end{cases}$$

$$-(G_n\eta)(t_\nu) = \begin{cases} z_0, & \nu = 0, \\ f(\eta_{\nu-1}), & \nu = 1(1)n. \end{cases}$$

The F_n are of the form (1.1.13), the discretization $\mathfrak{D}$ is stable with $S=1$ (see Section 1.2.4) and we have the situation of Corollary 2.2.6 since $(G_n \eta)(t_\nu)$ depends only on $\eta(t_{\nu-1})$. Since f of an IVP 1 satisfies the Lipschitz condition (2.1.6), the Euler method is stable for IVP 1, with stability bound e^L and stability threshold Re^{-L}. The size of L is now irrelevant except for the values of S and r—as indeed it should be.

2.3 Strong Stability of f.s.m.

In the previous sections we have exclusively considered the properties of discretizations of IVP 1 on a fixed finite interval. In this section we will attempt to analyze some of the phenomena which occur when the interval of integration increases beyond bound while the value n of the discretization parameter is kept fixed.

2.3.1 Perturbation of IVP 1

With initial value problems, the interesting aspects of perturbation sensitivity concern the growth or decay of the deviation of the solution of the perturbed problem from that of the unperturbed problem with increasing t, particularly the asymptotic behavior as $t \to \infty$. The perturbation theory for a system of first order ordinary differential equations has been elaborated to a considerable extent, the commonly used term for this subject matter is "stability theory". There are a number of comprehensive treatments, e.g. Hahn [1], [2], and many others[3]. A large part of the theory is concerned with the various types of asymptotic behavior and the conditions under which they occur.

According to our general exposition in Section 1.5.3 it will be our aim to find under which circumstances and to what extent the discretizations of an IVP 1 show the same perturbation sensitivity as the original problem. With respect to the assumptions on the IVP 1, we will usually restrict ourselves to the simplest and most fundamental case, viz. exponential stability (see Def. 2.3.2 below). Even so, we will find that much remains to be done towards a satisfactory analysis and it is hoped that further investigations will bring more insight.

In Section 2.1.1, we introduced the formal concept of a $\mathbb{T}$-sequence of IVP 1 (see Def. 2.1.3) in order to avoid consideration of discretizations on infinite intervals. This concept will be used throughout Section 2.3. Note that each first order system

$$(2.3.1) \qquad \begin{cases} y(0) - z_0 = 0, & \\ y' - f(t,y) = 0, & t \in [0,\infty), \end{cases}$$

[3] For convenience we quote only from Hahn [2].

with a unique solution z in $[0,\infty)$ defines an $\{\text{IVP }1\}_{\mathbb{T}}$ if f satisfies a *uniform* Lipschitz condition

(2.3.2)
$$\|f(t^1,x^1)-f(t^2,x^2)\| \le L\left\|\begin{pmatrix} t^1-t^2 \\ x^1-x^2 \end{pmatrix}\right\| \quad \text{for} \quad \left\|\begin{pmatrix} t^i \\ x^i \end{pmatrix}-\begin{pmatrix} t \\ z(t) \end{pmatrix}\right\| < R, \qquad i=1,2,$$

with constants L and R independent of $t\in[0,\infty)$. Since we will naturally assume that $f(t,z(t))$ remains bounded as $t\to\infty$ this implies a uniform bound:

(2.3.3) $$\|f(t,x)\| < M_f \quad \text{for } \|x-z(t)\| < R, \qquad t\in[0,\infty).$$

For many results we will furthermore assume that certain derivatives of f exist and are uniformly bounded for $\|x-z(t)\|<R$.

Together with the solution z of (2.3.1) we consider a continuous and piecewise differentiable function $\tilde{z}\colon [0,\infty)\to\mathbb{R}^s$ for which

$$\begin{cases} \tilde{z}(0)-z_0 = \delta_0, \\ \tilde{z}'(t)-f(t,\tilde{z}(t)) = \delta(t), \qquad t\in[0,\infty). \end{cases}$$

Then $e:=\tilde{z}-z$ satisfies

(2.3.4) $$\begin{cases} e(0)=\delta_0, \\ e'(t)-g(t,e(t)) = \delta(t), \qquad t\in[0,\infty), \end{cases}$$

where $g\colon [0,\infty)\times\mathbb{R}^s\to\mathbb{R}^s$ is defined by

(2.3.5) $$g(t,x) := f(t,z(t)+x)-f(t,z(t)).$$

It is the equation (2.3.4) which is treated in stability theory of differential equations. Its solution—unique under our assumptions—for a vanishing perturbation $(\delta_0,\delta(t))$ is $e(t)\equiv 0$, the "*equilibrium*" of (2.3.4).

Def. 2.3.1 (Hahn [2], Def. 56.1). The equilibrium of (2.3.4) is called *totally stable* if for each $\rho>0$ there exist $\overline{\delta}_0>0$ and $\overline{\delta}_1>0$ such that the solution e of (2.3.4) exists and satisfies

(2.3.6) $$\|e(t)\| < \rho \quad \text{for } t\in[0,\infty)$$

whenever

(2.3.7) $$\|\delta_0\| < \overline{\delta}_0 \quad \text{and} \quad \|\overline{\delta}(t)\| < \delta_1, \ t\in[0,\infty).$$

For convenience, we will also call the $\{\text{IVP }1\}_{\mathbb{T}}$ (2.3.1) totally stable if the equilibrium of (2.3.4) with g, given by (2.3.5), is totally stable.

Obviously, total stability expresses the fact that the initial value problem (2.3.1) is "properly posed" on the infinite interval $[0,\infty)$, i.e., that the solution of (2.3.1) depends continuously on its data z_0 and f (when a norm like $\|\delta_0\| + \max_{t\in[0,\infty)} \|\delta(t)\|$ is employed for the perturbation of the data and a maximum norm for the solution). Therefore, (2.3.1) has to be totally stable if a numerical solution of (2.3.1) may be expected to succeed at all.

In the stability theory of ordinary differential equations one normally considers—in place of the inhomogeneous problem (2.3.4)—the *family of homogeneous problems* with an initial perturbation at $t_0, 0 \le t_0 < \infty$:

$$\text{(2.3.8)} \qquad \begin{cases} e(t_0) = \delta_0, \\ e'(t) - g(t, e(t)) = 0, \quad t \in [t_0, \infty). \end{cases}$$

The solution of (2.3.8) will be denoted by $\bar{e}(t; \delta_0, t_0)$. By (2.3.5), it represents the effect of a perturbation of (2.3.1) which consists of displacing z by δ_0 at t_0 (Fig. 2.2).

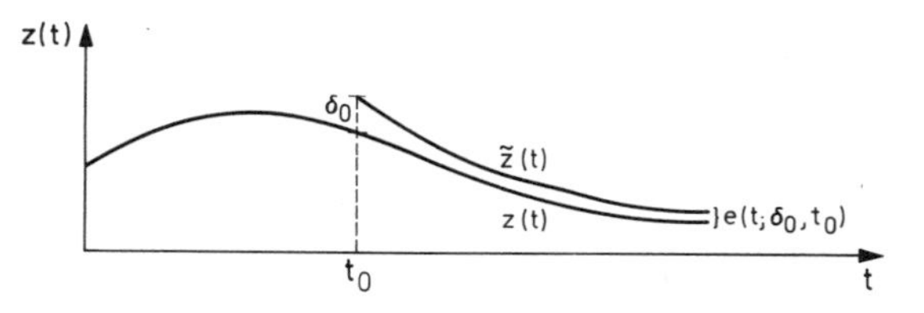

Fig. 2.2. Perturbation of z at t_0

Def. 2.3.2 (Hahn [2], Def. 26.2). If there exist positive constants a, μ, and r *independent of* t_0 such that $\bar{e}(t; \delta_0, t_0)$ exists and satisfies

$$\text{(2.3.9)} \qquad \|\bar{e}(t; \delta_0, t_0)\| \le a \|\delta_0\| \exp(-\mu(t - t_0)) \quad \text{for } t \ge t_0$$

whenever $\|\delta_0\| < r$, then the equilibrium $e_0 = 0$ of (2.3.8) is called *exponentially stable.* In this case we will also call the associated $\{\text{IVP 1}\}_{\mathbb{T}}$ (2.3.1) exponentially stable. We will use the term *μ-exponentially stable* ($\mu > 0$) to indicate that (2.3.9) holds with a specific value of μ.

Remark. Since the norms in (2.3.9) refer only to $\mathbb{R}^s$ they influence only the value of a but are immaterial for exponential stability.

Although several of the results which we will derive for discretizations of $\{\text{IVP 1}\}_{\mathbb{T}}$ also hold under slightly more general assumptions (viz. uniformly asymptotic stability) we will usually assume that our initial value problem (2.3.1) is exponentially stable since the technical details of many proofs become much simpler in the case of exponential stability. This is due to the following result:

Theorem 2.3.1 (Liapunov function, Hahn [2], Theorems 26.5 and 56.1). *The equilibrium of* (2.3.8) *is exponentially stable if and only if there* *exists a "Liapunov" function* $v\colon \mathbb{R}^s \times \mathbb{R} \to \mathbb{R}$ *for* (2.3.8) *which satisfies, for* $t \in [0, \infty)$, $\|x\| < r$,

$$a_1 \|x\|^2 \le v(x,t) \le a_2 \|x\|^2, \tag{2.3.10}$$

$$\dot{v}(x,t) := \frac{d}{d\tau} v(\bar{e}(\tau; x, t), \tau)\bigg|_{\tau=t} \le -a_3 \|x\|^2, \tag{2.3.11}$$

with positive constants a_1, a_2, a_3, *and* r *independent of* t.

Furthermore, if $\partial g(t,x)/\partial x$ *exists and is uniformly bounded for* $t \in [0, \infty)$, $\|x\| < r$, *then* v *may be chosen such that*

$$\left\| \frac{\partial v}{\partial x}(x,t) \right\| \le a_4 \|x\|, \quad \text{with } a_4 \text{ independent of } t. \tag{2.3.12}$$

Remark. v may be chosen as a quadratic form in x. If g is linear in x, both v and $\dot{v}$ may be assumed to be quadratic forms in x.

Some further properties of exponentially stable $\{\mathrm{IVP}\,1\}_{\mathbb{T}}$ are expressed in the following results:

Theorem 2.3.2 (Hahn [2], Theorem 56.4). *If the equilibrium of* (2.3.8) *is exponentially stable it is also totally stable in each interval* $[t_0, \infty)$, *with* $\bar{\delta}_0$ *and* $\bar{\delta}_1$ *independent of* t_0.

Theorem 2.3.3 (Hahn [2], Theorem 56.2). *For* $t \in [0, \infty)$, *let*

$$g(t,x) = g_0(t,x) + k(t,x) \tag{2.3.13}$$

and assume that, uniformly for $t \in [0, \infty)$ *and* $\|x\| < r$,

$$\|k(t,x)\| \le M \|x\|.$$

If the equilibrium of

$$\begin{cases} e(t_0) = \delta_0, \\ e'(t) - g_0(t, e(t)) = 0, \quad t \ge t_0, \end{cases} \tag{2.3.14}$$

is exponentially stable, with a Liapunov function v *satisfying* (2.3.11) *and* (2.3.12), *and if* $M < a_3/a_4$ *then the equilibrium of* (2.3.8) *is also exponentially stable.*

If f is *differentiable* w.r.t. x, with a uniformly bounded derivative, we may choose (cf. (2.3.5))

$$g_0(t,x) = f_y(t, z(t))\, x \tag{2.3.15}$$

and the condition of Theorem 2.3.3 can be satisfied for sufficiently small r if the remainder term $k(t,x) = o(\|x\|)$ *uniformly in* t. In the stability theory of ordinary differential equations, (2.3.14) with (2.3.15) is called the "first approximation" of (2.3.8).

For a *linear* problem

$$\begin{cases} e(t_0) = \delta_0, \\ e'(t) - g(t)e(t) = 0, \quad t \geq t_0, \end{cases} \tag{2.3.16}$$

a *Green's function* $\Gamma: [0,\infty) \times [0,\infty) \to \mathfrak{L}(\mathbb{R}^s \to \mathbb{R}^s)$ may be defined as the solution of the matrix differential equation

$$\Gamma(\tau, \tau) = I,$$

$$\frac{\partial}{\partial t} \Gamma(t, \tau) - g(t)\Gamma(t, \tau) = 0, \quad t \geq \tau \geq 0.$$

It permits the representation

$$e(t) = \Gamma(t, t_0)\delta_0 \tag{2.3.17}$$

for the solution of (2.3.16).

It is obvious from (2.3.17) that the equilibrium of (2.3.16) is exponentially stable if and only if there are positive constants a and μ independent of τ such that Γ satisfies

$$\|\Gamma(t, \tau)\| \leq a \exp(-\mu(t-\tau)), \quad t \geq \tau \geq 0. \tag{2.3.18}$$

If the matrix g in (2.3.16) is *constant* we have

$$\Gamma(t, \tau) = \exp(t-\tau)g, \quad t \geq \tau \geq 0, \tag{2.3.19}$$

and the stability behavior of (2.3.16) depends simply on the eigenvalues of g.

Def. 2.3.3. For a $s \times s$-matrix g with eigenvalues λ_σ, $\sigma = 1(1)s$, we define

$$r[g] := \max_{\sigma = 1(1)s} \operatorname{Re} \lambda_\sigma; \tag{2.3.20}$$

the eigenvalues with $\operatorname{Re} \lambda_\sigma = r[g]$ are called *dominant eigenvalues* of g.

Theorem 2.3.4. *The equilibrium of* (2.3.16) *with constant g is exponentially stable if and only if $r[g] < 0$. If none of the dominant eigenvalues of g possesses nonlinear elementary divisors the equilibrium is $-r[g]$-exponentially stable.*

Proof. A trivial consequence of (2.3.19) and (2.3.18). □

Def. 2.3.4 (see, e.g., Dahlquist [2]). W.r.t. a given l.u.b. norm the *logarithmic norm* of a quadratic matrix g is given by

$$\mu[g] := \lim_{h \to 0} \frac{\|I + hg\| - 1}{h}. \tag{2.3.21}$$

For various familiar norms in $\mathbb{R}^s$, the values of $\mu[g]$ are given as follows: Let $g=(g_{\mu\nu})$, μ row index, ν column index; then

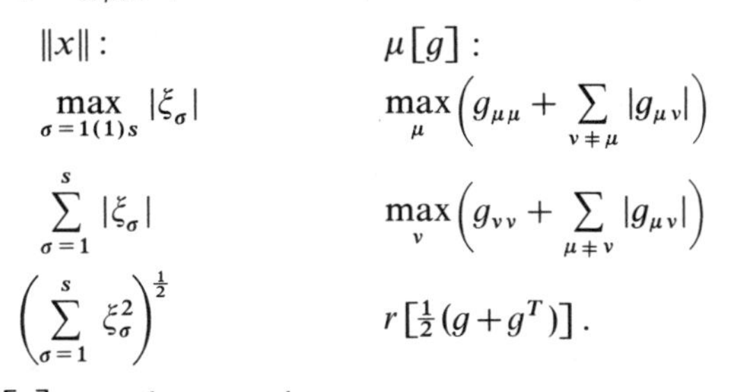

$\|x\|$:	$\mu[g]$:
$\max\limits_{\sigma=1(1)s} \lvert\xi_\sigma\rvert$	$\max\limits_{\mu}\Big(g_{\mu\mu}+\sum\limits_{\nu\neq\mu}\lvert g_{\mu\nu}\rvert\Big)$
$\sum\limits_{\sigma=1}^{s} \lvert\xi_\sigma\rvert$	$\max\limits_{\nu}\Big(g_{\nu\nu}+\sum\limits_{\mu\neq\nu}\lvert g_{\mu\nu}\rvert\Big)$
$\Big(\sum\limits_{\sigma=1}^{s} \xi_\sigma^2\Big)^{\frac{1}{2}}$	$r\big[\tfrac{1}{2}(g+g^T)\big]$.

Note that $\mu[g]$ may be negative.

Theorem 2.3.5 (Hahn [2], Theorem 58.4). *If there exists a constant $\mu>0$ such that g of* (2.3.16) *satisfies* (*w.r.t. some $\mathbb{R}^s$-norm*)

$$\mu[g(t)] \leq -\mu, \qquad t \geq 0, \tag{2.3.22}$$

then the equilibrium of (2.3.16) *is μ-exponentially stable.*

Example. For

$$g(t)=\begin{pmatrix} -5 & \dfrac{2}{1+t} \\ \dfrac{2t}{1+t} & -2 \end{pmatrix}$$

we obtain $\mu[g]=-1$ w.r.t. the Euclidean norm. Hence, (2.3.16) with this g is 1-exponentially stable.

2.3.2 Discretizations of $\{\text{IVP } 1\}_{\mathbb{T}}$

As indicated previously, we will formally use a sequence of discretization methods on longer and longer intervals to analyze the behavior of a f.s.m. for very large t. We denote the individual problems of an $\{\text{IVP } 1\}_{\mathbb{T}}$ by $(\text{IVP } 1)_T$, these being simply restrictions of (2.3.1) to the intervals $[0, T]$, $T \in \mathbb{T}$; cf. Def. 2.1.3 and the remark following it.

Def. 2.3.5. Consider an $\{\text{IVP } 1\}_{\mathbb{T}}$ and a sequence $\{\mathfrak{M}_T\}_{T\in\mathbb{T}}$ of f.s.m. $\mathfrak{M}_T$ applicable to $(\text{IVP } 1)_T$. If, for any $T_i, T_j \in \mathbb{T}$, $T_i < T_j$, the restriction of $\mathfrak{M}_{T_j}((\text{IVP } 1)_{T_j})$ to $[0, T_i]$ is identical with $\mathfrak{M}_{T_i}((\text{IVP } 1)_{T_i})$ then $\{\mathfrak{M}_T\}_{T\in\mathbb{T}}$ is called a $\mathbb{T}$-*sequence of f.s.m.*, or a $\mathfrak{M}_{\mathbb{T}}$, *applicable to the* $\{\text{IVP } 1\}_{\mathbb{T}}$. The sequence of discretizations $\mathfrak{D}_T = \mathfrak{M}_T((\text{IVP } 1)_T)$, $T \in \mathbb{T}$, is called a $\mathbb{T}$-*sequence of discretizations* and denoted by $\mathfrak{D}_{\mathbb{T}} = \mathfrak{M}_{\mathbb{T}}(\{\text{IVP } 1\}_{\mathbb{T}})$.

Note that each $\mathfrak{M}_T$ and $\mathfrak{D}_T$ is in itself a sequence $\{E_n, E_n^0, \Delta_n, \Delta_n^0, \varphi_n\}_{n\in\mathbb{N}'}$ and $\{E_n, E_n^0, F_n\}_{n\in\mathbb{N}'}$. The hypothesis of Def. 2.3.5 requires that $\mathbb{N}'$ is the same for each $T \in \mathbb{T}$ and that the meaning of the discretization parameter n is the same in each $\mathfrak{M}_T$ of an $\mathfrak{M}_{\mathbb{T}}$. In particular, if $\mathfrak{G}_{nT}$ denotes the grid $\mathfrak{G}_n$ of $\mathfrak{M}_T$ and $\mathfrak{D}_T$ then

$$\mathfrak{G}_{nT_i} \subset \mathfrak{G}_{nT_j} \quad \text{for any } T_i, T_j \in \mathbb{T}, \qquad T_i < T_j,$$

and $\mathfrak{G}_{nT_i}$ is equal to the restriction of $\mathfrak{G}_{nT_j}$ to $[0, T_i]$. Hence we may refer to the grid

$$\mathfrak{G}_n := \bigcup_{T \in \mathbb{T}} \mathfrak{G}_{nT} \subset [0, \infty), \qquad n \in \mathbb{N}',$$

as *the* grid $\mathfrak{G}_n$ of $\mathfrak{M}_{\mathbb{T}}$ and $\mathfrak{D}_{\mathbb{T}}$. Thus, *for fixed* $n \in \mathbb{N}'$, the expression "for all $t_\nu \geq t_{\nu_0}$" or "for all $\nu \geq \nu_0$" has a well-defined meaning with a given $\mathfrak{M}_{\mathbb{T}}$; it refers to the *infinitely many* points of $\mathfrak{G}_n$ which satisfy the inequality. Similarly, $h_\nu := t_\nu - t_{\nu-1}$ is defined in $\mathfrak{G}_n$ for all ν.

We will always assume that—in an obvious extension of Def. 2.1.5 to grids on $[0, \infty)$—the grid sequence $\{\mathfrak{G}_n\}_{n \in \mathbb{N}'}$ of an $\mathfrak{M}_{\mathbb{T}}$ is *linearly convergent*, i. e., that (2.1.11) holds *for all* ν with uniform constants c_1 and c_2. This implies that the steps h_ν in $\mathfrak{G}_n$ cannot become arbitrarily small or large as $\nu \to \infty$.

For reasons of simplicity, we will usually speak of *the* f.s.m. $\mathfrak{M}_{\mathbb{T}}$, discretization $\mathfrak{D}_{\mathbb{T}}$, etc. on $[0, \infty)$. However—no matter whether one uses our formal sophistication or simply an intuitive approach—one has to distinguish quite carefully between the *two types of limit processes* which are now possible as $\mathfrak{M}_{\mathbb{T}}$ and $\mathfrak{D}_{\mathbb{T}}$ are really sequences of sequences:

We may consider an arbitrarily large but *fixed interval* $[0, T]$, $T \in \mathbb{T}$, and let the discretization parameter n grow, i. e., let the stepsize decrease. This is the familiar limit process $n \to \infty$, $n \in \mathbb{N}'$, or $h \to 0$, which we have considered exclusively so far and which has been used to define such concepts as consistency, stability, convergence, etc.

However, what we are actually interested in now is the consideration of an arbitrary but *fixed value of* $n \in \mathbb{N}'$ and the investigation of what happens as we pass to larger and larger T in our $\mathbb{T}$-sequences, i. e., as we extend the grid $\mathfrak{G}_n$ to larger and larger intervals. This type of limit process will play the dominant role in our discussion of strong stability for f.s.m.; it is normally characterized by the phrase "for all ν".

According to Def. 2.1.8 and 2.1.9 and Def. 2.3.5, the difference equations of a discretization $\mathfrak{D}_{\mathbb{T}}$ generated by a k-step method $\mathfrak{M}_{\mathbb{T}}$ for some $\{\text{IVP } 1\}_{\mathbb{T}}$ have the form (for each $n \in \mathbb{N}'$)

$$(2.3.23) \quad \begin{cases} \eta(t_\nu) - s_{n\nu}(z_0) = 0, & \nu = 0(1)k-1, \\ \Phi_n(t_\nu, \eta(t_\nu), \ldots, \eta(t_{\nu-k})) = 0, & \nu \geq k, \end{cases} \qquad t_\nu \in \mathfrak{G}_n.$$

We are now indicating the dependence of the forward step procedure on ν by a separate argument t_ν rather than through a subscript (cf. (2.1.8)); this is analogous to denoting explicitly the argument t in (2.3.1).

By virtue of the hypothesis in Def. 2.3.5 we may again speak of *the* solution ζ_n of (2.3.23), i.e., the function $\mathbb{G}_n \to \mathbb{R}^s$ which arises from the $\zeta_{nT}: \mathbb{G}_{nT} \to \mathbb{R}^s$ as $T \to \infty$. Similarly, η_n, ε, etc. will denote functions $\mathbb{G}_n \to \mathbb{R}^s$ in the following.

The values of such functions at some $t_\nu \in \mathbb{G}_n$ will often be denoted by a subscript ν as usual. Furthermore, we will use the abbreviations

$$\begin{pmatrix} \eta_{\nu-1} \\ \vdots \\ \eta_{\nu-k} \end{pmatrix} =: \boldsymbol{\eta}_{\nu-1}, \qquad \Phi_n(t_\nu, \eta_\nu, \eta_{\nu-1}, \ldots, \eta_{\nu-k}) =: \Phi_n(t_\nu, \eta_\nu, \boldsymbol{\eta}_{\nu-1}), \quad \text{etc.}$$

when dealing with k-step methods, $k>1$. (We will not explicitly consider m-stage methods, $m>1$.) The norms for such quantities $\boldsymbol{\eta}_{\nu-1} \in \mathbb{R}^{ks}$ are assumed to be chosen such that

$$\|\eta_{\nu-\kappa}\|_{\mathbb{R}^s} \le \|\boldsymbol{\eta}_{\nu-1}\|_{\mathbb{R}^{ks}} \quad \text{for } \kappa = 1(1)k, \tag{2.3.24}$$

where $\|\cdots\|_{\mathbb{R}^s}$ is the $\mathbb{R}^s$-norm used in the particular context.

2.3.3 Exponential Stability for Difference Equations on $[0, \infty)$

We will now attempt to develop, for arbitrary but *fixed* $n \in \mathbb{N}'$, a "stability theory" for (2.3.23) which is analogous to that of Section 2.3.1. For $k=1$, some of the following developments have been suggested by Hahn ([1], [2]); for $k>1$ there appear to exist no systematic investigations.

In order to display the analogies more clearly we will treat (2.3.23) in the form (cf., e.g., (3.1.11) and (4.1.9))

$$\begin{cases} \boldsymbol{\eta}_{k-1} - \mathbf{s}(z_0) = 0, \\ \dfrac{\eta_\nu - \eta_{\nu-1}}{h_\nu} - \hat{\Phi}(t_\nu, \boldsymbol{\eta}_{\nu-1}) = 0, \qquad \nu \ge k, \qquad t_\nu \in \mathbb{G}_n; \end{cases} \tag{2.3.25}$$

we will assume that the "increment function" $\hat{\Phi}: \mathbb{R} \times \mathbb{R}^{ks} \to \mathbb{R}^s$ is defined and satisfies a *uniform* Lipschitz condition[4] for $t_\nu \in \mathbb{G}_n$, $\|\boldsymbol{\eta}_{\nu-1} - \zeta(t_{\nu-1})\| < R$. (It will not always be possible to express $\hat{\Phi}$ explicitly in terms of the function f of the $\{\text{IVP } 1\}_{\mathbb{T}}$ of which (2.3.25) is a discretization.)

Together with the solution $\zeta: \mathbb{G}_n \to \mathbb{R}^s$ of the unperturbed equation (2.3.25), the existence of which we assume, we consider a function $\tilde{\zeta}: \mathbb{G}_n \to \mathbb{R}^s$ which satisfies

$$\tilde{\zeta}(t_{k-1}) - \mathbf{s}(z_0) = \boldsymbol{\delta}_{k-1},$$

$$\frac{\tilde{\zeta}(t_\nu) - \tilde{\zeta}(t_{\nu-1})}{h_\nu} - \hat{\Phi}(t_\nu, \tilde{\zeta}(t_{\nu-1})) = \delta_\nu, \qquad \nu \ge k, \qquad t_\nu \in \mathbb{G}_n.$$

---

[4] The Lipschitz constant L may be of the form $L_1 n + L_2$ which is of no harm since n is fixed.

The difference $\varepsilon := \tilde{\zeta} - \zeta$ then satisfies

$$(2.3.26)\qquad \begin{cases} \boldsymbol{\varepsilon}_{k-1} = \boldsymbol{\delta}_{k-1}, \\ \dfrac{\varepsilon_\nu - \varepsilon_{\nu-1}}{h_\nu} - \Psi(t_\nu, \boldsymbol{\varepsilon}_{\nu-1}) = \delta_\nu, \quad \nu \geq k, \quad t_\nu \in \mathbb{G}_n, \end{cases}$$

where (cf. (2.3.5))

$$(2.3.27)\qquad \Psi(t_\nu, \boldsymbol{\varepsilon}_{\nu-1}) := \hat{\Phi}(t_\nu, \zeta(t_{\nu-1}) + \boldsymbol{\varepsilon}_{\nu-1}) - \hat{\Phi}(t_\nu, \zeta(t_{\nu-1})).$$

From the assumptions on $\hat{\Phi}$ we know that $\Psi: \mathbb{R} \times \mathbb{R}^{ks} \to \mathbb{R}^s$ is defined for $t_\nu \in \mathbb{G}_n$, $\|\boldsymbol{\varepsilon}_{\nu-1}\| < R$, and that it satisfies $(\xi^1, \xi^2 \in \mathbb{R}^{ks})$

$$(2.3.28)\qquad \|\Psi(t_\nu, \xi^1) - \Psi(t_\nu, \xi^2)\| \leq L \|\xi^1 - \xi^2\| \quad \text{for } \|\xi^i\| < R$$

uniformly for all $t_\nu \in \mathbb{G}_n$. As $\Psi(t_\nu, \mathbf{0}) = 0$ this implies also $\|\Psi(t_\nu, \xi)\| \leq L\|\xi\|$ for $\|\xi\| < R$.

The solution $\varepsilon \equiv 0$ of (2.3.26) for a vanishing perturbation δ is again called the *equilibrium* of (2.3.26).

Def. 2.3.6. The equilibrium of (2.3.26) is called *totally stable for n* if for each $\rho > 0\ (\rho \leq R)$ there exist positive constants $\bar{\delta}_0$ and $\bar{\delta}_1$ such that a unique solution of (2.3.26) exists which satisfies

$$\|\varepsilon_\nu\| < \rho \quad \text{for all } \nu$$

whenever

$$(2.3.29)\qquad \|\boldsymbol{\delta}_{k-1}\| < \bar{\delta}_0, \quad \|\delta_\nu\| < \bar{\delta}_1 \quad \text{for all } \nu.$$

In this case we will also call the difference equation (2.3.25) totally stable for that particular value of n.

As with an $\{\text{IVP } 1\}_{\mathbb{T}}$, the total stability of the difference equation (2.3.25) asserts that it is *properly posed* on the infinite grid $\mathbb{G}_n$ w.r.t. a norm on the perturbation δ like $\|\boldsymbol{\delta}_{k-1}\| + \sup_{\nu \geq k} \|\delta_\nu\|$. It is thus normally a prerequisite for a successful numerical treatment of (2.3.25).

In analogy with the development in Section 2.3.1 we will also consider the family of problems $(\nu_0 \geq k-1)$

$$(2.3.30)\qquad \begin{cases} \boldsymbol{\varepsilon}_{\nu_0} = \boldsymbol{\delta}_0, \\ \dfrac{\varepsilon_\nu - \varepsilon_{\nu-1}}{h_\nu} - \Psi(t_\nu, \boldsymbol{\varepsilon}_{\nu-1}) = 0, \quad \nu > \nu_0, \quad t_\nu \in \mathbb{G}_n; \end{cases}$$

we will denote the solution of (2.3.30) by $\bar{\varepsilon}_\nu(\boldsymbol{\delta}_0, t_{\nu_0})$, $\nu \geq \nu_0$.

Def. 2.3.7. If there exist positive constants a, μ, and r, *independent of* ν_0, such that $\bar{\varepsilon}_\nu(\boldsymbol{\delta}_0, t_{\nu_0})$ exists and satisfies

$$(2.3.31)\qquad \|\bar{\varepsilon}_\nu(\boldsymbol{\delta}_0, t_{\nu_0})\| \leq a\|\boldsymbol{\delta}_0\| \exp(-\mu(t_\nu - t_{\nu_0})) \quad \text{for all } \nu \geq \nu_0$$

whenever $\|\boldsymbol{\delta}_0\| < r$ then the equilibrium of (2.3.30) is called *exponentially stable for n* (or μ-exponentially stable if we wish to indicate the value of μ). In this case we will also call (2.3.25) (μ-)exponentially stable for n.

As with differential equations the concept of exponential stability is independent of the norms used in (2.3.31) (see Remark after Def. 2.3.2).

Theorem 2.3.6. *The equilibrium of* (2.3.30) *is exponentially stable for n if and only if there exists a Liapunov function* $v\colon \mathbb{R}^{ks} \times \mathbb{R} \to \mathbb{R}$ *which satisfies, for* $\|\xi\| < r$ *and all* $t_\nu \in \mathbb{G}_n$,

$$a_1 \|\xi\|^2 \le v(\xi, t_\nu) \le a_2 \|\xi\|^2, \tag{2.3.32}$$

$$\Delta v(\xi, t_\nu) := \frac{1}{h_{\nu+1}} [v(\overline{\boldsymbol{\varepsilon}}_{\nu+1}(\xi, t_\nu), t_{\nu+1}) - v(\xi, t_\nu)] \le -a_3 \|\xi\|^2 \tag{2.3.33}$$

with positive constants a_1, a_2, a_3, *and r independent of* t_ν.

Furthermore, v may be chosen such that

$$|v(\xi^1, t_\nu) - v(\xi^2, t_\nu)| \le a_4 \max(\|\xi^1\|, \|\xi^2\|) \|\xi^1 - \xi^2\|, \tag{2.3.34}$$

holds uniformly for all $t_\nu \in \mathbb{G}_n$ *and* $\|\xi^i\| < r$.

Proof. a) Assume the existence of a Liapunov function and let $V_\nu := v(\overline{\boldsymbol{\varepsilon}}_\nu(\boldsymbol{\delta}_0, t_{\nu_0}), t_\nu)$, then (2.3.33) and the definition of $\overline{\varepsilon}_\nu$ imply for $\nu > \nu_0$

$$V_\nu \le V_{\nu-1} - h_\nu a_3 \|\overline{\boldsymbol{\varepsilon}}_{\nu-1}(\boldsymbol{\delta}_0, t_{\nu_0})\|^2 \le V_{\nu-1}\left(1 - \frac{h_\nu a_3}{a_2}\right) \quad \text{by (2.3.32)}.$$

Without invalidating (2.3.32)—(2.3.33) we may increase a_2 and decrease a_3 in such a way that $1 - h_\nu a_3/a_2 > 0$ for all ν, due to the linear convergence relation (2.1.11) for $\mathbb{G}_n$. With these modified a_2, a_3 and c_1, c_2 from (2.1.11), we get

$$\begin{aligned} V_\nu &\le V_{\nu_0} \prod_{\lambda=\nu_0+1}^{\nu} \left(1 - \frac{h_\lambda a_3}{a_2}\right) \\ &\le V_{\nu_0} \exp\left(-\frac{c_1}{n}\frac{a_3}{a_2}(\nu - \nu_0)\right) \le V_{\nu_0} \exp\left(-\frac{c_1}{c_2}\frac{a_3}{a_2}(t_\nu - t_{\nu_0})\right). \end{aligned}$$

(2.3.32) now produces the desired estimate (2.3.31) with $a = \sqrt{a_2/a_1}$, $\mu = c_1 a_3 / 2 c_2 a_2$; the existence of $\overline{\varepsilon}_\nu(\delta_0, t_{\nu_0})$ is obtained recursively from this estimate if we assume $r \le \sqrt{a_1/a_2}\, R$.

b) Now assume the existence of $\overline{\varepsilon}_\nu$ and the exponential stability estimate (2.3.31) and define

$$v(\xi, t_\nu) := \frac{1}{n} \sum_{\lambda=\nu}^{\nu+m-1} \|\overline{\boldsymbol{\varepsilon}}_\lambda(\xi, t_\nu)\|^2; \tag{2.3.35}$$

the choice of m will be explained below.

The left-hand part of (2.3.32) is trivial for (2.3.35) as $\bar{\varepsilon}_\nu(\xi, t_\nu)=\xi$. On the other hand, by (2.3.31)

$$v(\xi, t_\nu) \le \frac{1}{n} a^2 \|\xi\|^2 \sum_{\lambda=\nu}^{\nu+m-1} \exp(-2\mu(t_\lambda - t_\nu))$$

$$\le \frac{1}{n} a^2 \|\xi\|^2 \sum_{\lambda=\nu}^{\nu+m-1} \exp\left(-2\mu c_1 \frac{\lambda-\nu}{n}\right) \quad \text{by (2.1.11)}$$

$$\le \frac{a^2}{n(1-\exp(-2\mu c_1/n))} \|\xi\|^2 .$$

From the definition of $\bar{\varepsilon}_\nu$ it follows that

$$\bar{\varepsilon}_\lambda(\bar{\varepsilon}_{\nu+1}(\xi, t_\nu), t_{\nu+1}) = \bar{\varepsilon}_\lambda(\xi, t_\nu)$$

so that

$$\Delta v(\xi, t_\nu) = \frac{1}{n h_{\nu+1}} \left[\sum_{\lambda=\nu+1}^{\nu+m} \|\bar{\varepsilon}_\lambda(\xi, t_\nu)\|^2 - \sum_{\lambda=\nu}^{\nu+m-1} \|\bar{\varepsilon}_\lambda(\xi, t_\nu)\|^2 \right]$$

$$= \frac{1}{n h_{\nu+1}} [\|\bar{\varepsilon}_{\nu+m}(\xi, t_\nu)\|^2 - \|\xi\|^2]$$

$$\le -\frac{1}{c_2}\left[1 - a^2 \exp\left(-\frac{2\mu m c_1}{n}\right)\right] \|\xi\|^2 \quad \text{by (2.3.31) and (2.1.11)}$$

if m is chosen such that the bracket is positive, which establishes (2.3.33).

c) While we could have taken $m=\infty$ for the purpose of b), the finiteness of m is essential for the derivation of (2.3.34). We abbreviate $\bar{\varepsilon}_\lambda(\xi^i, t_\nu)$ by $\bar{\varepsilon}^i_{\lambda\nu}$, $i=1,2$; then

$$|v(\xi^1, t_\nu) - v(\xi^2, t_\nu)| \le \frac{1}{n} \sum_{\lambda=\nu}^{\nu+m-1} \left| \|\bar{\varepsilon}^1_{\lambda\nu}\|^2 - \|\bar{\varepsilon}^2_{\lambda\nu}\|^2 \right|$$

$$= \frac{1}{n} \sum_{\lambda=\nu}^{\nu+m-1} (\|\bar{\varepsilon}^1_{\lambda\nu}\| + \|\bar{\varepsilon}^2_{\lambda\nu}\|) \left| \|\bar{\varepsilon}^1_{\lambda\nu}\| - \|\bar{\varepsilon}^2_{\lambda\nu}\| \right|$$

$$\le \frac{2}{n} \sum_{\lambda=\nu}^{\nu+m-1} \max(\|\bar{\varepsilon}^1_{\lambda\nu}\|, \|\bar{\varepsilon}^2_{\lambda\nu}\|) \|\bar{\varepsilon}^1_{\lambda\nu} - \bar{\varepsilon}^2_{\lambda\nu}\|.$$

Since we know that $\|\bar{\varepsilon}^i_{\lambda\nu}\| < R$ for all ν if $\|\xi^i\| < r$ (see end of part a) of the proof) we may use (2.3.30) and (2.3.28) to find that

$$\|\bar{\varepsilon}^1_{\lambda\nu} - \bar{\varepsilon}^2_{\lambda\nu}\| \le (1 + h_\lambda L) \|\bar{\varepsilon}^1_{\lambda-1,\nu} - \bar{\varepsilon}^2_{\lambda-1,\nu}\|$$

$$\le \prod_{\lambda'=\nu+1}^{\lambda} (1 + h_{\lambda'} L) \|\xi^1 - \xi^2\|$$

$$\le \exp\left(\frac{c_2}{n} L(\lambda - \nu)\right) \|\xi^1 - \xi^2\|.$$

With the estimate (2.3.31) for the $\|\bar{\varepsilon}^i_{\lambda\nu}\|$ we finally obtain

$$|v(\xi^1, t_\nu) - v(\xi^2, t_\nu)|$$
$$\leq \frac{2a}{n} \max(\|\xi^1\|, \|\xi^2\|) \|\xi^1 - \xi^2\| \sum_{\lambda=\nu}^{\nu+m-1} \exp(-\mu c_1 + Lc_2) \frac{\lambda - \nu}{n}$$

which is (2.3.34) with $a_4 = (2a/n) \sum_{\lambda=0}^{m-1} \exp(-\mu c_1 + Lc_2)\lambda/n$. □

Remark. In many cases, $v(\xi, t_\nu) = \|\xi\|^2$ or (2.3.31) holds with $a=1$ (these two assumptions imply each other) and Theorem 2.3.6 is almost trivial.

Theorem 2.3.7. *If the difference equation* (2.3.25) *is exponentially stable for n it is also totally stable for n.*

Proof. According to Theorem 2.3.6 there is a Liapunov function v satisfying (2.3.32)—(2.3.34). Denote the solution of (2.3.26) by $\tilde{\varepsilon}_\nu$ and let $\tilde{V}_\nu := v(\tilde{\varepsilon}_\nu, t_\nu)$. It suffices to show that

(2.3.36) $$\tilde{V}_\nu \leq \beta < a_1 \rho^2 \quad \text{for all } \nu$$

in order to conclude the assertion from (2.3.32) and (2.3.24).

To enable us to use (2.3.34) at $\tilde{\varepsilon}_\nu$ when we only know that $\|\tilde{\varepsilon}_{\nu-1}\| < \rho$, we choose $\overline{\delta}_1(\rho)$ such that, in any case,

(2.3.37) $$\left(1 + \frac{c_2}{n} L\right)\rho + \frac{c_2}{n}\overline{\delta}_1 \leq r,$$

cf. (2.3.26) and (2.3.28) (further restrictions on $\overline{\delta}_1$ are obtained below). Furthermore, the choice $\overline{\delta}_0 < \min(\rho, \sqrt{\beta/a_2})$ guarantees $\|\tilde{\varepsilon}_{k-1}\| = \|\overline{\delta}_{k-1}\| < \rho$ and $\tilde{V}_{k-1} < \beta$.

We now choose another constant β', $0 < \beta' < \beta$, and show that—for a suitable $\overline{\delta}_1$—

(2.3.38) $$\tilde{V}_\nu - \tilde{V}_{\nu-1} < \begin{cases} \beta - \beta', & \text{if } \tilde{V}_{\nu-1} \leq \beta', \\ 0, & \text{if } \tilde{V}_{\nu-1} > \beta'; \end{cases}$$

this establishes (2.3.36). By (2.3.33) and (2.3.34)

$$\tilde{V}_\nu - \tilde{V}_{\nu-1} = h_\nu \Delta v(\tilde{\varepsilon}_{\nu-1}, t_{\nu-1}) + [v(\tilde{\varepsilon}_\nu, t_\nu) - v(\overline{\varepsilon}_\nu(\tilde{\varepsilon}_{\nu-1}, t_{\nu-1}), t_\nu)]$$
$$\leq -h_\nu a_3 \|\tilde{\varepsilon}_{\nu-1}\|^2 + a_4 \max(\|\tilde{\varepsilon}_\nu\|, \|\overline{\varepsilon}_\nu(\tilde{\varepsilon}_{\nu-1}, t_{\nu-1})\|) h_\nu \|\delta_\nu\|$$

(where we have assumed that $\|(\delta_\nu, 0, \ldots, 0)^T\|_{\mathbb{R}^{ks}} = \|\delta_\nu\|_{\mathbb{R}^s}$). By virtue of (2.3.37), $a_4 r \overline{\delta}_1 < \min((\beta - \beta')n/c_2, a_3\beta'/a_2)$ guarantees (2.3.38). □

2.3.4 Exponential Stability of Neighboring Discretizations

We will now show that the exponential stability of (2.3.25) is retained when the increment function is changed by a sufficiently small amount. In (2.3.27), let

(2.3.39) $$\Psi(t_\nu, \xi) = \Psi^0(t_\nu, \xi) + X(t_\nu, \xi)$$

and assume that, uniformly for all $t_\nu \in \mathbb{G}_n$ and $\|\xi\| < R$,

$$\|\Psi^0(t_\nu, \xi)\| \le L_0 \|\xi\|, \qquad \|X(t_\nu, \xi)\| \le M \|\xi\|;$$

let $\hat{L} := \max(L_0, L)$, L from (2.3.28). Consider the family of problems ($\nu_0 \ge k-1$)

$$\text{(2.3.40)} \qquad \begin{cases} \varepsilon_{\nu_0} = \delta_0, \\ \dfrac{\varepsilon_\nu - \varepsilon_{\nu-1}}{h_\nu} - \Psi^0(t_\nu, \varepsilon_{\nu-1}) = 0, \qquad \nu > \nu_0, \qquad t_\nu \in \mathbb{G}_n; \end{cases}$$

the solution of (2.3.40) will be denoted by $\bar{\varepsilon}_\nu^0(\delta_0, t_{\nu_0})$, $\nu \ge \nu_0$.

Theorem 2.3.8. *If the equilibrium of either problem* (2.3.30) *or* (2.3.40) *is exponentially stable, with a Liapunov function v satisfying* (2.3.32) *to* (2.3.34), *then the equilibrium of the other problem is also exponentially stable provided that*

$$\text{(2.3.41)} \qquad M < \frac{a_3}{a_4\left(1 + \dfrac{c_2 \hat{L}}{n}\right)}.$$

Proof. The symmetry of the result follows immediately from the formulation of the assumptions. Assume that the hypotheses are satisfied for (2.3.30). Then

$$\begin{aligned} \Delta^0 v(\xi, t_\nu) &:= \frac{1}{h_{\nu+1}} [v(\bar{\varepsilon}^0_{\nu+1}(\xi, t_\nu), t_{\nu+1}) - v(\xi, t_\nu)] \\ &= \Delta v(\xi, t_\nu) + \frac{1}{h_{\nu+1}} [v(\bar{\varepsilon}^0_{\nu+1}(\xi, t_\nu), t_{\nu+1}) - v(\bar{\varepsilon}_{\nu+1}(\xi, t_\nu), t_{\nu+1})] \\ &\le -a_3 \|\xi\|^2 + \frac{a_4}{h_{\nu+1}} (1 + h_{\nu+1} \hat{L}) \|\xi\| \, h_{\nu+1} M \|\xi\|. \end{aligned}$$

Hence, (2.3.41) guarantees that $\Delta^0 v(\xi, t_\nu) \le -a_3^0 \|\xi\|^2$ with some $a_3^0 > 0$ so that v is also a Liapunov function for (2.3.40) and Theorem 2.3.6 implies the assertion. □

Corollary 2.3.9. *If Ψ is differentiable w.r.t. ξ and $\partial\Psi/\partial\xi =: \Psi'$ satisfies, uniformly for $t_\nu \in \mathbb{G}_n$ and $\|\xi\| < R$,*

$$\text{(2.3.42)} \qquad \|\Psi'(t_\nu, \xi) - \Psi'(t_\nu, \mathbf{0})\| \le L' \|\xi\|,$$

then the equilibrium of (2.3.30) *is exponentially stable if and only if the equilibrium of*

$$\text{(2.3.43)} \qquad \begin{cases} \varepsilon_{\nu_0} = \delta_0, \\ \dfrac{\varepsilon_\nu - \varepsilon_{\nu-1}}{h_\nu} - \Psi'(t_\nu, \mathbf{0}) \varepsilon_{\nu-1} = 0, \qquad \nu > \nu_0, \qquad t_\nu \in \mathbb{G}_n \end{cases}$$

is exponentially stable.

Proof. In (2.3.39), let $\Psi^0(t,\xi)=\Psi'(t_\nu,\mathbf{0})\xi$, this implies $L_0=L$. From $\Psi(t_\nu,\xi)=\left[\int_0^1 \Psi'(t_\nu,\lambda\xi)\,d\lambda\right]\xi$ (mean-value theorem) we obtain, with (2.3.42),

$$\|X(t_\nu,\xi)\|=\|\Psi(t_\nu,\xi)-\Psi'(t_\nu,\mathbf{0})\xi\|\leq\frac{L'}{2}\|\xi\|^2\leq\frac{L'R}{2}\|\xi\|.$$

Thus the Lipschitz-constant M for X_n can be made to satisfy (2.3.41) by choosing a sufficiently small R. □

Corollary 2.3.9 shows that it suffices to consider the exponential stability of a linearized version of (2.3.25) if the derivative of the increment function of (2.3.25) satisfies a Lipschitz condition uniformly for all t_ν in a neighborhood of the solution ζ.

If the linear operator $\Psi'(t_\nu,\mathbf{0})$ can be represented as the sum of a *constant* operator (i.e., independent of t_ν) and an operator whose norm is sufficiently small, then it may even suffice to establish the exponential of the difference equation with constant coefficients; an application of Theorem 2.3.8 will then confirm the exponential stability of the linearized difference equation and hence that of the original difference equation.

Example. $s=k=1$, $h=1/n=\text{const.}$

$$\text{(2.3.44)}\qquad \begin{cases} \eta_0-1=0, & \\ \dfrac{\eta_\nu-\eta_{\nu-1}}{h}+\sin\eta_{\nu-1}=0, & \nu>0. \end{cases}$$

In this simple case the exponential stability may, of course, be obtained in a straightforward manner; we will, however, employ the line of reasoning just explained.

Let ζ be the solution of (2.3.44); the function Ψ: $\mathbb{R}\times\mathbb{R}\to\mathbb{R}$ of (2.3.27) is given by

$$\Psi(t_\nu,\xi)=-[\sin(\zeta_{\nu-1}+\xi)-\sin\zeta_{\nu-1}],$$

its Lipschitz constant is $L=1$ for arbitrary R. Similarly, $\Psi'(t_\nu,\xi)=-\cos(\zeta_{\nu-1}+\xi)$ possesses the uniform Lipschitz constant $L'=1$, see (2.3.42). Thus, by Corollary 2.3.9, it suffices to consider the exponential stability of the equilibrium of

$$\text{(2.3.45)}\qquad \frac{\varepsilon_\nu-\varepsilon_{\nu-1}}{h}+(\cos\zeta_{\nu-1})\varepsilon_{\nu-1}=0.$$

We replace (2.3.45) by

$$\text{(2.3.46)}\qquad \frac{\varepsilon_\nu^0-\varepsilon_{\nu-1}^0}{h}+\frac{3}{4}\varepsilon_{\nu-1}^0=0$$

the exponential stability of which is trivial (for $h<\frac{8}{3}$). This yields $X(t_\nu,\xi)=(\frac{3}{4}-\cos\zeta_{\nu-1})\xi$ (cf. 2.3.39) and $M=\frac{1}{4}$ if we restrict $h\leq 2$ so that $|\zeta_\nu|\leq 1$ is immediate.

For (2.3.46), we may choose $v(\xi)=\xi^2$ (no dependence upon t_ν is necessary) which leads to

$$\Delta v(\xi)=\frac{1}{h}[(1-\tfrac{3}{4}h)^2-1]\xi^2=-\tfrac{3}{2}(1-\tfrac{3}{8}h)\xi^2\quad\text{and}$$

$$v(\xi^1)-v(\xi^2)\leq 2\max(|\xi^1|,|\xi^2|)|\xi^1-\xi^2|.$$

Furthermore, we may neglect $1+h\hat{L}$ in the denominator of (2.3.41) since both $|\bar{\varepsilon}|$ and $|\bar{\varepsilon}^0|$ cannot grow (cf. proof of Theorem 2.3.8). Hence the exponential stability of (2.3.46) implies that of (2.3.45) which in turn implies that of (2.3.44) if

$$\tfrac{1}{4}<\tfrac{3}{4}(1-\tfrac{3}{8}h) \quad \text{or} \quad h<\tfrac{16}{9}.$$

(Actually (2.3.44) is exponentially stable for $h<2$.)

2.3.5 Strong Exponential Stability

In Sections 2.3.2 to 2.3.4 we discussed the stability properties of certain k-step difference equations on $[0,\infty)$ irrespective of the associated $\{\text{IVP 1}\}_{\mathbb{T}}$. According to our discussion in Section 1.5.3, an ideal discretization method should generate discretizations which share precisely the perturbation sensitivity of the original problem for any $n\in\mathbb{N}'$; however, it was not clear how this general requirement should be concisely formulated. In the case of exponential stability we may do this as follows:

Def. 2.3.8. A f.s.m. $\mathfrak{M}_{\mathbb{T}}$ applicable to $\{\text{IVP 1}\}_{\mathbb{T}}$ is called *strongly exponentially stable* if it generates a discretization $\mathfrak{D}_{\mathbb{T}}$ which is exponentially stable for each $n\in\mathbb{N}'$ for each exponentially stable $\{\text{IVP 1}\}_{\mathbb{T}}$.

Strong exponential stability appears to be a very natural condition to impose on a f.s.m. which is to be used for the numerical integration of an exponentially stable $\{\text{IVP 1}\}_{\mathbb{T}}$ over long time intervals. Since the solution of such a system is totally stable according to Theorem 2.3.2, the effect of a sufficiently small persistent perturbation of the function $f(t,y)$ in (2.3.1) remains uniformly bounded over arbitrarily long intervals; see Def. 2.3.1. If we employ for the numerical integration of such a system a discretization which is not exponentially stable for the value of the discretization parameter used (i.e., for the step sizes used), then a persistent perturbation may cause a total error which grows without bound with the length of the interval of integration, irrespectively even of the effect of the local discretization error and the local computing error. Thus the numerical solution will become meaningless.

Unfortunately, it will turn out that strong exponential stability is a very severe requirement for f.s.m.; hence we will consider some relaxations of this requirement which seem reasonable from the point of view of applications.

One natural approach is to restrict the set of $\{\text{IVP 1}\}_{\mathbb{T}}$ under consideration:

Def. 2.3.9. Let J be a class of exponentially stable $\{\text{IVP 1}\}_{\mathbb{T}}$. A f.s.m. is called *strongly exponentially stable w.r.t. J* if for each $\{\text{IVP 1}\}_{\mathbb{T}}\in J$ it generates a discretization which is exponentially stable for each $n\in\mathbb{N}'$.

It will be an important objective to define large and meaningful classes J of exponentially stable $\{\text{IVP }1\}_{\mathbb{T}}$ such that there exist classes of f.s.m. which are strongly exponentially stable w.r.t. J.

Another natural restriction concerns the discretization parameter n; it may be combined with the restriction of Def. 2.3.9[5]:

Def. 2.3.10. A f.s.m. $\mathfrak{M}_{\mathbb{T}}$ is called *strongly exponentially stable (w.r.t. J) for sufficiently large n* if for each exponentially stable $\{\text{IVP }1\}_{\mathbb{T}}$ (from J) there exists an n_0 such that the discretization $\mathfrak{M}_{\mathbb{T}}(\{\text{IVP }1\}_{\mathbb{T}})$ is exponentially stable for $n \geq n_0$, $n \in \mathbb{N}'$.

It will turn out that large classes of f.s.m. are strongly exponentially stable for sufficiently large n. For these methods it will be important to have *criteria* to decide whether a *given n* is "sufficiently large" for a *given* $\{\text{IVP }1\}_{\mathbb{T}}$.

It has become customary also to characterize the absence of strong exponential stability for sufficiently large n:

Def. 2.3.11. A f.s.m. $\mathfrak{M}_{\mathbb{T}}$ is called *weakly stable* if each $\mathfrak{M}_T$ is stable in the sense of Def. 1.1.10 for each $(\text{IVP }1)_T$, $T \in \mathbb{T}$, but $\mathfrak{M}_{\mathbb{T}}(\{\text{IVP }1\}_{\mathbb{T}})$ is not even exponentially stable for arbitrarily large n for any exponentially stable $\{\text{IVP }1\}_{\mathbb{T}}$.

It is obvious that weakly stable f.s.m. are not suitable for integration over long time intervals; some such methods are, however, particularly valuable for integration over "normal" time intervals.

It would be desirable that the discretization of a μ-exponentially stable $\{\text{IVP }1\}_{\mathbb{T}}$ should also be μ-exponentially stable so that the effect of an individual perturbation of the discretization would remain bounded *relative* to the effect of the same perturbation in the differential equation. But this is asking too much except under very special circumstances.

Example. Euler method, equidistant grids, $h = 1/n$:

Any meaningful class of exponentially stable $\{\text{IVP }1\}_{\mathbb{T}}$ should at least contain all linear problems (2.3.1) with a constant matrix g whose eigenvalues lie in some given bounded region $B \subset \mathbb{C}_-$. Since $\bar{\varepsilon}_\nu(\delta_0, t_{\nu_0}) = (I + hg)^{\nu - \nu_0} \delta_0$ in this case (cf. (2.3.30)), it is sufficient and necessary for the exponential stability of the discretization that all eigenvalues of $I + hg$ are in the open unit disk D_0. Therefore we have to consider the location relative to the unit disk of the image B'_h of B under the mapping $\lambda \to 1 + h\lambda$:

For any $B \subset \mathbb{C}_-$ there are always values of h such that $B'_h \not\subset D_0$. This means that the Euler method is not strongly exponentially stable w.r.t. the above class J of exponentially stable $\{\text{IVP }1\}_{\mathbb{T}}$. On the other hand, as B was assumed *bounded* there is always an $h_0 > 0$ such that $B'_h \subset D_0$ for all $h < h_0$. Thus the Euler method is strongly exponentially stable w.r.t. J for sufficiently small h.

---

[5] The insertions in parantheses in Def. 2.3.10 serve this purpose.

2.3.6 Stability Regions

In the following, we will denote by J_c the class of *linear* $\{\text{IVP}\,1\}_{\mathbb{T}}$ *with constant coefficients*, and g will be the generic notation for the coefficient matrix. For an $\{\text{IVP}\,1\}_{\mathbb{T}}$ from J_c, the exponential stability of its discretization generated by a standard f.s.m. with equidistant grids depends only on the location of the eigenvalues of the matrix hg.

Def. 2.3.12. For a given f.s.m. $\mathfrak{M}_{\mathbb{T}}$ with equidistant grids, the region $\mathfrak{H}_0 \subset \mathbb{C}$ is called *region of absolute stability* (or: absolute stability region) if the discretization generated by $\mathfrak{M}_{\mathbb{T}}$ for an $\{\text{IVP}\,1\}_{\mathbb{T}} \in J_c$ is exponentially stable for the step h whenever all eigenvalues of the matrix hg are in $\mathfrak{H}_0$.

Of course, a region of absolute stability need not exist for each f.s.m.; in particular, $\mathfrak{H}_0$ is empty for a weakly stable f.s.m. (see Def. 2.3.11).

The knowledge of $\mathfrak{H}_0$ for a given discretization method provides a simple criterion of the type requested after Def. 2.3.10 at least for strong exponential stability w.r.t. J_c. This class is not very interesting from the practical point of view; but one may use Theorem 2.3.8 and Corollary 2.3.9 to extend the range of application. Results of this kind will be presented in Section 3.5.4.

Def. 2.3.13 (Dahlquist [3]). A f.s.m. $\mathfrak{M}_{\mathbb{T}}$ is called *A-stable* (absolutely stable) if it possesses a region $\mathfrak{H}_0$ of absolute stability and $\mathfrak{H}_0 \supset \mathbb{C}_-$.

Theorem 2.3.10. *An A-stable f.s.m. is strongly exponentially stable w.r.t.* J_c.

Proof. According to Theorem 2.3.4, for an exponentially stable $\{\text{IVP}\,1\}_{\mathbb{T}} \in J_c$ the eigenvalues of hg are in $\mathbb{C}_-$ for any $h>0$. ▯

Obviously, A-stability is a very desirable property for a f.s.m. which is to be used for the numerical integration of an IVP 1 over long time intervals.

Def. 2.3.14 (Widlund [1]). For $0<\alpha\leq\pi/2$, let $S(\alpha):=\{\gamma\in\mathbb{C}_-: |\arg\gamma-\pi|<\alpha\}$. A f.s.m. $\mathfrak{M}_{\mathbb{T}}$ is called *$A(\alpha)$-stable* if it possesses a region $\mathfrak{H}_0$ of absolute stability and $\mathfrak{H}_0 \supset S(\alpha)$. It is called *$A(0)$-stable* if there exists an $\alpha_0>0$ such that it is $A(\alpha_0)$-stable.

Obviously, if all eigenvalues of the matrix g are in $S(\alpha)$ and the $\{\text{IVP}\,1\}_{\mathbb{T}} \in J_c$ is discretized by an $A(\alpha)$-stable f.s.m. the discretization is exponentially stable for arbitrary h as $\gamma\in S(\alpha)$ implies $h\gamma\in S(\alpha)$. Note that $A(\pi/2)$-stability is equivalent to A-stability.

For a finer analysis of the error propagation in the solution of a discretization one may extend the idea of Def. 2.3.12 in the following manner:

Def. 2.3.15. For a given f.s.m. $\mathfrak{M}_{\mathbb{T}}$ with equidistant grids, the region $\mathfrak{H}_\mu \subset \mathbb{C}$ $(\mu > 0)$ is called the *region of μ-exponential stability* if the discretization by $\mathfrak{M}_{\mathbb{T}}$ for an $\{\text{IVP}\,1\}_{\mathbb{T}} \in J_c$ is μ/h-exponentially stable for the step h whenever all eigenvalues of hg are in $\mathfrak{H}_\mu$.

Obviously $\mathfrak{H}_0 = \bigcup_{\mu > 0} \mathfrak{H}_\mu$ for any f.s.m.

An $\{\text{IVP}\,1\}_{\mathbb{T}} \in J_c$ is μ-exponentially stable if $r[g] < -\mu < 0$; see Theorem 2.3.4. Hence we would hope that the regions $\mathfrak{H}_\mu$ of a f.s.m. with a favorable strong stability behavior would at least contain all H with an arbitrarily large negative real part. This demand represents a strengthening of the requirement of A-stability.

Def. 2.3.16 (Ehle [1]). A f.s.m. $\mathfrak{M}_{\mathbb{T}}$ is called *L-stable* if for each $\mu > 0$ there exists a $\mu' > 0$ such that $H_\mu \supset \mathbb{C}_{-\mu'} := \{H \in \mathbb{C}: \text{Re}\, H < -\mu'\}$.

Examples. 1. The absolute stability region $\mathfrak{H}_0$ of the Euler method is the disk $\{H \in \mathbb{C}: |H+1| < 1\}$; see the example at the end of Section 2.3.5. Clearly, the Euler method is neither A- nor $A(0)$-stable.

2. The *implicit Euler method* is the 1-step method with the forward step procedure

$$\frac{\eta_\nu - \eta_{\nu-1}}{h_\nu} - f(t_\nu, \eta_\nu) = 0\,.$$

For a problem from J_c we have (for equidistant grids)

$$\eta_\nu = (I - hg)^{-1}\, \eta_{\nu-1}\,,$$

so that the stability regions $\mathfrak{H}_\mu$ of the implicit Euler method are

$$\mathfrak{H}_\mu = \left\{ H \in \mathbb{C}: \left| \frac{1}{1-H} \right| < e^{-\mu} \right\}, \qquad \mu \geq 0\,.$$

Obviously $\mathfrak{H}_0 \supset \mathbb{C}_-$ so that the method is A-stable. Also $\mathfrak{H}_\mu \supset \mathbb{C}_{-\mu'}$ with $\mu' = e^\mu - 1$ so that it is even L-stable. (This shows that Def. 2.3.13 and 2.3.16 are meaningful.)

For large classes of k-step f.s.m. with equidistant grids the Equ. (2.1.13) generated for an $\{\text{IVP}\,1\}_{\mathbb{T}} \in J_c$ reduces essentially to a difference equation of some order $\geq k$ with constant coefficients. The growth of the solutions of such a discretization is fully determined by the zeros of the *characteristic polynomial* of the difference equation. If these zeros depend in a suitable fashion on the products of the step h with the eigenvalues of g then the stability regions of the f.s.m. may be determined from the characteristic polynomial; see (3.5.24) and Sections 4.6.2 and 5.5.1.

2.3.7 Stiff Systems of Differential Equations

The development of the concepts of the previous sections has been prompted largely by the observation that many standard f.s.m. perform very poorly on certain types of differential equations for which the name "stiff equations" has become customary.

Within the class J_c, the stiff systems are characterized by matrices g possessing eigenvalues with negative real parts of very large modulus. To see why systems of this kind present difficulties we consider the scalar $\{\mathrm{IVP}\,1\}_{\mathrm{T}}$ (cf. Dahlquist [3])

$$y(0)=y_0, \quad y'-gy-r(t)=0, \quad t\geq 0, \tag{2.3.47}$$

where r' exists and has small absolute values while $g \ll 0$. The solution of (2.3.47) is

$$z(t)=e^{tg}y_0+\int_0^t e^{(t-\tau)g}r(\tau)d\tau=e^{tg}(y_0+g^{-1}r(0))-g^{-1}\Big(r(t)-\int_0^t e^{(t-\tau)g}r'(\tau)d\tau\Big);$$

for all but very small t it will be determined solely by r; under our assumptions on r it will be slowly varying after the contribution from the initial value has decayed. Therefore, except for small t, one should be able to proceed with large steps in a discretization of (2.3.47) and still obtain a rather accurate approximation to the true solution.

However, for f.s.m. with a finite region of absolute stability hg will not be contained in $\mathfrak{H}_0$ when h is large. Perturbations introduced by the computational process may therefore *grow exponentially* and produce errors which become intolerable after a while. The only remedy is to use sufficiently small steps *throughout the whole integration* of (2.3.47) which is obviously very uneconomic considering the structure of the solution z.

Only if the f.s.m. is A-stable, or at least $A(0)$-stable (see Def. 2.3.14), may we increase h without running the risk of pushing hg out of $\mathfrak{H}_0$; then the discretization is exponentially stable with arbitrary steps (see Theorem 2.3.10). If the f.s.m. is even L-stable we may expect that the exponential decay of perturbations in the discretization is the more rapid the larger $-g$ is (cf. Def. 2.3.16).

A general IVP 1 will be considered as stiff if the differential equation possesses some particular solutions which show a rapid exponential decay while the true solution of the IVP 1 varies only moderately over most of the interval of integration. Of course, the stiffness may now be present only over certain portions of the time interval or with particular data only. Linear and non-linear IVP 1 which display this pattern of behavior occur in various fields of applications, notably in chemical engineering. For more information, see, e.g., Dahlquist et al. [1].

"Badly stiff" problems require very special types of f.s.m. for their effective numerical solution. Although considerable progress has been made in the construction of such methods within the past years, there is still much insight to be gained. In this treatise, we will devote no particular effort to a comprehensive analysis of methods for stiff problems as this seems premature at the present time. However, a good deal of our theory of discretizations on infinite intervals may be utilized in the analysis of such methods since, loosely speaking, stiff IVP1 show those phenomena on *short* intervals which occur naturally as the length of the interval tends to infinity.

Chapter 3

Runge-Kutta Methods

One-step methods (see Def. 2.1.8) form a particularly simple class of f.s.m. for IVP 1. Among these, a certain class of methods has commonly been associated with the names of C. Runge and W. Kutta and is widely used. These "Runge-Kutta methods" (RK-methods) are 1-step $m+1$-stage methods in the sense of Def. 2.1.10. However, only the final stage $\eta_{\nu-1}^{m+1}$ of the value $\boldsymbol{\eta}_{\nu-1}$ at $t_{\nu-1}$ enters into the computation of $\boldsymbol{\eta}_\nu$; also, only this final stage is normally taken as an approximation to the true solution. Thus, by formally disregarding the intermediate stages RK-methods may also be considered as 1-step 1-stage methods. We will use both interpretations depending on what is more convenient.

The "classical" RK-method is the 5-stage method of the last example at the end of Section 3.1.1. It is probably the most widely used individual discretization method. The extension of the concept "RK-method" to the generality presented in this section has only come about during the past decade. The theory of RK-methods is the best developed of all discretization methods.

3.1 RK-procedures

In this section we will consider the forward step procedures of RK-methods only. f will denote a mapping $\mathbb{R}^s \to \mathbb{R}^s$ (s a natural number) which satisfies a uniform Lipschitz condition in some domain $B_R(z) \subset \mathbb{R}^s$ (cf. (2.1.6)); further smoothness assumptions will be added as needed.

3.1.1 Characterization

Def. 3.1.1. Let A be a $(m+1) \times m$-matrix ($m+1$ rows, m columns), with elements $a_{\mu\lambda}$, $\mu=1(1)m+1$, $\lambda=1(1)m$. The 1-step $(m+1)$-stage forward step procedure (for terminology and notation, see Section 2.1.3)

(3.1.1) $$\Phi_\nu^\mu(\boldsymbol{\eta}_\nu,\boldsymbol{\eta}_{\nu-1}) = \frac{1}{h_\nu}(\eta_\nu^\mu - \eta_{\nu-1}^{m+1}) - \sum_{\lambda=1}^{m} a_{\mu\lambda} f(\eta_\nu^\lambda) = 0, \qquad \mu = 1(1)m+1,$$

is called an *m-stage*[1] *RK-procedure*, the matrix A its *generating matrix*.

If we let

(3.1.2) $$\mathbf{A} := \begin{pmatrix} a_{11}I & a_{12}I & \dots & a_{1m}I & 0 \\ a_{21}I & a_{22}I & \dots & a_{2m}I & 0 \\ \vdots & \vdots & & \vdots & \vdots \\ a_{m+1,1}I & a_{m+1,2}I & \dots & a_{m+1,m}I & 0 \end{pmatrix}, \qquad \mathbf{e} := \begin{pmatrix} I \\ I \\ \vdots \\ I \end{pmatrix},$$

where I and 0 are $s \times s$-matrices, we may use the notation introduced in Section 2.1.5 (see (2.1.30) and (2.1.31)) to write (3.1.1) in the form

(3.1.3) $$\Phi_\nu(\boldsymbol{\eta}_\nu,\boldsymbol{\eta}_{\nu-1}) = \frac{1}{h_\nu}(\boldsymbol{\eta}_\nu - \mathbf{e}\,\eta_{\nu-1}^{m+1}) - \mathbf{A}\,\mathbf{f}(\boldsymbol{\eta}_\nu) = 0\,.$$

Def. 3.1.2. If the elements $a_{\mu\lambda}$ of the generating matrix A satisfy

$$a_{\mu\lambda} = 0 \quad \text{for } \lambda \ge \mu, \qquad \mu = 1(1)m,$$

A is called *explicit*; otherwise it is called *implicit*. If $a_{\mu\lambda} = 0$ for $\lambda > \mu$ but $a_{\mu\mu} \neq 0$ for some μ, A is called *semi-implicit* (or semi-explicit).

Remarks. 1. Originally, only RK-procedures with explicit generating matrices were considered; the inclusion of implicit generating matrices has added much insight into the structure of RK-methods. In the following, a generating matrix may always be implicit unless stated otherwise.

2. For an implicit generating matrix A there may exist a permutation of the first m rows and columns such that the permuted matrix is explicit. In this case the implicitness is of a formal nature only. In Section 3.2.4 we will give a more satisfactory definition of explicitness and implicitness (Def. 3.2.8).

The intuitive interpretation of (3.1.1) is the following: The first m stages of (3.1.1) represent auxiliary computations to obtain the values $f(\eta_\nu^\lambda)$, $\lambda = 1(1)m$, which are then used to compute the value η_ν^{m+1} of the final stage. If A is explicit, the values η_ν^λ may be obtained recursively. Otherwise, some or all of the first m sets of s equations in (3.1.1) have to be solved simultaneously for the $\eta_\nu^\lambda \in \mathbb{R}^s$, $\lambda = 1(1)m$. The existence of a solution to these equations is guaranteed by

Lemma 3.1.1. *There is an* $h_0 > 0$ *such that, for* $|h_\nu| \le h_0$ *and* $\eta_{\nu-1}^{m+1} \in B_0 := B_{R/2}(z)$, (3.1.1), *resp.* (3.1.3) *has a unique solution* $\boldsymbol{\eta}_\nu \in (B_R(z))^{m+1}$.

Proof. By (2.1.6), f possesses the uniform Lipschitz constant L in $B_R(z)$. Thus the mapping

(3.1.4) $$F(\boldsymbol{\eta}; h, \xi) := \mathbf{e}\,\xi + h\,\mathbf{A}\,\mathbf{f}(\boldsymbol{\eta})$$

[1] We adhere to common usage in the literature w.r.t. counting the number of stages in this fashion; see also Example 1 at the end of Section 5.1.1.

satisfies for $\boldsymbol{\eta}^{(1)}, \boldsymbol{\eta}^{(2)} \in (B_R(z))^{m+1}$

$$\|F(\boldsymbol{\eta}^{(1)}; h, \xi) - F(\boldsymbol{\eta}^{(2)}; h, \xi)\|_{\mathbb{R}^{(m+1)s}} \leq h L \,\||A|\|\, \|\boldsymbol{\eta}^{(1)} - \boldsymbol{\eta}^{(2)}\|_{\mathbb{R}^{(m+1)s}};$$

$|A|$ is the matrix of the $|a_{\mu\lambda}|$, with an $(m+1)$st column of zeros, and $\|\cdot\|$ the l.u.b. norm for (2.1.32). We now choose h_0 such that

$$0 < h_0 < \frac{1}{L\||A|\|} \tag{3.1.5}$$

and that $F(\boldsymbol{\eta}; h, \xi) \in (B_R(z))^{m+1}$ for $\boldsymbol{\eta} \in (B_R(z))^{m+1}$, $|h| \leq h_0$ and $\xi \in B_0$ which is possible since f is bounded on $B_R(z)$. Then F is a contraction w.r.t. $\boldsymbol{\eta}$ on $(B_R(z))^{m+1}$ and

$$\boldsymbol{\eta}_\nu = F(\boldsymbol{\eta}_\nu; h_\nu, \eta_{\nu-1}^{m+1}) \tag{3.1.6}$$

has a unique solution in $(B_R(z))^{m+1}$ for $|h_\nu| \leq h_0$, $\eta_{\nu-1}^{m+1} \in B_0$. By (3.1.4), (3.1.6) is equivalent to (3.1.3). □

Remarks. 1. In our discussion of RK-procedures, h_0 and B_0 will always denote the quantities of Lemma 3.1.1 and the solution $\boldsymbol{\eta}_\nu$ of (3.1.3) is assumed to be the unique solution of (3.1.6) in $(B_R(z))^{m+1}$.

2. When (3.1.3) is implicit, "solving" it by means of a *truncated* iteration

$$\begin{cases} \boldsymbol{\eta}_\nu^{(i)} := F(\boldsymbol{\eta}_\nu^{(i-1)}; h_\nu, \eta_{\nu-1}^{m+1}), & i = 1(1)M, \\ \boldsymbol{\eta}_\nu := \boldsymbol{\eta}_\nu^{(M)}, \end{cases} \tag{3.1.7}$$

with fixed M and $\boldsymbol{\eta}_\nu^{(0)} = \boldsymbol{\eta}_\nu^{(0)}(h_\nu, \eta_{\nu-1}^{m+1})$, amounts to replacing the original RK-procedure by a higher stage explicit RK-procedure. Therefore (3.1.7) is *not* equivalent to (3.1.3) although it may often be acceptable as an algorithmic approximation of (3.1.3) (see also Section 3.5.3).

Examples. Generating matrices of commonly used RK-procedures:

$\begin{pmatrix} 0 \\ 1 \end{pmatrix}, \begin{pmatrix} 1 \\ 1 \end{pmatrix}$ Euler (polygon) method, implicit Euler method;

$\begin{pmatrix} 0 & 0 \\ \frac{1}{2} & 0 \\ 0 & 1 \end{pmatrix}, \begin{pmatrix} \frac{1}{2} \\ 1 \end{pmatrix}$ explicit, resp. implicit midpoint method;

$\begin{pmatrix} 0 & 0 \\ 1 & 0 \\ \frac{1}{2} & \frac{1}{2} \end{pmatrix}, \begin{pmatrix} 0 & 0 \\ \frac{1}{2} & \frac{1}{2} \\ \frac{1}{2} & \frac{1}{2} \end{pmatrix}$[2] explicit, resp. implicit trapezoidal method;

$\begin{pmatrix} 0 & 0 & 0 & 0 \\ \frac{1}{2} & 0 & 0 & 0 \\ 0 & \frac{1}{2} & 0 & 0 \\ 0 & 0 & 1 & 0 \\ \frac{1}{6} & \frac{1}{3} & \frac{1}{3} & \frac{1}{6} \end{pmatrix}$ classical RK-method, "RK 4".

[2] This slightly awkward notation results from our general formulation (3.1.1) for arbitrary RK-procedures.

3.1.2 Local Solution and Increment Function

Def. 3.1.3. For a given m-stage RK-procedure with generating matrix A, the mapping $\bar{\zeta}: \mathbb{R} \times \mathbb{R}^s \to \mathbb{R}^{(m+1)s}$ defined by

$$\bar{\zeta}(h,\xi) = \mathbf{e}\,\xi + h\,\mathbf{A}\,\mathbf{f}(\bar{\zeta}(h,\xi)) \tag{3.1.8}$$

is called the *local solution (through ξ)* of the RK-procedure.

By Lemma 3.1.1, the definition of $\bar{\zeta}(h,\xi)$ is valid and unique at least for $|h| \le h_0$, $\xi \in B_0$. The block-components of $\bar{\zeta}(h,\xi)$ will be denoted by $\bar{\zeta}^\mu(h,\xi) \in \mathbb{R}^s$, $\mu = 1(1)m+1$, see Section 2.1.5. Note that $\bar{\zeta}^\mu(0,\xi) = \xi$, $\mu = 1(1)m+1$.

Def. 3.1.4. For a given m-stage RK-procedure with generating matrix A, the mapping $\Psi: \mathbb{R} \times \mathbb{R}^s \to \mathbb{R}^s$ defined by

$$\Psi(h,\xi) := [\mathbf{A}\,\mathbf{f}(\bar{\zeta}(h,\xi))]^{m+1} = \begin{cases} \dfrac{\bar{\zeta}^{m+1}(h,\xi) - \xi}{h}, & h \neq 0, \\[2ex] \left(\sum\limits_{\lambda=1}^{m} a_{m+1,\lambda}\right) f(\xi), & h = 0, \end{cases} \tag{3.1.9}$$

is called the *increment function* of the RK-procedure.

As indicated in the opening remarks of Chapter 3, a RK-method may formally be considered as a 1-stage f.s.m. by restricting attention to the final stage η_ν^{m+1} of each step. The increment function serves exactly this purpose: Since the solution $\boldsymbol{\eta}_\nu$ of (3.1.3) is given by

$$\boldsymbol{\eta}_\nu = \bar{\zeta}(h_\nu, \eta_{\nu-1}^{m+1}) \tag{3.1.10}$$

so that in particular $\eta_\nu^{m+1} = \bar{\zeta}^{m+1}(h_\nu, \eta_{\nu-1}^{m+1})$, we can write the forward step procedure (3.1.3) in the 1-stage form

$$\Phi_\nu(\eta_\nu, \eta_{\nu-1}) = \frac{\eta_\nu - \eta_{\nu-1}}{h_\nu} - \Psi(h_\nu, \eta_{\nu-1}) = 0, \tag{3.1.11}$$

with $\eta_\nu = \eta_\nu^{m+1}$, $\nu = 0(1)n$. (3.1.11) resembles the forward step procedure of the Euler method, with $f(\eta_{\nu-1})$ replaced by $\Psi(h_\nu, \eta_{\nu-1})$. A closed expression for Ψ in terms of f can *only* be obtained for *explicit* generating matrices; thus the simplicity of (3.1.11) is deceptive.

Example. $\quad A = \begin{pmatrix} 0 & 0 \\ \frac{1}{2} & 0 \\ 0 & 1 \end{pmatrix}, \quad \Psi(h,\xi) = f\left(\xi + \frac{h}{2} f(\xi)\right).$

The smoothness properties of $\bar{\zeta}$ and Ψ depend on those of f:

Lemma 3.1.2. *$\bar{\zeta}$ and Ψ are Lipschitz continuous w.r.t. their second argument in $[-h_0, h_0] \times B_0$ (see Remark after Lemma 3.1.1). If f possesses* *p continuous (Frechet-) derivatives in $B_R(z)$ so do $\bar{\zeta}$ and Ψ in $[-h_0, h_0] \times B_0$. If $f^{(p)}$ is uniformly Lipschitz continuous in $B_R(z)$ so are the p-th derivatives of $\bar{\zeta}$ and Ψ w.r.t. their second arguments in $[-h_0, h_0] \times B_0$.*

Proof. Let L be the uniform Lipschitz constant of f in $B_R(z)$. By (3.1.8), we have for $|h| \le h_0$, $\xi_1, \xi_2 \in B_0$,

(3.1.12)
$$\|\bar{\zeta}^\mu(h,\xi_1) - \bar{\zeta}^\mu(h,\xi_2)\| \le \|\xi_1 - \xi_2\| + hL \sum_{\lambda=1}^{m} |a_{\mu\lambda}| \, \|\bar{\zeta}^\lambda(h,\xi_1) - \bar{\zeta}^\lambda(h,\xi_2)\| .$$

Since $h_0 L \| |A| \| < 1$ by (3.1.5), $[I - hL|A|]^{-1} = \sum_{\nu=0}^{\infty} (hL|A|)^\nu$ exists and has only positive elements for $|h| \le h_0$. Therefore, (3.1.12) implies

$$\|\bar{\zeta}(h,\xi_1) - \bar{\zeta}(h,\xi_2)\| \le \|[I - h_0 L|A|]^{-1}\| \, \|\xi_1 - \xi_2\| \le \frac{1}{1 - h_0 L \| |A| \|} \|\xi_1 - \xi_2\|.$$

The Lipschitz continuity of Ψ is then implied by (3.1.9).

Since the Lipschitz constant L of f is uniform in $B_R(z)$, L is also the uniform bound on the first (Frechet-)derivative of f in $B_R(z)$. Hence, the relations obtained by differentiating (3.1.8) determine the derivatives of $\bar{\zeta}$ uniquely for $|h| \le h_0$, $\xi \in B_0$, as is seen recursively by an argument like the one in the proof of Lemma 3.1.1. Similarly, the Lipschitz continuity of the derivatives of f carries over to those of $\bar{\zeta}$ as is seen, by induction with the order of the derivatives, through an argument analogous to the one involving (3.1.12). By (3.1.9), Ψ has the same smoothness properties as $\bar{\zeta}$. □

In order not to be distracted by secondary matters, we will often assume $f \in C^\infty(\mathbb{R}^s \to \mathbb{R}^s)$ in our further discussion of RK-procedures. In specific applications, Lemma 3.1.2 may then be used to determine the precise smoothness conditions necessary for the validity of certain results.

3.1.3 Elementary Differentials

Def. 3.1.5 (Butcher [2]). For a given mapping $f: \mathbb{R}^s \to \mathbb{R}^s$, the *elementary differentials* $F_i: \mathbb{R}^s \to \mathbb{R}^s$ of f and their *order* are recursively defined thus:

(i) f is the only elementary differential of f of order 1;

(ii) if $F_j, j = 1(1)J$, are elementary differentials of f of orders r_j resp. (not necessarily distinct) then

(3.1.13)
$$f^{(J)} F_1 F_2 \ldots F_J =: \{F_1 \ldots F_J\}$$

is an elementary differential of f of order $1 + \sum_{j=1}^{J} r_j$.

Here $f^{(J)}: \mathbb{R}^s \to \mathfrak{L}((\mathbb{R}^s)^J \to \mathbb{R}^s)$ is the J-th Frechet derivative of f: The application of $f^{(J)}(\xi_0)$ to the J arguments $\xi_1, \ldots, \xi_J$ ($\xi_j \in \mathbb{R}^s, j=0(1)J$) yields

$$f^{(J)}(\xi_0)\xi_1 \ldots \xi_J = \sum_{\sigma_1=1}^{s} \ldots \sum_{\sigma_J=1}^{s} \frac{\partial^J f(\xi_0)}{\partial \xi_{\sigma_1} \partial \xi_{\sigma_2} \ldots \partial \xi_{\sigma_J}} \xi_{1\sigma_1} \ldots \xi_{J\sigma_J}$$

where $\xi_{j\sigma_j}, \sigma_j = 1(1)s$, are the s components of ξ_j and $\partial \xi_{\sigma_j}$ signifies the partial derivative w.r.t. the σ_j-th component of the argument ξ_0 of f. Since the J-linear function $f^{(J)}(\xi_0)$ is symmetric in its arguments the sequence of the symbols in a bracket (3.1.13) is immaterial.

With the notation introduced in (3.1.13), each elementary differential of f is denoted by a string consisting of letters f and bracket symbols. Butcher further introduced the abbreviations

$$\underbrace{ff\ldots f}_{k} =: f^k,$$

$$\underbrace{\{\{\ldots\{}_{k} =: \{_k, \qquad \underbrace{\}\}\ldots\}}_{k} =: \}_k,$$

to make the strings shorter and easier to read. The sum of the exponents of f and of the left *or* right brackets gives the order of the differential.

Example. The elementary differentials of order 2 to 4:

(All sums run from 1 to s; the arguments of f and its derivatives are the *same* throughout one expression, viz. *the* argument of the elementary differential; f_σ, $\sigma = 1(1)s$, denotes the components of f, analogously for the derivatives.)

The only elementary differential of order 2 is

$$\{f\} = f'f = \sum_{\sigma} \frac{\partial f}{\partial \xi_\sigma} f_\sigma,$$

i.e., the σ-th component of the elementary differential $\{f\}$ of f at ξ is

$$\{f\}_\sigma(\xi) = \sum_{\sigma_1=1}^{s} \frac{\partial f_\sigma(\xi)}{\partial \xi_{\sigma_1}} f_{\sigma_1}(\xi).$$

There are two different elementary differentials of order 3:

$$\{_2 f\}_2 = f'(f'f) = \sum_{\sigma_1} \frac{\partial f}{\partial \xi_{\sigma_1}} \sum_{\sigma_2} \frac{\partial f_{\sigma_1}}{\partial \xi_{\sigma_2}} f_{\sigma_2},$$

$$\{f^2\} = f''(ff) = \sum_{\sigma_1} \sum_{\sigma_2} \frac{\partial^2 f}{\partial \xi_{\sigma_1} \partial \xi_{\sigma_2}} f_{\sigma_1} f_{\sigma_2}.$$

The different elementary differentials of order 4 are:

$$\{_3f\}_3 = f'(f'(f'f)) = \sum_{\sigma_1} \frac{\partial f}{\partial \xi_{\sigma_1}} \sum_{\sigma_2} \frac{\partial f_{\sigma_1}}{\partial \xi_{\sigma_2}} \sum_{\sigma_3} \frac{\partial f_{\sigma_2}}{\partial \xi_{\sigma_3}} f_{\sigma_3},$$

$$\{_2f^2\}_2 = f'(f''f^2) = \sum_{\sigma_1} \frac{\partial f}{\partial \xi_{\sigma_1}} \sum_{\sigma_2} \sum_{\sigma_3} \frac{\partial^2 f_{\sigma_1}}{\partial \xi_{\sigma_2} \partial \xi_{\sigma_3}} f_{\sigma_2} f_{\sigma_3},$$

$$\{\{f\}f\} = f''(f'f)f = \sum_{\sigma_1} \sum_{\sigma_2} \frac{\partial^2 f}{\partial \xi_{\sigma_1} \partial \xi_{\sigma_2}} \left(\sum_{\sigma_3} \frac{\partial f_{\sigma_1}}{\partial \xi_{\sigma_3}} f_{\sigma_3} \right) f_{\sigma_2},$$

$$\{f^3\} = f'''f^3 = \sum_{\sigma_1} \sum_{\sigma_2} \sum_{\sigma_3} \frac{\partial^3 f}{\partial \xi_{\sigma_1} \partial \xi_{\sigma_2} \partial \xi_{\sigma_3}} f_{\sigma_1} f_{\sigma_2} f_{\sigma_3}.$$

In Butcher [2], all elementary differentials up to order 8 have been listed; there are 115 different elementary differentials of order 8. The visualization and distinction of the elementary differentials is facilitated by the fact that there exists a one-to-one correspondence between the set of elementary differentials of f of order r and the set of rooted trees of order r (the order of a tree being the number of its nodes). This correspondence is brought about recursively in the following manner[3] (cf. Def. 3.1.5):

(i) $f \leftrightarrow \bullet$ (the only tree of order 1);

(ii) if the elementary differentials F_j of f, $j=1(1)J$, correspond to the trees τ_j, with r_j nodes resp., then

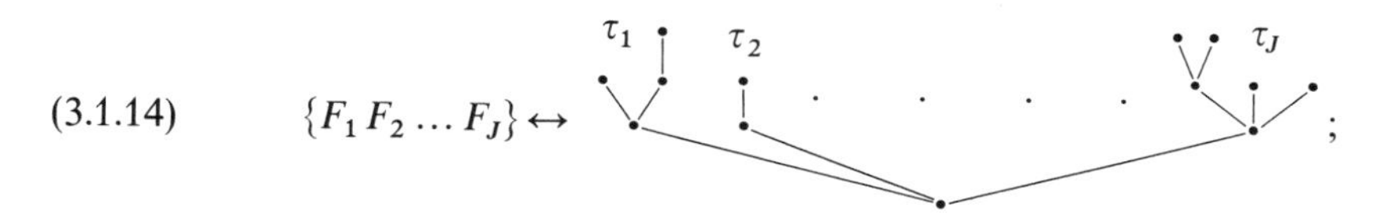

this tree obviously has $1 + \Sigma r_j$ nodes.

Example. The trees corresponding to the 4 different elementary differentials of order 4 are

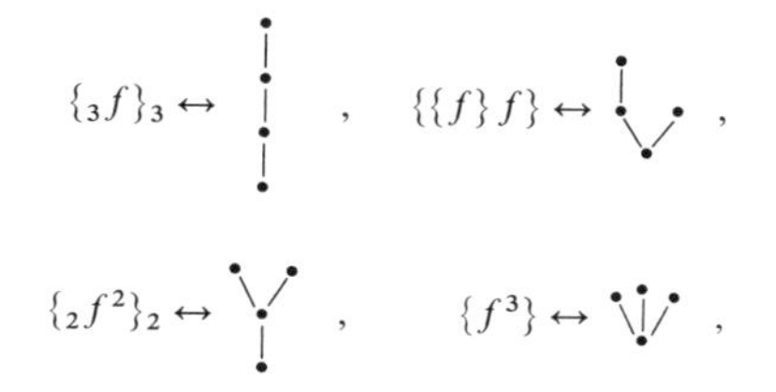

and there are no other rooted trees of order 4.

[3] For convenience, we will not handle rooted trees formally but represent them by their graphs.

If we arrange the n_r different elementary differentials of f of order r and the corresponding n_r rooted trees of order r into a *fixed sequence* for each $r \in \mathbb{N}$, both the set of elementary differentials and the set of rooted trees are mapped onto the set $\mathbb{D}$ of pairs of natural numbers (r, λ) with $\lambda \in \{1, 2, \ldots, n_r\}$, $r \in \mathbb{N}$. In this fashion, for a given $f: \mathbb{R}^s \to \mathbb{R}^s$, $F_\lambda^{(r)}$ denotes a specific elementary differential of order r for $\lambda \in \{1, \ldots, n_r\}$. On the other hand, the bracket product is defined for elements of $\mathbb{D}$ via (3.1.13).

The following are the values of n_r for $r = 1(1)10$ (see, e.g., Riordan [1]):

Table 3.1

r	1	2	3	4	5	6	7	8	9	10
n_r	1	1	2	4	9	20	48	115	286	719

If $s = 1$, several of the elementary differentials of order $r \geq 4$ become identical and the number of *different* elementary differentials is reduced considerably. Even when $s = 2$ but f is of the special form $\begin{pmatrix} 1 \\ f(\xi_1, \xi_2) \end{pmatrix}$— which occurs when *one* differential equation $y' = f(t, y)$ is written in the autonomous form $\begin{pmatrix} t \\ y \end{pmatrix}' = \begin{pmatrix} 1 \\ f(t, y) \end{pmatrix}$—, a reduction of the number of different elementary differentials occurs for $r \geq 5$.

On the other hand, the statement that for general s and f all n_r elementary differentials of a given order r are "different" can be made precise in the following way:

Lemma 3.1.3. *For each $(r_0, \lambda_0) \in \mathbb{D}$ there exist functions $f: \mathbb{R}^s \to \mathbb{R}^s$, $s \geq 2$, such that $F_{\lambda_0}^{(r_0)}(\xi) \not\equiv 0$ while $F_\lambda^{(r_0)}(\xi) \equiv 0$ for $\lambda \neq \lambda_0$, $\lambda \in \{1, \ldots, n_{r_0}\}$.*

The formal proof of this important fact contains numerous technicalities and we omit it; a proof of an equivalent result may be found in Section 5 of Butcher [4].

Example. For the second component of $f: \begin{pmatrix} t \\ y \end{pmatrix} \to \begin{pmatrix} 1 \\ g(t, y) \end{pmatrix}$ we have

$$\{_2\{f\} f\}_2 = g_y(g_{ty} + g_{yy} g)(g_t + g_y g) = (g_{ty} + g_{yy} g)(g_y g_t + g_y g_y g) = \{\{_2 f\}_2 f\}.$$

But for $f: \begin{pmatrix} \xi_1 \\ \xi_2 \end{pmatrix} \to \begin{pmatrix} \xi_2 \\ \frac{1}{2}\xi_1^2 \end{pmatrix}$ we have

$$\{_2\{f\} f\}_2 = \begin{pmatrix} \frac{1}{2}\xi_1^2 \xi_2 \\ 0 \end{pmatrix} \quad \text{while} \quad \{\{_2 f\}_2 f\} = \begin{pmatrix} 0 \\ \xi_1 \xi_2^2 \end{pmatrix}.$$

3.1.4 The Expansion of the Local Solution

For the purposes of the subsequent analysis we introduce the following notation:

For $a=(a_1,\dots,a_{m+1})^T\in\mathbb{R}^{m+1}$, $b=(b_1,\dots,b_s)^T\in\mathbb{R}^s$:

$$a\times b:=(a_1b_1,a_1b_2,\dots,a_1b_s,a_2b_1,\dots,a_2b_s,\\ \dots,a_{m+1}b_1,\dots,a_{m+1}b_s)^T\in\mathbb{R}^{(m+1)s}. \tag{3.1.15}$$

For $a,b\in\mathbb{R}^{m+1}$ we denote the componentwise product by

$$a\cdot b:=(a_1b_1,a_2b_2,\dots,a_{m+1}b_{m+1})^T\in\mathbb{R}^{m+1},\\ a^r:=a^{r-1}\cdot a=a\cdot a\cdot\dots\cdot a,\qquad r\in\mathbb{N}. \tag{3.1.16}$$

Furthermore we will assume that a zero column has been attached to every generating matrix A as an $(m+1)st$ column so that it becomes a *quadratic* $(m+1)\times(m+1)$-matrix.

We now define a mapping Φ from the Cartesian product of the set $\mathbb{D}$ (see Section 3.1.3) and the set of $(m+1)\times(m+1)$-matrices to $\mathbb{R}^{m+1}$ by (compare with Def. 3.1.5)

(i) $\Phi((1,1),A):=Ae$ where $e:=(1,\dots,1)^T\in\mathbb{R}^{m+1}$;

(ii) for $d_j\in\mathbb{D}$, $j=1(1)J$,

$$\Phi(\{d_1d_2\dots d_J\},A):=A\,\Phi(d_1,A)\cdot\Phi(d_2,A)\cdot\dots\cdot\Phi(d_J,A). \tag{3.1.17}$$

Due to the one-to-one correspondence between the set $\mathbb{D}$, the set of elementary differentials of f, and the set of rooted trees, we may take the first arguments of Φ from either set. We will also use the notation $\Phi_\lambda^{(r)}(A)$ for $\Phi((r,\lambda),A)$ if it is more convenient.

Example. Φ for the elementary differentials of order 4, cf. Section 3.1.3:

$$\Phi(\{_3f\}_3,A)=A(A(A(Ae)))=A^4e,$$
$$\Phi(\{_2f^2\}_2,A)=A(A(Ae\cdot Ae))=A^2(Ae)^2,$$
$$\Phi(\{\{f\}f\},A)=A(A(Ae)\cdot Ae)=A(A^2e\cdot Ae),$$
$$\Phi(\{f^3\},A)=A(Ae\cdot Ae\cdot Ae)=A(Ae)^3.$$

For the generating matrix of the classical RK4-procedure (see end of Section 3.1.1) we have after easy computations

$$\Phi(\{_3f\}_3,A)=\begin{pmatrix}0\\0\\0\\0\\\frac{1}{24}\end{pmatrix},\qquad \Phi(\{_2f^2\}_2,A)=\begin{pmatrix}0\\0\\0\\\frac{1}{8}\\\frac{1}{12}\end{pmatrix},$$

$$\Phi(\{\{f\}f\},A)=\begin{pmatrix}0\\0\\0\\\frac{1}{8}\\\frac{1}{8}\end{pmatrix},\qquad \Phi(\{f^3\},A)=\begin{pmatrix}0\\0\\\frac{1}{16}\\\frac{1}{16}\\\frac{1}{4}\end{pmatrix}.$$

Theorem 3.1.4. *For sufficiently smooth f, the local solution of the RK-procedure with generating matrix A satisfies*

$$(3.1.18)\qquad \frac{\partial^r}{\partial h^r}\bar{\zeta}(0,\xi)=r\sum_{\lambda=1}^{n_r}\beta_\lambda^{(r)}\Phi_\lambda^{(r)}(A)\times F_\lambda^{(r)}(\xi),\qquad r=1,2,\ldots,$$

where the $\beta_\lambda^{(r)}$ are non-vanishing constants independent of A and f.

Proof. By induction on r: For $r=1$, we have from (3.1.8)

$$\frac{\partial}{\partial h}\bar{\zeta}(0,\xi)=\mathbf{A}\,\mathbf{f}(\bar{\zeta}(0,\xi))=\mathbf{A}\,\mathbf{f}(\xi)=A\,e\times f(\xi)=\Phi_1^{(1)}(A)\times f(\xi)$$

(cf. the notational conventions (3.1.2), (2.1.31), and (3.1.15)) so that (3.1.18) holds with $\beta_1^{(1)}=1$.

We now assume that (3.1.18) holds for $r=1(1)r_0-1$, $r_0\geq 2$, so that

$$\begin{aligned}\bar{\zeta}(h,\xi)&=\mathbf{e}\xi+\sum_{r=1}^{r_0-1}\frac{h^r}{(r-1)!}\sum_{\lambda=1}^{n_r}\beta_\lambda^{(r)}\Phi_\lambda^{(r)}(A)\times F_\lambda^{(r)}(\xi)\\&\qquad+\frac{h^{r_0}}{r_0!}\frac{\partial^{r_0}}{\partial h^{r_0}}\bar{\zeta}(0,\xi)+O(h^{r_0+1})\\&=\mathbf{e}\xi+h\mathbf{A}\mathbf{f}\left(\mathbf{e}\xi+\sum_{r=1}^{r_0-1}\frac{h^r}{(r-1)!}\sum_{\lambda=1}^{n_r}\beta_\lambda^{(r)}\Phi_\lambda^{(r)}(A)\times F_\lambda^{(r)}(\xi)+O(h^{r_0})\right)\end{aligned}$$

and

$$(3.1.19)\qquad \begin{aligned}\frac{\partial^{r_0}}{\partial h^{r_0}}\bar{\zeta}(0,\xi)=r_0!\Bigg\{&\text{coefficient of } h^{r_0-1} \text{ in the expansion of}\\&\mathbf{A}\sum_{i=1}^{r_0-1}\frac{1}{i!}\mathbf{f}^{(i)}(\mathbf{e}\xi)\left(\sum_{r=1}^{r_0-1}\frac{h^r}{(r-1)!}\sum_{\lambda=1}^{n_r}\beta_\lambda^{(r)}\Phi_\lambda^{(r)}(A)\times F_\lambda^{(r)}(\xi)\right)^i\Bigg\}.\end{aligned}$$

An analysis of the right-hand side of (3.1.19) reveals that it is a sum of the r_0-th order differentials of f which are $\times$-premultiplied by the corresponding $\Phi_\lambda^{(r_0)}(A)$ and scalar factors $\beta_\lambda^{(r_0)}$ which depend only on the $\beta_\lambda^{(r)}$, $r<r_0$. We will carry this analysis through for $r_0=4$ and refer the reader to Butcher [2] for the general case:

$$\begin{aligned}&4!\Bigg\{\text{coefficient of } h^3 \text{ in}\\&\qquad\mathbf{A}\sum_{i=1}^{3}\frac{1}{i!}\mathbf{f}^{(i)}(\mathbf{e}\xi)\left(\sum_{r=1}^{3}\frac{h^r}{(r-1)!}\sum_{\lambda=1}^{n_r}\beta_\lambda^{(r)}\Phi_\lambda^{(r)}(A)\times F_\lambda^{(r)}(\xi)\right)^i\Bigg\}\\&=4!\Bigg\{\mathbf{A}\,\mathbf{f}'(\mathbf{e}\xi)\frac{1}{2!}(\beta_1^{(3)}\Phi_1^{(3)}(A)\times\{_2f\}_2(\xi)+\beta_2^{(3)}\Phi_2^{(3)}(A)\times\{f^2\}(\xi))\end{aligned}$$

$$+\mathbf{A}\frac{1}{2!}\mathbf{f}''(\mathbf{e}\,\xi)2(\beta_1^{(1)}\Phi_1^{(1)}(A)\times f(\xi))(\beta_1^{(2)}\Phi_1^{(2)}(A)\times\{f\}(\xi))$$

$$+\mathbf{A}\frac{1}{3!}\mathbf{f}'''(\mathbf{e}\,\xi)(\beta_1^{(1)}\Phi_1^{(1)}(A)\times f(\xi))^3\Big\}.$$

As

$$\mathbf{A}\,\mathbf{f}^{(i)}(\mathbf{e}\,\xi)(\Phi_{\lambda_1}^{(r_1)}(A)\times F_{\lambda_1}^{(r_1)}(\xi))\ldots(\Phi_{\lambda_i}^{(r_i)}(A)\times F_{\lambda_i}^{(r_i)}(\xi))$$
$$=A(\Phi_{\lambda_1}^{(r_1)}(A)\cdot\ldots\cdot\Phi_{\lambda_i}^{(r_i)}(A))\times f^{(i)}(\xi)F_{\lambda_1}^{(r_1)}(\xi)\ldots F_{\lambda_i}^{(r_i)}(\xi)$$

in our notation, this expression may be written as

$$4\{3\beta_1^{(3)}A\,\Phi_1^{(3)}(A)\times f'(\xi)\{_2f\}_2(\xi)+3\beta_2^{(3)}A\,\Phi_2^{(3)}(A)\times f'(\xi)\{f^2\}(\xi)$$
$$+6\beta_1^{(1)}\beta_1^{(2)}A(\Phi_1^{(1)}(A)\cdot\Phi_1^{(2)}(A))\times f''(\xi)f(\xi)\{f\}(\xi)$$
$$+(\beta_1^{(1)})^3A(\Phi_1^{(1)}(A))^3\times f'''(\xi)f(\xi)^3\}=:4\sum_{\lambda=1}^{n_4}\beta_\lambda^{(4)}\Phi_\lambda^{(4)}(A)\times F_\lambda^{(4)}(\xi)$$

according to (3.1.13) and (3.1.17) which is (3.1.18) for $r=4$. ▯

In Butcher [2], the general recursion for the $\beta_\lambda^{(r)}$ and their values for $r=1(1)8$ are to be found.

Naturally, (3.1.18) implies a similar expression for the h-derivatives at $(0,\xi)$ of the increment function of a RK-procedure. Define (Φ^{m+1} is the $(m+1)$th component of Φ)

(3.1.20) $$\hat{\Phi}_\lambda^{(r)}(A):=\hat{\Phi}((r,\lambda),A):=\Phi^{m+1}((r,\lambda),A)\in\mathbb{R}.$$

(This function $\hat{\Phi}(d,A)$, $d\in\mathbb{D}$, is written as $v_A(d)$ by Butcher in [3]; in his presentation d varies over the set of rooted trees.)

Corollary 3.1.5. *For sufficiently smooth f, the increment function of the RK-procedure with generating matrix A satisfies*

(3.1.21) $$\frac{\partial^r}{\partial h^r}\Psi(0,\xi)=\sum_{\lambda=1}^{n_{r+1}}\beta_\lambda^{(r+1)}\hat{\Phi}_\lambda^{(r+1)}(A)F_\lambda^{(r+1)}(\xi),\qquad r=0,1,2,\ldots.$$

Proof. From (3.1.9) in the form $\bar{\zeta}^{m+1}(h,\xi)=\xi+h\Psi(h,\xi)$ we obtain by differentiation

$$\frac{\partial^{r+1}}{\partial h^{r+1}}\bar{\zeta}^{m+1}(0,\xi)=(r+1)\frac{\partial^r}{\partial h^r}\Psi(0,\xi).$$

Since $[\Phi(d,A)\times F_\lambda^{(r)}(\xi)]^{m+1}=\hat{\Phi}(d,A)F_\lambda^{(r)}(\xi)$, (3.1.18) implies the assertion. ▯

3.1.5 The Exact Increment Function

The similarity between (3.1.8) and the integral equation equivalent to the differential equation (2.1.2) of an IVP 1 suggests the following definition:

Def. 3.1.6. For a given IVP 1, the mapping $\bar{z}: \mathbb{R} \times \mathbb{R}^s \to \mathbb{R}^s$ which is implicitly defined by

$$\bar{z}(h,\xi) = \xi + \int_0^h f(\bar{z}(\tau,\xi))\,d\tau \tag{3.1.22}$$

is called the *local solution (through ξ)* of the IVP 1.

The mapping $\Psi_\infty: \mathbb{R} \times \mathbb{R}^s \to \mathbb{R}^s$ defined by

$$\Psi_\infty(h,\xi) := \frac{1}{h}\int_0^h f(\bar{z}(\tau,\xi))\,d\tau = \begin{cases} \dfrac{\bar{z}(h,\xi)-\xi}{h}, & h \neq 0, \\ f(\xi), & h = 0, \end{cases} \tag{3.1.23}$$

is called the *exact increment function* of the IVP 1.

It is well-known from the theory of differential equations that Lemma 3.1.2 holds for $\bar{z}$ and Ψ_∞.

The local solution $\bar{z}(h,\xi)$ of the IVP 1 corresponds to the $(m+1)$st component $\bar{\zeta}^{m+1}(h,\xi)$ of a RK-procedure for that IVP 1. The analogy becomes evident if we replace the integral in (3.1.22) by a Riemann sum, say $\frac{1}{m}\sum_{\mu=0}^{m-1} f\left(\bar{z}\left(\frac{\mu}{m}h,\xi\right)\right)$. This reduces (3.1.22) to the definition of the local solution $\bar{\zeta}$ of the RK-procedure generated by the explicit m-stage generating matrix

$$\frac{1}{m}\begin{pmatrix} 0 & . & . & . & . & 0 \\ 1 & 0 & & & & . \\ 1 & 1 & 0 & & & . \\ : & & & \ddots & & . \\ 1 & . & . & . & 1 & 0 \\ 1 & . & . & . & . & 1 \end{pmatrix},$$

with $\bar{\zeta}^\mu(h,\xi) \approx \bar{z}\left(\frac{\mu-1}{m}h,\xi\right)$, $\mu = 1(1)m+1$. (This RK-procedure corresponds to solving the differential equation equivalent of (3.1.22) by the Euler method with step h/m.)

This analogy further suggests writing (3.1.22) in the form

$$\bar{z}(h,\xi) = \xi + h\left[\frac{1}{h}\int_0^h f(\bar{z}(\tau,\xi))\,d\tau\right]$$

and considering the "generating operator" $A_\infty^h: ([0,h] \to \mathbb{R}) \to ([0,h] \to \mathbb{R})$ defined by

$$(A_\infty^h x)(\tau) := \frac{1}{h}\int_0^\tau x(\tau')\,d\tau', \qquad \tau \in [0,h].$$

The function Φ of Section 3.1.4 may then be "extended" to such generating operators by the definition (cf. (3.1.17)):

(i) $\Phi((1,1), A_\infty^h) := A_\infty^h e$ where $e(\tau) \equiv 1$;

(ii) for $d_j \in \mathbb{D}$, $j = 1(1)J$,

$$\Phi(\{d_1 \dots d_J\}, A_\infty^h) := A_\infty^h [\Phi(d_1, A_\infty^h) \dots \Phi(d_J, A_\infty^h)], \tag{3.1.24}$$

where each $\Phi(d, A_\infty^h)$ is a function $[0, h] \to \mathbb{R}$ and the product in the bracket is the conventional product of the values for each τ.

It is easily seen from this definition that the values of the $\Phi((r,\lambda), A_\infty^h)(\tau)$ depend only on τ/h so that the numbers (cf. (3.1.20))

$$\hat{\Phi}_\lambda^{(r)}(A_\infty^h) := \hat{\Phi}((r,\lambda), A_\infty^h) := \Phi((r,\lambda), A_\infty^h)(h) \tag{3.1.25}$$

are independent of h and we may drop the superscript h. The fact that, with these $\hat{\Phi}_\lambda^{(r)}(A_\infty)$,

$$\frac{\partial^r}{\partial h^r} \Psi_\infty(0, \xi) = \sum_{\lambda=1}^{n_{r+1}} \beta_\lambda^{(r+1)} \hat{\Phi}_\lambda^{(r+1)}(A_\infty) F_\lambda^{(r+1)}(\xi), \qquad r = 0, 1, 2, \dots, \tag{3.1.26}$$

may finally be proved in a manner completely analogous to the proof of Theorem 3.1.4 and Corollary 3.1.5. We omit the details (see Butcher [2]).

Example. From the expressions for the $\Phi_\lambda^{(4)}$ obtained in the example before Theorem 3.1.4 we have

$$\Phi(\{_3 f\}_3, A_\infty^h)(\tau) = \frac{1}{h}\int_0^\tau \frac{1}{h}\int_0^{\tau'} \frac{1}{h}\int_0^{\tau''} \frac{1}{h}\int_0^{\tau'''} d\sigma\, d\tau'''\, d\tau''\, d\tau' = \frac{1}{24}\left(\frac{\tau}{h}\right)^4,$$

$$\Phi(\{_2 f^2\}_2, A_\infty^h)(\tau) = \frac{1}{h}\int_0^\tau \frac{1}{h}\int_0^{\tau'} \left(\frac{1}{h}\int_0^{\tau''} d\tau'''\right)^2 d\tau''\, d\tau' = \frac{1}{12}\left(\frac{\tau}{h}\right)^4,$$

$$\Phi(\{\{f\} f\}, A_\infty^h)(\tau) = \frac{1}{h}\int_0^\tau \left(\frac{1}{h}\int_0^{\tau'} \frac{1}{h}\int_0^{\tau''} d\tau'''\, d\tau'' \cdot \frac{1}{h}\int_0^{\tau'} d\tau''\right) d\tau' = \frac{1}{8}\left(\frac{\tau}{h}\right)^4,$$

$$\Phi(\{f^3\}, A_\infty^h)(\tau) = \frac{1}{h}\int_0^\tau \left(\frac{1}{h}\int_0^{\tau'} d\tau''\right)^3 d\tau' = \frac{1}{4}\left(\frac{\tau}{h}\right)^4,$$

so that the $\hat{\Phi}_\lambda^{(4)}(A_\infty)$ take the values $\frac{1}{24}$, $\frac{1}{12}$, $\frac{1}{8}$, and $\frac{1}{4}$. As seen from (3.1.20) and the example in Section 3.1.4, the values of the $\hat{\Phi}_\lambda^{(4)}(A)$ for the generating matrix A of the classical RK4-method are identical to those of the $\hat{\Phi}_\lambda^{(4)}(A_\infty)$.

Theorem 3.1.6.

$$\hat{\Phi}_\lambda^{(r)}(A_\infty) = \frac{1}{\gamma_\lambda^{(r)}}, \qquad \lambda = 1(1)n_r, \qquad r = 1, 2, \dots, \tag{3.1.27}$$

where the $\gamma_\lambda^{(r)}$ *are integers recursively defined by*

(i) $\gamma_1^{(1)} := 1$;

(ii) *for* $(r,\lambda)=\{(r_1,\lambda_1)\dots(r_J,\lambda_J)\}$, $(r_j,\lambda_j)\in\mathbb{D}$, $j=1(1)J$,

(3.1.28) $$\gamma_\lambda^{(r)} := r\prod_{j=1}^{J}\gamma_{\lambda_j}^{(r_j)}.$$

Proof. We prove by induction

(3.1.29) $$\Phi((r,\lambda),A_\infty^h)(\tau) = \frac{1}{\gamma_\lambda^{(r)}}\left(\frac{\tau}{h}\right)^r, \qquad (r,\lambda)\in\mathbb{D},$$

which is equivalent to the assertion by (3.1.25). For $r=1$, we have $\Phi((1,1),A_\infty^h)(\tau)=1/h\int_0^\tau d\tau'=\tau/h$. Let $(r_0,\lambda)=\{(r_1,\lambda_1)\dots(r_J,\lambda_J)\}$ where necessarily $r_j<r_0$; by (3.1.24)

$$\Phi((r_0,\lambda),A_\infty^h)=A_\infty^h\prod_{j=1}^{J}\Phi((r_j,\lambda_j),A_\infty) = \frac{1}{h}\int_0^\tau \frac{1}{\prod_j \gamma_{\lambda_j}^{(r_j)}}\left(\frac{\tau'}{h}\right)^{\Sigma r_j} d\tau'$$

$$= \frac{1}{r_0\prod_j\gamma_{\lambda_j}^{(r_j)}}\left(\frac{\tau}{h}\right)^{r_0} \quad \text{since } r_0=1+\sum_j r_j \quad \text{by Def. 3.1.5.} \qquad \square$$

The following is a table of the $\gamma_\lambda^{(r)}$ for $r=1(1)5$. Values for $r=6,7,8$ may be found in Butcher [2].

Table 3.2

r	(r,λ)	$\gamma_\lambda^{(r)}$	r	(r,λ)	$\gamma_\lambda^{(r)}$
1	f	1	5	$\{_4 f\}_4$	120
			5	$\{_3 f^2\}_3$	60
2	$\{f\}$	2	5	$\{_2\{f\}f\}_2$	40
			5	$\{_2 f^3\}_2$	20
3	$\{_2 f\}_2$	6	5	$\{\{_2 f\}_2 f\}$	30
3	$\{f^2\}$	3	5	$\{\{f^2\}f\}$	15
4	$\{_3 f\}_3$	24	5	$\{\{f\}^2\}$	20
4	$\{_2 f^2\}_2$	12	5	$\{\{f\}f^2\}$	10
4	$\{\{f\}f\}$	8	5	$\{f^4\}$	5
4	$\{f^3\}$	4			

3.2 The Group of RK-schemes

3.2.1 RK-schemes

If we regard a RK-method as a 1-step *one*-stage method with the forward step procedure (3.1.11) it is clear that two RK-procedures (3.1.3) which define the same increment function Ψ for any given function

$f:\mathbb{R}^s \to \mathbb{R}^s$ are equivalent. The essential information about the sequence of mappings $\{\varphi_n\}_{n\in\mathbb{N}'}$ of a RK-discretization method (cf. Def. 1.1.2) is contained in the mapping ψ which maps the given Lipschitz-continuous function f of an IVP 1 into the corresponding increment function of (3.1.11). It is thus natural to focus our attention on these mappings ψ.

Def. 3.2.1. A mapping ψ which assigns an increment function $\Psi: \mathbb{R} \times \mathbb{R}^s \to \mathbb{R}^s$ to each Lipschitz-continuous function $f:\mathbb{R}^s \to \mathbb{R}^s$ and which may be generated via (3.1.8)/(3.1.9) through a generating matrix A is called *a RK-scheme*. The set of all RK-schemes will be denoted by $\mathfrak{R}$.

Def. 3.2.2. Two generating matrices which generate the same RK-scheme are called *equivalent*. Also the RK-procedures generated by equivalent generating matrices are called equivalent.

Obviously, a RK-scheme represents a class of equivalent generating matrices or a class of equivalent RK-procedures.

Example. The generating matrices

$$\begin{pmatrix} 0 & 0 & 0 \\ 0 & 0 & \frac{2}{3} \\ \frac{1}{3} & 0 & 0 \\ \frac{1}{4} & \frac{3}{4} & 0 \end{pmatrix}, \quad \begin{pmatrix} 0 & 0 & 0 & 0 \\ \frac{1}{3} & 0 & 0 & 0 \\ 0 & \frac{1}{3} & 0 & \frac{1}{3} \\ \frac{1}{3} & 0 & 0 & 0 \\ \frac{1}{4} & 0 & \frac{3}{4} & 0 \end{pmatrix}, \quad \begin{pmatrix} 0 & \frac{2}{3} & 0 & 0 \\ 0 & 0 & \frac{1}{3} & 0 \\ 0 & 0 & 0 & 0 \\ 0 & \frac{2}{3} & 0 & 0 \\ \frac{1}{4} & 0 & \frac{1}{4} & \frac{1}{2} \end{pmatrix}$$

are equivalent to the explicit generating matrix $\begin{pmatrix} 0 & 0 & 0 \\ \frac{1}{3} & 0 & 0 \\ 0 & \frac{2}{3} & 0 \\ \frac{1}{4} & 0 & \frac{3}{4} \end{pmatrix}$;

they all generate the RK-scheme

$$\psi: f \to \Psi(h,\xi) = \frac{1}{4} f(\xi) + \frac{3}{4} f\left(\xi + \frac{2}{3} h f\left(\xi + \frac{h}{3} f(\xi)\right)\right).$$

By (3.1.9), $\Psi(0,\xi) = \left(\sum_{\lambda=1}^{m} a_{m+1,\lambda}\right) f(\xi)$. The value of $\sum_{\lambda} a_{m+1,\lambda}$ thus has to be the same for equivalent generating matrices (see also Example above).

Def. 3.2.3. Let $A = (a_{\mu\lambda})$ be any generating matrix of the RK-scheme ψ. The number

$$\alpha(\psi) = \sum_{\lambda=1}^{m} a_{m+1,\lambda} \tag{3.2.1}$$

is called the *step factor* of ψ.

Remark. It will turn out that the last component of the local solution $\bar{\zeta}(h,\xi)$ of a RK-procedure associated with the RK-scheme ψ approximates the true solution of $y' = f(y)$ through (t_0, ξ) at $t_0 + \alpha(\psi) h$ which explains the term "step factor".

We now introduce two *operations* with RK-schemes:

1. A *multiplication by a real number*: $\mathbb{R} \times \mathfrak{R} \to \mathfrak{R}$. For $\rho \in \mathbb{R}, \psi \in \mathfrak{R}$:

$$(\rho\psi[f])(h,\xi) := \rho(\psi[f])(\rho h, \xi), \tag{3.2.2}$$

or: If ψ generates the increment function $\Psi(h,\xi)$ for a given f then $\rho\psi$ generates the increment function $\rho\Psi(\rho h, \xi)$.

2. A *composition:* $\mathfrak{R} \times \mathfrak{R} \to \mathfrak{R}$. For $\psi_1, \psi_2 \in \mathfrak{R}$:

$$(\psi_1\psi_2[f])(h,\xi) := \psi_1[f](h,\xi) + \psi_2[f](h, \xi + h(\psi_1[f])(h,\xi)), \tag{3.2.3}$$

or: If ψ_i generates the increment function $\Psi_i(h,\xi)$, $i=1,2$, then $\psi_1\psi_2$ generates the increment function $\Psi_1(h,\xi) + \Psi_2(h, \xi + h\Psi_1(h,\xi))$.

To verify that these operations do not lead out of $\mathfrak{R}$ we have to show that $\rho\psi$, resp. $\psi_1\psi_2$ can be generated by a generating matrix: Let A, A_1, A_2 be generating matrices for the RK-schemes ψ, ψ_1, ψ_2, resp.; then

(3.2.4) $\rho\psi$ is generated by ρA,

(3.2.5) $\psi_1\psi_2$ is generated by $A_{12} := \left(\begin{array}{c|c} A_1 & 0 \\ \hline LA_1 & A_2 \end{array}\right) \updownarrow m_1 + m_2 + 1$, with column blocks of widths $\leftarrow m_1 \rightarrow$ and $\leftarrow m_2 \rightarrow$

where $L := \begin{pmatrix} 0 & \dots & 0 & 1 \\ \vdots & & & \vdots \\ 0 & \dots & 0 & 1 \end{pmatrix} \updownarrow m_2$ (with $\leftarrow m_1 + 1 \rightarrow$ columns), LA_1 reproduces the last row of A_1.

Proof of (3.2.4). With an obvious notation for the local solution and increment function of the RK-procedure generated by ρA, we have from (3.1.8)

$$(\rho\bar{\zeta})(h,\xi) = \mathbf{e}\xi + h\rho\mathbf{A}\mathbf{f}((\rho\bar{\zeta})(h,\xi))$$

which implies $(\rho\bar{\zeta})(h,\xi) = \bar{\zeta}(\rho h, \xi)$. Hence,

$$(\rho\Psi)(h,\xi) = \frac{(\rho\bar{\zeta})^{m+1}(h,\xi) - \xi}{h} = \rho\frac{\bar{\zeta}^{m+1}(\rho h,\xi) - \xi}{\rho h} = \rho\Psi(\rho h, \xi).$$

Proof of (3.2.5). From

$$\bar{\zeta}_i(h,\xi) = \mathbf{e}\xi + h\mathbf{A}_i\mathbf{f}(\bar{\zeta}_i(h,\xi)), \qquad i = 1, 2, 12$$

and the form (3.2.5) of A_{12} we have

$$\bar{\zeta}^{\mu}_{12}(h,\xi) = \begin{cases} \bar{\zeta}^{\mu}_1(h,\xi), & \mu = 1(1)m_1, \\ \bar{\zeta}^{\mu-m_1}_3(h,\xi), & \mu = m_1 + 1(1)m_1 + m_2 + 1, \end{cases}$$

where

$$\bar{\zeta}_3(h,\xi) = \mathbf{e}(\xi + h[\mathbf{A}_1 \mathbf{f}(\bar{\zeta}_1(h,\xi))]^{m_1+1}) + h\mathbf{A}_2 \mathbf{f}(\bar{\zeta}_3(h,\xi))$$
$$= \mathbf{e}(\xi + h\Psi_1(h,\xi)) + h\mathbf{A}_2 \mathbf{f}(\bar{\zeta}_3(h,\xi))$$

so that $\bar{\zeta}_3(h,\xi) = \bar{\zeta}_2(h, \xi + h\Psi_1(h,\xi))$ and

$$\Psi_{12}(h,\xi) = \frac{\bar{\zeta}_{12}^{m_1+m_2+1}(h,\xi) - \xi}{h}$$
$$= \frac{\bar{\zeta}_2^{m_2+1}(h, \xi + h\Psi_1(h,\xi)) - \xi - h\Psi_1(h,\xi)}{h} + \Psi_1(h,\xi)$$
$$= \Psi_1(h,\xi) + \Psi_2(h, \xi + h\Psi_1(h,\xi)).$$

Example. Let ψ_1 be generated by $\begin{pmatrix} 0 \\ 1 \end{pmatrix}$, ψ_2 by $\begin{pmatrix} 1 \\ 1 \end{pmatrix}$.

Then the RK-scheme $\frac{1}{2}\psi_1\psi_2$ is generated by $\frac{1}{2}\begin{pmatrix} 0 & 0 \\ 1 & 1 \\ 1 & 1 \end{pmatrix} = \begin{pmatrix} 0 & 0 \\ \frac{1}{2} & \frac{1}{2} \\ \frac{1}{2} & \frac{1}{2} \end{pmatrix}$,

similarly $\frac{1}{2}\psi_2\psi_1$ is generated by $\frac{1}{2}\begin{pmatrix} 1 & 0 \\ 1 & 0 \\ 1 & 1 \end{pmatrix} = \begin{pmatrix} \frac{1}{2} & 0 \\ \frac{1}{2} & 0 \\ \frac{1}{2} & \frac{1}{2} \end{pmatrix} \equiv \begin{pmatrix} \frac{1}{2} \\ 1 \end{pmatrix}$.

Cf. the example at end of Section 3.1.1.

Theorem 3.2.1. *The operations* (3.2.2) *and* (3.2.3) *have the following properties* ($\rho, \sigma \in \mathbb{R}, \psi, \psi_1, \psi_2 \in \mathfrak{R}$):

1. $\rho(\sigma\psi) = (\rho\sigma)\psi$,
   $0\psi = \psi_0 : f \to 0$ *for each* f,
   $\rho\psi_0 = \psi_0$ *for each* ρ,
   $\alpha(\rho\psi) = \rho\alpha(\psi)$;
2. $(\psi_1\psi_2)\psi_3 = \psi_1(\psi_2\psi_3)$,
   $\psi\psi_0 = \psi_0\psi = \psi$ *for each* $\psi \in \mathfrak{R}$,

but: $\psi_1\psi_2 \neq \psi_2\psi_1$ *generally (thus* $\mathfrak{R}$ *is a non-commutative semigroup w.r.t. composition),*
   $\alpha(\psi_1\psi_2) = \alpha(\psi_1) + \alpha(\psi_2)$;

3. $\rho(\psi_1\psi_2) = (\rho\psi_1)(\rho\psi_2)$,

but: $\underbrace{\psi \ldots \psi}_{\mu\text{-times}} \neq \mu\psi$ *and* $(-\psi)\psi \neq \psi_0$ *generally (where* $-\psi := (-1)\psi$*).*

Proof. 1. The relations follow trivially from (3.2.2), (3.2.4), and (3.2.1).
2. The associativity of the composition is seen thus:

$$\begin{aligned}\Psi_{(12)3}(h,\xi) &= \Psi_{12}(h,\xi)+\Psi_3(h,\xi+h\Psi_{12}(h,\xi))\\ &= \Psi_1(h,\xi)+\Psi_2(h,\xi+h\Psi_1(h,\xi))\\ &\quad +\Psi_3(h,\xi+h\Psi_1(h,\xi)+h\Psi_2(h,\xi+h\Psi_1(h,\xi)))\\ &= \Psi_1(h,\xi)+\Psi_{23}(h,\xi+h\Psi_1(h,\xi))=\Psi_{1(23)}(h,\xi)\,.\end{aligned}$$

(3.2.3) implies immediately the unit property of $\psi_0: f\to 0$ w.r.t. composition. The non-commutativity of the composition is established by simple counterexamples: Take ψ_1 and ψ_2 of the Example before this Theorem and $f\begin{pmatrix}\xi_1\\ \xi_2\end{pmatrix}=\begin{pmatrix}1\\ g(\xi_1)\end{pmatrix}$. Then

$$\tfrac{1}{2}(\psi_1\psi_2)[f](h,\xi)=\begin{pmatrix}1\\ \dfrac{g(\xi_1)+g(\xi_1+h)}{2}\end{pmatrix} \quad \text{while}$$

$$\tfrac{1}{2}(\psi_2\psi_1)[f](h,\xi)=\begin{pmatrix}1\\ g\left(\xi_1+\dfrac{h}{2}\right)\end{pmatrix}.$$

The additivity of α follows from (3.2.5) and (3.2.1).

3. The distributivity of the scalar multiplication over the composition follows easily from (3.2.2) and (3.2.3):

$$\begin{aligned}\rho(\psi_1\psi_2)[f](h,\xi) &= \rho\psi_{12}(\rho h,\xi)\\ &= \rho\psi_1(\rho h,\xi)+\rho\psi_2(\rho h,\xi+\rho h\psi_1(\rho h,\xi))\\ &= (\rho\psi_1)(\rho\psi_2)[f](h,\xi)\,.\end{aligned}$$

The *non*-validity of, e.g., $\psi\psi=2\psi$ and $(-\psi)\psi=0$ is again established by counterexamples: For ψ_1 as above, $\psi_1^2[f](h,\xi)=f(\xi)+f(\xi+hf(\xi))$ while $2\psi_1[f](h,\xi)=2f(\xi)$. Similarly, $(-\psi_1)\psi_1[f](h,\xi)=-f(\xi)+f(\xi-hf(\xi))\neq 0$. □

The *meaning* of the two operations on $\mathfrak{R}$ is obvious:

1. The scalar multiplication simply extends the step h in the local solution $\bar{\zeta}(h,\xi)$ by the scalar factor ρ so that the new RK-procedure advances the approximate solution of the IVP 1 by ρh.

2. The composition $\psi_1\psi_2$ simply defines a RK-procedure the solution of which is obtained by first solving the RK-procedure of ψ_1 and then the one of ψ_2. (In this interpretation the associativity becomes trivial.)

3.2.2 Inverses of RK-schemes

Theorem 3.2.2. *$\mathfrak{R}$ is a group w.r.t. the composition* (3.2.3). *If A is a generating matrix of $\psi \in \mathfrak{R}$ then the matrix*

$$A_- = RA \quad \text{where } R := \left(\begin{array}{ccc|c} 1 & \ddots & 0 & -1 \\ 0 & & 1 & \vdots \\ \hline 0 & .. & 0 & -1 \end{array}\right) \begin{matrix} \uparrow \\ m+1 \\ \downarrow \end{matrix} \qquad (3.2.6)$$

$$\leftarrow m+1 \rightarrow$$

generates the inverse RK-scheme ψ^{-1}.

Proof. Consider the local solution $\bar{\zeta}_-$ of the RK-procedure generated by A_-: From (3.2.6) and (3.1.8)

$$\bar{\zeta}^{\mu}_-(h,\xi) = \xi + h \sum_{\lambda=1}^{m} (a_{\mu\lambda} - a_{m+1,\lambda}) f(\bar{\zeta}^{\lambda}_-(h,\xi)), \qquad \mu = 1(1)m, \qquad (3.2.7)$$

$$\bar{\zeta}^{m+1}_-(h,\xi) = \xi + h \sum_{\lambda=1}^{m} (-a_{m+1,\lambda}) f(\bar{\zeta}^{\lambda}_-(h,\xi)),$$

so that

$$\Psi_-(h,\xi) = -\sum_{\lambda=1}^{m} a_{m+1,\lambda} f(\bar{\zeta}^{\lambda}_-(h,\xi))$$

is the increment function of that RK-procedure. Thus, (3.2.7) may be written as

$$\bar{\zeta}^{\mu}_-(h,\xi) = \xi + h\Psi_-(h,\xi) + h \sum_{\lambda=1}^{m} a_{\mu\lambda} f(\bar{\zeta}^{\lambda}_-(h,\xi)), \qquad \mu = 1(1)m. \qquad (3.2.8)$$

By an argument similar to that in the proof of Lemma 3.1.1 one obtains $h_0^- > 0$ and $B_0^- \subset B_R(z)$ such that (3.2.8) implies, at least for $|h| \le h_0^-$ and $\xi \in B_0^-$,

$$\bar{\zeta}^{\mu}_-(h,\xi) = \bar{\zeta}^{\mu}(h, \xi + h\Psi_-(h,\xi)), \qquad \mu = 1(1)m,$$

and

$$\Psi_-(h,\xi) = -\sum_{\lambda=1}^{m} a_{m+1,\lambda} f(\bar{\zeta}^{\lambda}(h, \xi + h\Psi_-(h,\xi))) = -\Psi(h, \xi + h\Psi_-(h,\xi)).$$

Thus the increment function Ψ_- satisfies

$$\Psi_-(h,\xi) + \Psi(h, \xi + h\Psi_-(h,\xi)) \equiv 0. \qquad (3.2.9)$$

On the other hand, let

$$\begin{aligned} \Psi_0(h,\xi) &:= \Psi(h,\xi) + \Psi_-(h, \xi + h\Psi(h,\xi)) \\ &= \Psi(h,\xi) - \Psi(h, \xi + h\Psi(h,\xi) + h\Psi_-(h, \xi + h\Psi(h,\xi))) \quad \text{by (3.2.9)} \\ &= \Psi(h,\xi) - \Psi(h, \xi + h\Psi_0(h,\xi)). \end{aligned}$$

The obvious solution $\Psi_0(h,\xi)\equiv 0$ has to be unique at least for sufficiently small $|h|$ since Ψ satisfies a Lipschitz condition w.r.t. its second argument by Lemma 3.1.2. Thus

$$\Psi(h,\xi)+\Psi_-(h,\xi+h\Psi(h,\xi))\equiv 0\,. \tag{3.2.10}$$

According to (3.2.3), (3.2.9) and (3.2.10), $\psi^{-1}: f\to\Psi_-$ is the inverse element of $\psi: f\to\Psi$. Since ψ was arbitrary, each $\psi\in\mathfrak{R}$ possesses an inverse in $\mathfrak{R}$ so that $\mathfrak{R}$ is a group. □

Remark. The intuitive meaning of ψ^{-1} is the following: If η_ν has been computed from $\eta_{\nu-1}$ by a RK-procedure (3.1.11) associated with ψ for step h then $\eta_{\nu-1}$ is recovered from η_ν by a RK-procedure associated with ψ^{-1} for the same step h. Or: ψ^{-1} "undoes" what ψ has done and vice versa.

The following relations hold for the inverse $(\rho\in\mathbb{R}, \psi\in\mathfrak{R})$:

$$(\rho\psi)^{-1}=\rho(\psi^{-1})\,, \tag{3.2.11}$$

$$\alpha(\psi^{-1})=-\alpha(\psi)\,, \tag{3.2.12}$$

but:

$$\psi^{-1}\neq-\psi \quad \text{generally}.$$

Proof of (3.2.11).

$$\begin{aligned}(\rho\psi)^{-1}(\rho\psi)&=\psi_0=\rho\psi_0 \quad \text{(Theorem 3.2.1, 1)}\\ &=\rho(\psi^{-1}\psi)=(\rho\psi^{-1})(\rho\psi) \quad \text{(Theorem 3.2.1, 3)}.\end{aligned}$$

Proof of (3.2.12).

$$\alpha(\psi_0)=\alpha(\psi^{-1}\psi)=\alpha(\psi^{-1})+\alpha(\psi)=0 \quad \text{(Theorem 3.2.1, 1 and 2)}.$$

The non-validity of $\psi^{-1}=-\psi$ can be seen from Example 1 below.

While it is true that $R^2=I$ (see (3.2.6)) so that

$$(A_-)_-=A\,, \tag{3.2.13}$$

A_- is, of course, not the only generating matrix for ψ^{-1} but only an element from a class of equivalent generating matrices.

Examples. 1. For $A=\begin{pmatrix}0\\1\end{pmatrix}$ we have $A_-=\begin{pmatrix}-1\\-1\end{pmatrix}=-\begin{pmatrix}1\\1\end{pmatrix}$; i.e., the inverse of the RK-scheme of the Euler method is the "backward" copy of the RK-scheme of the implicit Euler-method which is intuitively clear. Since the explicit and the implicit Euler-method are not equivalent this shows that $\psi^{-1}\neq-\psi$ in general.

2. For $A=\begin{pmatrix}\frac{1}{2}\\1\end{pmatrix}$, which generates the RK-scheme of the implicit midpoint rule, we have $A_-=\begin{pmatrix}-\frac{1}{2}\\-1\end{pmatrix}=-A$ so that $\psi^{-1}=-\psi$ does hold for that scheme.

Def. 3.2.4. A RK-scheme is called *symmetric* if

(3.2.14) $$\psi^{-1} = -\psi .$$

Each RK-procedure associated with a symmetric RK-scheme is also called symmetric.

It is very easy to generate symmetric RK-schemes due to

Theorem 3.2.3. *For an arbitrary $\psi \in \mathfrak{R}$, the RK-schemes $\psi(-\psi^{-1})$ and $(-\psi^{-1})\psi$ are symmetric.*

Proof. Both $\psi(-\psi^{-1})$ and $(-\psi^{-1})\psi$ satisfy (3.2.14):

$$\begin{aligned}[\psi(-\psi^{-1})]^{-1} &= (-\psi^{-1})^{-1}\psi^{-1} = ((-\psi)^{-1})^{-1}\psi^{-1} && \text{by (3.2.11)}\\ &= (-\psi)\psi^{-1} = -\psi(-\psi^{-1}) && \text{by Theorem 3.2.1, 3.}\\ [(-\psi^{-1})\psi]^{-1} &= \psi^{-1}(-\psi) = -(-\psi^{-1})\psi && \text{analogously.} \quad \square\end{aligned}$$

Remark. If $\alpha(\psi)=1$ we obtain a symmetric scheme with step factor 1 by forming $\frac{1}{2}\psi(-\psi^{-1})$ or $\frac{1}{2}(-\psi^{-1})\psi$.

Example. Take ψ generated by $\begin{pmatrix}0\\1\end{pmatrix}$. The schemes $\frac{1}{2}\psi(-\psi^{-1})$ and $\frac{1}{2}(-\psi^{-1})\psi$ are generated by the matrices

$$\frac{1}{2}\begin{pmatrix}0 & 0\\ 1 & 1\\ 1 & 1\end{pmatrix} = \begin{pmatrix}0 & 0\\ \frac{1}{2} & \frac{1}{2}\\ \frac{1}{2} & \frac{1}{2}\end{pmatrix} \quad \text{and} \quad \frac{1}{2}\begin{pmatrix}1 & 0\\ 1 & 0\\ 1 & 1\end{pmatrix} \equiv \begin{pmatrix}\frac{1}{2}\\ 1\end{pmatrix}$$

(cf. the example in Section 3.2.1).

3.2.3 Equivalent Generating Matrices

Def. 3.2.5. For $k \le m/2$, let $\Lambda_{mk} = \{\lambda_1, \dots, \lambda_k\}$ and $\Lambda'_{mk} = \{\lambda'_1, \dots, \lambda'_k\}$, $\Lambda_{mk} \cap \Lambda'_{mk} = \emptyset$, be two subsets of $\{1, \dots, m\}$. The m-stage generating matrix A *permits the reduction* $R((\lambda_1, \lambda'_1), \dots, (\lambda_k, \lambda'_k))$ if the elements of A satisfy

(3.2.15) $$\begin{cases} a_{\lambda_i\lambda_j} + a_{\lambda_i\lambda'_j} = a_{\lambda'_i\lambda_j} + a_{\lambda'_i\lambda'_j}, & i,j = 1(1)k,\\ a_{\lambda_i\lambda} = a_{\lambda'_i\lambda}, \quad \lambda \notin \Lambda_{mk} \cup \Lambda'_{mk}, & i = 1(1)k.\end{cases}$$

Theorem 3.2.4. *If the m-stage generating matrix A permits the reduction $R((\lambda_1, \lambda'_1), \dots, (\lambda_k, \lambda'_k))$ then A is equivalent to the $(m-k)$-stage generating matrix $\hat{A}$ with the elements $\hat{a}_{\mu\lambda}$, $\mu \in \{1, \dots, m+1\} - \Lambda'_{mk}$, $\lambda \in \{1, \dots, m\} - \Lambda'_{mk}$, where*

(3.2.16) $$\hat{a}_{\mu\lambda} = \begin{cases} a_{\mu\lambda}, & \lambda \notin \Lambda_{mk},\\ a_{\mu\lambda_i} + a_{\mu\lambda'_i}, & \lambda = \lambda_i \in \Lambda_{mk}.\end{cases}$$

Proof. Let $\hat{\bar{\zeta}}(h,\xi)=(\hat{\bar{\zeta}}^\mu(h,\xi), \mu\in\{1,\dots,m+1\}-\Lambda'_{mk})$ be the local solution of the RK-procedure generated by $\hat{A}$. Then the local solution $\bar{\zeta}$ of the RK-procedure generated by A satisfies

$$(3.2.17)\qquad \bar{\zeta}^\mu(h,\xi)=\begin{cases}\hat{\bar{\zeta}}^\mu(h,\xi) & \text{for } \mu\in\{1,\dots,m+1\}-\Lambda'_{mk}\\ \hat{\bar{\zeta}}^{\lambda_i}(h,\xi) & \text{for } \mu=\lambda'_i\in\Lambda'_{mk}\end{cases}$$

as is seen by substituting (3.2.17) into (3.1.8) and taking into account the relations (3.2.15) and the definition of $\hat{\bar{\zeta}}$. Hence

$$\bar{\zeta}^{m+1}(h,\xi)=\hat{\bar{\zeta}}^{m+1}(h,\xi)\,. \qquad \square$$

Def. 3.2.6. A generating matrix which permits no reduction is called *irreducible*.

Example.

$$A=\begin{pmatrix}\frac{1}{4} & \frac{1}{4}-\frac{\sqrt{3}}{6} & 0 & 0\\ \frac{1}{4}+\frac{\sqrt{3}}{6} & \frac{1}{4} & 0 & 0\\ \frac{1}{6} & \frac{1}{3} & -\frac{1}{12} & \frac{1}{12}+\frac{\sqrt{3}}{6}\\ \frac{1}{6} & \frac{1}{3} & -\frac{1}{12}-\frac{\sqrt{3}}{6} & \frac{1}{12}\\ \frac{1}{6} & \frac{1}{3} & \frac{1}{6} & \frac{1}{3}\end{pmatrix}$$

permits the reduction $R((1,4),(2,3))$ which leads to

$$\hat{A}=\begin{pmatrix}\frac{1}{4} & \frac{1}{4}-\frac{\sqrt{3}}{6}\\ \frac{1}{4}+\frac{\sqrt{3}}{6} & \frac{1}{4}\\ \frac{1}{2} & \frac{1}{2}\end{pmatrix}\equiv A\,.$$

$\hat{A}$ is irreducible as the only possible reduction $R((1,2))$ is not feasible.

For an $m\times m$ permutation matrix P let $P'=\left(\begin{array}{ccc|c} & P & & 0\\ & & & \vdots\\ & & & 0\\ \hline 0 & .. & 0 & 1\end{array}\right)$.

Theorem 3.2.5. *If two m-stage generating matrices A and $\tilde{A}$ satisfy*

$$(3.2.18)\qquad \tilde{A}=P'\,A\,P^T$$

then $\tilde{A}\equiv A$.

Proof. From (3.1.8) we have, with our notational conventions,

$$\mathbf{P}'\bar{\zeta}(h,\xi)=\mathbf{e}\,\xi+h\mathbf{P}'\mathbf{A}\mathbf{P}'^T\mathbf{f}(\mathbf{P}'\bar{\zeta}(h,\xi))$$

so that $\tilde{\zeta}(h,\xi) := \mathbf{P}'\bar{\zeta}(h,\xi)$ satisfies

$$\tilde{\zeta}(h,\xi) = \mathbf{e}\xi + h\tilde{\mathbf{A}}\mathbf{f}(\tilde{\zeta}(h,\xi))$$

and hence is the local solution of the RK-procedure generated by $\tilde{A}$. By the definition of P', $\tilde{\zeta}^{m+1}(h,\xi) = \bar{\zeta}^{m+1}(h,\xi)$. □

The converse of Theorem 3.2.5 is true for irreducible generating matrices:

Theorem 3.2.6. *If two m-stage generating matrices are equivalent and irreducible then one is a permutation* (3.2.18) *of the other.*

Proof. If A and $\tilde{A}$ are irreducible there must be functions $f: \mathbb{R}^s \to \mathbb{R}^s$ such that all components of the local solution $\bar{\zeta}$ are different and the same holds for $\tilde{\zeta}$. (If $\bar{\zeta}^{\mu_1}(h,\xi) \equiv \bar{\zeta}^{\mu_2}(h,\xi)$ for arbitrary f then A admits the reduction $R((\mu_1,\mu_2))$!) $\tilde{A} \equiv A$ requires that $\tilde{\zeta}^{m+1}(h,\xi) \equiv \bar{\zeta}^{m+1}(h,\xi)$ for arbitrary f, hence each of the m first components of $\tilde{\zeta}$ must be identical to one of the m first components of $\bar{\zeta}$ which implies that there exists an $m \times m$ permutation matrix P such that $\tilde{\zeta}(h,\xi) = \mathbf{P}'\bar{\zeta}(h,\xi)$. Hence, $\tilde{\zeta}$ satisfies

$$\tilde{\xi}(h,\xi) = \mathbf{e}\xi + h\mathbf{P}'\mathbf{A}\mathbf{P}'^T\mathbf{f}(\tilde{\zeta}(h,\xi)) = \mathbf{e}\xi + h\tilde{\mathbf{A}}\mathbf{f}(\tilde{\zeta}(h,\xi))$$

for arbitrary f which implies $P'AP^T = \tilde{A}$. □

Example. In the example after Def. 3.2.2, the first matrix reduces to the last one through the permutation $P = \begin{pmatrix} 1 & 0 & 0 \\ 0 & 0 & 1 \\ 0 & 1 & 0 \end{pmatrix}$. The third one permits the reduction $R((1,4))$, the resulting matrix $\hat{A}$ is related to the last matrix through the permutation $P = \begin{pmatrix} 0 & 0 & 1 \\ 0 & 1 & 0 \\ 1 & 0 & 0 \end{pmatrix}$.

We are now in a position to characterize the generating matrices of symmetric RK-schemes:

Theorem 3.2.7. *An irreducible m-stage generating matrix A generates a symmetric RK-scheme if and only if there is a one-to-one mapping of* $\{1,\dots,m\}$ *onto itself (denoted by ') such that the elements* $a_{\mu\lambda}$ *of A satisfy, for* $\mu,\lambda = 1(1)m$,

$$a_{\mu\lambda} + a_{\mu'\lambda'} = a_{m+1,\lambda} = a_{m+1,\lambda'}\,. \tag{3.2.19}$$

Proof. a) Assume that A generates a symmetric RK-scheme then $-A \equiv A_-$ by Def. 3.2.4. Furthermore, it is easily checked that both $-A$ and A_- have to be irreducible if A is irreducible. Hence, by Theorem 3.2.6, there must be an $m \times m$ permutation matrix P such that

$$-A = P'A_-P^T. \tag{3.2.20}$$

Let μ' be the column index of the element in the μ-th row of P which is 1. Then (3.2.20) implies (see (3.2.6))

$$-a_{\mu\lambda} = -a_{m+1,\lambda'} + a_{\mu'\lambda'}, \qquad \mu, \lambda = 1(1)m,$$
$$-a_{m+1,\lambda} = -a_{m+1,\lambda'}, \qquad \lambda = 1(1)m.$$

b) If the relations (3.2.19) hold for A with some correspondence $\mu \leftrightarrow \mu'$ then it is easily established that $-A = P' A_- P^T$ and hence $-A \equiv A_-$ by Theorem 3.2.5. One simply has to reverse the argument given in a). □

Example.

$$A = \begin{pmatrix} \frac{1}{4} & \frac{1}{4} - \frac{\sqrt{3}}{6} \\ \frac{1}{4} + \frac{\sqrt{3}}{6} & \frac{1}{4} \\ \frac{1}{2} & \frac{1}{2} \end{pmatrix} \text{ satisfies (3.2.19) with } 1' = 2,\ 2' = 1.$$

We finally show that the form $\psi(-\psi^{-1})$ is characteristic for a symmetric RK-scheme, cf. Theorem 3.2.3:

Theorem 3.2.8. *For each symmetric RK-scheme ψ there exist RK-schemes $\hat{\psi}$ such that $\hat{\psi}(-\hat{\psi}^{-1}) = \psi$. In particular, if $A = (a_{\mu\lambda})$ is an irreducible m-stage generating matrix for ψ then the m-stage generating matrix $\hat{A} = (\hat{a}_{\mu\lambda})$ with*

$$(3.2.21) \qquad \hat{a}_{\mu\lambda} = \begin{cases} a_{\mu\lambda}, & \mu = 1(1)m, \\ \frac{1}{2} a_{m+1,\lambda}, & \mu = m+1, \end{cases} \qquad \lambda = 1(1)m,$$

is a generating matrix for a $\hat{\psi}$.

Proof. Consider the $2m$-stage generating matrix $\hat{A}'$ of $\hat{\psi}(-\hat{\psi}^{-1})$ formed from $\hat{A}$ and $-\hat{A}_-$ by (3.2.5). By Theorem 3.2.7, there exists a correspondence $\mu \leftrightarrow \mu'$, $\mu, \mu' \in \{1, \ldots, m\}$ such that (3.2.19) holds for the elements of A. $\hat{A}' = (\hat{a}'_{\mu\lambda})$ permits the reduction $R((1, m+1'), \ldots, (m, m+m'))$ as we have, for $\mu, \lambda = 1(1)m$,

$$\begin{aligned}
\hat{a}'_{\mu\lambda} + \hat{a}'_{\mu,m+\lambda'} &= \hat{a}_{\mu\lambda} = a_{\mu\lambda} && \text{by (3.2.5) and (3.2.21)} \\
&= a_{m+1,\lambda} - a_{\mu'\lambda'} && \text{by (3.2.19)} \\
&= \tfrac{1}{2} a_{m+1,\lambda} + \tfrac{1}{2} a_{m+1,\lambda'} - a_{\mu'\lambda'} && \text{by (3.2.19)} \\
&= \hat{a}_{m+1,\lambda} + \hat{a}_{m+1,\lambda'} - \hat{a}_{\mu'\lambda'} && \text{by (3.2.21)} \\
&= \hat{a}'_{m+\mu',\lambda} + \hat{a}'_{m+\mu',m+\lambda'} && \text{by (3.2.5)}
\end{aligned}$$

and the definition of $\hat{A}'$ so that (3.2.15) holds. But the reduced equivalent form of $\hat{A}'$ becomes A by (3.2.16) and (3.2.21). □

Example. Take A of the example following Theorem 3.2.7.

$$\hat{A}=\begin{pmatrix} \frac{1}{4} & \frac{1}{4}-\frac{\sqrt{3}}{6} \\ \frac{1}{4}+\frac{\sqrt{3}}{6} & \frac{1}{4} \\ \frac{1}{4} & \frac{1}{4} \end{pmatrix} \quad \text{gives} \quad -\hat{A}_{-}=\begin{pmatrix} 0 & \frac{\sqrt{3}}{6} \\ -\frac{\sqrt{3}}{6} & 0 \\ \frac{1}{4} & \frac{1}{4} \end{pmatrix}$$

and

$$\hat{A}'=\begin{pmatrix} \frac{1}{4} & \frac{1}{4}-\frac{\sqrt{3}}{6} & 0 & 0 \\ \frac{1}{4}+\frac{\sqrt{3}}{6} & \frac{1}{4} & 0 & 0 \\ \frac{1}{4} & \frac{1}{4} & 0 & \frac{\sqrt{3}}{6} \\ \frac{1}{4} & \frac{1}{4} & -\frac{\sqrt{3}}{6} & 0 \\ \frac{1}{4} & \frac{1}{4} & \frac{1}{4} & \frac{1}{4} \end{pmatrix}$$

permits the reduction $R((1,4),(2,3))$ and reduces to A (cf. example following Def. 3.2.6).

3.2.4 Explicit and Implicit RK-schemes

We want to characterize those equivalence classes of generating matrices which contain explicit generating matrices (see Def. 3.1.2). In the present context, $\bar{A}$ will denote the upper $m \times m$ quadratic submatrix of an m-stage generating matrix A, b^T will denote the last row of A, and $e:=(1,\ldots,1)^T \in \mathbb{R}^m$.

The local solution $\bar{\zeta}$ of a RK-procedure (3.1.3) for $s=1$ and $f(\xi)\equiv\xi$ satisfies

$$\bar{\zeta}^{\mu}(h,\xi)=\xi+h\sum_{\lambda=1}^{m} a_{\mu\lambda}\bar{\zeta}^{\lambda}(h,\xi), \qquad \mu=1(1)m+1, \tag{3.2.22}$$

which implies ($\xi\in\mathbb{R}$!)

$$\bar{\zeta}^{m+1}(h,\xi)=\xi[1+hb^T(I-h\bar{A})^{-1}e]. \tag{3.2.23}$$

The right-hand side of (3.2.23) is a rational function of h so that it may be extended to complex values of h except at finitely many poles. Furthermore, since $\xi(b^T(I-h\bar{A})^{-1}e)$ is the increment function for the present case, (3.2.23) must be independent of the particular generating matrix but only depend on the RK-scheme.

Def. 3.2.7. For a given RK-scheme ψ, the rational function

$$\gamma_{\psi}(h):=1+hb^T(I-h\bar{A})^{-1}e \tag{3.2.24}$$

where A is any generating matrix of ψ, is called the *growth function* of ψ. (The term "growth function" will be justified in Section 3.5.3.)

Theorem 3.2.9. *Let A be a generating matrix of the RK-scheme ψ then*

$$\gamma_\psi(h) = \frac{\det(I - h\bar{A}_-)}{\det(I - h\bar{A})}, \qquad h \in \mathbb{C}. \tag{3.2.25}$$

Proof. Cramer's rule applied to (3.2.22), with $\xi = 1$, expresses $\bar{\zeta}^{m+1}(h, 1)$ as a quotient of determinants. The denominator is clearly $\det(I - h\bar{A})$, the numerator is

$$\begin{vmatrix} 1 - h a_{11} & -h a_{12} & \dots & -h a_{1m} & 1 \\ -h a_{21} & 1 - h a_{22} & & -h a_{2m} & 1 \\ \dots & \dots & \dots & \dots & \dots \\ -h a_{m1} & -h a_{m2} & \dots & 1 - h a_{mm} & 1 \\ -h a_{m+1,1} & \dots & \dots & -h a_{m+1,m} & 1 \end{vmatrix} = \det(I - h\bar{A}_-)$$

as is seen by subtraction of the last row from the first m rows (see (3.2.6)). □

Remark. While each RK-scheme possesses a unique growth function, different RK-schemes may have the same growth function.

Theorem 3.2.10. *The growth function satisfies* $(\rho \in \mathbb{R}, \psi, \psi_1, \psi_2 \in \mathfrak{R})$

$$\gamma_{\rho\psi}(h) = \gamma_\psi(\rho h), \tag{3.2.26}$$

$$\gamma_{\psi_1\psi_2}(h) = \gamma_{\psi_1}\gamma_{\psi_2}(h), \tag{3.2.27}$$

$$\gamma_{\psi^{-1}}(h) = [\gamma_\psi(h)]^{-1}. \tag{3.2.28}$$

Proof. (3.2.26) follows immediately from (3.2.24) and (3.2.4). To see that (3.2.27) holds we recall that a RK-procedure corresponding to $\psi_1\psi_2$ consists of the application of a RK-procedure of ψ_2 after one of ψ_1. Thus, to obtain (3.2.23) for $\psi_1\psi_2$ we evaluate the bracket in (3.2.23) for ψ_2 but replace ξ by $\bar{\zeta}^{m+1}(h, \xi)$ for ψ_1 which produces (3.2.27) for $\xi = 1$.

(3.2.28) follows from (3.2.27) with $\psi_1 = \psi$, $\psi_2 = \psi^{-1}$ since $\gamma_{\psi_0}(h) \equiv 1$ (take $\rho = 0$ in (3.2.26)). It also follows from (3.2.25) since $(A_-)_- = A$ by (3.2.13). □

Example. Let ψ_E be the Euler scheme generated by $\binom{0}{1}$; Theorem 3.2.9 gives $\gamma_{\psi_E}(h) = 1 + h$. For the RK-scheme $\frac{1}{2}\psi_E(-\psi_E^{-1})$ of the implicit trapezoidal method (see the example at the end of Section 3.2.2) we have by Theorem 3.2.10

$$\gamma(h) = \left(1 + \frac{1}{2}h\right)\left(1 + \frac{1}{2}(-h)\right)^{-1} = \frac{1 + h/2}{1 - h/2}.$$

Since the right-hand side of (3.2.27) is commutative, we obtain the same growth function for the RK-scheme $\frac{1}{2}(-\psi_E^{-1})\psi_E$ of the implicit midpoint method.

We are now in a position to give a definition of explicitness which does not refer to a particular generating matrix:

Def. 3.2.8. A RK-scheme ψ is called *explicit* if γ_ψ is a polynomial; otherwise it is called *implicit*.

Remark. As observed in Remark 2 following Def. 3.1.2, an implicit generating matrix may be equivalent to an explicit one and hence generate an explicit RK-scheme. According to Theorem 3.2.9, $\det(I-h\bar{A})$ must reduce to 1 or be a factor of $\det(I-h\bar{A}_{-})$ in such cases. An implicit RK-scheme cannot possess any explicit generating matrix while an explicit RK-scheme does possess an explicit generating matrix.

Examples.

1. $A=\begin{pmatrix}0 & 0 & 0\\ 0 & 0 & \frac{2}{3}\\ \frac{1}{3} & 0 & 0\\ \frac{1}{4} & \frac{3}{4} & 0\end{pmatrix}$ yields $\det(I-h\bar{A})=1$ (see the example after Def. 3.2.2),

2. $A=\begin{pmatrix}0 & 0\\ 1 & -1\\ 1 & -1\end{pmatrix}$ which generates ψ_0 (by the reduction $R((1,2))$) has $\det(I-h\bar{A})$ $=\det(I-h\bar{A}_{-})=1+h$, hence $\gamma(h)\equiv 1$.

Corollary 3.2.11. *If ψ is an explicit RK-scheme, ψ^{-1} is implicit (with the trivial exception of $\psi_0=\psi_0^{-1}$). Each symmetric RK-scheme (except ψ_0) is implicit.*

Proof. Immediate consequence of Def. 3.2.8, (3.2.28), and Def. 3.2.4. □

Remark. By Corollary 3.2.11, the consideration of *implicit* RK-schemes was a prerequisite for obtaining inverses to RK-schemes and symmetric RK-schemes.

Corollary 3.2.12. *The growth function of a symmetric RK-scheme satisfies*

$$\gamma(-h)=\frac{1}{\gamma(h)}. \tag{3.2.29}$$

Proof. Immediate consequence of Def. 3.2.4, (3.2.26), and (3.2.28). □

Remark. (3.2.29) is not a sufficient condition for the symmetry of a RK-scheme. A simple counter-example shows this:

$A=\begin{pmatrix}1/6 & 1/6\\ 1/4 & 1/3\\ 1/3 & 2/3\end{pmatrix}$ generates a RK-scheme with the growth function $\gamma(h)=(1+h/2+h^2/72)/(1-h/2+h^2/72)=\gamma(-h)^{-1}$. But $-A$ is not equivalent to A_{-} as can be easily established.

3.2.5 Symmetric RK-procedures

Def. 3.2.9. A RK-procedure is called *symmetric* if it is generated by a generating matrix of a symmetric RK-scheme.

Since symmetric RK-procedures are important in various contexts we will give two important characterizations for them.

Theorem 3.2.13. *A symmetric RK-procedure with increment function* Ψ *can be written in the form* (*cf.* (3.1.11))

$$\frac{\eta_\nu - \eta_{\nu-1}}{h_\nu} - \frac{1}{2}\left[\Psi(h_\nu, \eta_{\nu-1}) + \Psi(-h_\nu, \eta_\nu)\right] = 0\,. \tag{3.2.30}$$

Proof. Since, by Def. 3.2.4, $\psi(-\psi) = \psi_0$ for the associated symmetric RK-scheme ψ, we have by (3.2.2) and (3.2.3)

$$\Psi(h_\nu, \eta_{\nu-1}) - \Psi(-h_\nu, \eta_{\nu-1} + h_\nu \Psi(h_\nu, \eta_{\nu-1})) \equiv 0$$

which implies $\Psi(h_\nu, \eta_{\nu-1}) = \Psi(-h_\nu, \eta_\nu)$ by (3.1.11) so that (3.2.30) is equivalent to (3.1.11). □

The converse of Theorem 3.2.13 takes the following form:

Theorem 3.2.14. *If a RK-procedure can be written as*

$$\frac{\eta_\nu - \eta_{\nu-1}}{h_\nu} - X(h_\nu, \eta_{\nu-1}, \eta_\nu) = 0 \tag{3.2.31}$$

with a mapping $X: \mathbb{R} \times \mathbb{R}^s \times \mathbb{R}^s \to \mathbb{R}^s$ *(depending on f of the IVP 1) which is Lipschitz-continuous in its second and third argument for Lipschitz-continuous f and which satisfies*

$$X(h, \xi_1, \xi_2) = X(-h, \xi_2, \xi_1) \tag{3.2.32}$$

then it is symmetric.

Proof. The increment function Ψ of the RK-procedure (3.2.31) satisfies (cf. (3.1.11))

$$\Psi(h, \xi) = X(h, \xi, \xi + h\Psi(h, \xi))$$

which implies

$$-\Psi(-h, \xi) = -X(-h, \xi, \xi + h(-\Psi(-h, \xi)))\,. \tag{3.2.33}$$

On the other hand, (3.2.31) and (3.2.32) imply

$$\frac{\eta_{\nu-1} - \eta_\nu}{h_\nu} + X(h_\nu, \eta_{\nu-1}, \eta_\nu) = \frac{\eta_{\nu-1} - \eta_\nu}{h_\nu} + X(-h_\nu, \eta_\nu, \eta_{\nu-1}) = 0$$

so that the increment function Ψ_- of the RK-scheme inverse to the one associated with (3.2.31) satisfies

(3.2.34) $$\Psi_-(h,\xi) = -X(-h,\xi,\xi + h\,\Psi_-(h,\xi)).$$

As X is Lipschitz-continuous w.r.t. its third argument, (3.2.33) and (3.2.34) define the same function (at least for sufficiently small h) so that $-\Psi(-h,\xi) = \Psi_-(h,\xi)$ which proves the assertion by Def. 3.2.4 and (3.2.2). □

As symmetric RK-schemes are always implicit by Corollary 3.2.11 so that the increment functions of symmetric RK-procedures cannot be explicitly expressed in terms of the function f of the IVP 1, the form (3.2.31) is usually more convenient and intuitive for the formulation of a symmetric RK-procedure.

Example. The implicit trapezoidal procedure may be written as

$$\frac{\eta_\nu - \eta_{\nu-1}}{h_\nu} - \frac{1}{2}[f(\eta_{\nu-1}) + f(\eta_\nu)] = 0$$

as is evident from the generating matrix $\begin{pmatrix} 0 & 0 \\ \frac{1}{2} & \frac{1}{2} \\ \frac{1}{2} & \frac{1}{2} \end{pmatrix}$.

The implicit midpoint procedure may be written as

$$\frac{\eta_\nu - \eta_{\nu-1}}{h_\nu} - f\left(\frac{\eta_{\nu-1} + \eta_\nu}{2}\right) = 0$$

since the first stage η_ν^1 satisfies

$$\eta_\nu^1 = \eta_{\nu-1} + \tfrac{1}{2}h_\nu f(\eta_\nu^1) = \eta_\nu - \tfrac{1}{2}h_\nu f(\eta_\nu^1)$$

so that $\eta_\nu^1 = \frac{1}{2}(\eta_{\nu-1} + \eta_\nu)$.

3.3 RK-methods and Their Orders

3.3.1 RK-methods

In most of the following considerations, we will treat RK-methods as *one-stage* one-step methods with forward step procedures (3.1.11) (cf. the introductory remarks of Chapter 3).

Def. 3.3.1. For any RK-scheme ψ the *uniform RK-method based upon* ψ is the f.s.m. applicable to IVP 1 defined by $(n \in \mathbb{N}')$:

$$E_n = E_n^0 = (\mathbb{G}_n \to \mathbb{R}^s), \text{ with norms (2.2.2) and (2.2.3) resp.,}$$

$$(\Delta_n y)(t_\nu) = y(t_\nu),$$

$$\Delta_n^0 \begin{pmatrix} d_0 \\ d(t) \end{pmatrix} (t_\nu) = \begin{cases} d_0, & \nu = 0, \\ d(t_{\nu-1}), & \nu = 1(1)n,^4 \end{cases} \tag{3.3.1}$$

$$t_\nu \in \mathbb{G}_n .$$

$$[\varphi_n(F)\eta](t_\nu) = \begin{cases} \eta_0 - z_0, & \nu = 0, \\ \dfrac{\eta_\nu - \eta_{\nu-1}}{h_\nu} - \psi[f](h_\nu, \eta_{\nu-1}), & \nu = 1(1)n, \end{cases} \tag{3.3.2}$$

In a *non-uniform RK-method*, (3.3.2) is replaced by

$$[\varphi_n(F)\eta](t_\nu) = \begin{cases} \eta_0 - z_0, & \nu = 0, \\ \dfrac{\eta_\nu - \eta_{\nu-1}}{h_\nu} - \psi_{n\nu}[f](h_\nu, \eta_{\nu-1}), & \nu = 1(1)n, \end{cases} \tag{3.3.2. a}$$

where the $\psi_{n\nu}$, $\nu = 1(1)n$, $n \in \mathbb{N}'$, are RK-schemes specified by the method.

Remark. The determination of the grids $\mathbb{G}_n$ which is part of a f.s.m. is left open at present. As stated in Section 2.1.2, we assume linearly convergent grid sequences $\{\mathbb{G}_n\}_{n \in \mathbb{N}'}$ throughout.

RK-methods are stable discretization methods in the sense of Section 1.1.4:

Theorem 3.3.1. *Each uniform RK-method $\mathfrak{M}$ is stable for an IVP 1, with stability bound $S = c_2 e^{c_2 L_0}$ and stability threshold $R/2S$ where c_2 is the grid constant*[5] *of (2.1.11), L_0 is the Lipschitz constant of $\psi[f] = \Psi$ w.r.t. its second argument in $[-h_0, h_0] \times B_0$, and R is from (2.1.6). (It may be necessary to restrict $\mathbb{N}'$ to $\bar{\mathbb{N}}' := \{n \in \mathbb{N}' : n \geq c_2/h_0\}$.)*

Proof. For $n \in \mathbb{N}'$, let

$$(F_n \eta)(t_\nu) := \begin{cases} \eta_0, & \nu = 0, \\ \dfrac{\eta_\nu - \eta_{\nu-1}}{h_\nu}, & \nu = 1(1)n, \end{cases}$$

$$(G_n \eta)(t_\nu) := \begin{cases} z_0, & \nu = 0, \\ \Psi(h_\nu, \eta_{\nu-1}), & \nu = 1(1)n. \end{cases}$$

[4] For symmetric RK-methods, see (3.4.13).

[5] Without loss of generality we assume $c_2 \geq 1$.

The F_n are of the form (1.1.13); from

$$\|\eta\|_{E_n} = \max_\nu \|\eta_\nu\| \le \|(F_n\eta)(t_0)\| + \sum_{\nu=1}^{n} h_\nu \|(F_n\eta)(t_\nu)\|$$

$$\le c_2 \left[\|(F_n\eta)(t_0)\| + \frac{1}{n}\sum_{\nu=1}^{n} \|(F_n\eta)(t_\nu)\|\right] = c_2 \|F_n\eta\|_{E_n^0}$$

we have the stability of $\{E_n, E_n^0, F_n\}_{n\in\mathbb{N}'}$ with a stability bound $c_2 \ge 1$. The G_n satisfy

$$\|(G_n\eta_\nu^{(1)})(t_\nu) - (G_n\eta_\nu^{(2)})(t_\nu)\| = \|\Psi(h_\nu, \eta_{\nu-1}^{(1)}) - \Psi(h_\nu, \eta_{\nu-1}^{(2)})\| \le L_0 \|\eta_{\nu-1}^{(1)} - \eta_{\nu-1}^{(2)}\|$$

for $|h_\nu| \le h_0$, $\eta_{\nu-1}^{(1)}, \eta_{\nu-1}^{(2)} \in B_0$, see Lemma 3.1.2.

Hence—after we have reduced $\mathbb{N}'$ to $\bar{\mathbb{N}}'$ if necessary—we may apply Corollary 2.2.7 with $\eta_n = \Delta_n z$, $R = \frac{1}{2}$ of the R used in (2.1.6) (cf. Lemma 3.1.1) to obtain the stability of $\{E_n, E_n^0, F_n - G_n\}_{n\in\bar{\mathbb{N}}'} = \mathfrak{M}(\text{IVP } 1)$ at $\{\Delta_n z\}$. The values for the stability bound and threshold follow from (2.2.23) as $(G_n\eta)(t_\nu)$ depends only on $\eta_{\nu-1}$. □

Corollary 3.3.2. *If, for a given IVP 1, the increment functions* $\Psi_{n\nu} = \psi_{n\nu}[f]$ *occuring in a non-uniform RK-method* $\mathfrak{M}$ *possess a common uniform Lipschitz constant* L_0 *in the region* $[-h_0, h_0] \times B_0$ *then* $\mathfrak{M}$ *is stable for the IVP 1 with stability bound and threshold as in Theorem* 3.3.1.

Proof. Under the assumptions made the proof of Theorem 3.3.1 may be repeated literally. □

Theorem 3.3.3. *A RK-method* $\mathfrak{M}$ *which is consistent of some order* $p \ge 1$ *with a given IVP 1* $\mathfrak{P}$ *generates a discretization* $\mathfrak{M}(\mathfrak{P})$ *which possesses solution elements* ζ_n *for all sufficiently large* $n \in \mathbb{N}'$. *Furthermore,* $\mathfrak{M}$ *is convergent for* $\mathfrak{P}$ *of order p.*

Proof. With Def. 2.1.1 of an IVP 1, the assertion is an immediate consequence of Theorems 3.1.1 resp. 3.1.2, 1.2.3, and 1.2.4. □

Theorem 3.3.3 permits us to restrict our attention to a large extent to the analysis of the *local* discretization error of RK-methods.

3.3.2 The Order of Consistency

By Def. 3.1.6, the true solution z of an IVP 1 satisfies $z(t_{\nu-1} + h_\nu) = \bar{z}(h_\nu, z(t_{\nu-1}))$ so that

$$\frac{z(t_\nu) - z(t_{\nu-1})}{h_\nu} = \frac{\bar{z}(h_\nu, z(t_{\nu-1})) - z(t_{\nu-1})}{h_\nu} = \Psi_\infty(h_\nu, z(t_{\nu-1}))$$

and the expression for the local discretization error of a uniform RK-method becomes (see (1.1.7), (3.3.1), (3.3.2))

$$(3.3.3)\qquad l_n(t_\nu)=\begin{cases}0, & \nu=0\,,\\ \Psi_\infty(h_\nu,z(t_{\nu-1}))-\Psi(h_\nu,z(t_{\nu-1})), & \nu=1(1)n\,.\end{cases}$$

In a non-uniform RK-method $\Psi_{n\nu}$ takes the place of Ψ. Obviously, the local discretization error is of order $O(n^{-p})$ if the expansions of $\Psi_\infty(h,\xi)$ and $\Psi(h,\xi)$ in powers of h exist and agree up to the h^{p-1}-term and both remainder terms are of order $O(h^p)$.

The coefficients of the expansion of Ψ for a given generating matrix A have been found in Corollary 3.1.5. Since the increment function Ψ is the same for all generating matrices of the same RK-scheme ψ, the quantities $\hat{\Phi}_\lambda^{(r)}(A)$ of (3.1.20) are really mappings from $\mathfrak{R}$ to $\mathbb{R}$ for each $(r,\lambda)\in\mathbb{D}$.

Def. 3.3.2. A RK-scheme ψ is called *of generalized order p* if (cf. Theorem 3.1.6)

$$(3.3.4)\qquad \hat{\Phi}_\lambda^{(r)}(\psi)=\frac{\alpha(\psi)^r}{\gamma_\lambda^{(r)}},\qquad \lambda=1(1)n_r\,,\qquad r=1(1)p\,;$$

it is called *of order p* if it is of generalized order p and $\alpha(\psi)=1$. ψ is called of *exact* (generalized) order p if (3.3.4) holds for no larger value of p.

All generating matrices of ψ and their associated RK-procedures are also called of (exact) (generalized) order p if ψ is of (exact) (generalized) order p.

Remark. By (3.1.20) and (3.2.1) we have for each ψ

$$(3.3.5)\qquad \hat{\Phi}_1^{(1)}(\psi)=\sum_{\lambda=1}^{m}a_{m+1,\lambda}=\alpha(\psi)=\frac{\alpha(\psi)}{\gamma_1^{(1)}}$$

where the $a_{m+1,\lambda}$ are from any generating matrix of ψ. Hence, *each* ψ is of generalized order 1.

Lemma 3.3.4. *If ψ is of generalized order p and $\alpha(\psi)\neq 0$, then $\dfrac{1}{\alpha(\psi)}\psi$ is of order p.*

Proof. From Theorem 3.2.1,1 we have $\alpha(\alpha(\psi)^{-1}\psi)=1$. From (3.2.4) and the definition (3.1.17) of $\Phi(d,A)$ it is seen by induction that

$$(3.3.6)\qquad \Phi((r,\lambda),\rho A)=\rho^r\,\Phi((r,\lambda),A),\qquad \rho\in\mathbb{R}\,,$$

which implies via (3.1.20)

$$(3.3.7)\qquad \hat{\Phi}_\lambda^{(r)}\left(\frac{1}{\alpha(\psi)}\psi\right)=\frac{1}{\alpha(\psi)^r}\hat{\Phi}_\lambda^{(r)}(\psi)=\frac{1}{\gamma_\lambda^{(r)}}\,.\qquad\square$$

Theorem 3.3.5. *If and only if ψ is of generalized order p and $\alpha(\psi) \neq 0$, the uniform RK-method based upon the RK-scheme $(\alpha(\psi)^{-1}\psi)$ is consistent of order p with each IVP$^{(p)}$ 1.*

Proof. a) By Def. 2.1.2, f from an IVP$^{(p)}$ 1 possesses a uniformly Lipschitz-continuous p-th derivative in $B_R(z)$ so that $\Psi = (\alpha(\psi)^{-1}\psi)[f]$ as well as Ψ_∞ for f possess Lipschitz-continuous p-th derivatives w.r.t. h in $[-h_0, h_0] \times B_0$ by Lemma 3.1.2 and the remark after Def. 3.1.6. Hence (3.1.21) and (3.1.26)/(3.1.27) imply, for $|h| \leq h_0$ and $\xi \in B_0$,

$$\Psi_\infty(h,\xi) - \Psi(h,\xi) = \sum_{r=0}^{p-1} \frac{h^r}{r!} \frac{\partial^r}{\partial h^r}[\Psi_\infty(0,\xi) - \Psi(0,\xi)] + O(h^p)$$

$$(3.3.8) \qquad = \sum_{r=1}^{p-1} \frac{h^r}{r!} \sum_{\lambda=1}^{n_{r+1}} \beta_\lambda^{(r+1)} \left[\frac{1}{\gamma_\lambda^{(r+1)}} - \hat{\Phi}_\lambda^{(r+1)}\left(\frac{1}{\alpha(\psi)}\psi\right)\right] F_\lambda^{(r+1)}(\xi) + O(h^p)$$

$$= O(h^p) \quad \text{by (3.3.7)},$$

under our assumptions on ψ. Since $z(t_\nu) \in B_0$ for all ν, this implies $\|l_n\|_{E_n^0} = O(n^{-p})$ by (3.3.3), (3.3.1), and (2.2.3).

b) If we assume that ψ is only of exact generalized order $p_0 < p$, then there is at least one pair $(p_0+1, \lambda_0) \in \mathbb{D}$ such that

$$\hat{\Phi}_{\lambda_0}^{(p_0+1)}(\psi) \neq \frac{\alpha(\psi)^{p_0+1}}{\gamma_{\lambda_0}^{(p_0+1)}}.$$

According to Lemma 3.1.3, we may choose $f: \mathbb{R}^s \to \mathbb{R}^s$ such that $F_{\lambda_0}^{(p_0+1)}(\xi) \not\equiv 0$ but $F_\lambda^{(p_0+1)}(\xi) \equiv 0$ for $\lambda \neq \lambda_0$, $(p_0+1, \lambda) \in \mathbb{D}$. Since all $\beta_\lambda^{(r)} > 0$—which is seen from the proof of Theorem 3.1.4—there remains exactly one term in the coefficient of h^{p_0} in (3.3.8) which does not vanish so that $\|l_n\|$ becomes of exact order $O(h^{p_0})$. □

Remark. The remark after Def. 3.3.2 implies that for *each RK-scheme* with $\alpha(\psi) \neq 0$ the uniform RK-method based upon $(1/\alpha(\psi))\psi$ is consistent with each IVP 1, and hence also *convergent for each IVP 1* by Theorem 3.3.3. Thus, uniform RK-methods based upon RK-schemes with step-factor 1 form a very well-behaved class of discretization methods for the numerical treatment of IVP 1.

Corollary 3.3.6. *If all RK-schemes $\psi_{n\nu}$ of a non-uniform RK-method $\mathfrak{M}$ are of order p then $\mathfrak{M}$ is consistent of order p with each IVP$^{(p)}$ 1.*

Proof. Completely analogous to part a) of proof of Theorem 3.3.5. □

Remark. If Spijker's norm is used for the spaces E_n^0 a non-uniform RK-method may be consistent of order p with an IVP$^{(p)}$ 1 although some or all of its constituent RK-schemes are of an exact order smaller than p (see Example 1 in Section 2.2.4).

It is intuitively clear that the operations in $\mathfrak{R}$ should not decrease the order of the involved operands. For "scalar multiplication", this is an immediate consequence of (3.3.6), (3.3.4) and $\alpha(\rho\psi)=\rho\alpha(\psi)$ (Theorem 3.2.1, 1). Furthermore:

Theorem 3.3.7. *The RK-schemes of generalized order p form a subgroup* $\mathfrak{R}_p \subset \mathfrak{R}$ *w.r.t. composition.*

Proof. The unit scheme ψ_0 is of any generalized order p since $\alpha(\psi_0)=0$ and $\hat{\Phi}_\lambda^{(r)}(\psi_0)=0$ for all $(r,\lambda)\in\mathbb{D}$; thus $\psi_0\in\mathfrak{R}_p$ for each p. To show that $\psi_1\psi_2\in\mathfrak{R}_p$ if ψ_1 and ψ_2 are in $\mathfrak{R}_p$ we observe the following fact about the exact increment function: By the definition of the local solution of an IVP 1 (see Def. 3.1.6) we have $\bar{z}((\rho_1+\rho_2)h,\xi)=\bar{z}(\rho_2 h,\bar{z}(\rho_1 h,\xi))$ so that, by (3.1.23),

$$
\begin{aligned}
(\rho_1+\rho_2)\Psi_\infty((\rho_1+\rho_2)h,\xi) &= \frac{\bar{z}((\rho_1+\rho_2)h,\xi)-\xi}{h} \\
&= \frac{\bar{z}(\rho_1 h,\xi)-\xi}{h}+\frac{\bar{z}(\rho_2 h,\bar{z}(\rho_1 h,\xi))-\bar{z}(\rho_1 h,\xi)}{h} \qquad (3.3.9)\\
&=\rho_1\Psi_\infty(\rho_1 h,\xi)+\rho_2\Psi_\infty(\rho_2 h,\xi+\rho_1 h\Psi_\infty(\rho_1 h,\xi)).
\end{aligned}
$$

Now let $\Psi_{12}=(\psi_1\psi_2)[f]$, with f from an IVP$^{(p)}$ 1, $\rho_1=\alpha(\psi_1)$, $\rho_2=\alpha(\psi_2)$; then, by (3.3.9) and (3.2.3),

$$
\begin{aligned}
&(\rho_1+\rho_2)\Psi_\infty((\rho_1+\rho_2)h,\xi)-\Psi_{12}(h,\xi)\\
&=\rho_1\Psi_\infty(\rho_1 h,\xi)+\rho_2\Psi_\infty(\rho_2 h,\xi+\rho_1 h\Psi_\infty(\rho_1 h,\xi))\\
&\quad -\Psi_1(h,\xi)-\Psi_2(h,\xi+h\Psi_1(h,\xi))\\
&=[\rho_1\Psi_\infty(\rho_1 h,\xi)-\Psi_1(h,\xi)]\\
&\quad +[\rho_2\Psi_\infty(\rho_2 h,\xi+\rho_1 h\Psi_\infty(\rho_1 h,\xi))-\Psi_2(h,\xi+\rho_1 h\Psi_\infty(\rho_1 h,\xi))]\\
&\quad +[\Psi_2(h,\xi+\rho_1 h\Psi_\infty(\rho_1 h,\xi))-\Psi_2(h,\xi+h\Psi_1(h,\xi))].
\end{aligned}
$$

By (3.1.26)/(3.1.27), in the expansion of $\rho\Psi_\infty(\rho h,\xi)$ the $1/\gamma_\lambda^{(r)}$ are replaced by $\rho^r/\gamma_\lambda^{(r)}$. Therefore we find, as in part a) of the proof of Theorem 3.3.5 that the first and the second bracket are of order $O(h^p)$. The third bracket is $O(h^{p+1})$ due to the Lipschitz continuity of Ψ_2 and the result about Ψ_1. From $(\rho_1+\rho_2)\Psi_\infty((\rho_1+\rho_2)h,\xi)-\Psi_{12}(h,\xi)=O(h^p)$ for f from an IVP$^{(p)}$ 1, $\xi\in B_0$ and sufficiently small h we may conclude that $\psi_1\psi_2$ is of generalized order p as in part b) of the proof of Theorem 3.3.5.

Similarly, from

$$
\begin{aligned}
0=&-\rho\Psi_\infty(-\rho h,\xi)+\rho\Psi_\infty(\rho h,\xi-\rho h\Psi_\infty(-\rho h,\xi))\\
&-\Psi_-(h,\xi)-\Psi(h,\xi+h\Psi_-(h,\xi))\\
=&[-\rho\Psi_\infty(-\rho h,\xi)-\Psi_-(h,\xi)]\\
&+[\rho\Psi_\infty(\rho h,\xi-\rho h\Psi_\infty(-\rho h,\xi))-\Psi(h,\xi-\rho h\Psi_\infty(-\rho h,\xi))]\\
&+[\Psi(h,\xi-\rho h\Psi_\infty(-\rho h,\xi))-\Psi(h,\xi+h\Psi_-(h,\xi))]
\end{aligned}
$$

we obtain $\|-\rho\,\Psi_\infty(-\rho h,\xi)-\Psi_-(h,\xi)\|=O(h^p)$ which implies $\psi^{-1}\in\mathfrak{R}_p$. □

Remark. If ψ_1 and ψ_2 are of exact generalized order p, $\psi_1\psi_2$ may be of an exact generalized order greater than p.

Corollary 3.3.8. *The growth function γ_ψ of a RK-scheme ψ of order p satisfies*

$$\gamma_\psi(h)=e^h+O(h^{p+1}) \quad as\ |h|\to 0\,. \tag{3.3.10}$$

Proof. According to Def. 3.2.7, $\gamma_\psi(h)=\bar{\zeta}^{m+1}(h,1)$ where $\bar{\zeta}$ is the local solution of a RK-procedure of ψ for the scalar differential equation $y'=y$. From (3.1.9) and (3.1.23) for $f(y)=y$ we have

$$\Psi_\infty(h,1)-\Psi(h,1)=\frac{e^h-1}{h}-\frac{\gamma_\psi(h)-1}{h}=O(h^p)$$

which implies the assertion. □

Remark. (3.3.10) is a necessary but not a sufficient condition for ψ to be of order p.

3.3.3 Construction of High-order RK-procedures

According to Theorems 3.3.5 and 3.3.3, the generating matrix of the RK-procedure of a uniform RK-method $\mathfrak{M}$ has to satisfy

$$\hat{\Phi}_\lambda^{(r)}(A)=\frac{1}{\gamma_\lambda^{(r)}},\quad \lambda=1(1)n_r,\quad r=1(1)p\,, \tag{3.3.11}$$

in order that $\mathfrak{M}$ be convergent of order p for each IVP$^{(p)}$ 1. The number $N_p=\sum_{r=1}^{p} n_r$ of conditions (3.3.11) increases rapidly with p as can be seen from Table 3.1 in Section 3.1.3. Since most of the conditions (3.3.11) are non-linear in the elements $a_{\mu\lambda}$ of A (see the recursive definition of the $\hat{\Phi}_\lambda^{(r)}$), there is no general procedure for solving (3.3.11) for the $a_{\mu\nu}$ except in the simplest cases. Also, the number of free parameters in A necessary for the existence of a solution to (3.3.11) for a given p cannot be immediately related to N_p except for small p.

The problem of constructing m-stage generating matrices of order p with a maximal ratio p/m has challenged many mathematicians during recent years. The most systematic approach is based on the theory developed by Butcher [3]; we will attempt to indicate some of its considerations. The objective is to satisfy the conditions (3.3.11) for a given value of p with a generating matrix subject to given restrictions (number of stages, explicitness, etc.).

First we note that for the degenerate case of a mere *quadrature*, i.e., for the IVP 1

$$y_1' - 1 = 0, \qquad y_1(0) = 0,$$
$$y_2' - f(y_1) = 0, \qquad y_2(0) = 0,$$

all elementary differentials of an order $r > 1$ vanish except $\{f^{r-1}\} = \begin{pmatrix} 0 \\ f^{(r-1)}(y_1) \end{pmatrix}$. Hence, only the corresponding terms remain in the expansion (3.3.8) and the conditions (3.3.11) reduce to

$$\text{(3.3.12)} \qquad \begin{cases} \hat{\Phi}(f, A) = 1, \\ \hat{\Phi}(\{f^{r-1}\}, A) = \hat{\Phi}(\{f^{r-1}\}, A_\infty) = \dfrac{1}{r}, \qquad r = 2(1)p. \end{cases}$$

If we use the common abbreviations (cf. the beginning of Section 3.2.4)

$$\bar{A} := \begin{pmatrix} a_{11} & \dots & a_{1m} \\ \vdots & & \vdots \\ a_{m1} & \dots & a_{mm} \end{pmatrix}, \qquad \begin{aligned} b &:= (a_{m+1,1}, \dots, a_{m+1,m})^T \in \mathbb{R}^m, \\ c &:= \bar{A} e = \left(\sum_{\lambda=1}^{m} a_{1\lambda}, \dots, \sum_{\lambda=1}^{m} a_{m\lambda} \right)^T \in \mathbb{R}^m, \end{aligned}$$

these basic conditions (3.3.12) take the simple form

$$\text{(3.3.13)} \qquad b^T c^{r-1} = \frac{1}{r}, \qquad r = 1(1)p,$$

where c^r denotes componentwise powers as in (3.1.16), $c^0 = e$.

The remaining conditions (3.3.11) may be greatly or even completely reduced to (3.3.13) if the generating matrix A satisfies certain relations which hold for the integral operator A_∞^h. These are (cf. Section 3.1.5):

$$\text{(3.3.14)} \qquad r A_\infty^h (A_\infty^h e)^{r-1}(\tau) = (A_\infty^h e)^r(\tau), \qquad r = 2, 3, \dots,$$

$$\text{(3.3.15)} \qquad A_\infty^h [r(A_\infty^h e)^{r-1}(A_\infty^h x) + (A_\infty^h e)^r x - x](h) = 0, \qquad r = 1, 2, \dots,$$

where x is an arbitrary function from $C[0, h]$. (3.3.14) is nothing but the integration rule for powers; (3.3.15) is the rule of partial integration:

$$\frac{1}{h} \int_0^h \left[r \left(\frac{\tau}{h} \right)^{r-1} \frac{1}{h} \int_0^\tau x(\tau') d\tau' + \left(\frac{\tau}{h} \right)^r x(\tau) \right] d\tau - \frac{1}{h} \int_0^h x(\tau) d\tau = 0.$$

Thus one should aim at having A satisfy relations of the type

$$\text{(3.3.16)} \qquad r \bar{A} c^{r-1} = c^r, \qquad r = 2, 3, \dots,$$

$$\text{(3.3.17)} \qquad b^T [r c^{r-1} \cdot \bar{A} x + c^r \cdot x - x] = 0, \qquad x \in \mathbb{R}^m, \qquad r = 1, 2, \dots.$$

Example. The 8 conditions (3.3.11) for order 4 are

(1)	$b^T e = 1$,	(4.1)	$b^T \bar{A}^2 c = \frac{1}{24}$,
(2)	$b^T c = \frac{1}{2}$,	(4.2)	$b^T \bar{A} c^2 = \frac{1}{12}$,
(3.1)	$b^T \bar{A} c = \frac{1}{6}$,	(4.3)	$b^T (\bar{A} c) \cdot c = \frac{1}{8}$,
(3.2)	$b^T c^2 = \frac{1}{3}$,	(4.4)	$b^T c^3 = \frac{1}{4}$.

The quadrature conditions (3.3.13) are (1), (2), (3.2), and (4.4). If A satisfies (3.3.16) for $r=2$, (3.1) reduces to (3.2), (4.1) to (4.2), and (4.3) to (4.4). If, furthermore, A satisfies (3.3.17) for $r=1$, the remaining non-quadrature condition (4.2) is reduced (4.4) and (3.2) by taking $x = c^2$. The same effect is achieved if A satisfies (3.3.16) for $r=3$.

While each condition (3.3.16) or (3.3.17) amounts to m conditions on A, our example shows that these conditions need not always hold in full generality. Also, if A is to be explicit the first component of (3.3.16) is satisfied trivially while the second cannot be satisfied at all; but this defect can be remedied by choosing $a_{m+1,2} = 0$ in b.

Naturally, the satisfaction of (3.3.16) and (3.3.17) for some low values of r is not a necessary condition for the satisfaction of (3.3.11). But proceeding in this fashion makes the attempt to satisfy the large number of non-linear equations more transparent and feasible. Of course, for larger values of p a great deal of ingenuity and skill has to be applied. No further details are discussed here, but the reader is referred to the relevant literature.

Example. Construction of an implicit 2-stage RK-procedure of order 4:

The quadrature conditions are

$$\sum_{\mu=1}^{2} a_{3\mu} c_\mu^{r-1} = \frac{1}{r}, \qquad r = 1(1)4.$$

As is well-known, these 4 equations possess a unique solution corresponding to the 2-point Gauss-Legendre quadrature rule: $a_{31} = a_{32} = \frac{1}{2}$, $c_{1,2} = \frac{1}{2} \mp \sqrt{3}/6$.

Condition (3.3.16) for $r=2$ determines the $a_{\mu\lambda}$, $\mu, \lambda = 1, 2$, uniquely from the linear system

$$\begin{aligned} a_{\mu 1} + a_{\mu 2} &= c_\mu, \\ a_{\mu 1} c_1 + a_{\mu 2} c_2 &= \tfrac{1}{2} c_\mu^2, \end{aligned} \qquad \mu = 1, 2,$$

which yields $a_{11} = a_{22} = \frac{1}{4}$, $a_{12} = \frac{1}{4} - \sqrt{3}/6$, $a_{21} = \frac{1}{4} + \sqrt{3}/6$. These values also satisfy Condition (3.3.17) for $r=1$ which requires that

$$a_{11} + a_{21} = c_2, \qquad a_{12} + a_{22} = c_1.$$

Thus the $a_{\mu\lambda}$ generate a RK-procedure of order 4 (see the discussion in the Example above). Note that we have satisfied 8 conditions (3.3.11) with 6 parameters $a_{\mu\lambda}$, $\mu = 1(1)3$, $\lambda = 1, 2$.

As observed in Section 3.1.3, for *one* differential equation $y' = f(t, y)$ the number of different elementary differentials is reduced for orders $r \geq 5$. Therefore, a generating matrix A has to satisfy fewer than N_p conditions in order to generate a RK-method which is consistent of order $p \geq 5$ with a scalar differential equation. We will not analyze these simplifications. Some authors do consider the simplified set of conditions, however, so that one has to be careful.

3.3.4 Attainable Order of m-stage RK-procedures

As we have seen in the previous section, the conditions (3.3.11) are highly non-linear as well as interdependent so that a count of the number of conditions and available parameters is of little value. The fact that RK-procedures of arbitrarily high orders exist at all is seen rather easily:

Theorem 3.3.9. *There exist (explicit) RK-procedures of arbitrarily high orders.*

Proof. a) There exist quadrature rules of arbitrary order p, i.e., such that for sufficiently smooth functions $x \in ([0,h] \to \mathbb{R}^s)$

$$\frac{1}{h}\int_0^h x(\tau)\,d\tau = \sum_{\mu=1}^{m} \beta_\mu x(\gamma_\mu h) + O(h^p). \tag{3.3.18}$$

(They may be obtained, e.g., from polynomial interpolation.)

b) We proceed by induction: The Euler-procedure is an (explicit) RK-procedure of order 1. Assume that we have found an (explicit) RK-procedure of order $p-1$, with increment function Ψ_{p-1}. Then the (explicit) increment function

$$\Psi_p(h,\xi) := \sum_{\mu=1}^{m} \beta_\mu f(\xi + \gamma_\mu h \Psi_{p-1}(\gamma_\mu h, \xi))$$

satisfies (see (3.1.23))

$$\Psi_\infty(h,\xi) - \Psi_p(h,\xi) = \left[\frac{1}{h}\int_0^h f(\bar{z}(\tau,\xi))\,d\tau - \sum_{\mu=1}^{m} \beta_\mu f(\bar{z}(\gamma_\mu h,\xi))\right]$$
$$+ \sum_{\mu=1}^{m} \beta_\mu [f(\bar{z}(\gamma_\mu h,\xi)) - f(\xi + \gamma_\mu h \Psi_{p-1}(\gamma_\mu h,\xi))].$$

The first bracket is $O(h^p)$ by (3.3.18); the subsequent sum is of the same order since

$$\bar{z}(\gamma_\mu h,\xi) - \xi - \gamma_\mu h \Psi_{p-1}(\gamma_\mu h,\xi) = \gamma_\mu h[\Psi_\infty(\gamma_\mu h,\xi) - \Psi_{p-1}(\gamma_\mu h,\xi)] = O(h^p)$$

by the induction assumption on Ψ_{p-1}. Thus, $\Psi_\infty(h,\xi) - \Psi_p(h,\xi) = O(h^p)$ for sufficiently smooth f which implies order p for the RK-procedure with increment function Ψ_p. □

Remark. It is obvious that the RK-procedures constructed in the above proof are not efficient (in the sense of a large p/m) for $p > 2$.

There exists no general result on the highest attainable order with a given number m of stages for *explicit* RK-procedures (i.e., RK-procedures of the traditional form). It is well-known that there exist explicit m-stage procedures of order m for $m=1(1)4$ but no explicit 5-stage RK-procedure of order 5 could be found. In his paper [5], Butcher *proved* that the conditions (3.3.11) are contradictory for explicit generating matrices A

$$\text{for } m \le p \quad \text{if } p \ge 5,$$
$$\text{for } m \le p+1 \quad \text{if } p \ge 7.$$

Furthermore explicit generating matrices with the next larger number of stages have actually been constructed for $p=5,6,7$.

Thus for the highest attainable order p of an explicit m-stage RK-procedure we have the following table:

Table 3.3

m	1	2	3	4	5	6	7	8	9	10
p	1	2	3	4	4	5	6	6	7	?

Whether there exists an explicit 10-stage RK-procedure of order 8 seems to be still unknown.

If we admit *implicit* RK-procedures the construction of Theorem 3.3.9 can be made more effective:

Theorem 3.3.10. *Each quadrature rule with m nodes which is of order at least* $m+1$ *leads to an implicit m-stage RK-procedure which is of order at least* $m+1$.

Proof. Take the quadrature rule (3.3.18), with $p \ge m+1$, and let for $\mu=1(1)m$

$$(3.3.19) \qquad L_\mu(\tau) := \frac{\prod_{\lambda=1}^{m}{}' (\tau-\gamma_\lambda h)}{\prod_{\lambda=1}^{m}{}' (\gamma_\mu-\gamma_\lambda) h}, \qquad a_{\mu\lambda} := \frac{1}{h}\int_0^{\gamma_\mu h} L_\lambda(\tau)\,d\tau, \qquad \lambda=1(1)m,$$

and $a_{m+1,\lambda} := \beta_\lambda$, $\lambda=1(1)m$. From the theory of polynomial interpolation we have for sufficiently smooth f

$$(3.3.20) \quad f(\bar{z}(\tau,\xi)) = \sum_{\lambda=1}^{m} L_\lambda(\tau) f(\bar{z}(\gamma_\lambda h,\xi)) + \frac{1}{m!}\prod_{\lambda=1}^{m}(\tau-\gamma_\lambda h)\frac{d^m}{dh^m} f(\bar{z}(\tilde{\tau},\xi)),$$

where $\tilde{\tau}$ is an "intermediate" value, which implies

$$\bar{z}(\gamma_\mu h,\xi) = \xi + \int_0^{\gamma_\mu h} f(\bar{z}(\tau,\xi))\,d\tau \qquad \text{see (3.1.22)}$$

$$(3.3.21) \qquad = \xi + h\sum_{\lambda=1}^{m} a_{\mu\lambda} f(\bar{z}(\gamma_\lambda h,\xi)) + O(h^{m+1}) \quad \text{by (3.3.20) and (3.3.19).}$$

A comparison of (3.3.21) with (3.1.8) shows—the details are omitted—that

$$\bar{z}(\gamma_\mu h, \xi) = \bar{\zeta}^\mu(h, \xi) + O(h^{m+1}), \qquad \mu = 1(1)m, \tag{3.3.22}$$

where $\bar{\zeta}$ is the local solution of the RK-procedure with the coefficients $a_{\mu\lambda}$. Hence the associated increment function Ψ satisfies

$$\begin{aligned}\Psi_\infty(h,\xi) - \Psi(h,\xi) &= \left[\frac{1}{h}\int_0^h f(\bar{z}(\tau,\xi))\,d\tau - \sum_{\mu=1}^m \beta_\mu f(\bar{z}(\gamma_\mu h, \xi))\right] \\ &\quad + \sum_{\mu=1}^m \beta_\mu [f(\bar{z}(\gamma_\mu h,\xi)) - f(\bar{\zeta}^\mu(h,\xi))] \\ &= O(h^{m+1}) \quad \text{by the assumption on (3.3.18) and by (3.3.22)}\end{aligned}$$

which proves the assertion. □

Actually, one may reach considerably higher orders for special quadrature rules and larger m:

Theorem 3.3.11 (*Butcher* [6]). *For each m there exists an implicit m-stage RK-procedure of order $2m$.*

The proof of this result proceeds basically along the same lines as our construction of a 2-stage 4th order RK-procedure in Section 3.3.3: The values of the components of b and $c \in \mathbb{R}^m$ which satisfy the quadrature conditions (3.3.13) for $p = 2m$ are taken from the Gauss-Legendre quadrature. Then (3.3.16) for $r = 1(1)m$ becomes a linear system for the $a_{\mu\lambda}$ which determines them uniquely. The more difficult part of the proof is to show that these $a_{\mu\lambda}$ also satisfy (3.3.17) for $r = 1(1)m$. This establishes the assertion since the satisfaction of (3.3.16) *and* (3.3.17) for $r = 1(1)m$ reduces all conditions (3.3.11) for $p = 2m$ to the quadrature conditions for $2m$ as has been established by Butcher [6].

For $m = 1, 2, 3,$ one obtains the following generating matrices

$$\begin{pmatrix} \frac{1}{2} \\ 1 \end{pmatrix}, \quad \begin{pmatrix} \frac{1}{4} & \frac{1}{4} - \frac{\sqrt{3}}{6} \\ \frac{1}{4} + \frac{\sqrt{3}}{6} & \frac{1}{4} \\ \frac{1}{2} & \frac{1}{2} \end{pmatrix}, \quad \begin{pmatrix} \frac{5}{36} & \frac{2}{9} - \frac{w}{15} & \frac{5}{36} - \frac{w}{30} \\ \frac{5}{36} + \frac{w}{24} & \frac{2}{9} & \frac{5}{36} - \frac{w}{24} \\ \frac{5}{36} + \frac{w}{30} & \frac{2}{9} + \frac{w}{15} & \frac{5}{36} \\ \frac{5}{18} & \frac{4}{9} & \frac{5}{18} \end{pmatrix} \quad \text{with } w = \sqrt{15}.$$

It is obvious that these matrices satisfy the condition of Theorem 3.2.7 for the symmetry of the generated RK-scheme.

Theorem 3.3.12. *The optimal order RK-procedures of Theorem* 3.3.11 *are symmetric for each m.*

Proof. a) The coefficients $a_{\mu\lambda}$ of the optimal order generating matrices are given by (3.3.19), where the γ_μ are taken from the Gauss-Legendre quadrature rule (3.3.18) with m nodes: For $r=1(1)m$,

$$\sum_{\lambda=1}^{m} L_\lambda(\tau)(\gamma_\lambda h)^{r-1}=\tau^{r-1},$$

$$\frac{1}{h}\sum_{\lambda=1}^{m}\int_0^{\gamma_\mu h} L_\lambda(\tau)\,d\tau\,(\gamma_\lambda h)^{r-1}=\frac{1}{rh}(\gamma_\mu h)^r,$$

or (cf. (3.3.19))

$$\sum_{\lambda=1}^{m} a_{\mu\lambda}\gamma_\lambda^{r-1}=\frac{1}{r}\gamma_\mu^r,\qquad \mu=1(1)m,\qquad r=1(1)m.$$

These are exactly the conditions (3.3.16) which have been used to determine the $a_{\mu\lambda}$ for the optimal order generating matrices (see the outline of the proof of Theorem 3.3.11).

b) Let the γ_μ be ordered according to increasing magnitude. Then $\gamma_\mu=1-\gamma_{m-\mu}$, $\mu=1(1)m$, which implies (see 3.3.19)

$$L_\lambda(\tau)=L_{m-\lambda}(h-\tau),\qquad \int_0^{\gamma_\mu h} L_\lambda(\tau)\,d\tau=\int_{(1-\gamma_\mu)h}^{h} L_{m-\lambda}(\tau)\,d\tau,$$

so that

$$a_{\mu\lambda}+a_{m-\mu,m-\lambda}=\frac{1}{h}\int_0^{\gamma_\mu h} L_\lambda(\tau)\,d\tau+\frac{1}{h}\int_0^{(1-\gamma_\mu)h} L_{m-\lambda}(\tau)\,d\tau$$

$$=\int_0^h L_{m-\lambda}(\tau)\,d\tau=\int_0^h L_\lambda(\tau)\,d\tau.$$

The last two quantities are nothing but $\beta_{m-\lambda}=a_{m+1,m-\lambda}$ and $\beta_\lambda=a_{m+1,\lambda}$, resp. so that the symmetry conditions (3.2.19) of Theorem 3.2.7 hold with $\mu'=m-\mu$, $\mu=1(1)m$. □

By Theorem 3.2.9, the growth functions associated with the above m-stage implicit RK-procedures of order $2m$ are rational functions with numerators and denominators of a degree not larger than m. At the same time they have to satisfy

$$\gamma(h)=e^h+O(h^{2m+1})$$

by Corollary 3.3.8. Hence these growth functions must be the *diagonal Padé-approximations* of e^h; for $m=1,2,3$, we have

$$\gamma(h) = \frac{1+h/2}{1-h/2},\ \frac{1+h/2+h^2/12}{1-h/2+h^2/12},\ \frac{1+h/2+h^2/10+h^3/120}{1-h/2+h^2/10-h^3/120}.$$

Note that these growth functions also satisfy the necessary symmetry condition (3.2.29) in accordance with Corollary 3.2.12.

Generating matrices for RK-schemes whose growth functions are from the upper or lower co-diagonals in the Padé-table of e^h have been constructed by Butcher [7] and Ehle [1].

3.3.5 Effective Order of RK-schemes

The power of the approach of Section 3.2 comes to light in the following development of Butcher [8]:

Def. 3.3.3. A RK-scheme ψ is called *of effective order* $\bar{p}$ if there exists a RK-scheme χ such that the RK-scheme $\chi\psi\chi^{-1}$ is of order $\bar{p}$. Each generating matrix of ψ is also called of effective order $\bar{p}$ and so are the associated RK-procedures.

It is trivial that each RK-scheme of order p is of effective order p (take $\chi=\psi_0$).

Theorem 3.3.13. *Let the RK-scheme ψ be of effective order $\bar{p}$ and let $\bar{\psi}:=\chi\psi\chi^{-1}$ be of order $\bar{p}$; furthermore, let $\psi_s:=\chi\psi$ and $\psi_e:=\psi\chi^{-1}$. Then, for each $IVP^{(\bar{p})}$ 1, the non-uniform RK-method $\mathfrak{M}$ (see Def. 3.3.1) with equidistant grids based upon the RK-schemes*

$$\psi_{n1}=\psi_s; \quad \psi_{n\nu}=\psi,\ \nu=2(1)n-1; \quad \psi_{nn}=\psi_e \tag{3.3.23}$$

generates a discretization the solution sequence $\{\zeta_n\}$ of which satisfies

$$\|\zeta_n(1)-z(1)\|=O(h^{\bar{p}}).$$

Proof. Since

$$\psi_s\psi^{n-2}\psi_e=\chi\psi^n\chi^{-1}=(\chi\psi\chi^{-1})^n=\bar{\psi}^n, \tag{3.3.24}$$

the increment function $\Psi^{(n)}$ of the RK-scheme $\psi_s\psi^{n-2}\psi_e$ is identical to the increment function $\bar{\Psi}^{(n)}$ of $\bar{\psi}^n$ for any $f:\mathbb{R}^s\to\mathbb{R}^s$ (at least for sufficiently small $|h|$ and suitable ξ). Hence, for any IVP 1 and sufficiently small h,

$$\zeta_n(1)=z_0+h\,\Psi^{(n)}(h,z_0)=z_0+h\,\bar{\Psi}^{(n)}(h,z_0)$$

is the value that would have been obtained from the uniform RK-method based upon the RK-scheme $\bar{\psi}$. The assertion now follows from Theorems 3.3.3 and 3.3.5 and from our assumption on $\bar{\psi}$. □

Naturally, Def. 3.3.3 and Theorem 3.3.13 become meaningful only if there exist RK-schemes whose effective order $\bar{p}$ is greater than their order p, and if this order $\bar{p}$ could not have been attained by a RK-scheme of comparable simplicity.

By Theorem 3.2.10, the growth functions of ψ and $\bar{\psi}=\chi\psi\chi^{-1}$ are the same. If $\bar{\psi}$ is to be of order $\bar{p}$ this common growth function γ_ψ of ψ and $\bar{\psi}$ must satisfy

$$\gamma_\psi(h)=e^h+O(h^{\bar{p}+1})$$

by Corollary 3.3.8. If ψ is generated by an explicit m-stage generating matrix, γ_ψ is a polynomial of degree m by Theorem 3.2.9; hence, there cannot exist an explicit m-stage RK-procedure of effective order $\bar{p}>m$. Thus, the concept of effective order seems uninteresting for $m=1(1)4$. (Cf., however, Section 3.4.3!)

However, for $m=5$, Butcher has constructed a family of *explicit* 5-stage RK-procedures of effective order 5 in [8]. Furthermore, the associated "starting" and "ending" schemes ψ_s and ψ_e (see (3.3.23)) can also be generated by explicit 5-stage generating matrices. Thus each forward step of the non-uniform RK-method defined by (3.3.23) consists of an explicit 5-stage RK-procedure and yet the value of the solution of the discretization at the end of the interval differs from the value of the true solution by a quantity of order $O(h^5)$ for an arbitrary IVP$^{(5)}$ 1. This could not have been achieved by a uniform RK-method with a 5-stage explicit RK-procedure according to Section 3.3.4. Similarly, there exist explicit 6-stage RK-procedures of effective order 6[6].

In the example given by Butcher [8], the RK-scheme ψ of (3.3.23) is of order 4 while ψ_s and ψ_e are only of order 3 each. From (3.3.3) and our definition (2.2.3) of the norm in E_n^0 it follows that the non-uniform RK-method (3.3.23) is convergent of order 4, i.e., that $\|\zeta_n-\Delta_n z\|_{E_n}=O(h^4)$.

Actually the 5-th order convergence is restricted to the value at $t=1$ for (3.3.23); but one may also obtain a 5-th order correct approximation at any other gridpoint (except t_1) simply by using the RK-procedure associated with ψ_e to arrive at this point. (The continuation has to be done by $\psi_s\psi\ldots$ if the effect is not to be lost for the remainder of the interval.) With a *cyclic* use of ψ_s and ψ_e every even grid point would enjoy the 5-th order convergence.

According to the second expression $\chi\psi^n\chi^{-1}$ in (3.3.24), the computation of $\zeta_n(1)$ may also be regarded as follows:

Apply a RK-procedure associated with the RK-scheme χ to the initial value z_0; this amounts to a certain modification (depending upon h and f) of the initial value z_0 to a new initial value $\bar{z}_0$.

---

[6] Private communication by J. C. Butcher.

Apply the *uniform* RK-method based upon ψ to the modified IVP 1: $y(0)=\bar{z}_0$, $y'=f(y)$, to obtain an approximation $\bar{\zeta}_n(1)$, with an error of order $O(h^p)$.

Modify $\bar{\zeta}_n(1)$ by applying a RK-procedure associated with χ^{-1} to obtain $\zeta_n(1)$ the error of which is of order $O(h^{\bar{p}})$ only.

In this interpretation, Butcher's result seems even more startling at first sight; but it will help us to understand the analytic meaning of effective order in Section 3.4.3.

3.4 Analysis of the Discretization Error

3.4.1 The Principal Error Function

Def. 3.4.1. For a given RK-scheme ψ of exact order p and a given IVP$^{(p+1)}$ 1, the mapping $\varphi: \mathbb{R}^s \to \mathbb{R}^s$ defined by

$$(3.4.1) \qquad \varphi(\xi) := \frac{1}{p!} \sum_{\lambda=1}^{n_{p+1}} \beta_\lambda^{(p+1)} \left[\hat{\Phi}_\lambda^{(p+1)}(\psi) - \frac{1}{\gamma_\lambda^{(p+1)}} \right] F_\lambda^{(p+1)}(\xi)$$

is called the *principal error function* of ψ for that IVP 1.

This definition coincides with the meaning of the term in most of the current literature (see, e.g., Henrici [1]). It is an immediate consequence of Def. 3.4.1 that (see (3.1.21) and (3.1.26)—(3.1.27)) for an m-stage RK-procedure of order p and an IVP$^{(p+1)}$ 1

$$(3.4.2) \qquad \bar{\zeta}^{m+1}(h,\xi) - \bar{z}(h,\xi) = h^{p+1}\varphi(\xi) + O(h^{p+2}),$$

$$(3.4.3) \qquad \Psi(h,\xi) - \Psi_\infty(h,\xi) = h^p \varphi(\xi) + O(h^{p+1}).$$

Lemma 3.4.1. *Consider the uniform RK-method $\mathfrak{M}$ based upon the RK-scheme ψ of order p, with a coherent grid sequence with step size function θ. The local discretization error l_n of $\mathfrak{M}$ for an IVP$^{(p+1)}$ 1 satisfies*

$$(3.4.4) \qquad \begin{aligned} l_n(t_\nu) &= -n^{-p}\theta(t_{\nu-1})^p \varphi(z(t_{\nu-1})) + O(n^{-(p+1)}) \\ &= -n^{-p}\theta(t_{\nu-1})^p \varphi(\zeta_n(t_{\nu-1})) + O(n^{-(p+1)}), \end{aligned} \qquad \nu = 1(1)n.$$

Proof. For $\nu=1(1)n$, we have from (3.3.3)

$$\begin{aligned} l_n(t_\nu) &= \Psi_\infty(h_\nu, z(t_{\nu-1})) - \Psi(h_\nu, z(t_{\nu-1})) \\ &= -h_\nu^p \varphi(z(t_{\nu-1})) + O(h^{p+1}) \quad \text{by (3.4.3)} \\ &= -n^{-p}\theta(t_{\nu-1})^p \varphi(z(t_{\nu-1})) + O(n^{-(p+1)}) \quad \text{by (2.1.36).} \end{aligned}$$

Since $F_\lambda^{(p+1)}: \mathbb{R}^s \to \mathbb{R}^s$ is Lipschitz continuous in $B_R(z)$ for an IVP$^{(p+1)}$ 1 by Def. 2.1.2, $\varphi(z(t_{\nu-1}))$ may be replaced by $\varphi(\zeta_n(t_{\nu-1}))$ due to Theorem 3.3.3. □

Remark. In the terminology of our general discussion in Section 1.3, $-n^{-p}\Delta_n^0(\theta^p\varphi(z))$ is the principal term of the local discretization error of $\mathfrak{M}$ for the IVP$^{(p+1)}$ 1 (cf. Def. 1.3.5).

From the expression (3.4.1) for the principal error function, a general classification of the RK-schemes of a certain order p according to the size of their local discretization error can be obtained only for $p=1$: Here the sum in (3.4.1) consists only of one term so that the modulus of the coefficient of $F_1^{(2)}$ is a quantitative measure for the quality of ψ. For $p\geq 2$, the superiority of one scheme of order p over another can only be decided for a given IVP$^{(p+1)}$ 1: Even if *all* coefficients of the $F_\lambda^{(p+1)}$ in (3.4.1) are smaller in modulus for one scheme ψ_1 than for another scheme ψ_2, a favorable cancellation may still render ψ_2 superior to ψ_1 for a particular IVP$^{(p+1)}$ 1.

On the other hand, it is only for very restrictive classes of IVP 1 that the number of non-vanishing elementary differentials is reduced so drastically that it becomes worthwhile to look for a RK-scheme which is particularly adapted to this situation. E.g., for linear systems with constant coefficients

$$y'-gy=0, \quad g \text{ a } s\times s \text{ matrix}, \tag{3.4.5}$$

it is easily seen that all r-th order differentials vanish except $\{_{r-1}f\}_{r-1}(\xi)=g^r\xi$.

Lemma 3.4.2. *For an arbitrary linear IVP 1 with constant coefficients, the uniform RK-method based upon the RK-scheme ψ is convergent of order p if and only if the growth function γ_ψ of ψ satisfies*

$$\gamma_\psi(h)=e^h+O(h^{p+1}).$$

Proof. For (3.4.5), the local solution $\bar\zeta$ of the discretization is obtained from the linear system of equations (cf. (3.1.8))

$$\bar\zeta^\mu(h,\xi)=\xi+hg\sum_{\lambda=1}^m a_{\mu\lambda}\bar\zeta^\lambda(h,\xi), \qquad \mu=1(1)m+1.$$

The elimination proceeds exactly as in Section 3.2.4 for $y'=y$ and yields (see proof of Theorem 3.2.9)

$$\det(I-hg\bar A)\bar\zeta^{m+1}(h,\xi)=\det(I-hg\bar A_-)\xi, \tag{3.4.6}$$

where $g\bar A$ is the $ms\times ms$-matrix

$$\begin{pmatrix} a_{11}g & \dots & a_{1m}g \\ \vdots & & \vdots \\ a_{m1}g & \dots & a_{mm}g \end{pmatrix}$$

(analogously for $g\bar{A}_-$). Thus, if we interpret the application of the rational function γ to a quadratic non-singular matrix in the usual manner, we obtain from (3.2.25) and (3.4.6)

$$\text{(3.4.7)} \qquad \bar{\zeta}^{m+1}(h,\xi)=\gamma_\psi(hg)\xi \quad \text{and} \quad \Psi(h,\xi)=\frac{1}{h}[\gamma_\psi(hg)-I]\xi\,.$$

Since $\Psi_\infty(h,\xi)=1/h[e^{hg}-I]\xi$ for (3.4.5), the assertion follows from (3.3.3) and Theorem 3.3.3 in a straightforward manner. □

Example. The RK-procedures generated by the explicit m-stage generating matrices $(m\geq 1)$

$$\begin{pmatrix} 0 & 0 & \cdot & \cdot & \cdot & \cdot & 0 \\ \frac{1}{m} & 0 & & & & & \cdot \\ 0 & \frac{1}{m-1} & \cdot & & & & \cdot \\ \cdot & & \cdot & \cdot & & & \cdot \\ \cdot & & & \cdot & \cdot & & \cdot \\ \cdot & & & & \cdot & \frac{1}{2} & 0 \\ 0 & \cdot & \cdot & \cdot & \cdot & 0 & 1 \end{pmatrix}$$

lead to RK-methods which are convergent of order m for an IVP 1 of type (3.4.5), for each integer m. (They are of order m for $m=1,2$ only.)

However, for linear IVP 1 with constant coefficients—if one wants to solve them numerically by a RK-method at all—the concept of stages looses its meaning: According to (3.4.7) one simply takes any rational approximation $\gamma(t)$ to e^t, computes the matrix $\gamma(hg)$, and proceeds with the "RK-procedure" $\eta_\nu=\gamma(hg)\eta_{\nu-1}$, $\nu=1(1)n$.

For general linear systems $y'-g(t)y=0$ the principal error function is a linear transform of ξ:

Lemma 3.4.3. *For* $f:\mathbb{R}^{s+1}\to\mathbb{R}^{s+1}$ *of the form*

$$\text{(3.4.8)} \qquad f\begin{pmatrix} t \\ \xi \end{pmatrix}=\begin{pmatrix} 1 \\ g(t)\xi \end{pmatrix}, \qquad \xi\in\mathbb{R}^s, \qquad g(t) \ \text{a}\ s\times s\ \text{matrix}$$

the principal error function of an arbitrary RK-scheme of order p is of the form

$$\text{(3.4.9)} \qquad \varphi\begin{pmatrix} t \\ \xi \end{pmatrix}=\begin{pmatrix} 0 \\ \hat{g}(t)\xi \end{pmatrix}, \qquad \hat{g}(t)\ \text{a}\ s\times s\text{-matrix}.$$

Proof. All elementary differentials of f of order $r\geq 2$ are of the form (3.4.9) as is seen by induction on their order r. Clearly, $\{f\}\begin{pmatrix} t \\ \xi \end{pmatrix}=\begin{pmatrix} 0 \\ (g'(t)+g^2(t))\xi \end{pmatrix}$ is of that form. If it is true for $r<r_0$, consider $(r_0,\lambda)=\{(r_1,\lambda_1),\ldots,(r_J,\lambda_J)\}$, $J\geq 1$. The vanishing of the first component is trivial. Of the derivatives $\partial^J f_2/\partial t^j\partial\xi^{J-j}$ occuring in the second

component of $F_\lambda^{(r_0)}$ (see (3.1.13)), only $\partial^J f_2/\partial t^J$ and $\partial^J f_2/\partial t^{J-1}\partial\xi$ do not vanish identically; the first one gives a contribution $g^{(J)}(t)\xi$ only if $(r_j,\lambda_j)=(1,1)$, $j=1(1)J$; the second one contributes terms of the form $g^{(J-1)}(t)g_{\lambda_i}^{(r_j)}\xi$ where $g_{\lambda_i}^{(r_j)}\xi$ is the second component of $F_{\lambda_i}^{(r_j)}(\xi)$. Thus $F_\lambda^{(r_0)}$ is of the form (3.4.9). □

E.B. Shanks [1] constructs generating matrices of order p such that many of the coefficients in (3.4.1) vanish and the *remaining ones are small*. Theoretically, one could make them arbitrarily small; but the elements of the generating matrices then tend to infinity with alternating signs so that there is a practical limit due to round-off amplification.

Example. The following is an example of an explicit 5-stage "almost 5th-order" generating matrix taken from Shanks [1]:

$$\begin{pmatrix} 0 & 0 & 0 & 0 & 0 \\ \frac{1}{10} & 0 & 0 & 0 & 0 \\ -\frac{3}{20} & \frac{9}{20} & 0 & 0 & 0 \\ \frac{27}{16} & -\frac{45}{16} & \frac{15}{8} & 0 & 0 \\ -\frac{64}{9} & \frac{115}{9} & -\frac{490}{81} & \frac{112}{81} & 0 \\ \frac{5}{54} & 0 & \frac{250}{567} & \frac{32}{81} & \frac{1}{14} \end{pmatrix}.$$

The generated RK-scheme satisfies 7 of the 9 conditions (3.3.11) for $p=5$, its principal error function becomes

$$\varphi(\xi)=\tfrac{1}{1440}[\{\{f^2\}f\}(\xi)-\{_3f^2\}_3(\xi)].$$

Thus the RK-method with this generating matrix is consistent of order 5 for all IVP$^{(5)}$ 1 for which either $\{\{f^2\}f\}$ and $\{_3f^2\}_3$ both vanish at $z(t)$ or are identical; for many other IVP$^{(5)}$ 1 it should have a rather small 4th-order discretization error. (Note the large moduli of the elements in the fifth row!)

3.4.2 Asymptotic Expansion of the Discretization Error

The theory of Section 1.3 is applicable to arbitrary uniform RK-methods for sufficiently smooth IVP 1.

Theorem 3.4.4. *Consider the uniform RK-method* $\mathfrak{M}:=\{E_n,E_n^0,\Delta_n,\Delta_n^0,\varphi_n\}_{n\in\mathbb{N}'}$ *of Def.* 3.3.1 *based upon a RK-scheme of order* $p\geq 1$, *with a coherent grid sequence with stepsize function* θ. *The sequence* $\{\Lambda_n\}_{n\in\mathbb{N}'}$ *of mappings* $\Lambda_n: E\to E^0$ *defined by*

$$(3.4.10)\qquad \Lambda_n y=\begin{pmatrix} 0 \\ \dfrac{y(t+\theta(t)/n)-y(t)}{\theta(t)/n}-\psi[f](\theta(t)/n,y(t))-y'(t)+f(y(t))\end{pmatrix}$$

is a local error mapping of $\mathfrak{M}$ *for the IVP 1 with the function* f *(cf. Def.* 1.3.1).

Furthermore, for an $IVP^{(J+1)}$ 1 this local error mapping possesses an asymptotic expansion to order J (cf. Def. 1.3.2) *the second component of which is given by*

$$(3.4.11) \qquad (\Lambda_n y)(t) = \sum_{j=1}^{J} \frac{\theta(t)^j}{n^j j!} \left[\frac{y^{(j+1)}(t)}{j+1} - \sum_{\lambda=1}^{n_{j+1}} \beta_\lambda^{(j+1)} \hat{\Phi}_\lambda^{(j+1)}(\psi) F_\lambda^{(j+1)}(y(t)) \right] + O(n^{-(J+1)})$$

for $y \in D_J := \{y \in E$: $y^{(J+1)}$ exists and is Lipschitz continuous, $y(t) \in B_0$[7]$\}$.

The asymptotic expansion (3.4.11) *is (J,p)-smooth for each $p \geq 1$ (cf. Def.* 1.3.4), *the associated domains $D_{J,k}$ are*

> $D_{J,k} := \{e \in E$: *e possesses (piecewise) a Lipschitz-continuous $(J-k+1)$st derivative. (The derivatives of e may possess step discontinuities whenever θ does.)*$\}$.

Proof. It is evident that (3.4.10) satisfies (1.3.1): For $y \in E$,

$$[\Delta_n^0(F + \Lambda_n)y](t_v) = \frac{y(t_{v-1} + h_v) - y(t_{v-1})}{h_v} - \Psi(h_v, y(t_{v-1}))$$
$$= [F_n \Delta_n y](t_v), \qquad v = 1(1)n.$$

The asymptotic expansion (3.4.11) follows immediately from the expansion (3.1.21) for $\Psi = \psi[f]$ and the smoothness assumptions on f since $y(t)$ is in the domain of the $F_\lambda^{(r)}$.

Furthermore, the $F_\lambda^{(j+1)}$ are $(J-j)$-times differentiable and the highest derivatives are Lipschitz-continuous in $B_R(z)$ as an elementary differential of order $j+1$ contains at most a j-th derivative of f. If $y(t) \in B_0$ then $\left(y(t) + \sum_{k=p}^{J} (1/n^k) e_k(t)\right) \in B_R(z)$ for sufficiently large n. Finally,

$$\frac{1}{n^k} \frac{e_k(t+1/n) - e_k(t)}{1/n} = \frac{1}{n^k} \sum_{j=0}^{J-k} \frac{1}{n^j} \frac{e_k^{(j+1)}(t)}{(j+1)!} + O(n^{-(J+1)})$$

holds if e_k possesses a Lipschitz continuous $(J-k+1)$st derivative. Thus the conditions for (J,p)-smoothness of (3.4.11) are satisfied, for any $p \geq 1$.

Naturally, if $\theta(t)$ possesses step discontinuities at finitely many $t \in [0,1]$, the other terms in the expansion (3.4.11) may also be discontinuous at these points; this explains the definition of $D_{J,k}$. □

---

[7] See Remark 1 after Lemma 3.1.1.

Theorem 3.4.5. *Consider the RK-method* $\mathfrak{M}$ *of Theorem* 3.4.4. *For an* $IVP^{(J+1)}$ *1*, $J \geq p$, *the global discretization error of* $\mathfrak{M}$ *possesses an asymptotic expansion to order J* (*cf. Def.* 1.3.3).

Proof. We have to check the various hypotheses of Theorem 1.3.1:

(i) The required stability follows from Theorem 3.3.1.

(ii) The required consistency and the asymptotic expansion of order J for the local error mapping follow from Theorems 3.3.5 and 3.4.4.

(iii) The smoothness requirements follow from Def. 2.1.2 and Theorem 3.4.4.

(iv) The solvability of the linearized IVP 1 is trivial under our smoothness assumptions.

It remains to check whether the e_j defined by (1.3.9)/(1.3.10) are in the domains $D_{J,j}$ defined in Theorem 3.4.4: The second components of the operators occuring in (1.3.9) and (1.3.10) have the following form in our case (see (2.1.5) and (3.4.11)):

$$[F'(z)e](t) = e'(t) - f'(z(t))e(t),$$

$$[F^{(m)}(z)e^m](t) = -f^{(m)}(z(t))e(t)^m, \qquad m \geq 2;$$

$$[\lambda_j(z)](t) = \frac{\theta(t)^j}{j!}\left[\frac{z^{(j+1)}(t)}{j+1} - \sum_{\lambda=1}^{n_{j+1}} \beta_\lambda^{(j+1)}\hat{\Phi}_\lambda^{(j+1)}(\psi)F_\lambda^{(j+1)}(z(t))\right],$$

$$[\lambda_j'(z)e](t) = \frac{\theta(t)^j}{j!}\left[\frac{e^{(j+1)}(t)}{j+1} - \sum_{\lambda=1}^{n_{j+1}} \beta_\lambda^{(j+1)}\hat{\Phi}_\lambda^{(j+1)}(\psi)(F_\lambda^{(j+1)})'(z(t))e(t)\right],$$

$$[\lambda_j^{(m)}(z)e^m](t) = -\frac{\theta(t)^j}{j!}\sum_{\lambda=1}^{n_{j+1}} \beta_\lambda^{(j+1)}\hat{\Phi}_\lambda^{(j+1)}(\psi)(F_\lambda^{(j+1)})^{(m)}(z(t))e(t)^m, \qquad m \geq 2.$$

For each $r \geq 1$, $F_\lambda^{(r)}$ contains at most an $(r-1)$st derivative of f and hence possesses $(J-r+1)$ Lipschitz continuous derivatives in $B_R(z)$ so that we can determine the smoothness of each of the terms occuring in (1.3.9) and (1.3.10).

For $j=p$, the right hand side of (1.3.10) consists simply of $-\lambda_p z$ and is thus (piecewise) $J-p$ times Lipschitz-continuously differentiable. The solution of (1.3.10), i.e., of

$$e_p(0) = 0, \qquad e_p' - f'(z(t))e_p = -[\lambda_p z](t),$$

is therefore in $D_{J,p}$. As in the example in Section 1.3.2, one can now proceed by induction to show that the right hand side of (1.3.10) is in $D_{J,j+1}$ and thus $e_j \in D_{J,j}$ for each $j = p(1)J$. This completes the proof. □

Remark. Although Theorem 3.4.5 permits the recursive construction of the e_j from (1.3.9)/(1.3.10)—assuming knowledge of z—its main importance lies in the assertion of the *existence* of an asymptotic expansion for the global discretization error as a basis for the application of the Richardson extrapolation technique.

Normally, none of the $e_j, j=p(1)J$, will vanish identically. However, we know from Theorem 1.3.2 that the expansion will be in *even* powers of n^{-1} if the expansion of the local error mapping has this property.

Theorem 3.4.6. *In the situation of Theorem 3.4.5, if the RK-method $\mathfrak{M}$ is based upon a symmetric RK-scheme ψ the asymptotic expansion of the global discretization error is in even powers of $1/n$.*

Proof. According to Theorem 3.2.13, if $\mathfrak{M}$ is based on a symmetric RK-scheme, the mappings F_n of the discretization $\mathfrak{M}$(IVP 1) may be written in the form ($\eta \in E_n$)

$$(3.4.12)\qquad [F_n\eta](t_\nu) = \begin{cases} \eta_0 - z_0, & \nu=0, \\ \dfrac{\eta_\nu - \eta_{\nu-1}}{h_\nu} - \dfrac{1}{2}[\Psi(h_\nu, \eta_{\nu-1}) + \Psi(-h_\nu, \eta_\nu)], & \nu=1(1)n. \end{cases}$$

Since the solution sequence $\{\zeta_n\}$ of $\mathfrak{M}$ (IVP 1) and its asymptotic expansion obviously cannot depend on the choice of $\{\Delta_n^0\}$ in the formal definition of $\mathfrak{M}$ we may change the definition of the Δ_n^0 without affecting the structure of the asymptotic expansion of $\zeta_n - \Delta_n z$ (cf. the discussion at the end of Section 1.3.3). If we replace the previous definition (3.3.1) of Δ_n^0 by

$$(3.4.13)\qquad \Delta_n^0 \begin{pmatrix} d_0 \\ d(t) \end{pmatrix}(t_\nu) = \begin{cases} d_0, & \nu=0, \\ d\left(\dfrac{t_{\nu-1}+t_\nu}{2}\right), & \nu=1(1)n, \end{cases}$$

we find that

(3.4.14)

$$\frac{y\left(t+\frac{\theta(t)}{2n}\right) - y\left(t-\frac{\theta(t)}{2n}\right)}{\frac{\theta(t)}{n}} - \frac{1}{2}\left[\Psi\left(\frac{\theta(t)}{n}, y\left(t-\frac{\theta(t)}{2n}\right)\right) + \Psi\left(-\frac{\theta(t)}{n}, y\left(t+\frac{\theta(t)}{2n}\right)\right)\right] - y'(t) + f(y(t))$$

becomes the second component of the local error mapping of $\mathfrak{M}$ (based upon a symmetric ψ) for IVP 1. Furthermore, it can be shown as in the proof of Theorem 3.4.4 that this error mapping possesses an asymptotic expansion to order J which is (J,p)-smooth.

Obviously, (3.4.14) is invariant w.r.t. the replacement of n by $-n$, hence its expansion in powers of $1/n$ can only contain even powers of $1/n$. An application of Theorem 1.3.2 completes the proof. □

Theorem 3.4.6 establishes the distinguishing feature of "symmetric RK-methods", i.e., uniform RK-methods based upon symmetric RK-schemes.

Examples. 1. Implicit trapezoidal method (cf. end of Section 3.2.5). For equidistant grids, i.e., $\theta(t)\equiv 1$, and $y\in D^{(J+1)}$, the second component (3.4.14) of the local error mapping takes the form $(h=1/n)$

$$\frac{y(t+h/2)-y(t-h/2)}{h}-\frac{1}{2}[f(y(t-h/2))+f(y(t+h/2))]-y'(t)+f(y(t))$$

$$=\sum_{j=1}^{\left[\frac{J}{2}\right]}\frac{h^{2j}}{2^{2j}(2j)!}\left[\frac{y^{(2j+1)}(t)}{2j+1}-\frac{d^{2j}}{dt^{2j}}f(y(t))\right].$$

2. The optimal order implicit m-stage RK-schemes of Theorem 3.3.11 are symmetric as asserted by Theorem 3.3.12. Thus the uniform RK-methods based upon these schemes possess asymptotic expansions in even powers of $1/n$ for the global discretization error.

Theorems 3.4.5 and 3.4.6 provide a firm foundation for the application of the Richardson extrapolation technique to uniform RK-methods with coherent grids for sufficiently smooth IVP 1. Note that the use of equidistant grids is *not* a prerequisite as is often presumed. It is only necessary that corresponding stepsize changes are made at the same t in each of the computations whose results are combined by Richardson extrapolation.

3.4.3 The Principal Term of the Global Discretization Error

Consider the application of a uniform RK-method $\mathfrak{M}$ based upon a RK-scheme ψ of order p with a coherent grid sequence with stepsize function θ to an IVP$^{(p+1)}$ 1. By Theorem 3.4.5, there exists an asymptotic expansion of the global discretization error

(3.4.15) $$\varepsilon_n=\zeta_n-\Delta_n z=n^{-p}\Delta_n e_p+O(n^{-(p+1)}),\qquad n\in\mathbb{N}',$$

where the function $e_p\in E$ is determined by the IVP 1

(3.4.16) $$e_p(0)=0$$
$$e_p'-f'(z(t))e_p=\theta(t)^p\varphi(z(t)),\qquad t\in[0,1],$$

according to (1.3.16), (2.1.5), and (3.4.4).

For sufficiently large n, i.e. for sufficiently small steps, the qualitative and quantitative behavior of the global discretization error of $\mathfrak{M}$ for a given IVP$^{(p+1)}$ 1 is largely determined by the function e_p according to (3.4.15). If approximate values of the principal error function at the $\zeta_n(t_\nu)$ can be found in conjunction with the computation of the $\zeta_n(t_\nu)$

themselves, a numerical integration of (3.4.16), with $z(t_\nu)$ replaced by $\zeta_n(t_\nu)$, normally provides a good estimation for ε_n. The problem of obtaining approximate values of the local discretization error or of the principal error function will be discussed in the subsequent section.

The other general approach to an estimation of ε_n is to solve in parallel the discretization for n and $n/2$ (or some other integer part of n) and to use polynomial Richardson extrapolation with $r=1$ at selected points of the wider grid. This results in (cf. Section 1.4.2)

$$n^{-p} e_p(t) = \frac{1}{2^p - 1} (\zeta_{\frac{n}{2}}(t) - \zeta_n(t)) + O(n^{-(p+1)}). \tag{3.4.17}$$

Both approaches may be used either to estimate the error of ζ_n or to improve the approximation to z, cf. the "paradox" discussed in Section 1.5.4.

For a scalar differential equation $y' - g(t)y = 0$, Henrici ([1], Section 2.2–9) uses the result of Lemma 3.4.3 to show that e_p satisfies

$$e_p(t) = \left(\int_0^t \theta(\tau)^p \hat{g}(\tau) d\tau \right) z(t), \qquad t \in [0,1], \tag{3.4.18}$$

where θ is again the stepsize function and $\hat{g}$ is from (3.4.9). That (3.4.18) satisfies (3.4.16) with $\varphi(z(t)) = \hat{g}(t) z(t)$ (see Lemma 3.4.3) is an immediate consequence of the form of the differential equation: $f'(z(t)) = g(t)$. The result extends to systems of the form (3.4.8) only if g commutes with the integral in (3.4.18) as is easily seen.

In simple cases, (3.4.18) may be used to predict points where $e_p = 0$ so that ζ_n is an $O(h^{p+1})$-approximation to z at these points. This may be important for Richardson extrapolation, cf. Section 1.5.4. An example is treated in Section 2.2–9 of Henrici [1].

Another special possibility for the estimation of the global discretization error arises when the principal error function φ is of such a structure that (3.4.16) becomes an *exact differential equation*; see Stetter [5]. (A system of s differential equations is called exact if it may be generated from

$$Q(t, y(t)) = \text{const.}, \qquad Q: \mathbb{R} \times \mathbb{R}^s \to \mathbb{R}^s, \tag{3.4.19}$$

by differentiation:

$$\frac{\partial Q}{\partial t}(t, y(t)) + \frac{\partial Q}{\partial y}(t, y(t)) y'(t) = 0. \tag{3.4.20}$$

The solution of (3.4.20) is obviously given by (3.4.19).)

Lemma 3.4.7. *For a given IVP 1 if there exists a function* $\Phi: \mathbb{R}^s \to \mathbb{R}^s$ *such that*

(3.4.21) $$\varphi(\xi) = \Phi'(\xi) f(\xi) - f'(\xi)\Phi(\xi), \qquad \xi \in B_0,$$

then the solution of

(3.4.22) $$\begin{aligned} &e_p(0) = \Phi(z_0), \\ &e_p' - f'(z(t))\, e_p = \varphi(z(t)), \qquad t \in [0,1], \end{aligned}$$

is given by

(3.4.23) $$e_p(t) = \Phi(z(t))$$

where z is the true solution of the IVP 1.

Proof. Introduction of (3.4.23) into the left-hand side of (3.4.22) yields

$$\frac{d}{dt}\Phi(z(t)) - f'(z(t))\Phi(z(t)) = \Phi'(z(t)) f(z(t)) - f'(z(t))\Phi(z(t))$$

since $z'(t) = f(z(t))$. □

Theorem 3.4.8. *Assume that the condition* (3.4.21) *of Lemma* 3.4.7 *is satisfied for the principal error function of some RK-scheme* ψ *of order* p *for some* $IVP^{(p+1)}$ *1 and that the mapping* Φ *is Lipschitz-continuous in* B_0.

If we modify the uniform RK-method $\mathfrak{M}$ *based upon* ψ, *with equidistant grids, by replacing its starting procedure* $\eta_0 = z_0$ *by*

(3.4.24) $$\eta_0 - z_0 - h^p \Phi(z_0) = 0$$

then the global discretization error of this new method $\mathfrak{M}'$ *for the* $IVP^{(p+1)}$ *1 satisfies*

(3.4.25) $$\varepsilon_n(t_\nu) = h^p \Phi(\zeta_n(t_\nu)) + O(h^{p+1}).$$

Proof. The second component of the local error mapping of $\mathfrak{M}'$ is identical to that of $\mathfrak{M}$ while the first component becomes $-h^p\Phi(z_0) = -(1/n^p)\Phi(z_0)$. Thus, the first component of λ_p in the asymptotic expansion of the local error mapping does not vanish but equals $-\Phi(z_0)$ so that (3.4.22) becomes the defining IVP 1 for the function e_p in (3.4.15) and $e_p(t_\nu) = \Phi(z(t_\nu))$ by (3.4.23). The replacement of $z(t_\nu)$ by $\zeta_n(t_\nu) = z(t_\nu) + O(h^p)$ does not invalidate (3.4.15) because of the Lipschitz-continuity of Φ. □

Under the assumptions of Theorem 3.4.8 we may convert the discretization method $\mathfrak{M}'$ which is convergent of order p for the $\text{IVP}^{(p+1)}$ 1 into a method of order $p+1$ by forming

(3.4.26) $$\bar{\zeta}_n(t_\nu) := \zeta_n(t_\nu) - h^p \Phi(\zeta_n(t_\nu)), \qquad t_\nu \in \mathbb{G}_n, \qquad n \in \mathbb{N}',$$

where ζ_n is the solution of $\mathfrak{M}'$ (IVP$^{(p+1)}$ 1). (In practice one will compute the $\bar{\zeta}_n(t_\nu)$ at selected grid points only.) It is clear from (3.4.25) that

$$\bar{\zeta}_n(t_\nu) = z(t_\nu) + O(h^{p+1}). \tag{3.4.27}$$

The whole approach becomes interesting because of the fact that there exist principal error functions which satisfy (3.4.21) for an *arbitrary* (sufficiently smooth) IVP 1: We need only take for Φ some linear combination of elementary differentials of f of order p, then $\Phi' f$ as well as $f' \Phi$ are linear combinations of elementary differentials of order $p+1$ so that

$$\Phi'(\xi) f(\xi) - f'(\xi) \Phi(\xi) = \sum_{\lambda=1}^{n_{p+1}} \alpha_\lambda^{(p+1)} F_\lambda^{(p+1)}(\xi). \tag{3.4.28}$$

Thus, if a RK-scheme of order p is chosen such that (3.4.1) matches (3.4.28), this RK-scheme satisfies the assumptions of Theorem 3.4.8 for any IVP$^{(p+1)}$ 1. The fact that there exist RK-schemes of order p with prescribed values for the $\hat{\Phi}_\lambda^{(p+1)}(\psi)$ has been shown by Butcher [3, Theorem 6.9]. However, most of these RK-schemes will have more complicated generating matrices than some schemes of order $p+1$ so that only special cases are of interest. (Note that (3.4.28) contains only n_p free parameters, viz. those of Φ.)

It is obvious that Theorem 3.4.8 provides the link between our analysis of the global discretization error and Butcher's RK-methods *of effective order* $p+1$ which were based on a combination $\chi \psi^n \chi^{-1}$ of RK-schemes, with ψ of order p and $\alpha(\chi)=0$ (see Section 3.3.5): If ψ has the principal error function (3.4.21) while $\chi[f](h,\xi) = h^{p-1} \Phi(\xi) + O(h^p)$ and $\chi^{-1}[f](h,\xi) = -h^{p-1} \Phi(\xi) + O(h^p)$ we have exactly the situation of Theorem 3.4.8:

$\eta_0' = z_0 + h\chi[f](h, z_0) = z_0 + h^p \Phi(z_0) + O(h^{p+1})$ becomes the starting value for our method $\mathfrak{M}'$ while

$$\bar{\zeta}_n(1) = \zeta_n(1) + h\chi^{-1}[f](h, \zeta_n(1)) = \zeta_n(1) - h^p \Phi(\zeta_n(1)) + O(h^{p+1})$$

effects the subtraction of $h^p e_p(1)$ so that $\bar{\zeta}_n(1)$ has an error of order $O(h^{p+1})$ only; see (3.4.26) and (3.4.27).

We can even see now that Butcher's "algebraic" approach is slightly more restrictive than it need be; it would have been sufficient in Def. 3.3.3 to require that there are two RK-schemes χ_s and χ_e such that $\chi_s \psi \chi_e$ is of order $\bar{p}$ and $\chi_s \chi_e = \bar{\psi}_0$ where $\bar{\psi}_0[f] = O(h^{\bar{p}-1})$ for any f. The following example is to indicate the wider scope of Theorem 3.4.8 in comparison with Theorem 3.3.13; it is not meant to be of practical significance.

Example. The RK-scheme generated by the explicit 3-stage generating matrix

$$\begin{pmatrix} 0 & 0 & 0 \\ \frac{1}{3} & 0 & 0 \\ 0 & \frac{1}{2} & 0 \\ 0 & 0 & 1 \end{pmatrix}$$

is of order 2 and has the principal error function $-\frac{1}{24}\{f^2\}$. Since $\{f^2\}=\{f\}'f-f'\{f\}$, we have (3.4.21) valid with $\Phi=-\frac{1}{24}\{f\}$ so that $\Phi(z(t))=-\frac{1}{24}z''(t)$.

The starting value $\eta_0'=z_0-(h^2/24)z''(0)$ may be computed by obtaining $z''(0)$ from the IVP 1 or by a RK-procedure $\eta_0'=z_0+(\alpha/24)[f(z_0)-f(z_0+(h/\alpha)f(z_0))]$ with any $\alpha\neq 0$. One can also construct a RK-procedure which produces $\zeta_n(h)=z(h)-(h^2/24)z''(h)+O(h^3)$, see Stetter [5].

For the correction at some t_ν, we may estimate $(h^2/24)z''(t)$ by forming (cf. (3.4.15))

$$\begin{aligned}\tfrac{1}{24}&[\zeta_n(t_{\nu+1})-2\zeta_n(t_\nu)+\zeta_n(t_{\nu-1})]\\ &=\tfrac{1}{24}[h^2z''(t_\nu)+h^2(e_2(t_{\nu+1})-2e_2(t_\nu)+e_2(t_{\nu-1}))+O(h^3)]\\ &=\frac{h^2}{24}z''(t_\nu)+O(h^3)\quad \text{as } e_2 \text{ is differentiable for an IVP}^{(3)}\,1.\end{aligned}$$

Thus the formation of the second difference of the values of ζ_n provides either a check on the accuracy or a means for improvement, as one desires.

3.4.4 Estimation of the Local Discretization Error

The two main purposes for which values of the local discretization error may be used are

a) to determine approximately the global discretization error (see Section 1.5.2 and beginning of Section 3.4.3),

b) to monitor the stepsize according to various stepsize control strategies.

In both cases one is scarcely interested in rigid bounds which would have to use bounds on derivatives of f and normally grossly overestimate the error. Instead, all practical algorithms content themselves with an estimation of the local discretization error in which only terms of the lowest power in the stepsize are considered.

The quantity normally used for stepsize control purposes is the generated contribution to the global discretization error in the current step, referred to a step of *unit* length, i.e. (see Lemma 3.4.1)

$$\text{(3.4.29)}\qquad \frac{1}{h_\nu}[\bar{\zeta}^{m+1}(h_\nu,\zeta_{\nu-1})-\bar{z}(h_\nu,\zeta_{\nu-1})]=\begin{cases}-l_n(t_\nu)\,[1+O(h_\nu^p)]\,,\\ h_\nu^p\,\varphi(z(t_{\nu-1}))+O(h_\nu^{p+1})\,,\\ h_\nu^p\,\varphi(\zeta_{\nu-1})+O(h_\nu^{p+1})\,.\end{cases}$$

The following simple but effective approach to the estimation of (3.4.29) has been suggested by various authors:

According to the theory of linear multistep schemes in Section 4.1 we may, for $k \geq [p/2]+1$, find coefficients $\alpha_\kappa, \beta_\kappa\ \kappa=0(1)k$, such that

$$
(3.4.30)\quad \begin{cases} \sum_{\kappa=0}^{k} \alpha_\kappa = 0\,, \qquad \sum_{\kappa=0}^{k} \beta_\kappa = 1\,, \\ \sum_{\kappa=0}^{k} [\alpha_\kappa y(t-kh+\kappa h) - h\beta_\kappa y'(t-kh+\kappa h)] = \begin{cases} O(h^{p+2}) \\ O(h^r) \end{cases} \\ \qquad \text{if } y \text{ possesses} \begin{cases} p+2 \\ r<p+2 \end{cases} \quad \text{Lipschitz continuous derivatives in } (t-kh,t). \end{cases}
$$

Theorem 3.4.9. *Let $\zeta_n \in E_n$ be the solution of the discretization of an $IVP^{(p+2)}1$ generated by a uniform RK-method of order $p \geq 1$, with a grid $\mathbb{G}_n$. If the steps in $\mathbb{G}_n$ are constant in $[t_{\nu-k}, t_\nu]$ then*

(3.4.31)

$$
h_\nu^{p+1}\varphi(z(t_{\nu-1})) = \sum_{\kappa=0}^{k} [\alpha_\kappa \zeta_n(t_{\nu-k+\kappa}) - h_\nu \beta_\kappa f(\zeta_n(t_{\nu-k+\kappa}))] + O(h_\nu^{p+2})
$$

where the $\alpha_\kappa, \beta_\kappa$ are from (3.4.30).

Proof. By Theorem 3.4.5, with $J=p+1$,

$$
\zeta_n(t_\nu) = z(t_\nu) + \frac{1}{n^p} e_p(t_\nu) + \frac{1}{n^{p+1}} e_{p+1}(t_\nu) + O(n^{-(p+2)})\,;
$$

z, e_p, and e_{p+1} possess $p+2, 2$, and 1 Lipschitz-continuous derivatives in $(t_{\nu-k}, t_\nu)$. Hence we obtain for the right hand side of (3.4.31)

$$
\begin{aligned}
&\sum_{\kappa=0}^{k} [\alpha_\kappa z(t_{\nu-k+\kappa}) - h_\nu \beta_\kappa f(z(t_{\nu-k+\kappa}))] \\
&\quad + \frac{1}{n^p} \sum_{\kappa=0}^{k} [\alpha_\kappa e_p(t_{\nu-k+\kappa}) - h_\nu \beta_\kappa f'(z(t_{\nu-k+\kappa}))\, e_p(t_{\nu-k+\kappa})] \\
&\quad + \frac{1}{n^{p+1}} \sum_{\kappa=0}^{k} \alpha_\kappa e_{p+1}(t_{\nu-k+\kappa}) + O(n^{-(p+2)}) \\
&= \frac{1}{n^p} \sum_{\kappa=0}^{k} [\alpha_\kappa e_p(t_{\nu-k+\kappa}) - h_\nu \beta_\kappa e_p'(t_{\nu-k+\kappa})] + h_\nu^{p+1} \sum_{\kappa=0}^{k} \beta_\kappa \varphi(z(t_{\nu-k+\kappa})) \\
&\quad + \frac{1}{n^{p+1}} \left(\sum_{\kappa=0}^{k} \alpha_\kappa \right) e_{p+1}(t_\nu) + O(n^{-(p+2)}) \\
&= h_\nu^{p+1} \varphi(z(t_{\nu-1})) + O(n^{-(p+2)})
\end{aligned}
$$

by virtue of (3.4.30), (3.4.16), the smoothness of z, e_p, and e_{p+1}, and the Lipschitz continuity of φ for an $\text{IVP}^{(p+2)}1$ (see (3.4.1)). □

Since values of $f(\zeta_n(t_\nu))$ are normally computed anyway (they *must* be computed in an explicit RK-procedure), the application of Theorem 3.4.9 requires simply that the values of $\zeta_n(t_\nu)$ and $f(\zeta_n(t_\nu))$ are saved over the subsequent k steps. If the $f(\zeta_n(t_\nu))$ do not occur in the RK-procedure used, as may be the case with some implicit RK-procedures, the expression (3.4.30) can be modified such that available f-values may be used. Thus the evaluation of the right-hand side of (3.4.31) is very "cheap" computationally.

The only drawback of the procedure of Theorem 3.4.9 is that it can be used only after k steps of equal length have been made. (One could adjust (3.4.30) to unequal steps but the derivatives of e_p are no longer continuous in this case, see (3.4.16).) Thus it cannot be used if it is desirable to monitor and possibly change the stepsize after each step.

Example. $p=3,\ k=2,\ h_{\nu-1}=h_\nu$:

$$h_\nu^4\,\varphi(z(t_{\nu-1}))\approx\frac{1}{2}\left[\zeta_n(t_\nu)-\zeta_n(t_{\nu-2})\right]-\frac{h_\nu}{6}\left[f(\zeta_n(t_\nu))+4f(\zeta_n(t_{\nu-1}))+f(\zeta_n(t_{\nu-2}))\right].$$

The following different approach is more in the spirit of RK-methods: For a given p-th order m-stage RK-method, supplement the computation of $\zeta_\nu=\zeta_n(t_\nu)$ by the computation of a value $\bar{\zeta}_\nu$ through a RK-procedure of order greater than p, with increment function $\bar{\Psi}$, so that (for an IVP$^{(p+1)}$ 1)

$$\bar{\zeta}_\nu=\zeta_{\nu-1}+h_\nu\bar{\Psi}(h_\nu,\zeta_{\nu-1})=\bar{z}(h_\nu,\zeta_{\nu-1})+O(h_\nu^{p+2}),$$

$$\text{(3.4.32)}\qquad h_\nu^{p+1}\,\varphi(z(t_{\nu-1}))=\zeta_\nu-\bar{z}(h_\nu,\zeta_{\nu-1})+O(h_\nu^{p+2})=\zeta_\nu-\bar{\zeta}_\nu+O(h_\nu^{p+2}).$$

The approach is practical if the m-stage, resp. $\bar{m}$-stage generating matrices A and $\bar{A}$ of the RK-procedures *are almost identical.* Such *pairs* of RK-procedures, with $\bar{a}_{\mu\nu}=a_{\mu\nu}$ for $\mu=1(1)m$, have been constructed by England [1] and, more systematically, by Fehlberg [1], [2], [3]. The latter gives examples of explicit RK-procedure pairs with the following values for m and $\bar{m}$

p	1	2	3	4	5	6	7	8
m	1	2	3, 4	4, 5	6	8	11	15
$\bar{m}$	2	3	5	6	8	10	13	17

The use of $m=p+1$ for $p=3$ and 4 makes it possible to have very small values for the coefficients of the elementary differentials in (3.4.1). E.g., for $p=4$, the coefficients of the principal error function of the RK-procedure determined by A in Fehlberg's $m=5$, $\bar{m}=6$ pair are only 1—15% of those of the classical RK 4. Thus the use of this method might be economical (in spite of $m=5$) even if $\bar{\zeta}_\nu$ is computed only infrequently.

One may even find pairs such that

$$\bar{a}_{\mu\nu}=a_{\mu\nu}, \qquad \mu=1(1)m+1, \qquad \bar{m}=m+1.$$

In this case, no additional function evaluations are needed for the computation of $\bar{\zeta}_\nu$ (if not the current step is rejected by the stepsize control) since $\zeta_{\nu+1}^0 = \zeta_\nu^{m+1}$.

Example. From Fehlberg [3]; $p=2$, $m=3$, $\bar{m}=4$:

$$A, \bar{A} = \left(\begin{array}{ccc|c} 0 & 0 & 0 & 0 \\ \frac{1}{4} & 0 & 0 & 0 \\ -\frac{189}{800} & \frac{729}{800} & 0 & 0 \\ \frac{241}{891} & \frac{1}{33} & \frac{650}{891} & 0 \\ \hline \frac{533}{2106} & 0 & \frac{800}{1053} & -\frac{1}{78} \end{array}\right).$$

The principal error function of the RK-scheme generated by A is

$$\varphi(\xi) = \tfrac{1}{2112}[\{f^2\} - \{_2 f\}_2](\xi).$$

The classical but computationally expensive approach to the estimation of the local discretization error is a local version of Richardson extrapolation: Subdivide the step from $t_{\nu-1}$ to t_ν into 2 steps of length $h_\nu/2$ each and compute $\bar{\zeta}_\nu$ with the original RK-procedure (i.e. take $\bar{\Psi} = (1/2\psi)^2[f]$):

Theorem 3.4.10. *Under the conditions of Theorem* 3.4.9, *if*

$$\bar{\zeta}_\nu := \zeta_{\nu-1} + \frac{h_\nu}{2}\left[\Psi\left(\frac{h_\nu}{2}, \zeta_{\nu-1}\right) + \Psi\left(\frac{h_\nu}{2}, \zeta_{\nu-1} + \frac{h_\nu}{2}\Psi\left(\frac{h_\nu}{2}, \zeta_{\nu-1}\right)\right)\right]$$

then

$$h_\nu^{p+1}\varphi(z(t_{\nu-1})) = \frac{2^p}{2^p-1}(\zeta_\nu - \bar{\zeta}_\nu) + O(h_\nu^{p+2}). \tag{3.4.33}$$

Proof. By Theorem 3.4.5, $J=p+1$, we have

$$\zeta_\nu = z(t_\nu) + \frac{1}{n^p} e_p(t_\nu) + \frac{1}{n^{p+1}} e_{p+1}(t_\nu) + O(n^{-(p+2)}),$$

$$\bar{\zeta}_\nu = z(t_\nu) + \frac{1}{n^p} \bar{e}_p(t_\nu) + \frac{1}{n^{p+1}} \bar{e}_{p+1}(t_\nu) + O(n^{-(p+2)}),$$

where, by (3.4.16),

$$(e_p - \bar{e}_p)(t_{\nu-1}) = 0,$$

$$(e_p - \bar{e}_p)' - f'(z(t))(e_p - \bar{e}_p) = \left(1 - \frac{1}{2^p}\right)\theta(t_{\nu-1})^p \varphi(z(t)), \qquad t \in [t_{\nu-1}, t_\nu];$$

this implies

$$(e_p - \bar{e}_p)(t_\nu) = h_\nu\left(1 - \frac{1}{2^p}\right)\theta(t_{\nu-1})^p \varphi(z(t_{\nu-1})) + O(h_\nu^2).$$

$(e_{p+1} - \bar{e}_{p+1})(t_\nu) = O(h_\nu)$ by a similar consideration. □

3.5 Strong Stability of RK-methods

In the definition of an $\{\mathrm{IVP}\,1\}_{\mathbb{T}}$ we distinguished the role of t from that of the dependent variables as t grows without bound, cf. Def. 2.1.3 and Section 2.3.1. Accordingly, we will in this section regard the increment function of a RK-method—or, more precisely, of a $\mathbb{T}$-sequence of RK-methods, cf. Section 2.3.2—as a function $\Psi: \mathbb{R} \times [0, \infty) \times \mathbb{R}^s \to \mathbb{R}^s$ and denote its values by $\Psi(h, t, \xi)$. Naturally, these values are defined by (3.1.9) via the interpretation (2.1.2) of a non-autonomous system. Similarly, the domains $B_R(z)$ and B_0 in $\mathbb{R}^s$ which we have used previously (see, e.g., Section 3.1.1) now become domains in $[0, \infty) \times \mathbb{R}^s$ by means of the same process, and we now have $B_R(z) := \{(\tau, \xi): \tau \geq 0,\ \|(\tau, \xi) - (t, z(t))\| < R,\ t \in [0, \infty)\}$.

Since our $\mathbb{T}$-sequences consist of ordinary IVP 1, resp. RK-methods, it follows from Lemma 3.1.1 and the uniformity of the Lipschitz condition (2.3.1) that Ψ is well-defined for $|h| \leq h_0$, $(t, \xi) \in B_0$. Similarly, any smoothness assumptions on f in $B_R(z)$ imply analogous smoothness properties for Ψ in $[-h_0, h_0] \times B_0$ as stated in Lemma 3.1.2, and the derivatives of Ψ are *uniformly* bounded in this domain if the same is true for the derivatives of f in $B_R(z)$.

3.5.1 Strong Stability for Sufficiently Large n

We first demonstrate that for a properly posed $\{\mathrm{IVP}\,1\}_{\mathbb{T}}$ it is possible by means of RK-methods to compute approximations to z which are arbitrarily good uniformly for all $t \geq 0$.

Theorem 3.5.1. *Consider a totally stable $\{\mathrm{IVP}\,1\}_{\mathbb{T}}$ and an arbitrary uniform RK-method $\mathfrak{M}$ of (minimal) order 1; let ζ_n be the solution of $\mathfrak{M}_{\mathbb{T}}(\{\mathrm{IVP}\,1\}_{\mathbb{T}})$ for $n \in \mathbb{N}'$. For each given $\rho < R/2$, $\rho > 0$, there is an $n_0(\rho)$ such that ζ_n exists and*

$$\|\zeta_n(t_\nu) - z(t_\nu)\| < \rho \quad \text{for all } t_\nu \in \mathbb{G}_n \tag{3.5.1}$$

for each $n \geq n_0$, $n \in \mathbb{N}'$.

Proof. At first take n_0 sufficiently large for $\Psi(h_\nu, t_\nu, \xi)$ to be defined for $(t_\nu, \xi) \in B_0$ for $n \geq n_0$. Consider the continuous and piecewise differentiable (viz. piecewise linear) function $\tilde{z}_n$ defined by

$$\tilde{z}_n(t) := \zeta_n(t_{\nu-1}) + (t - t_{\nu-1})\,\Psi(h_\nu, t_{\nu-1}, \zeta_n(t_{\nu-1})) \tag{3.5.2}$$

for $t \in [t_{\nu-1}, t_\nu)$, where $t_\nu \in \mathbb{G}_n$, $n \geq n_0$, $n \in \mathbb{N}'$.

Trivially $\tilde{z}_n(0)=z_0$ and in each interval $(t_{\nu-1}, t_\nu)$

$$\begin{aligned}\tilde{z}_n'(t) &= \Psi(h_\nu, t_{\nu-1}, \zeta_n(t_{\nu-1}))\\ &= f(t,\tilde{z}_n(t)) + [f(t_{\nu-1},\zeta_n(t_{\nu-1})) - f(t,\tilde{z}_n(t))]\\ &\quad + [\Psi(h_\nu, t_{\nu-1}, \zeta_n(t_{\nu-1})) - \Psi(0, t_{\nu-1}, \zeta_n(t_{\nu-1}))]\\ &=: f(t,\tilde{z}_n(t)) + \delta_n(t);\end{aligned}$$

we have used the fact that order 1 implies $\Psi(0,t,\xi)=f(t,\xi)$ (see Theorem 3.3.5). According to the assumption of total stability (Def. 2.3.1) there exists a $\overline{\delta}_1(\rho)$ such that $\|\delta_n(t)\| < \overline{\delta}_1(\rho)$ implies the existence of $\tilde{z}_n$ and $\|\tilde{z}_n(t)-z(t)\| < \rho$ for all t which is equivalent to (3.5.1).

By (3.5.2) and (2.3.1),

$$\begin{aligned}&\|f(t_{\nu-1},\zeta_n(t_{\nu-1})) - f(t,\tilde{z}_n(t))\|\\ &\le L\left\|\begin{pmatrix} t_{\nu-1}-t \\ \zeta_n(t_{\nu-1})-\tilde{z}_n(t)\end{pmatrix}\right\| < L\left\|\begin{pmatrix} h_\nu \\ h_\nu\,\Psi(h_\nu, t_{\nu-1},\zeta_n(t_{\nu-1}))\end{pmatrix}\right\|\\ &\le M_1 h_\nu \le M_1 c_2/n \quad \text{(see (2.1.11))},\end{aligned}$$

since Ψ is bounded by $\|A\|_{\max} M_f$ in $[-h_0, h_0]\times B_0$ according to (3.1.9) and (2.3.3), A being the generating matrix of the RK-method. By (3.1.9) and (3.1.8) we obtain, using maximum norms throughout,

$$\begin{aligned}&\|\Psi(h_\nu, t_{\nu-1},\zeta_n(t_{\nu-1})) - \Psi(0, t_{\nu-1},\zeta_n(t_{\nu-1}))\|\\ &\le \|A\|\, L\, \|\overline{\zeta}(h_\nu, t_{\nu-1},\zeta_n(t_{\nu-1})) - \mathbf{e}\,\zeta_n(t_{\nu-1})\|\\ &\le \|A\|\, L h_\nu \|A\|\, M_f \le M_2 c_2/n\,.\end{aligned}$$

Hence it is sufficient to have $n_0(\rho) > (M_1+M_2)c_2/\overline{\delta}_1(\rho)$ to achieve (3.5.1). (It is clear that ζ_n cannot leave B_0 since $\rho < R/2$.) □

Remark. The proof applies equally to non-uniform RK-methods if we replace $\|A\|$ by $\max_i(\|A_i\|)$ where A_i are the generating matrices of the various RK-schemes used.

The uniform Lipschitz-continuity and boundedness of f in $B_R(z)$ is essential: Consider the scalar differential equation

$$y' + t\,y = 0 \tag{3.5.3}$$

which is exponentially stable, since

$$|\overline{e}(t;\delta_0,t_0)| \le \delta_0\, e^{\frac{1}{2}}\, e^{-(t-t_0)} \quad \text{for } t \ge t_0,$$

and hence totally stable. However, the Euler discretization of (3.5.3) gives

$$\eta_\nu = (1 - h_\nu t_{\nu-1})\,\eta_{\nu-1}$$

and $\lim_{\nu\to\infty} |\eta_\nu| = \infty$ even for arbitrarily small h (except when the grid is chosen to produce $h_\nu t_{\nu-1} = 1$ at a certain ν). Note that (3.5.3) does *not* define an $\{\text{IVP } 1\}_{\mathbb{T}}$ according to our assumptions!

Corollary 3.5.2. *In the situation of Theorem* 3.5.1, *consider a computational solution* $\tilde{\zeta}_n$ *of* $\mathfrak{M}_{\mathbb{T}}(\{\text{IVP } 1\}_{\mathbb{T}})$, *cf. Section* 1.5.1. *For each* $\rho < R/2$, $\rho > 0$, *there exists a* $\tilde{\delta}(n, \rho)$ *such that*

$$\|\tilde{\zeta}_n(t_\nu) - z(t_\nu)\| < \rho \quad \text{for all } t_\nu \in \mathbb{G}_n$$

whenever $n > n_0(\rho)$ *(see Theorem* 3.5.1) *and the local computing error* ρ_n *satisfies* $\|\rho_n(t_\nu)\| < \tilde{\delta}(n, \rho)$ *for all* ν.

Proof. We simply take

$$\tilde{z}_n(t) := \tilde{\zeta}_n(t_{\nu-1}) + (t - t_{\nu-1})[\Psi(h_\nu, t_{\nu-1}, \tilde{\zeta}_n(t_{\nu-1})) + \rho_n(t_\nu)]$$

and follow the same argument as in the proof of Theorem 3.5.1. □

Remarks. 1. Our (implicit) assumption that the computing error does not affect the t_ν is unrealistic when $t_\nu \to \infty$. However, if f does not depend upon t, or depends sufficiently slightly on it for large t, this assumption may be discarded.

2. Corollary 3.5.2 could also have been formulated thus: In the situation of Theorem 3.5.1, the difference equations of $\mathfrak{M}_{\mathbb{T}}(\{\text{IVP } 1\}_{\mathbb{T}})$ are totally stable for sufficiently large n.

Theorem 3.5.3. *Consider an exponentially stable* $\{\text{IVP } 1\}_{\mathbb{T}}$ *with differentiable* f *and a uniform Lipschitz condition for* f_y *in* $B_R(z)$. *For each uniform RK-method* $\mathfrak{M}$ *of (minimal) order* 1 *there exists an* n_0 *such that the difference equations of* $\mathfrak{M}_{\mathbb{T}}(\{\text{IVP } 1\}_{\mathbb{T}})$ *are exponentially stable for each* $n \geq n_0$, $n \in \mathbb{N}'$.

Proof. For some $n \in \mathbb{N}'$ which is sufficiently large for ζ_n to exist and for $(t_\nu, \zeta_n(t_\nu))$ to be in B_0 for all ν (cf. Theorem 3.5.1), let $\{\varepsilon_\nu\}$ be the solution of

$$\frac{\varepsilon_\nu - \varepsilon_{\nu-1}}{h_\nu} - [\Psi(h_\nu, t_{\nu-1}, \zeta_n(t_{\nu-1}) + \varepsilon_{\nu-1}) - \Psi(h_\nu, t_{\nu-1}, \zeta_n(t_{\nu-1}))] = 0 \tag{3.5.4}$$

for some given initial perturbation δ_0 at t_{ν_0}. The linear interpolant e_n of the ε_ν satisfies in each interval $(t_{\nu-1}, t_\nu)$, $\nu > \nu_0$,

$$\begin{aligned} e_n'(t) &= \Psi(h_\nu, t_{\nu-1}, \zeta_n(t_{\nu-1}) + \varepsilon_{\nu-1}) - \Psi(h_\nu, t_{\nu-1}, \zeta_n(t_{\nu-1})) \\ &= g(t, e_n(t)) + k_n(t, e_n(t)), \end{aligned} \tag{3.5.5}$$

with (see (2.3.5))

$$\begin{aligned} k_n(t, x) = {} & [\Psi(h_\nu, t_{\nu-1}, \zeta_n(t_{\nu-1}) + x_{\nu-1}) - \Psi(h_\nu, t_{\nu-1}, \zeta_n(t_{\nu-1}))] \\ & - [f(t, z(t) + x) - f(z(t))]. \end{aligned}$$

Here v is chosen such that $t \in [t_{v-1}, t_v)$ and x_{v-1} is a function of both t and x defined implicitly by

$$x = x_{v-1} + (t - t_{v-1})[\Psi(h_v, t_{v-1}, \zeta_n(t_{v-1}) + x_{v-1}) - \Psi(h_v, t_{v-1}, \zeta_n(t_{v-1}))].$$

It is easy to show by the method used in the proof of Lemma 3.1.1 that x_{v-1} is well defined for $h_v \leq h_0$ and $\|x\| < r$, with r sufficiently small, and that there is a constant K such that

$$\|x_{v-1} - x\| \leq h_v K \|x\|. \tag{3.5.6}$$

We now rewrite $k_n(t,x)$ in the following form[8]:

$$\begin{aligned} k_n(t,x) &= \int_0^1 \Psi'(h_v, t_{v-1}, \zeta_n(t_{v-1}) + \vartheta x_{v-1}) x_{v-1} \, d\vartheta - \int_0^1 f'(t, z(t) + \vartheta x) x \, d\vartheta \\ &= \int_0^1 \Psi'(h_v, t_{v-1}, \zeta_n(t_{v-1}) + \vartheta x_{v-1})(x_{v-1} - x) \, d\vartheta \\ &+ \int_0^1 [\Psi'(h_v, t_{v-1}, \zeta_n(t_{v-1}) + \vartheta x_{v-1}) - \Psi'(0, t_{v-1}, \zeta_n(t_{v-1}) + \vartheta x_{v-1})] x \, d\vartheta \\ &+ \int_0^1 [f'(t_{v-1}, \zeta_n(t_{v-1}) + \vartheta x_{v-1}) - f'(t, z(t) + \vartheta x)] x \, d\vartheta; \end{aligned}$$

the existence, uniform boundedness, and uniform Lipschitz continuity of Ψ' in $[-h_0, h_0] \times B_0$ follow from the remarks at the start of Section 3.5, and we have used $\Psi'(0,t,x) = f'(t,x)$ for consistent RK-methods.

From (3.5.6) it is clear that the norm of the first member of $k_n(t,x)$ may be estimated uniformly by $K_1 h_v \|x\|$; an estimate $K_2 h_v \|x\|$ for the norm of the second member is obtained from (3.1.8), (3.1.9), and the assumed Lipschitz condition in a manner similar to that used for the second member of $\delta_n(t)$ in the proof of Theorem 3.5.1 (we omit the details). For the third member, the norm is smaller than

$$L' \left\| \begin{pmatrix} t_{v-1} - t \\ \zeta_n(t_{v-1}) - z(t) + \vartheta(x_{v-1} - x) \end{pmatrix} \right\| \|x\|;$$

but $|t_{v-1} - t| < h_v$, $\|\zeta_n(t_{v-1}) - z(t)\| = o(1)$ by Theorem 3.5.1, and $\|\vartheta(x_{v-1} - x)\| = O(h_v)$ by (3.5.6).

Hence we do possess an estimate $\|k_n(t,x)\| \leq M_n \|x\|$ uniformly for $t \in [t_{v0}, \infty)$, $t_{v0} \geq 0$, and $\|x\| < r$ where M_n can be made arbitrarily small by taking n sufficiently large and r sufficiently small. By Theorem 2.3.3, it therefore follows that the equilibrium of (3.5.5) is exponentially stable for sufficiently large n. By the construction of (3.5.5) this trivially implies the exponential stability of the equilibrium of (3.5.4) (see Def. 2.3.7). □

[8] The prime denotes differentiation w.r.t. the last argument.

Remark. The remark after Theorem 3.5.1 applies again.

Theorem 3.5.3 shows that each consistent RK-method is strongly exponentially stable for sufficiently large n w.r.t. the class of $\{\text{IVP } 1\}_{\mathbb{T}}$ with a uniformly bounded and Lipschitz continuous derivative f_y, cf. Def. 2.3.10. This class is rather large and contains most of the differential equations which arise in practice; however, it does not include problems which may get arbitrarily stiff like the one discussed after Theorem 3.5.1.

With RK-discretizations the constant a in the exponential stability estimate (2.3.31) will actually be 1 in many cases. For a linear IVP 1 this happens when $\Psi'(h_\nu, t_{\nu-1}, \zeta_n(t_{\nu-1}))$ satisfies

$$\|I + h_\nu \Psi'(h_\nu, t_{\nu-1}, \zeta_n(t_{\nu-1}))\| \le e^{-\mu h_\nu}, \quad \mu > 0, \tag{3.5.7}$$

w.r.t. the norm used in (2.3.31). If $\Psi'(h_\nu, t_{\nu-1}, \zeta_n(t_{\nu-1}))$ is a normal matrix and the Euclidean norm is used, the satisfaction of (3.5.7) depends solely on the position of the eigenvalues of Ψ'. We will return to this subject in Section 3.5.5.

Corollary 3.5.4. *In the situation of Theorem* 3.5.3, *if* $a=1$ *in the exponential stability estimate* (2.3.31) *for* (3.5.4), *or if* (3.5.4) *is linear, then there exists a constant* S *such that, for each* $n \ge n_0$, $n \in \mathbb{N}'$, *the solution* $\tilde{\zeta}_n$ *of*

$$\begin{cases} \tilde{\zeta}_n(0) - z_0 = \delta_0, \\ \dfrac{\tilde{\zeta}_n(t_\nu) - \tilde{\zeta}_n(t_{\nu-1})}{h_\nu} - \Psi(h_\nu, t_{\nu-1}, \tilde{\zeta}_n(t_{\nu-1})) = \delta_\nu, \quad \nu > 0, \quad t_\nu \in \mathbb{G}_n, \end{cases} \tag{3.5.8}$$

satisfies

$$\|\tilde{\zeta}_n(t_\nu) - \zeta_n(t_\nu)\| \le S \left[\sup_\nu \|\delta_\nu\| + \|\delta_0\|\right] \tag{3.5.9}$$

if the right-hand side of (3.5.9) *remains less than* r.

Proof. Denote the solution of (3.5.4), with an initial perturbation ξ at t_{ν_0}, by $\varepsilon_\nu^{\nu_0}(\xi)$. Then

$$\tilde{\zeta}_n(t_\nu) = \zeta_n(t_\nu) + h_\nu \delta_\nu + \varepsilon_\nu^{\nu-1}(h_{\nu-1}\delta_{\nu-1} + \varepsilon_{\nu-1}^{\nu-2}(h_{\nu-2}\delta_{\nu-2} + \cdots + \varepsilon_2^1(h_1\delta_1 + \varepsilon_1^0(\delta_0))\ldots)).$$

With $\|\varepsilon_\nu^{\nu-1}(\xi)\| \le \|\xi\| \, e^{-\mu h_\nu}$ we obtain, for each $\nu \ge 0$,

$$\|\tilde{\zeta}_n(t_\nu) - \zeta_n(t_\nu)\| \le \sum_{\lambda=1}^{\nu} h_\lambda \|\delta_\lambda\| \exp\left(-\mu \sum_{\lambda'=\lambda+1}^{\nu} h_{\lambda'}\right) + \|\delta_0\| \exp(-\mu t_\nu)$$

which implies (3.5.9), with $S = \max((1 - \exp(-\mu c_1/n_0))^{-1} c_2/n_0, 1)$.

In the linear case, we may trivially superpose the effects of the various δ_ν. □

Of course, (3.5.9) is simply a quantitative version of the fact that (3.5.8) is totally stable which follows from Theorem 3.5.3 via Theorem 2.3.7, cf. Def. 2.3.6. It is, however, interesting to note that (3.5.9) is an extension of the usual stability estimate on a finite interval to the infinite interval $[0, \infty)$; compare, e.g., Def. 1.1.10 (r was called the stability threshold there). For discretizations of an arbitrary $(\text{IVP 1})_T$ the size of S in (3.5.9) depends on the length T of the interval and grows without bound as $T \to \infty$ for any fixed finite n. RK-discretizations of an exponentially stable $\{\text{IVP 1}\}_{\mathbb{T}}$ are distinguished by having S uniformly bounded as $T \to \infty$ for each sufficiently large n.

3.5.2 Strong Stability for Arbitrary n

In the previous section we have shown that it is possible to compute arbitrarily accurate approximations to the true solution of a totally stable $\{\text{IVP 1}\}_{\mathbb{T}}$ by means of a RK-method, provided that the discretization parameter n is taken sufficiently large, i.e. the steps are sufficiently small. With a truly strongly stable discretization method $\mathfrak{M}$, however, a computational solution $\tilde{\zeta}_n$ of $\mathfrak{M}_{\mathbb{T}}(\{\text{IVP 1}\}_{\mathbb{T}})$ should remain within a prescribed vicinity of the solution z of the totally stable $\{\text{IVP 1}\}_{\mathbb{T}}$ for all $t_\nu \in \mathbb{G}_n$ if the local discretization error and the local computing error are sufficiently small, *independently of n*.

From the attempts to solve numerically stiff systems of differential equations it is well-known that most RK-methods do not possess this property, not even for linear IVP 1 with constant coefficients (see the following section for a detailed analysis). Instead, one normally has stability restrictions on the size of the steps as well as accuracy restrictions, and the former may be more severe than the latter.

Furthermore, as soon as our steps become larger than the quantity h_0 defined in Section 3.1.1 we can no longer be sure of the *existence of a unique solution* of the discretization in $B_R(z)$. Thus we have to assume that the definition of the local solution $\bar{\zeta}(h, t, \xi)$ (see Def. 3.1.3) of the RK-procedure for the IVP 1 under consideration may be extended to $|h| > h_0$ for $(t, \xi) \in B_{R'}(z)$, $0 < R' \leq R$. In what follows, Ψ will denote the increment function defined via (3.1.9) by this fixed though not necessarily unique extension of the local solution.

If we succeed in establishing the total stability of the equilibrium of the difference equation

$$(3.5.10) \qquad \frac{\varepsilon_\nu - \varepsilon_{\nu-1}}{h_\nu} - [\Psi(h_\nu, t_{\nu-1}, z(t_{\nu-1}) + \varepsilon_{\nu-1}) - \Psi(h_\nu, t_{\nu-1}, z(t_{\nu-1}))] = 0$$

for arbitrary $n \in \mathbb{N}'$, the situation is quite simple:

Theorem 3.5.5. *Consider an* $\{\mathrm{IVP}\ 1\}_{\mathbb{T}}$ *with solution* z *such that the equilibrium of* (3.5.10) *is totally stable for arbitrary* $n \in \mathbb{N}'$, *where* Ψ *is the (extended) increment function of some RK-method* $\mathfrak{M}$. *Let* $\bar{\delta}_0(\rho)$ *and* $\bar{\delta}_1(\rho)$, $0<\rho<r$, *be the functions from the definition of total stability. Then, for any given* $n \in \mathbb{N}'$, *a computational solution* $\tilde{\zeta}_n$ *of* $\mathfrak{M}_{\mathbb{T}}(\{\mathrm{IVP}\ 1\}_{\mathbb{T}})$ *exists and satisfies*

$$\|\tilde{\zeta}_n(t_\nu)-z(t_\nu)\| < \rho \quad \text{for all } t_\nu \in \mathbb{G}_n$$

if the local discretization error l_n *and the local computing error* ρ_n *satisfy*

$$\|l_n(t_\nu)-\rho_n(t_\nu)\| < \bar{\delta}_1(\rho) \quad \text{for all } t_\nu \in \mathbb{G}_n,\ \nu>0,$$

and if $\|\tilde{\zeta}_n(0)-z_0\| < \bar{\delta}_0(\rho)$.

Proof. An immediate consequence of

$$\frac{\tilde{\zeta}_n(t_\nu)-\tilde{\zeta}_n(t_{\nu-1})}{h_\nu} - \Psi(h_\nu, t_{\nu-1}, \tilde{\zeta}_n(t_{\nu-1})) = \rho_n(t_\nu),$$

$$\frac{z(t_\nu)-z(t_{\nu-1})}{h_\nu} - \Psi(h_\nu, t_{\nu-1}, z(t_{\nu-1})) = l_n(t_\nu),$$

and the definition of total stability. □

For an $\{\mathrm{IVP}\ 1\}_{\mathbb{T}} \in J_c$ (linear with constant coefficients, see beginning of Section 2.3.6) the exponential stability of the $\{\mathrm{IVP}\ 1\}_{\mathbb{T}}$ implies that of (3.5.10) if the RK-method is A-stable, cf. Def. 2.3.13 and Theorem 2.3.10. To see that not even L-stability of the RK-method is sufficient to guarantee the exponential stability of (3.5.10) for arbitrary n for an exponentially stable $\{\mathrm{IVP}\ 1\}_{\mathbb{T}} \notin J_c$, consider a simple counter-example:

The linear differential equation

$$y' - (-\tfrac{1}{2} + \cos 2\pi t) y = 0 \tag{3.5.11}$$

is exponentially stable; for an initial perturbation δ_0 at t_0 we have

$$|\bar{e}(t;\delta_0,t_0)| \le e^{\frac{1}{\pi}} |\delta_0| \exp(-\tfrac{1}{2}(t-t_0)), \qquad t \ge t_0,$$

independently of t_0. On the other hand, the discretization of (3.5.11) by the implicit Euler method (which is L-stable, see Example 2 in Section 2.3.6) gives, with a constant step $h=1$,

$$\eta_\nu - \eta_{\nu-1} - \tfrac{1}{2}\eta_\nu = 0 \quad \text{or} \quad \eta_\nu = 2\eta_{\nu-1}$$

the equilibrium of which is violently unstable.

Very little is known about the exponential stability of (3.5.10) for $\{\mathrm{IVP}\ 1\}_{\mathbb{T}} \notin J_c$. In the following, we give a few supplementary conditions on $\{\mathrm{IVP}\ 1\}_{\mathbb{T}}$ which lead to exponential stability of (3.5.10) for the *implicit*

Euler discretization with arbitrary large steps. In view of Corollary 2.3.9 and the fact that—for the implicit Euler method—the discretization of the linearized differential equation is a "neighbor" of the linearization of the difference equation, we restrict ourselves to linear differential equations

(3.5.12) $$y' - g(t)y = 0\,, \qquad t \geq 0\,,$$

so that (3.5.10) has the form

(3.5.13) $$\frac{1}{h_\nu}\{[I - h_\nu g(t_\nu)]\eta_\nu - \eta_{\nu-1}\} = 0\,, \qquad t_\nu \in \mathbb{G}_n\,.$$

As usual, it is assumed that the grid sequence $\{\mathbb{G}_n\}_{n\in\mathbb{N}'}$ is linearly convergent, i.e., that (2.1.11) holds.

Theorem 3.5.6. *If $g(t)$ in* (3.5.12) *is a normal matrix for all t and satisfies (see Def.* 2.3.3)

(3.5.14) $$r[g(t)] \leq -\mu < 0 \quad \textit{for all} \quad t \geq 0\,,$$

then the equilibrium of (3.5.13) *is exponentially stable for arbitrary n.*

Proof. The effect of a perturbation δ_0 at t_{ν_0} is given by

$$\bar{\varepsilon}_\nu(\delta_0, t_{\nu_0}) = \prod_{\lambda=\nu_0+1}^{\nu} (I - h_\lambda g(t_\lambda))^{-1}\delta_0\,.$$

As the Euclidean norm $\|\cdot\|_E$ of a normal matrix is its spectral radius we have by our assumption (3.5.14)

$$\begin{aligned}
\|\bar{\varepsilon}_\nu(\delta_0, t_{\nu_0})\|_E &\leq \prod_{\lambda=\nu_0+1}^{\nu} \|(I - h_\lambda g(t_\lambda))^{-1}\|_E \|\delta_0\|_E \\
&\leq \prod_{\lambda=\nu_0+1}^{\nu} \frac{1}{1+h_\lambda \mu}\|\delta_0\|_E \leq \left(1 + \mu\frac{c_1}{n}\right)^{-(\nu-\nu_0)} \|\delta_0\|_E \\
&\leq e^{-\mu' c_1 \frac{\nu-\nu_0}{n}} \|\delta_0\|_E \leq e^{-\mu' \frac{c_1}{c_2}(t_\nu - t_{\nu_0})} \|\delta_0\|_E\,,
\end{aligned}$$

where c_1 and c_2 are from (2.1.11) and $\mu' > 0$ has been chosen such that $\exp(\mu' c_1/n) \leq 1 + \mu c_1/n$ for all $n \in \mathbb{N}'$. The use of a different norm in $\mathbb{R}^s$ can at most introduce a fixed constant into the final estimate. □

For the other two results on the exponential stability of (3.5.13) we need an adaptation of Theorem 2.3.6:

Corollary 3.5.7. *Assume that* (2.3.30) *is a one-step difference equation and that it may be solved for $\nu < \nu_0$ as well as for $\nu > \nu_0$. Then the equilibrium of* (2.3.30) *is exponentially stable for n if and only if there exists a*

Liapunov function v: $\mathbb{R}^s \times \mathbb{R} \to \mathbb{R}$ *which satisfies, for* $\|\xi\| < r$ *and all* $t_\nu \in \mathbb{G}_n$, (2.3.32) *and*

$$\text{(3.5.15)} \quad \nabla v(\xi, t_\nu) := \frac{1}{h_\nu}\left[v(\xi, t_\nu) - v(\bar{\varepsilon}_{\nu-1}(\xi, t_\nu), t_{\nu-1})\right] \leq -a_3 \|\xi\|^2,$$

with positive constants a_1, a_2, a_3, *and* r *independent of* t_ν. *Furthermore,* v *may be chosen such that* (2.3.34) *holds.*

Proof. a) Let (as in the proof of Theorem 2.3.6) $V_\nu := v(\bar{\varepsilon}_\nu(\delta_0, t_{\nu 0}), t_\nu)$; then we have by (3.5.15) and (2.3.32)

$$\begin{aligned} V_{\nu-1} &\geq V_\nu + h_\nu a_3 \|\bar{\varepsilon}_\nu(\delta_0, t_{\nu 0})\|^2 \geq \left(1 + \frac{h_\nu a_3}{a_2}\right) V_\nu \\ &\geq \left(1 + \frac{c_1}{n}\frac{a_3}{a_2}\right) V_\nu \geq \exp\frac{\mu'}{n} V_\nu \end{aligned}$$

with a suitably chosen $\mu' > 0$. This implies exponential stability as in the proof of Theorem 2.3.6.

Parts b) and c) of the proof of Theorem 2.3.6 may be adapted similarly. See also Driver [1]. □

Theorem 3.5.8. *If the Liapunov function* v *of* (3.5.12) *may be chosen as a constant quadratic form*

$$\text{(3.5.16)} \quad v(\xi, t) = \xi^T G \xi, \quad G = G^T \text{ a constant, positive definite } s \times s\text{-matrix,}$$

then the equilibrium of (3.5.13) *is exponentially stable for arbitrary* n.

Proof. From the fact that (3.5.16) is a Liapunov function for (3.5.12) it follows (cf. Theorem 2.3.1) that

$$\dot{v}(\xi, t_\nu) = 2\,\xi^T G g(t_\nu)\,\xi \leq -a_3 \|\xi\|^2 \quad \text{for all } t_\nu.$$

With the same v, we obtain for (3.5.13)

$$\begin{aligned} \nabla v(\xi, t_\nu) &= \frac{1}{h_\nu}\left[\xi^T G \xi - \xi^T (I - h_\nu g(t_\nu))^T G (I - h_\nu g(t_\nu))\,\xi\right] \\ &= \dot{v}(\xi, t_\nu) - h_\nu \xi^T g(t_\nu)^T G g(t_\nu)\,\xi \leq -a_3 \|\xi\|^2 \quad \text{for all } t_\nu \end{aligned}$$

due to the positive definiteness of G. □

Remark. Note that the proofs of Theorems 3.5.6 and 3.5.8 do not use the upper bound on $\|g(t_\nu)\|$ which is guaranteed by (2.3.1). Thus the implicit Euler method is exponentially stable also for $y' + ty = 0$ with arbitrary steps. See the example after Theorem 3.5.1.

Theorem 3.5.9. *For some constant matrix g_0, let*

$$\frac{1}{h_\nu}\{(I-h_\nu g_0)\eta_\nu-\eta_{\nu-1}\}=0\,,\qquad t_\nu\in\mathbb{G}_n\,, \tag{3.5.17}$$

be exponentially stable. Let a_3, a_4 be the constants of Corollary 3.5.7 *for* (3.5.17) *and* $\hat{L}:=\max(\|g_0\|, \max_\nu \|g(t_\nu)\|)$. *If $g(t)$ of* (3.5.12) *satisfies*

$$\|g(t_\nu)-g_0\|<\frac{a_3}{a_4(1+c_2\hat{L}/n)}\quad \textit{for all } t_\nu \tag{3.5.18}$$

then (3.5.13) *is exponentially stable for n.*

Proof. Use ∇v in place of Δv in the proof of Theorem 2.3.8 applied to the pair of difference equations (3.5.13) and (3.5.17). □

With regard to an extension of Theorems 3.5.6, 3.5.8 and 3.5.9 to other A-stable RK-methods we observe the following:

Theorem 3.5.6 immediately extends to A-stable RK-methods which, like the implicit midpoint rule, use $g(t)$ at *one* t-value only in each step. It does not hold for the trapezoidal method; see the counter-example at the end of Section 3.5.4.

In his paper on the stability behavior of trapezoidal discretizations on $[0,\infty)$, Dahlquist ([3]) essentially derives Theorem 3.5.5 for the trapezoidal method and states assumption (3.5.16) on the Liapunov function of the (linearized) differential equation as a sufficient condition for the applicability of Theorem 3.5.5. The simplicity of our proof of Theorem 3.5.8 depends largely on the special structure of the implicit Euler method.

Theorem 3.5.9 is simply a special version of the general theorem on the exponential stability of neighboring discretizations (Theorem 2.3.8); the simplicity of the formulation is again a consequence of the special form of (3.5.13).

3.5.3 Stability Regions of RK-methods

In Section 2.3.6, we defined stability regions to characterize the stability behavior of f.s.m. when applied to $\{\text{IVP } 1\}_{\mathbb{T}}\in J_c$ (linear equations with constant coefficients). For RK-methods with equidistant grids it is quite easy to find these regions.

Theorem 3.5.10. *Each consistent uniform RK-method $\mathfrak{M}_{\mathbb{T}}$ with equidistant grids possesses a non-empty region of absolute stability; it is given by*

$$\mathfrak{H}_0:=\{H\in\mathbb{C}: |\gamma_\psi(H)|<1\}\subset\mathbb{C} \tag{3.5.19}$$

where $\gamma_\psi: \mathbb{C} \to \mathbb{C}$ *is the growth function of the generating RK-scheme* ψ *of* $\mathfrak{M}_{\mathbb{T}}$.

Proof. For an $\{\text{IVP } 1\}_{\mathbb{T}}$ from J_c with $f(y) = gy$, g a constant $s \times s$-matrix, the RK-procedure generated by the RK-scheme ψ reduces to (cf. (3.4.7))

$$\frac{1}{h}[\eta_\nu - \gamma_\psi(hg)\eta_{\nu-1}] = 0 \quad \text{for all } \nu > 0. \tag{3.5.20}$$

If all eigenvalues λ_σ of hg are in $\mathfrak{H}_0$ then there exists a $\mu > 0$ such that

$$|\gamma_\psi(h\lambda_\sigma)| < e^{-\mu} < 1, \qquad \sigma = 1(1)s. \tag{3.5.21}$$

If g is similar to a diagonal matrix so is $\gamma_\psi(hg)$ and (3.5.21) immediately implies the μ/h-exponential stability of (3.5.20). If g has non-linear elementary divisors, the norm of $[\gamma_\psi(hg)]^\nu$ behaves like $\nu^r e^{-\mu\nu}$ as $\nu \to \infty$, with r a natural number; thus, there exists a constant a such that $\|[\gamma_\psi(hg)]^\nu\| \le a e^{-\mu'\nu}$, with $0 < \mu' < \mu$.

It remains to show that $\mathfrak{H}_0$ cannot be empty. For this purpose we observe that (3.2.24) implies

$$\gamma_\psi(H) = 1 + b^T e H + U(H^2) \quad \text{as } H \to 0 \tag{3.5.22}$$

and that $b^T e = 1$ for a consistent RK-method, cf. Theorem 3.3.5. Hence, γ_ψ conformally maps a neighborhood of 0 into a neighborhood of 1 and there is an open domain in the left halfplane, with 0 on its boundary, which is mapped into the open unit disk. □

Corollary 3.5.11. *For a uniform RK-method* $\mathfrak{M}_{\mathbb{T}}$ *with equidistant grids, the regions* $\mathfrak{H}_\mu$ *of* μ*-exponential stability are given by*

$$\mathfrak{H}_\mu := \{H \in \mathbb{C} : |\gamma_\psi(H)| < e^{-\mu}\}, \qquad \mu > 0; \tag{3.5.23}$$

some of these regions may be empty.

Proof. As in the proof of Theorem 3.5.10 the fact that all eigenvalues of hg are in some $\mathfrak{H}_\mu$ implies $\|[\gamma_\psi(hg)]^\nu\| \le a e^{-\mu\nu}$. (If we would admit the equality sign in (3.5.23), this estimate would not hold for matrices g with non-linear elementary divisors.) □

Remarks. 1. Theorem 3.5.10 and Corollary 3.5.11 show that the term *growth function* was well chosen in Def. 3.2.7.

2. For a RK-method, the *characteristic polynomial* introduced at the end of Section 2.3.6 is simply

$$\varphi(z; H) := z - \gamma_\psi(H). \tag{3.5.24}$$

Due to the simple structure of (3.5.23) and of γ_ψ the stability regions of RK-methods are easy to construct. If a region $\mathfrak{H}_\mu$ for a RK-method $\mathfrak{M}_{\mathbb{T}}$ contains points H which satisfy $\operatorname{Re} H > -\mu$ this implies the existence of $\{\mathrm{IVP}\,1\}_{\mathbb{T}}$ whose discretizations generated by $\mathfrak{M}_{\mathbb{T}}$ have solutions which decay faster than those of the original problem, or which decay exponentially although those of the original problem do not. As this indicates a lack of approximation rather than a lack of stability the definition of the stability regions is often modified into

$$\hat{\mathfrak{H}}_\mu := \{H \in \mathbb{C}: \operatorname{Re} H < -\mu,\ |\gamma_\psi(H)| < e^{-\mu}\}, \qquad \mu \geq 0. \tag{3.5.25}$$

Fig. 3.1 displays some regions $\mathfrak{H}_0$ for the explicit RK-methods with $m = p = 1(1)4$, cf. Section 3.3.4.

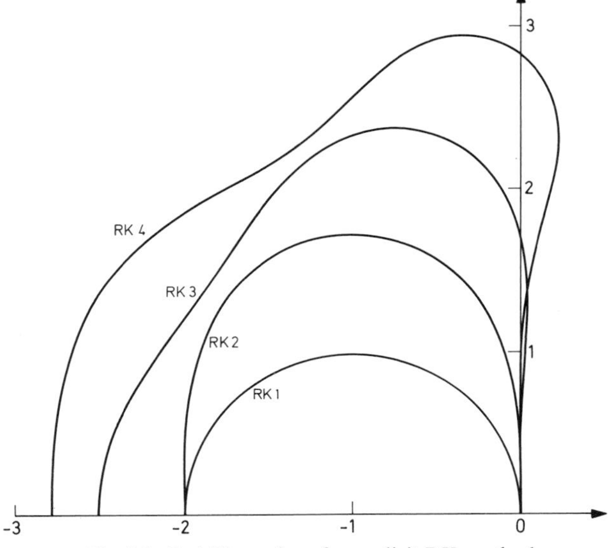

Fig. 3.1. Stability regions for explicit RK-methods

According to Theorem 2.3.10, it would be most desirable to have $\hat{\mathfrak{H}}_0 \equiv \mathbb{C}_-$ (i.e., A-stability, see Def. 2.3.13). Unfortunately, implicitness of the RK-method is a necessary condition for A-stability.

Theorem 3.5.12. *There is no explicit RK-method which is A-stable.*

Proof. According to Def. 3.2.8, γ_ψ is a polynomial for an explicit RK-scheme ψ; thus $|\gamma_\psi(H)| < 1$ cannot hold for all $H \in \mathbb{C}_-$ as would be required for A-stability. □

Naturally, most implicit RK-methods are not A-stable either. Examples of A-stable RK-methods are the implicit trapezoidal and the implicit midpoint methods, with $\gamma(H) = (1 + H/2)/(1 - H/2)$, and the implicit Euler method with $\gamma(H) = 1/(1 - H)$. (The implicit Euler method is even L-stable as was observed in Example 2 following Def. 2.3.16.)

Furthermore, the maximal order symmetric implicit RK-methods of Section 3.3.4 are A-stable according to the following result of Ehle [1]:

Theorem 3.5.13 *(Ehle). In the Padé-table of e^z let $r_{n,m}(z)$ be the entry with an n-th degree polynomial in the numerator and an m-th degree polynomial in the denominator. For each $m \geq 1$, $r_{m,m}$, $r_{m,m+1}$, and $r_{m,m+2}$ map $\mathbb{C}_-$ into the unit disk.*

In view of the computational difficulties which arise with implicit RK-methods a search for explicit RK-methods with *large* regions of absolute stability seems appropriate. For an explicit RK-scheme ψ, γ_ψ is a polynomial whose maximal degree is the number m of stages in an irreducible representation of ψ, cf. Section 3.2.3. On the other hand, $\gamma_\psi(H) = e^H + O(H^{p+1})$ for a RK-scheme of order p according to Corollary 3.3.8. Thus we must have $m > p$ if there are to be any parameters available to increase the stability regions beyond the standard regions defined by $\gamma_\psi(H) = \sum_{\rho=0}^{p} \frac{H^\rho}{\rho!}$, cf. Fig. 3.1. We then have the problem of choosing $c_\rho \in \mathbb{R}$, $\rho = p+1(1)m$, in

$$\gamma_\psi(H) = \sum_{\rho=0}^{p} \frac{H^\rho}{\rho!} + c_{p+1} H^{p+1} + \cdots + c_m H^m \tag{3.5.26}$$

such that $|\gamma_\psi(H)| < 1$ (resp. $e^{-\mu}$) in large and suitably shaped subdomains of $\mathbb{C}_-$.

If only the extension of $\mathfrak{H}_0$ (resp. $\mathfrak{H}_\mu$) along the negative real axis is considered as a criterion, the problem of selecting the c_ρ, $\rho > p$, may be solved numerically by a modification of the Remes algorithm for determining optimal approximations. In this fashion, Riha [1] has obtained the following strict lower bounds for $H_0 := -\min_{H \in \mathfrak{H}_0} \operatorname{Re} H$ for given p and m:

Table 3.4

p \ m	2	3	4	5	6	7	8
1	8.00	18.00	32.00	50.00	72.00	98.00	128.00
2	—	6.26	12.05	19.46	28.50	39.19	51.52
3	—	—	6.03	10.54	16.05	22.56	30.08
4	—	—	—	6.06	9.97	14.59	19.93
5	—	—	—	—	6.26	9.81	13.91
6	—	—	—	—	—	6.51	9.83

In the case $p=1$, the limit polynomial whose modulus remains ≤ 1 on a maximal interval $[-H_0, 0]$ is obtained by a transformation of the

m-th Chebyshev polynomial T_m: $\gamma(H)=T_m(1+H/m^2)$. Thus H_0 attains the value $2m^2$.

Although the values given in Table 3.4 cannot be reached when a substantial width of $\mathfrak{H}_0$ in the imaginary direction is required, it is obvious that the stability regions may be considerably enlarged by introducing more stages than would be required to attain order p.

After the construction of a suitable polynomial γ_ψ there remains the task of determining the coefficients of an explicit m-stage generating matrix A which satisfies the order conditions (3.3.11) and produces γ_ψ via (3.2.25). Since $m>p$ there would appear to be sufficiently many free parameters for this goal to be achieved, see e.g. Metzger [1].

Actually one may attempt to impose further restrictions on the coefficients of A which lead to a saving in computational effort. An interesting goal is to construct A such that it represents a sequence of several independent RK-steps with only a few stages each. For this purpose we factor the polynomial γ_ψ into r polynomials γ_{ψ_ρ} of degrees m_ρ, $\rho=1(1)r$; naturally, $\sum m_\rho = m$. According to (3.2.27), this corresponds to a decomposition of ψ into $\psi_1 \dots \psi_r$ and to a decomposition of the associated m-stage RK-procedure into r successive RK-procedures with m_ρ stages each. We are thus led to consider non-uniform RK-methods which use a sequence of r different RK-procedures in a *cyclic* fashion.

Example. Consider a RK-method which consists of using Euler steps of length ah and $(1-a)h$ alternatingly. (For $a=\frac{1}{2}$ we have the uniform Euler method, with step $h/2$.) Obviously, we have the generating matrix

$$A=\begin{pmatrix} 0 & 0 \\ a & 0 \\ a & 1-a \end{pmatrix} \quad \text{and} \quad \gamma_\psi(H)=1+H+a(1-a)H^2.$$

A short calculation shows that $|\gamma_\psi(H)|<1$ for $H\in(-1/(a(1-a)),0)$, with $a\in(1/2-1/\sqrt{8},\ 1/2+1/\sqrt{8})$. Thus we may nearly double the extension of $\mathfrak{H}_0$ along the negative real axis through a deviation from uniform grids (represented by $a=\frac{1}{2}$).

Although this example is not very practical, it seems that sizeable improvements may be obtained in this way for larger values of m. A more detailed analysis of the situation is in progress.

3.5.4 Use of Stability Regions for General $\{\text{IVP } 1\}_{\mathbb{T}}$

As we saw in Section 3.5.1, RK-methods produce exponentially stable discretizations to exponentially stable $\{\text{IVP } 1\}_{\mathbb{T}}$ if the steps h_ν are sufficiently small. We would like to be able to make use of the stability regions discussed in the previous section to tell just *how* small they have to be for a given $\{\text{IVP } 1\}_{\mathbb{T}}$ and a given method. The ideal result would be the following:

(R) Let $g_\nu := f_y(t_\nu, \zeta_\nu)$, $\nu \geq 0$, where ζ_ν are the values of the solution of the discretization. If all eigenvalues of $h_{\nu+1} g_\nu$ are in $\mathfrak{H}_0$ for each $\nu \geq 0$ then the discretization is exponentially stable.

Unfortunately, (R) does not hold even in the simplest case of a scalar linear differential equation and of a discretization with equidistant grids, as soon as there are variable coefficients. Here is a simple *counter-example:*

All RK-schemes $\psi(a) = \begin{pmatrix} 0 & 0 \\ 1/(2a) & 0 \\ 1-a & a \end{pmatrix}$, $a \neq 0$, have the same growth function $\gamma_\psi(H) = 1 + H + H^2/2$, hence $|\gamma_\psi(H)| < 1$ for $H \in (-2, 0)$. The RK-method based upon $\psi(a)$ discretizes $y' - g(t)y = 0$, $y \in \mathbb{R}$, into

$$\eta_\nu = \left[1 + (1-a)hg(t_{\nu-1}) + ahg(t_{\nu-1+\frac{1}{2a}}) + \frac{h^2}{2} g(t_{\nu-1}) g(t_{\nu-1+\frac{1}{2a}})\right] \eta_{\nu-1}.$$

For fixed $a \in (-1, 0)$ and a given step h, assume that $g(t)$ is such that

$$hg(t_{\nu-1}) = a, \qquad hg(t_{\nu-1+\frac{1}{2a}}) = -(1-a).$$

Then

$$\eta_\nu = [1 + \tfrac{1}{2}(-a)(1-a)]\eta_{\nu-1} \tag{3.5.27}$$

and there is exponential instability although both $hg(t_{\nu-1})$ and $hg(t_{\nu-1+1/2a})$ are in $\mathfrak{H}_0$ throughout.

If the $\{\text{IVP } 1\}_{\mathbb{T}}$ under consideration is non-linear, a further difficulty concerning the validity of (R) arises from the inherent *linearization*: In general, if we linearize the error equation (3.5.10) with an increment function Ψ formed for a non-linear $\{\text{IVP } 1\}_{\mathbb{T}}$, the result is different from the error equation obtained for the linearized $\{\text{IVP } 1\}_{\mathbb{T}}$ even if we refer both linearizations to the same function, say z.

Example. Take the usual 2-stage RK-procedure

$$\frac{\eta_\nu - \eta_{\nu-1}}{h} - f\left(\eta_{\nu-1} + \frac{h}{2} f(\eta_{\nu-1})\right) = 0. \tag{3.5.28}$$

The linearization of (3.5.10) at z becomes $(z_\nu := z(t_\nu))$:

$$\frac{\varepsilon_\nu - \varepsilon_{\nu-1}}{h} - f'\left(z_{\nu-1} + \frac{h}{2} f(z_{\nu-1})\right)\left[I + \frac{h}{2} f'(z_{\nu-1})\right]\varepsilon_{\nu-1} = 0 \tag{3.5.29}$$

while the error equation for the application of (3.5.28) to $y' - f'(z(t))y = 0$ is

$$\frac{\varepsilon_\nu - \varepsilon_{\nu-1}}{h} - f'(z_{\nu-\frac{1}{2}})\left[I + \frac{h}{2} f'(z_{\nu-1})\right]\varepsilon_{\nu-1} = 0. \tag{3.5.30}$$

Although the difference between (3.5.29) and (3.5.30) is minute for sufficiently small h, it may be crucial in a decision about the admissible size of h. Naturally, if we replace $z_{\nu-1}$ by $\zeta_{\nu-1}$ in (3.5.29), as stipulated in (R), another discrepancy is introduced.

In the literature there seem to be no significant results giving side-conditions under which (*R*) would become valid. Actually, our Theorems 3.5.6, 3.5.8, and 3.5.9 for the implicit Euler method may be interpreted as such results, as $\mathfrak{H}_0 \supset \mathbb{C}_-$ for this RK-method. An extension of a modified form of Theorem 3.5.6 to arbitrary one-stage RK-methods is possible:

Theorem 3.5.14. *Consider a one-stage RK-method* $\mathfrak{M}_{\mathbb{T}}$ *based upon the RK-scheme* $\psi(a) = \binom{a}{1}$, *a arbitrary, with grids* $\mathbb{G}_n$, $n \in \mathbb{N}'$. *Denote the regions of* μ*-exponential stability for* $\psi(a)$ *by* $\mathfrak{H}_\mu$. *For a given* $\{\text{IVP } 1\}_{\mathbb{T}}$ *with uniformly Lipschitz continuous* f_y *and for a fixed* $n \in \mathbb{N}'$, *if* $g_\nu := f_y(t_{\nu-1+a}, \zeta_\nu^{(1)})$ *is a normal matrix whose eigenvalues satisfy*

$$\lambda_\sigma[h_\nu g_\nu] \in \mathfrak{H}_{h_\nu \mu}, \qquad \mu > 0, \quad \textit{for all } t_\nu \in \mathbb{G}_n, \tag{3.5.31}$$

then the discretization $\mathfrak{M}_{\mathbb{T}}(\{\text{IVP } 1\}_{\mathbb{T}})$ *is* μ*-exponentially stable for n.*

Proof. From

$$\eta_\nu^{(1)} := \eta_{\nu-1} + a h_\nu f(t_{\nu-1+a}, \eta_\nu^{(1)}),$$
$$\eta_\nu := \eta_{\nu-1} + h_\nu f(t_{\nu-1+a}, \eta_\nu^{(1)}),$$

we have $\eta_\nu^{(1)} = (1-a)\eta_{\nu-1} + a\eta_\nu$ and the RK-procedure may be written as

$$\frac{\eta_\nu - \eta_{\nu-1}}{h_\nu} - f(t_{\nu-1+a}, (1-a)\eta_{\nu-1} + a\eta_\nu) = 0,$$

the linearization of which at ζ is

$$\frac{\varepsilon_\nu - \varepsilon_{\nu-1}}{h_\nu} - f_y(t_{\nu-1+a}, (1-a)\zeta_{\nu-1} + a\zeta_\nu)[(1-a)\varepsilon_{\nu-1} + a\varepsilon_\nu] = 0 \tag{3.5.32}$$

or

$$\varepsilon_\nu = (I - a h_\nu g_\nu)^{-1}(I + (1-a)h_\nu g_\nu)\varepsilon_{\nu-1} = \gamma_{\psi(a)}(h_\nu g_\nu)\varepsilon_{\nu-1}$$

by the definition (3.2.25) of $\gamma_{\psi(a)}$.

According to Corollary 3.5.11, our assumption (3.5.31) and the normality of g_ν imply

$$\|\gamma_{\psi(a)}(h_\nu g_\nu)\|_E < e^{-h_\nu \mu} \quad \text{for all } \nu$$

so that

$$\begin{aligned} \|\bar{\varepsilon}_\nu(\delta_0, t_{\nu_0})\|_E &\le \prod_{\lambda=\nu_0+1}^{\nu} \|\gamma_{\psi(a)}(h_\lambda g_\lambda)\|_E \|\delta_0\|_E \\ &< e^{-\mu \sum h_\lambda} \|\delta_0\|_E = e^{-\mu(t_\nu - t_{\nu_0})} \|\delta_0\|_E. \end{aligned} \tag{3.5.33}$$

According to Corollary 2.3.9 it is sufficient to prove the exponential stability of (3.5.32). □

Naturally, if we demand only $\lambda_\sigma[h_\nu g_\nu] \in \mathfrak{H}_0$ in (3.5.31) we still obtain exponential stability. For $a \geq \frac{1}{2}$, we may then replace (3.5.31) by (3.5.14) as is easily seen. Thus the *implicit midpoint method* ($a=\frac{1}{2}$) shares the pleasant property of the implicit Euler method of leading to an exponentially stable discretization for arbitrarily large steps in the situation of Theorem 3.5.6. In addition, $\psi(\frac{1}{2})$ is of order 2 while all other RK-schemes $\psi(a)$ are of order 1 only.

Actually, the implicit midpoint rule is superior in its strong stability to the *implicit trapezoidal method* which is often considered as the prototype of a strongly stable method of order 2, cf. for example Dahlquist [3]. To prove that claim we show that Theorem 3.5.6 cannot be extended to this RK-method.

It is true that the implicit trapezoidal method generates an exponentially stable discretization if the $g_\nu := f_y(t_\nu, \zeta_\nu)$ are normal matrices satisfying $\operatorname{Re}\lambda_\sigma[g_\nu] \leq -\mu < 0$ for all ν and all σ and if an arbitrary but *constant step* is used: In place of (3.5.32) we obtain

$$\frac{\varepsilon_\nu - \varepsilon_{\nu-1}}{h} - \frac{1}{2}\left[g_\nu \varepsilon_\nu + g_{\nu-1}\varepsilon_{\nu-1}\right] = 0,$$

$$\bar{\varepsilon}_\nu(\delta_0, t_{\nu_0}) = \left(\prod_{\lambda=\nu_0+1}^{\nu}\left[I - \frac{h}{2} g_\lambda\right]^{-1}\left[I + \frac{h}{2} g_{\lambda-1}\right]\right)\delta_0$$

$$= \left[I - \frac{h}{2} g_\nu\right]^{-1}\left(\prod_{\lambda=\nu_0+1}^{\nu-1}\left[I + \frac{h}{2} g_\lambda\right]\left[I - \frac{h}{2} g_\lambda\right]^{-1}\right)\left[I + \frac{h}{2} g_{\nu_0}\right]\delta_0,$$

and the same reasoning as in the proof of Theorem 3.5.14 can be applied. Since the $\|g_\nu\|$ are uniformly bounded the extra factors $[I - (h/2) g_\nu]^{-1}$ and $[I + (h/2) g_{\nu_0}]$ introduce only a fixed factor into the estimate corresponding to (3.5.33). (The example of Gourlay [1] does not contradict this result since an exponentially stable $\{\text{IVP } 1\}_{\mathbb{T}}$ cannot possess the property assumed by Gourlay over an infinite interval.)

On the other hand, if we admit *variable steps* we may construct counter-examples in which the implicit trapezoidal method fails to produce an exponentially stable discretization even for a scalar exponentially stable $\{\text{IVP } 1\}_{\mathbb{T}}$:

Take $s=1$ so that $g_\nu \in \mathbb{R}$; choose

$$g_\nu = \begin{cases} g_1, & \nu \text{ odd} \\ g_2, & \nu \text{ even} \end{cases}, \qquad h_\nu = \begin{cases} h_1, & \nu \text{ odd} \\ h_2, & \nu \text{ even} \end{cases}.$$

Then, for even $\nu-\nu_0$,

$$\bar{\varepsilon}_\nu(\delta_0, t_{\nu_0}) = \left[\frac{\left(1+\frac{h_2}{2}g_1\right)\left(1+\frac{h_1}{2}g_2\right)}{\left(1-\frac{h_2}{2}g_2\right)\left(1-\frac{h_1}{2}g_1\right)}\right]^{\frac{\nu-\nu_0}{2}} \delta_0$$

and the modulus of the bracket may be made greater than 1 for negative g_1, g_2 by choosing $h_1 < h_2$, $|g_1| \gg 0$, $|g_2| \approx 0$. Thus, exponential instability arises although both $h_\nu g_\nu$ and $h_\nu g_{\nu-1}$ are in $\mathfrak{H}_0$ for all ν.

3.5.5 Suggestion for a General Approach

A more general approach to our problem of finding side-conditions to make (R) valid is the following: We replace the linear difference equation resulting from a linearization of the error equation by

(3.5.34) $$\varepsilon_\nu = \gamma_\psi(h_\nu g_j)\varepsilon_{\nu-1} \quad \text{for} \quad \nu_{j-1} < \nu \leq \nu_j .$$

Here the ν_j, $j=1(1)J$, form a *finite* increasing sequence and the g_j are J fixed $s\times s$-matrices. If each of the difference equations with constant coefficients, $j=1(1)J$,

(3.5.35) $$\varepsilon_\nu = \gamma_\psi(h_\nu g_j)\varepsilon_{\nu-1} \quad \text{for all} \quad \nu > 0 ,$$

is μ_j-exponentially stable, then (3.5.34) is μ-exponentially stable with $\mu := \min_j \mu_j$. The exponentially stability of the difference equations (3.5.35) may immediately be checked with the aid of stability regions.

The original linearized error equation is then interpreted as a neighboring difference equation of (3.5.34) and Theorem 2.3.8 is used to establish exponential stability if the two difference equations are sufficiently close. Condition (2.3.41) of Theorem 2.3.8 will normally lead to a restriction of the variation of f_y along the solution of the discretization. If the exponent μ for (3.5.34) is large (small) the admissible variation of f_y will be large (small).

One may also immediately use the exponential stability estimate (2.3.31) for (3.5.34) to derive suitable restrictions on the variation of f_y. If $a=1$ in (2.3.31) one can even admit infinitely many different values of g_j, i.e., one can use a suitable g_j for each step. In this case the restrictions on the variation of f_y may become quite moderate and rather transparent. We explain this strategy by a simple example:

Consider the 2-stage RK-method

(3.5.36) $$\frac{\eta_\nu - \eta_{\nu-1}}{h_\nu} - f\left(t_{\nu-1} + \frac{h_\nu}{2},\ \eta_{\nu-1} + \frac{h_\nu}{2} f(t_{\nu-1}, \eta_{\nu-1})\right) = 0 , \qquad \nu > 0 .$$

The linearized form of the associated error equation is

$$(3.5.37)\qquad \frac{\varepsilon_\nu - \varepsilon_{\nu-1}}{h_\nu} - g_{\nu-\frac{1}{2}}\left[I + \frac{h_\nu}{2} g_{\nu-1}\right] \varepsilon_{\nu-1} = 0\,, \qquad \nu > 0\,,$$

where

$$g_{\nu-1} := f_y(t_{\nu-1}, \zeta_{\nu-1})\,,$$

$$g_{\nu-\frac{1}{2}} := f_y\left(t_{\nu-1} + \frac{h_\nu}{2},\ \zeta_{\nu-1} + \frac{h_\nu}{2} f(t_{\nu-1}, \zeta_{\nu-1})\right).$$

We write (3.5.37) in the form

$$\varepsilon_\nu = \left[I + h_\nu g_{\nu-\frac{1}{2}}\left(I + \frac{h_\nu}{2} g_{\nu-\frac{1}{2}}\right)\right] \varepsilon_{\nu-1} + \left[\frac{h_\nu^2}{2} g_{\nu-\frac{1}{2}}(g_{\nu-1} - g_{\nu-\frac{1}{2}})\right] \varepsilon_{\nu-1}\,.$$

If $g_{\nu-1/2}$ is a normal matrix and if all eigenvalues of $h_\nu g_{\nu-1/2}$ are in $\mathfrak{H}_\mu$, we know that the Euclidean norm of the first bracket may be bounded by $e^{-\mu}$. Hence we have to show that, for some $\mu' > 0$,

$$\frac{h_\nu^2}{2} \|g_{\nu-\frac{1}{2}}\|_E \, \|g_{\nu-1} - g_{\nu-\frac{1}{2}}\|_E \leq e^{-\mu'} - e^{-\mu}\,,$$

in order to establish the exponential stability of (3.5.37).

Thus, it would appear that we ought to supplement the plot of the stability region $\mathfrak{H}_0$ for (3.5.36) by also plotting the boundaries of various regions $\mathfrak{H}_\mu$ for (3.5.36), marking each of these curves with the value of $1 - e^{-\mu}$, see Fig. 3.2. Such a plot could then be used as follows:

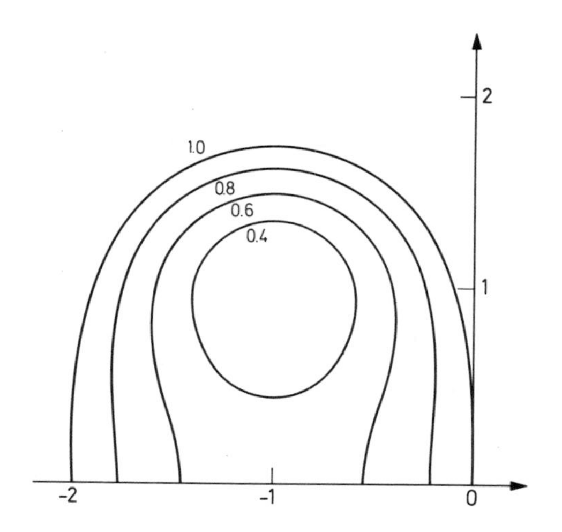

Fig. 3.2. Stability regions $\mathfrak{H}_\mu$ for (3.5.36) with values of $s(\mu) = 1 - e^{-\mu}$

At each step of (3.5.36) we determine the eigenvalues λ_σ of $h_\nu g_{\nu-1/2}$ and the smallest region $\mathfrak{H}_\mu$ which contains all these eigenvalues. (If they are not all in $\mathfrak{H}_0$ we have to choose a smaller step h_ν.) If the boundary of this region is marked with the number $s(\mu)$ we test whether

$$\frac{h_\nu^2}{2}\,\|g_{\nu-\frac{1}{2}}\|_E\,\|g_{\nu-1}-g_{\nu-\frac{1}{2}}\|_E<s(\mu)\,. \tag{3.5.38}$$

If this criterion holds throughout, the difference equation (3.5.36) must have been exponentially stable and we must have remained close to its exact solution in our computation if we have kept the local errors sufficiently small.

If $g_{\nu-1/2}$ is not normal and $a>1$ in the estimate (2.3.31) for (3.5.37) with constant coefficients, then we have to be much more restrictive: We now have to choose a *finite* number of g_j and write (3.5.37) in the form

$$\begin{aligned}\varepsilon_\nu=&\left[I+h_\nu g_j\left(I+\frac{h_\nu}{2}g_j\right)\right]\varepsilon_{\nu-1}\\&+\left[h_\nu(g_{\nu-\frac{1}{2}}-g_j)+\frac{h_\nu^2}{2}(g_{\nu-\frac{1}{2}}g_{\nu-1}-g_j^2)\right]\varepsilon_{\nu-1}\,,\qquad \nu_{j-1}<\nu\leq\nu_j\,.\end{aligned}$$

If all eigenvalues of $h_\nu g_j$ are in $\mathfrak{H}_\mu$ for all ν and the corresponding j, we have to require that the appropriate norm of the second bracket be smaller than $s(\mu)$.

Chapter 4

Linear Multistep Methods

The structure of general m-stage k-step methods in the sense of Def. 2.1.8 and 2.1.10 is so complex that we will deal in this chapter only with the special class of one-stage k-step methods whose forward-step procedures consist simply of a *linear* combination of values of η_μ and $f(\eta_\mu)$ at $k+1$ consecutive gridpoints t_μ, $\mu=\nu-k(1)\nu$. No re-substitutions into f (which formed the essence of RK-methods) will be permitted. Even with these limitations, this situation is sufficiently interesting and by no means trivial. Many of the results obtained will serve as background material in the treatment of more general classes of multistep methods in Chapter 5.

4.1 Linear k-step Schemes

In Section 4.1, we will analyze individual linear multistep procedures, their construction, and their local approximation properties for IVP 1 without reference to their use in a discretization method.

4.1.1 Characterization

In this Chapter 4, we will consider one-stage k-step f.s.m. for IVP 1 with forward step procedures of the form (cf. Section 2.1.3)

(4.1.1) $$\Phi_{n\nu}(\eta_\nu, \ldots, \eta_{\nu-k}) = \sum_{\kappa=0}^{k} [\alpha_{n\nu\kappa}\eta_{\nu-k+\kappa} - \beta_{n\nu\kappa} f(\eta_{\nu-k+\kappa})] = 0, \qquad \nu \geq k.$$

The dependence on ν of the coefficients of the general procedure (4.1.1) may arise from two sources:

a) The grid $\mathbb{G}_n$ may not be equidistant;

b) varying difference procedures may be employed as ν changes.

If a constant step $h=1/n$ is used and the discretization scheme does not change with v, the dependence of the coefficients in (4.1.1) on v vanishes and the one on n becomes trivial:

$$\alpha_{n\nu\kappa} = \frac{\alpha_\kappa}{h}, \qquad \beta_{n\nu\kappa} = \beta_\kappa .$$

Def. 4.1.1. The forward step procedure, with $f\colon \mathbb{R}^s \to \mathbb{R}^s$ from an IVP 1,

(4.1.2)
$$\Phi(h;\eta_\nu,\dots,\eta_{\nu-k}) = \frac{1}{h}\sum_{\kappa=0}^{k} [\alpha_\kappa \eta_{\nu-k+\kappa} - h\beta_\kappa f(\eta_{\nu-k+\kappa})] = 0, \qquad \nu \ge k,$$

with

$$\sum_{\kappa=0}^{k} \alpha_\kappa = 0, \qquad \alpha_k \neq 0, \qquad \alpha_0^2 + \beta_0^2 > 0, \tag{4.1.3}$$

is called a *linear k-step procedure* and the $2\times(k+1)$-matrix

$$M = \begin{pmatrix} \alpha_0 & \dots & \alpha_k \\ \beta_0 & \dots & \beta_k \end{pmatrix}$$

its *generating matrix.*

Remark. The condition $\sum_{\kappa=0}^{k} \alpha_\kappa = 0$ is clearly necessary if (4.1.2) is to approximate an IVP 1 (consider $f\equiv 0, y\equiv 1$); it is convenient to include it in the definition of a linear multistep procedure. The other two restrictions in (4.1.3) serve obvious purposes: (4.1.2) has to be solvable for η_ν (at least for $f\equiv 0$) and k should be uniquely defined.

With each linear k-step procedure (4.1.2) and its generating matrix we may associate the polynomials[1]

$$\rho(x) := \sum_{\kappa=0}^{k} \alpha_\kappa x^\kappa, \qquad \sigma(x) := \sum_{\kappa=0}^{k} \beta_\kappa x^\kappa . \tag{4.1.4}$$

Def. 4.1.2. Two linear multistep procedures (4.1.2)—not necessarily with the same step numbers—are called *equivalent* if their associated polynomials $\rho_i, \sigma_i, i=1,2,$ satisfy

$$\frac{\rho_1(x)}{\sigma_1(x)} \equiv \frac{\rho_2(x)}{\sigma_2(x)} \tag{4.1.5}$$

(in the sense of identity of rational functions). In this case the generating matrices of the multistep procedures are also called equivalent.

[1] Cf. Def. 4.1.6.

Theorem 4.1.1. *Let a linear k_1-step procedure $\Phi_1=0$ and a k_2-step procedure $\Phi_2=0$ be equivalent. For an arbitrary fixed function $f: \mathbb{R}^s \to \mathbb{R}^s$ and an arbitrary fixed $h>0$, consider a sequence $\{\eta_\nu\}_{\nu \geq 0}$, $\eta_\nu \in \mathbb{R}^s$, which satisfies*

$$\Phi_1(h; \eta_\nu, \ldots, \eta_{\nu-k_1})=0 \quad \text{for } \nu \geq k_1 .$$

If $\{\eta_\nu\}$ also satisfies

$$\text{(4.1.6)} \qquad \Phi_2(h; \eta_\nu, \ldots, \eta_{\nu-k_2})=0 \quad \text{for } \nu=\nu_0(1)\nu_0+k_1+k_2, \qquad \nu_0 \geq k_2 ,$$

then $\{\eta_\nu\}$ satisfies $\Phi_2=0$ for all $\nu \geq \nu_0$.

Proof. Let the polynomials ρ_0 and σ_0 be the greatest common divisors of ρ_1, ρ_2 and σ_1, σ_2 resp. and denote by $\Phi_0(h; \eta_\nu, \ldots, \eta_{\nu-k_0})=0$ the linear k_0-step procedure formed with the coefficients of ρ_0 and σ_0. Due to (4.1.5) there exist polynomials c_1 and c_2, of degrees $l_1=k_1-k_0$ and $l_2=k_2-k_0$ resp., which are relatively prime to each other and satisfy

$$\begin{aligned} \rho_1(x)&=c_1(x)\rho_0(x), & \rho_2(x)&=c_2(x)\rho_0(x), \\ \sigma_1(x)&=c_1(x)\sigma_0(x), & \sigma_2(x)&=c_2(x)\sigma_0(x). \end{aligned}$$

This implies, with $c_i(x) =: \sum_{\lambda=0}^{l_i} c_\lambda^{(i)} x^\lambda$, $i=1,2$,

$$\text{(4.1.7)} \qquad \begin{cases} \Phi_1(h; \eta_\nu, \ldots, \eta_{\nu-k_1}) = \sum_{\lambda=0}^{l_1} c_\lambda^{(1)} \Phi_0(h; \eta_{\nu-l_1+\lambda}, \ldots, \eta_{\nu-k_0-l_1+\lambda}), \\ \Phi_2(h; \eta_\nu, \ldots, \eta_{\nu-k_2}) = \sum_{\lambda=0}^{l_2} c_\lambda^{(2)} \Phi_0(h; \eta_{\nu-l_2+\lambda}, \ldots, \eta_{\nu-k_0-l_2+\lambda}). \end{cases}$$

According to (4.1.7), a sequence $\{\eta_\nu\}$ may satisfy $\Phi_1=0$ for all $\nu \geq k_1$ either by satisfying $\Phi_0=0$ for all $\nu \geq k_0$ or by generating values $\xi_\nu := \Phi_0(h; \eta_\nu, \ldots, \eta_{\nu-k_0}) \not\equiv 0$ which satisfy

$$\sum_{\lambda=0}^{l_1} c_\lambda^{(1)} \xi_{\nu-l_1+\lambda}=0 \quad \text{for all } \nu \geq k_1 .$$

In the first case $\{\eta_\nu\}$ also satisfies $\Phi_2=0$ by (4.1.7). In the second case, however, the ξ_ν cannot satisfy $\sum_{\lambda=0}^{l_2} c_\lambda^{(2)} \xi_{\nu-l_2+\lambda}=0$ for all $\nu \geq k_2$ since the polynomials c_1 and c_2 have no zeros in common which implies that the solution sets of the linear difference equations with coefficients $c_\lambda^{(1)}$ and $c_\lambda^{(2)}$ resp. are disjoint.

Thus, if the number l of consecutive gridpoints at which the validity of (4.1.6) is required is taken sufficiently large, (4.1.6) can be satisfied only in the first case. It may be established that $l=\max(k_1, k_2)+k_0+1 \leq k_1+k_2+1$ serves the purpose; the details are omitted. □

Remark. Theorem 4.1.1 shows that two different but equivalent linear multistep procedures generate the same sequence $\{\eta_\nu\}$ from given initial values.

By (4.1.5), if the associated polynomials ρ and σ of a linear multistep procedure possess one or more common zeros there exists an equivalent procedure with a smaller step number. On the other hand, all equivalent multistep procedures without such common zeros must have the same step number k which is the minimal step number in the equivalence class. Since 1 is always a zero of ρ by (4.1.3), it cannot be a zero of σ for such procedures so that $\sum_\kappa \beta_\kappa \neq 0$. Thus for each equivalence class there exist a unique linear k-step procedure and a corresponding generating matrix with

$$\sum_{\kappa=0}^{k} \beta_\kappa = 1 . \tag{4.1.8}$$

Def. 4.1.3. An equivalence class ψ of generating matrices for linear multistep procedures is called a *linear k-step scheme* where k is the minimal step number in ψ.

Thus a linear k-step scheme ψ may be viewed as mapping a function $f: \mathbb{R}^s \to \mathbb{R}^s$ from an IVP 1 into the function Φ of a linear multistep procedure (4.1.2) with coefficients from one of the equivalent generating matrices in ψ. For the sake of definiteness, we will assume—if not explicitly stated otherwise—that

$$\psi[f](h; \xi_k, \ldots, \xi_0) = \frac{1}{h} \sum_{\kappa=0}^{k} [\alpha_\kappa \xi_\kappa - h\beta_\kappa f(\xi_\kappa)]$$

has the coefficients of the unique k-step representative satisfying (4.1.8) and we will often call these *the* coefficients of ψ. In particular, we may assume without loss of generality that the polynomials ρ and σ associated with $\psi[f]$ *have no common zero.*

With Def. 4.1.1—4.1.3 we have attempted—as we did with RK-methods—to distinguish conceptually between the actual forward step *procedure* by which the solution of the discretization is computed for a given IVP 1, and the k-step *scheme* which contains the essential information about the construction of that procedure independently of a particular function f.

It would have been tempting to write (4.1.2) in the form

(4.1.9)

$$\frac{\eta_\nu - \eta_{\nu-1}}{h} - \left[-\frac{1}{\alpha_k h}\left(\sum_{\kappa=0}^{k-1} \alpha_\kappa \eta_{\nu-k+\kappa} + \alpha_k \eta_{\nu-1} - h \sum_{\kappa=0}^{k} \beta_\kappa f(\eta_{\nu-k+\kappa})\right)\right] = 0$$

so that the expression in brackets would have been the exact analog of the increment function of a RK-method, cf. (3.1.11). However, this would have meant a deviation from all the existing literature and the advantages (if any) would not have justified this innovation. We suggest that the interested reader formulate some of the well-known linear k-step procedures in the form (4.1.9) and try to parallel some of the developments of Chapter 3 for the "increment functions" of these procedures.

Def. 4.1.4. A linear k-step scheme ψ and its associated linear k-step procedure are called *explicit* or a *predictor* if $\beta_k=0$, otherwise *implicit* or a *corrector*.

Remark. Note that the condition $\beta_k=0$ is invariant under equivalence so that it is indeed a property of the scheme ψ.

An explicit linear k-step procedure (4.1.2) may be solved for η_ν immediately ($\alpha_k \neq 0$!). An implicit linear k-step procedure defines a unique η_ν if h is sufficiently small:

Lemma 4.1.2. *There is an* $h_0>0$ *such that* (4.1.2) *has a unique solution* $\eta_\nu \in B_R(z)$ (see (2.1.6)) *for* $|h|<h_0$ *and* $\eta_{\nu-k+\kappa} \in B_0 := B_{R/2a}(z)$, $\kappa=0(1)k-1$, *where* R *is from* (2.1.6) *and* $a := \dfrac{1}{|\alpha_k|}\sum_{\kappa=0}^{k-1} |\alpha_\kappa|$.

Proof. By (2.1.6), f possesses the uniform Lipschitz constant L in $B_R(z)$, $R>0$. Thus, the mapping

$$F(\eta; h, \xi_{k-1}, \ldots, \xi_0) := h \frac{\beta_k}{\alpha_k} f(\eta) - \frac{1}{\alpha_k} \sum_{\kappa=0}^{k-1} [\alpha_\kappa \xi_\kappa - h \beta_\kappa f(\xi_\kappa)] \tag{4.1.10}$$

possesses a Lipschitz constant smaller than 1 w.r.t. η if $\eta \in B_R(z)$ and $|h|<|\alpha_k|/L|\beta_k|$. Furthermore, since f is bounded in $B_R(z)$ by some M_f, we have

$$F(\eta; h, \xi_{k-1}, \ldots, \xi_0) \in B_R(z) \quad \text{for } \eta \in B_R(z)$$

if $|h|<R|\alpha_k|/(2M_f \sum|\beta_\kappa|)$ and $\xi_\kappa \in B_0$ as defined in the Lemma, $\kappa=0(1)k-1$. (Note that $\sum_{\kappa=0}^{k-1} \alpha_\kappa = -\alpha_k$ due to (4.1.3).)

Now (4.1.2) is equivalent to $\eta_\nu = F(\eta_\nu; h, \eta_{\nu-1}, \ldots, \eta_{\nu-k})$ and the assertion holds with

$$h_0 = \min\left(\frac{1}{L}\left|\frac{\alpha_k}{\beta_k}\right|, \frac{R}{2M_f} \frac{|\alpha_k|}{\sum_\kappa |\beta_\kappa|}\right), \tag{4.1.11}$$

since F is a contraction w.r.t. its first variable on $B_R(z)$. □

Remarks. 1. In our discussion of linear k-step procedures, h_0 and B_0 will always denote the quantities of Lemma 4.1.2 and the solution η_ν of (4.1.2) is assumed to be the unique solution in $B_R(z)$.

2. When (4.1.2) is implicit, "solving" it by means of a *truncated* iteration (with a suitable $\eta_\nu^{(0)}$)

$$\text{(4.1.12)} \qquad \begin{cases} \eta_\nu^{(i)} := F(\eta_\nu^{(i-1)}; h, \eta_{\nu-1}, \ldots, \eta_{\nu-k}), & i=1(1)m, \\ \eta_\nu := \eta_\nu^{(m)}, \end{cases}$$

with m fixed, amounts to replacing (4.1.2) by an explicit $(m+1)$-stage k-step procedure. Therefore, (4.1.12) is *not* equivalent to (4.1.2) although it may often be acceptable as an algorithmic approximation to (4.1.2).

Multistage k-step procedures of the type (4.1.12) are commonly called *predictor-corrector procedures*, their theory will be taken up in Section 5.2. Throughout Chapter 4 we will assume that η_ν is the *exact* solution of (4.1.2).

The coefficients α_κ and β_κ may be chosen in various ways such that (4.1.2) approximates $y'-f(y)$ for $\eta_\nu = y(t_\nu)$. Several natural approaches which lead to important classes of linear k-step schemes have been discussed thoroughly in Henrici [1], Section 5.1, and in the older literature. A few examples are given below; for details the reader is referred to the literature. Other special classes of linear k-step schemes will appear later at the appropriate places. A general approach to the construction of linear k-step schemes is presented in Section 4.1.3 below.

Examples. 1. *Adams-schemes.* The Adams-schemes are characterized by[2]

$$\alpha_k = 1, \quad \alpha_{k-1} = -1, \quad \alpha_\kappa = 0, \quad \kappa = 0(1)k-2;$$

for the choice of the β_κ see Theorem 4.1.9. Tables of the coefficients β_κ are to be found, e. g., in Henrici [1], Section 5.1-2.

For example, the 4-step explicit and implicit Adams procedures are defined by the generating matrices

$$\tfrac{1}{24}\begin{pmatrix} 0 & 0 & 0 & -24 & 24 \\ 9 & 37 & -59 & 55 & 0 \end{pmatrix} \text{ and } \tfrac{1}{720}\begin{pmatrix} 0 & 0 & 0 & -720 & 720 \\ -19 & 106 & -264 & 646 & 251 \end{pmatrix}.$$

Note that the Adams procedures are automatically of the form (4.1.9); the explicit and implicit 1-step Adams procedures are identical to the Euler and trapezoidal RK-procedures.

2. *Differentiation schemes.* The differentiation schemes are characterized by[2]

$$\beta_k = 1, \quad \beta_\kappa = 0, \quad \kappa = 0(1)k-1;$$

for the choice of the α_κ see Theorem 4.1.9. A table of the coefficients can be found in Liniger [1].

---

[2] The coefficients are those of the unique representative satisfying (4.1.8).

The 4-step differentiation procedure is given by

$$\frac{1}{12}\begin{pmatrix} 3 & -16 & 36 & -48 & 25 \\ 0 & 0 & 0 & 0 & 12 \end{pmatrix}.$$

The 1-step differentiation procedure is identical to the implicit Euler RK-procedure.

3. *The two-step midpoint and Simpson schemes.* These linear 2-step schemes are given by[2]

$$\alpha_0 = -\tfrac{1}{2}, \quad \alpha_1 = 0, \quad \alpha_2 = +\tfrac{1}{2},$$

$$\beta_0 = \begin{cases} 0 \\ \frac{1}{6} \end{cases}, \quad \beta_1 = \begin{cases} 1 \\ \frac{2}{3} \end{cases}, \quad \beta_2 = \begin{cases} 0 & \text{(midpoint)} \\ \frac{1}{6} & \text{(Simpson)} \end{cases}.$$

They are related to the numerical integration rules denoted by the same names in an obvious manner.

4.1.2 The Order of Linear k-step Schemes

So far we have not referred the values η_ν in (4.1.2) to the points t_ν of a particular grid. (Of course, we viewed h as a parameter characterizing the steps $h_{\nu-k+\kappa} = t_{\nu-k+\kappa} - t_{\nu-k+\kappa-1}$, $\kappa = 1(1)k$; see Lemma 4.1.2.) To avoid duplication of various developments we will admit variable steps h_ν in the remainder of this section.

We will assume that

(4.1.13) $$h_{\nu-k+\kappa} = r_\kappa h, \quad r_\kappa > 0, \quad \kappa = 1(1)k,$$

so that

(4.1.14) $$t_{\nu-k+\kappa} = t_{\nu-k} + \bar{r}_\kappa h, \quad \kappa = 0(1)k,$$

where

(4.1.15) $$\bar{r}_\kappa := \sum_{\lambda=1}^{\kappa} r_\lambda, \quad \kappa = 1(1)k; \quad \bar{r}_0 := 0.$$

Def. 4.1.5. In the situation of (4.1.13)—(4.1.15), the r_κ will be called *step ratios*, $r := (r_1, \ldots, r_k)^T \in \mathbb{R}^k$ the *step ratio vector*. $r = e := (1, \ldots, 1)^T$ will refer to the equidistant case where $\bar{r}_\kappa = \kappa$, $\kappa = 0(1)k$.

Def. 4.1.6. For a given linear k-step scheme ψ and a given step ratio vector $r \in \mathbb{R}^k$, the linear difference-differential operators $L_h[\psi, r]$: $C^{(1)}[0,1] \to C[0,1]$, $h > 0$, defined by[3]

(4.1.16) $$(L_h[\psi, r]y)(t) := \frac{1}{h} \sum_{\kappa=0}^{k} [\alpha_\kappa y(t + \bar{r}_\kappa h) - h\beta_\kappa y'(t + \bar{r}_\kappa h)]$$

---

[2] The coefficients are those of the unique representative satisfying (4.1.8).

[3] α_κ and β_κ are the coefficients of that generating matrix in ψ with minimal k and $\sum \beta_\kappa = 1$, cf. the discussion after Def. 4.1.3.

are called *associated difference operators* of ψ for r. The functions

(4.1.17) $$\rho[r](x) := \sum_{\kappa=0}^{k} \alpha_\kappa x^{\bar{r}_\kappa}, \qquad \sigma[r](x) := \sum_{\kappa=0}^{k} \beta_\kappa x^{\bar{r}_\kappa}$$

are the *associated functions* of ψ for r.

If r is not indicated (or clear from the context) $r=e$ is assumed, $\rho[e]=:\rho$ and $\sigma[e]=:\sigma$ are called *associated polynomials* of ψ. The linear k-step scheme with associated polynomials ρ and σ will be denoted by $\langle\rho,\sigma\rangle$.

Remark. Note that (4.1.3) implies

(4.1.18) $$\rho[r](1)=0 \quad \text{for any } r;$$

furthermore, (4.1.8) implies

(4.1.19) $$\sigma[r](1)=1 \quad \text{for any } r.$$

A handy and widely used abbreviation for the sums which occur in (4.1.2) and (4.1.16) is the following: For any function y from $\mathbb{R}$ or a subset of $\mathbb{R}$ to some domain X let

$$T_h y(t) := y(t+h).$$

Then, in the situation of (4.1.13)—(4.1.15), (4.1.2) may be written as

$$\frac{1}{h}\{\rho[r](T_h)\eta_{\nu-k} - h\sigma[r](T_h)f(\eta_{\nu-k})\}=0$$

and the associated difference operators of $\psi=\langle\rho,\sigma\rangle$ are simply

(4.1.20) $$L_h[\psi,r] = \frac{1}{h}(\rho[r](T_h) - h\sigma[r](T_h)\partial),$$

with $\partial y(t) := y'(t)$ for a differentiable function $y\colon \mathbb{R}\to X$.

The forward step procedure (4.1.2) generated by a linear k-step scheme $\langle\rho,\sigma\rangle$ for a given IVP 1 will represent a good local approximation to the IVP 1 in the situation of (4.1.13)—(4.1.15) if $L_h[\langle\rho,\sigma\rangle,r]y$ is small for sufficiently smooth functions y:

Def. 4.1.7. A linear k-step scheme $\langle\rho,\sigma\rangle$ is called *of order p for r* if its associated difference operators for r satisfy, for any $y\in D^{(p)}$ (see (2.1.7))

(4.1.21) $$(L_h[\langle\rho,\sigma\rangle,r]y)(t) = O(h^p) \quad \text{as } h\to 0.$$

$\langle\rho,\sigma\rangle$ is said to be *of order p* if it is order p for e. It is said to be of *exact* order p (for r) if p is the greatest integer such that (4.1.21) holds.

If (4.1.21) holds, the difference operators L_h themselves are also said to be of (exact) order p for r and so are the linear multistep procedures generated by $\psi=\langle\rho,\sigma\rangle$.

Theorem 4.1.3. *Each of the following conditions is necessary and sufficient that the linear k-step scheme $\langle\rho,\sigma\rangle$ is of order p for r:*

(4.1.22) $$\sum_{\kappa=0}^{k}\left(\frac{\alpha_\kappa \bar{r}_\kappa^{\mu+1}}{\mu+1}-\beta_\kappa \bar{r}_\kappa^{\mu}\right)=0 \quad \textit{for } \mu=0(1)p-1;$$

(4.1.23) $$\frac{1}{h}\rho[r](e^h)-\sigma[r](e^h)=O(h^p) \quad \textit{as } h\to 0;$$

(4.1.24) $$\frac{\rho[r](x)}{\log x}-\sigma[r](x)=O((x-1)^p) \quad \textit{as } x\to 1,$$

where $\log x$ *is the principal value of the logarithm function;*

(4.1.25) $L_h[\langle\rho,\sigma\rangle,r]$ *annihilates all polynomials of degree p.*

Proof. Taylor expansion of the right hand side of (4.1.16) yields

(4.1.26)
$$\begin{aligned}(L_h[\langle\rho,\sigma\rangle,r]y)(t) &= \frac{1}{h}\sum_{\kappa=0}^{k}\left[\alpha_\kappa\sum_{\mu=0}^{p}\frac{(\bar{r}_\kappa h)^\mu}{\mu!}y^{(\mu)}(t)-\beta_\kappa h\sum_{\mu=0}^{p-1}\frac{(\bar{r}_\kappa h)^\mu}{\mu!}y^{(\mu+1)}(t)\right] + O(h^p)\\ &=\sum_{\mu=0}^{p-1}\sum_{\kappa=0}^{k}\left(\frac{\alpha_\kappa \bar{r}_\kappa^{\mu+1}}{\mu+1}-\beta_\kappa \bar{r}_\kappa^{\mu}\right)\frac{h^\mu}{\mu!}y^{(\mu+1)}(t)+O(h^p)\end{aligned}$$

due to (4.1.3); this proves (4.1.22).

The expansion of (4.1.23) in powers of h yields

$$\frac{1}{h}\sum_{\kappa=0}^{k}\left[\alpha_\kappa e^{\bar{r}_\kappa h}-h\beta_\kappa e^{\bar{r}_\kappa h}\right]=\sum_{\mu=0}^{p-1}\sum_{\kappa=0}^{k}\left(\frac{\alpha_\kappa \bar{r}_\kappa^{\mu+1}}{\mu+1}-\beta_\kappa \bar{r}_\kappa^{\mu}\right)\frac{h^\mu}{\mu!}+O(h^p)$$

and a comparison with (4.1.26) proves (4.1.23).

(4.1.24) arises from (4.1.23) by the transformation $e^h=x$, $h=\log x$, $O(h)=O((x-1))$.

(4.1.25) implies and is implied by (4.1.22) according to (4.1.26). □

Remark. (4.1.23) shows that the conditions for order p may also be expressed in terms of the derivatives of ρ and σ at 1:

(4.1.27)
$$\begin{aligned}\rho'[r](1)=\sigma[r](1)&=1 && \text{for } p=1,\\ \rho''[r](1)+\rho'[r](1)&=2\sigma'[r](1) && \text{for } p=2,\\ \rho'''[r](1)+3\rho''[r](1)+\rho'[r](1)&=3(\sigma''[r](1)+\sigma'[r](1)) && \text{for } p=3,\\ &\text{etc.}\end{aligned}$$

Corollary 4.1.4. *All equivalent linear multistep procedures* (4.1.2) *for which the associated polynomial* ρ *does not possess a multiple zero at* 1 *are of the same exact order.*

Proof. Let $\psi = \langle \rho_0, \sigma_0 \rangle$ be an equivalence class of generating matrices for linear multistep procedures. By Def. 4.1.2 and the distinction of the representative matrix, an arbitrary generating matrix from ψ has associated polynomials $\rho(x) = c(x)\rho_0(x)$, $\sigma(x) = c(x)\sigma_0(x)$. If $c(1) \neq 0$ the assertion follows from (4.1.23). $c(1) = 0$ implies that ρ has a multiple zero at 1 which we have excluded. □

Remark. Corollary 4.1.4 shows that order is an invariant of the equivalence class so that it was correct to define the order of a linear k-step *scheme* in Def. 4.1.7. The restriction $\rho'(1) = \sum_\kappa \kappa \alpha^\kappa \neq 0$ is natural (and should perhaps have been included in the definition of a linear multistep procedure) since (4.1.2) cannot approximate an IVP 1 otherwise.

The equivalence of (4.1.21) and (4.1.23) is, of course, due to the fact that e^h is the generating function for the coefficients of the Taylor-expansion. We will later need the following

Lemma 4.1.5. *Let ρ, σ, and τ be polynomials, $\rho(1) = 0$, $\tau(1) \neq 0$. If the sequence of coefficients $c_\mu, \mu > 0$, is defined by*

$$\frac{\rho(e^h) - h\sigma(e^h)}{h\tau(e^h)} = \sum_{\mu=0}^{\infty} c_\mu h^\mu \tag{4.1.28}$$

then, for $y \in D^{(p)}$,

$$\frac{1}{h}[\rho(T_h)y(t) - h\sigma(T_h)y'(t)] = \tau(T_h) \sum_{\mu=0}^{p-1} c_\mu h^\mu y^{(\mu+1)}(t) + O(h^p). \tag{4.1.29}$$

Proof. Taylor expansion of the two sides of (4.1.29) leads to the same relations between the coefficients of ρ, σ, τ, and the c_μ as does expansion in powers of h of the two sides of

$$\rho(e^h) - h\sigma(e^h) = h\tau(e^h) \sum_{\mu=0}^{\infty} c_\mu h^\mu. \quad \square$$

Def. 4.1.8. For a linear k-step scheme $\langle \rho, \sigma \rangle$ which is of exact order p for r, the quantity

$$C[r] := \frac{1}{p!} \sum_{\kappa=1}^{k} \left(\frac{\alpha_\kappa \overline{r}_\kappa^{p+1}}{p+1} - \beta_\kappa \overline{r}_\kappa^p \right) \tag{4.1.30}$$

is called *error constant* of $\langle \rho, \sigma \rangle$ *for r*, $C := C[e]$ is called *error constant* of $\langle \rho, \sigma \rangle$.

Corollary 4.1.6. *The associated difference operator for r of a linear k-step scheme ψ of exact order p for r satisfies, for any $y \in D^{(p+1)}$ and arbitrary but fixed μ,*

$$\begin{aligned}(L_h[\psi, r]y)(t) &= C[r]h^p y^{(p+1)}(t + \mu h) + O(h^{p+1}) \\ &= C[r]h^p \sigma[r](T_h) y^{(p+1)}(t) + O(h^{p+1}).\end{aligned} \tag{4.1.31}$$

Proof. (4.1.26) with $p+1$ in place of p yields (4.1.31) with t as the argument of $y^{(p+1)}$. But $y^{(p+1)}(t+\mu h)=y^{(p+1)}(t)+O(h)$ for $y\in D^{(p+1)}$; furthermore, $\sigma[r](1)=1$ by (4.1.19). □

Remark. As in Theorem 4.1.3 it is seen that the following are equivalent definitions of $C[r]$:

$$C[r]=\lim_{h\to 0}\frac{1}{h^{p+1}}[\rho[r](e^h)-h\sigma[r](e^h)]\,; \tag{4.1.32}$$

$$C[r]=\lim_{x\to 1}\frac{1}{(x-1)^p}\left[\frac{\rho[r](x)}{\log x}-\sigma[r](x)\right]. \tag{4.1.33}$$

Example. 2-step midpoint and Simpson schemes, see Example 3 at the end of Section 4.1.1:

a) The associated polynomials of the midpoint scheme are

$$\rho(x)=\tfrac{1}{2}(x^2-1),\qquad \sigma(x)=x.$$

The conditions (4.1.27) hold for $p=1$ and 2 but not for $p=3$, hence the midpoint scheme is of exact order 2 (for equal steps).

If we use a general step ratio vector $r=(r_1,r_2)$ we can retain at least order 1 for r with a scheme of the same general structure if we choose $\rho[r](x)=(x^{r_1+r_2}-1)/(r_1+r_2)$, $\sigma[r](x)=x$.

b) The associated polynomials of the Simpson scheme are

$$\rho(x)=\tfrac{1}{2}(x^2-1)=(x-1)[1+\tfrac{1}{2}(x-1)],$$
$$\sigma(x)=\tfrac{1}{6}(x^2+4x+1)=1+(x-1)+\tfrac{1}{6}(x-1)^2.$$

Furthermore,

$$\frac{1}{\log x}=\left[\sum_{\mu=1}^{\infty}(-1)^{\mu+1}\frac{(x-1)^\mu}{\mu}\right]^{-1}$$
$$=\frac{1}{x-1}[1+\tfrac{1}{2}(x-1)-\tfrac{1}{12}(x-1)^2+\tfrac{1}{24}(x-1)^3-\tfrac{19}{720}(x-1)^4+\cdots]$$

so that

$$\frac{\rho(x)}{\log x}-\sigma(x)=[1+\tfrac{1}{2}(x-1)][1+\tfrac{1}{2}(x-1)-\tfrac{1}{12}(x-1)^2+\tfrac{1}{24}(x-1)^3-\tfrac{19}{720}(x-1)^4+\cdots]$$
$$-[1+(x-1)+\tfrac{1}{6}(x-1)^2]=-\tfrac{1}{180}(x-1)^4+O((x-1)^5).$$

Thus, by (4.1.33) the Simpson scheme is of order 4 and its error constant $C=-\frac{1}{180}$.

4.1.3 Construction of Linear k-step Schemes of High Order

According to Theorem 4.1.3, the coefficients of a linear k-step scheme of order p (for r) have to satisfy p linear relations (4.1.22). We may therefore except that the attainable order of linear k-step schemes is equal to the number of free parameters which are, in view of the restrictions (4.1.18)—(4.1.19),

$$\left.\begin{matrix}2k-1\\ 2k\end{matrix}\right\}\text{ for an }\begin{matrix}\text{explicit}\\ \text{implicit}\end{matrix}\ k\text{-step scheme}.$$

As a basis for a simple construction method for high order multistep schemes we will display yet another relation between the coefficients and the order of a linear k-step scheme:

Lemma 4.1.7. *For a given linear k-step scheme ψ, the differential-difference operators $L_h[\psi,r]$ are of order p for r if and only if the rational function of x*

(4.1.34) $$\psi[r](x) := \frac{1}{h}\sum_{\kappa=0}^{k}\left[\frac{\alpha_\kappa}{x-\bar{r}_\kappa h} - h\frac{\beta_\kappa}{(x-\bar{r}_\kappa h)^2}\right], \qquad x\in\mathbb{C},$$

satisfies

(4.1.35) $$\psi[r](x)=O(|x|^{-(p+2)}) \quad as \ |x|\to\infty.$$

Proof. For any polynomial y and fixed $t\in\mathbb{R}$ we have

(4.1.36) $$(L_h[\psi,r]y)(t) = \frac{1}{2\pi i}\oint_{|x|=R} \psi[r](x-t)y(x)dx$$

for $R>\max_\kappa |t+\bar{r}_\kappa h|$ as follows from (4.1.34) and the Residual Theorem of complex function theory. If $\psi[r](x-t)y(x)=O(|x|^{-2})$ as $|x|$ becomes large the value of the integral is $O(R^{-1})$; since it is independent of R for sufficiently large R it has to vanish. Thus, (4.1.35) implies the vanishing of L_h for all polynomials of degree not larger than p which is equivalent to order p by (4.1.25).

If the expansion of $\psi[r](x)$ in powers of $1/x$ contains a term x^{-p-1+q}, $q\geq 0$, (4.1.36) cannot vanish for $y(x)=x^{p-q}$. This shows the necessity of (4.1.35). □

Theorem 4.1.8. *For each $k\geq 1$ and each step ratio vector $r\in\mathbb{R}^k$ there exists a unique implicit linear k-step scheme of order $2k$ for r and a unique explicit linear k-step scheme of order $2k-1$ for r.*

Proof. In the case of the corrector, we define the coefficients α_κ and β_κ, $\kappa=0(1)k$, by the partial fraction decomposition (4.1.34) of

(4.1.37) $$\psi[r](x)=K h^{2k}\prod_{\kappa=0}^{k}(x-\bar{r}_\kappa h)^{-2}=O(|x|^{-(2k+2)})$$

so that the order of the generated linear k-step scheme is $2k$ by Lemma 4.1.7. Since

$$\frac{\beta_\kappa}{K} = \prod_{\lambda\neq\kappa}{}' (\bar{r}_\kappa-\bar{r}_\lambda)^{-2}>0, \qquad \kappa=0(1)k,$$

we can choose K such that $\sum_\kappa \beta_\kappa=1$. Furthermore, by (4.1.34),

(4.1.38) $$\frac{1}{h}\alpha_\kappa = \operatorname*{Res}_{x=\bar{r}_\kappa h}\psi[r](x), \qquad \kappa=0(1)k;$$

since $\oint \psi[r](x)dx=0$ over a sufficiently large circle about the origin, the sum of the residuals (4.1.38) vanishes; this establishes (4.1.3).

In the case of the predictor, we take analogously

$$\psi[r](x)=Kh^{2k}(x-\bar{r}_k h)^{-1}\prod_{\kappa=0}^{k-1}(x-\bar{r}_\kappa h)^{-2} \tag{4.1.39}$$

to obtain a partial fraction decomposition (4.1.34) with $\beta_k=0$ and an order $p=2k-1$ for the associated linear k-step scheme. $\square$

Example. $k=2$; $r=(r_1,r_2)$, $t_{\nu-1}=t_{\nu-2}+r_1h$, $t_\nu=t_{\nu-2}+(r_1+r_2)h$. From (4.1.39) we obtain after some manipulation

$$\alpha_2=\frac{r_1^3}{n_\alpha},\qquad \alpha_1=\frac{-r_1^3+3r_2^2r_1+2r_2^3}{n_\alpha},\qquad \alpha_0=-\frac{r_2^2(3r_1+2r_2)}{n_\alpha},$$

$$\beta_2=0,\qquad \beta_1=\frac{r_1^2+2r_1r_2+r_2^2}{n_\beta},\qquad \beta_0=\frac{r_1r_2+r_2^2}{n_\beta},$$

with $n_\alpha:=r_1r_2n_\beta$, $n_\beta:=r_1^2+3r_1r_2+2r_2^2$, for the coefficients of the explicit linear 2-step scheme which is of order 3 for r. For $r=e=(1,1)$, the well-known values appear:

$$\alpha_2=\tfrac{1}{6},\quad \alpha_1=\tfrac{4}{6},\quad \alpha_0=-\tfrac{5}{6},\quad \beta_1=\tfrac{4}{6},\quad \beta_0=\tfrac{2}{6}.$$

(4.1.37) yields the coefficients of the implicit linear 2-step scheme of order 4 for r:

$$\alpha_2=2\frac{(r_1+2r_2)r_1^2}{r_2(r_1+r_2)n},\qquad \alpha_1=-2\frac{(r_1-r_2)(r_1+r_2)^2}{r_1r_2n},\qquad \alpha_0=-2\frac{(2r_1+r_2)r_2^2}{r_1(r_1+r_2)n},$$

$$\beta_2=\frac{r_1^2}{n},\qquad \beta_1=\frac{(r_1+r_2)^2}{n},\qquad \beta_0=\frac{r_2^2}{n},$$

with $n:=r_1^2+r_2^2+(r_1+r_2)^2$. For $r=e$, we recover the coefficients of the Simpson-scheme which had been found to be of order 4 in the example at the end of Section 4.1.2.

If either the coefficients α_κ or the β_κ of a linear k-step scheme are specified, the remaining free parameters may be fully utilized to satisfy the order conditions (4.1.22):

Theorem 4.1.9. *Let a step ratio vector $r\in\mathbb{R}^k$ be given. For each set of values α_κ, $\kappa=0(1)k$, with $\sum_{\kappa=0}^{k}\alpha_\kappa=0$, $\sum_{\kappa=0}^{k}\alpha_\kappa\bar{r}_\kappa=1$, there exist unique values β_κ, $\kappa=0(1)k'\le k$, such that the linear k-step scheme with these coefficients and $\beta_\kappa=0$, $\kappa>k'$, is of order $k'+1$ for r.*

Also for each set of values β_κ, $\kappa=0(1)k$, with $\sum_{\kappa=0}^{k}\beta_\kappa=1$, there exist unique values α_κ, $\kappa=0(1)k$, such that the linear k-step scheme with these coefficients is of order k for r.

Proof. In the case of specified α_κ, we form the expansion

$$\frac{\rho[r](x)}{\log x}=\sum_{\mu=0}^{\infty}c_\mu(x-1)^\mu.$$

In view of (4.1.24) we would then like to choose the β_κ in

$$\sigma[r](x) = \sum_{\kappa=0}^{k'} \beta_\kappa x^{\overline{r}_\kappa} = \sum_{\kappa=0}^{k'} \beta_\kappa (1+(x-1))^{\overline{r}_\kappa} = \sum_{\mu=0}^{\infty} \left(\sum_{\kappa=0}^{k'} \beta_\kappa \binom{\overline{r}_\kappa}{\mu} \right) (x-1)^\mu$$

such that

$$\text{(4.1.40)} \qquad \sum_{\kappa=0}^{k'} \binom{\overline{r}_\kappa}{\mu} \beta_\kappa = c_\mu , \qquad \mu = 0(1)k' .$$

The determinant of the linear system (4.1.40) for the β_κ is a multiple of the Vandermonde determinant of the $\overline{r}_\kappa$ which are distinct by (4.1.15) and $r_\kappa > 0$. Thus the β_κ are uniquely determined and generate a k-step scheme of order $k'+1$ by (4.1.24). The assumptions on the α_κ imply $c_0 = 1 = \sum_\kappa \beta_\kappa$.

If the β_κ are given we form analogously

$$\sigma[r](x) \log x = \sum_{\mu=1}^{\infty} d_\mu (x-1)^\mu$$

and determine the α_κ from

$$\sum_{\kappa=0}^{k} \binom{\overline{r}_\kappa}{\mu} \alpha_\kappa = \begin{cases} 0, & \mu = 0, \\ d_\mu , & \mu = 1(1)k . \end{cases}$$

The assertion about the order follows again from (4.1.24). □

Example. Construction of an optimal order 3-step Adams predictor and corrector for the step ratio vector $(1, 1, r_3)$:

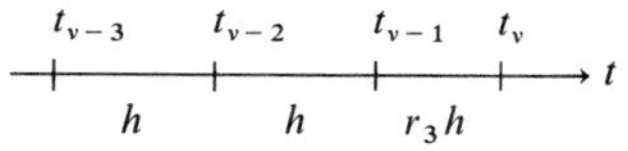

$t_{\nu-3}$ $t_{\nu-2}$ $t_{\nu-1}$ t_ν t

h h $r_3 h$

Fig. 4.1. Section of a non-equidistant grid

We start with $\alpha_3 = -\alpha_2 = 1/r_3$, $\alpha_1 = \alpha_0 = 0$, to have $\rho[r](1) = 0$ and $\rho'[r](1) = 1$ and expand

$$\rho[r](x) = \frac{1}{r_3}[x^{2+r_3} - x^2] = \frac{1}{r_3}[(1+(x-1))^{2+r_3} - (1+(x-1))^2]$$

$$= (x-1)\left[1 + \frac{3+r_3}{2}(x-1) + \frac{2+3r_3+r_3^2}{2}(x-1)^2 + \frac{-2-r_3+2r_3^2+r_3^3}{24}(x-1)^3 + \cdots\right]$$

which yields (see the example at the end of Section 4.1.2)

$$\frac{\rho(x)}{\log x} = 1 + \left(2 + \frac{r_3}{2}\right)(x-1) + \left(1 + \frac{3}{4} r_3 + \frac{r_3^2}{6}\right)(x-1)^2 + \frac{r_3}{24}(4 + 4r_3 + r_3^2)(x-1)^3 + \cdots$$

For the predictor scheme, we have simply $\sigma(x) = \beta_0 + \beta_1 x + \beta_2 x^2$ and a short calculation yields

$$\beta_2 = 1 + \frac{3}{4} r_3 + \frac{r_3^2}{6}, \quad \beta_1 = -r_3\left(1 + \frac{r_3}{3}\right), \quad \beta_0 = r_3\left(\frac{1}{4} + \frac{r_3}{6}\right).$$

For the corrector scheme, we have to convert

$$\sigma(x)=\beta_0+\beta_1 x+\beta_2 x^2+\beta_3 x^{2+r_3}$$

into an expansion in powers of $(x-1)$ and compare coefficients to obtain

$$\beta_3=\frac{2+r_3}{4(1+r_3)},\quad \beta_2=\frac{1}{2}+\frac{r_3}{4}+\frac{r_3^2}{24},\quad \beta_1=-\frac{r_3^2(4+r_3)}{12(1+r_3)},\quad \beta_0=\frac{r_3^2}{24}.$$

For $r_3=1$, we recover, of course, the well-known values of the standard 3-step Adams-schemes.

Fig. 4.1 indicates the use of linear k-step schemes of a reasonably high order for a step ratio vector different from e: If the *step is changed* these schemes may be used on the section of the grid with unequally spaced gridpoints (cf. Theorem 4.2.1).

4.2 Uniform Linear k-step Methods

4.2.1 Characterization, Consistency

Def. 4.2.1. For any linear k-step scheme $\psi=\langle\rho,\sigma\rangle$ a *uniform linear k-step method based on* ψ is a f.s.m. applicable to IVP 1 defined by ($n\in\mathbb{N}'$)

$E_n=E_n^0=(\mathbb{G}_n\to\mathbb{R}^s)$, where $\{\mathbb{G}_n\}_{n\in\mathbb{N}'}$ is a coherent gridsequence, with norms (2.2.2) and (2.2.3) resp.;

$$\left.\begin{array}{ll}
(\Delta_n y)(t_\nu)=y(t_\nu); & \\
(4.2.1)\quad \Delta_n^0\begin{pmatrix} d_0\\ d(t)\end{pmatrix}(t_\nu)=\begin{cases} d_0+\int_0^{t_\nu} d(t)\,dt, & \nu=0(1)k-1,\\ \sigma(T_{h_\nu})d(t_{\nu-k}), & \nu=k(1)n;\end{cases} & \\
(4.2.2)\quad [\varphi_n(F)\eta](t_\nu)=\begin{cases}\eta_\nu-s_{n\nu}(z_0), & \nu=0(1)k-1.\\ \psi[f](h_\nu;\eta_\nu,\dots,\eta_{\nu-k}), & \nu=k(1)n;\end{cases} &
\end{array}\right\}\ t_\nu\in\mathbb{G}_n.$$

If the value of the stepsize function of the coherent gridsequence $\{\mathbb{G}_n\}$ changes at t_ν (cf. Def. 2.1.11) the method has to specify the procedure to be used for the computation of $\eta_{\nu+\kappa}$, $\kappa=1(1)k-1$.

Remark. Note that (4.2.1) satisfies the condition in Def. 1.1.2 since $|t_\nu|\to 0$ for $\nu=1(1)k-1$ as $n\to\infty$ and the weighted sum $\sigma(T_h)d(t_{\nu-k})$ satisfies (4.1.8). The above definition of the Δ_n^0 will be adapted to particular needs in some contexts. (4.2.1) will be used whenever no other specification is indicated.

In a coherent gridsequence the number of possible *changes in stepsize* remains constant as n increases; see Def. 2.1.11. Therefore the character of the special forward step procedures which are used at step changes is of little importance as long as certain consistency requirements are satisfied (see Theorems 4.2.1 and 4.2.2 below). Without much loss of generality we will assume that linear k-step procedures (4.1.2) are used. This includes all procedures which involve polynomial interpolation of the η_ν and $f(\eta_\nu)$ values.

Similarly, the precise character of the starting procedure of a uniform linear k-step method is often of little importance as long as it satisfies certain consistency requirements (see Theorems 4.2.1 and 4.2.2, and Lemma 2.2.1 and the remark following it). Of course, the specification of a particular starting procedure will be considered an intrinsic part of the specification of a particular uniform linear k-step method in accordance with our general concepts.

It is obvious that Def. 4.2.1 does not include the so-called predictor-corrector methods, cf. also Remark 2 after Lemma 4.1.2. At present we are considering only k-step methods which use a *single* linear k-step procedure at each step to obtain the value of the solution of the discretization at the next gridpoint.

Example. The so-called extrapolation or Gragg-Bulirsch-Stoer method employs the following uniform linear 2-step method (see Section 6.3.2):

The starting procedure is

$$\eta_0 - z_0 = 0, \qquad \eta_1 - [z_0 + h f(z_0)] = 0;$$

the forward step procedure is based on the 2-step midpoint scheme:

$$\psi[f](h_\nu; \eta_\nu, \eta_{\nu-1}, \eta_{\nu-2}) = \frac{\eta_\nu - \eta_{\nu-2}}{2h_\nu} - f(\eta_{\nu-1}) = 0;$$

if the step changes at $t_{\nu-1}$ this is replaced by

$$\frac{\eta_\nu - \eta_{\nu-2}}{h_{\nu-1} + h_\nu} - f(\eta_{\nu-1}) = 0.$$

A uniform linear k-step method approximates any IVP 1 arbitrarily well as $n \to \infty$ under very mild assumptions:

Theorem 4.2.1. *A uniform linear k-step method based upon ψ is consistent with an IVP 1 at any $y \in E$ if*

(i) *its starting procedure is consistent with the IVP 1;*

(ii) *its linear k-step scheme ψ is (at least) of order 1;*

(iii) *any linear multistep scheme is used at step changes.*

Proof. We have to show that, for $y \in E$ (see (2.1.3)),

$$\lim_{n \to \infty} \| F_n \Delta_n y - \Delta_n^0 F y \|_{E_n^0} = 0.$$

Let $\delta_\nu := [F_n \Delta_n y - \Delta_n^0 F y](t_\nu)$, $t_\nu \in \mathbb{G}_n$. By (4.2.2), (2.1.5), and (4.2.1) we have for $\nu = 0(1)k-1$

$$\delta_\nu = y(t_\nu) - s_{n\nu}(z_0) - \left[y(0) - z_0 + \int_0^{t_\nu} (y'(\tau) - f(y(\tau)))\, d\tau \right]$$

$$= -\left[s_{n\nu}(z_0) - z_0 - \int_0^{t_\nu} f(y(\tau))\, d\tau \right]$$

and (i) implies $\lim_{n\to\infty} \max_{\nu=0(1)k-1} \|\delta_\nu\| = 0$ due to (2.2.7) and $t_\nu = O(n^{-1})$.

At any $t_\nu \in \mathbb{G}_n$, $\nu \geq k$, which is not within $k-1$ steps after a step change we have, by (4.2.2) and (4.2.1),

$$\delta_\nu = \frac{1}{h_\nu} [\rho(T_{h_\nu}) y(t_{\nu-k}) - h_\nu \sigma(T_{h_\nu}) f(y(t_{\nu-k}))] - \sigma(T_{h_\nu})[y'(t_{\nu-k}) - f(y(t_{\nu-k}))]$$

$$= (L_{h_\nu}[\psi] y)(t_{\nu-k}) = O(h_\nu) \quad \text{by (ii) and (4.1.21).}$$

At step changes, only the $O(h_\nu^{-1})$-term will drop out due to (4.1.18) if the linear k-step schemes used are not even of order 1. But in the formation of

$$\|\delta\|_{E_n^0} = \max_{\nu=0(1)k-1} \|\delta_\kappa\| + \frac{1}{n} \sum_{\nu=k}^{n} \|\delta_\nu\|$$

these finitely many terms are multiplied by $1/n$ so that nevertheless $\|\delta\|_{E_n^0} \to 0$ as $n \to \infty$. □

Remark. The above proof shows why it is natural to define the second component of $\Delta_n^0 d$ by means of $\sigma(T_h)$; see (4.2.1).

What is really wanted is, of course, a high order of consistency at the true solution z of a sufficiently smooth IVP 1.

Theorem 4.2.2. *A uniform linear k-step method based upon ψ is consistent of order p with an* IVP$^{(p)}$1 *if*

(i) *its starting procedure is of order p;*

(ii) *its linear k-step scheme ψ is of order p;*

(iii) *linear multistep schemes of order $p-1$ for the appropriate step ratio vector are used at step changes.*

Proof. As $Fz = 0$, we have only to check the behavior of $\|F_n \Delta_n z\|_{E_n^0}$ as $n \to \infty$; see Def. 1.1.5:

$$\text{(4.2.3)} \quad F_n \Delta_n z(t_\nu) = \begin{cases} z(t_\nu) - s_{n\nu}(z_0), & \nu = 0(1)k-1, \\ (L_{h_\nu}[\psi] z)(t_{\nu-k}), & \nu \geq k \text{ “normal”}, \\ (L_{h_\nu}[\tilde{\psi}_\nu, \tilde{r}_\nu] z)(t_{\nu-k}), & \nu \geq k \text{ after step changes,} \end{cases}$$

where L_h is the associated difference operator (Def. 4.1.6), $\tilde{\psi}_\nu$ is the particular scheme used at t_ν because of the step change and $\tilde{r}_\nu$ is the step ratio vector arising from the step change.

The assertion follows immediately from the assumptions, with (2.2.6), (4.1.21), and the norm definition (2.2.3). □

Example. As above.

We have $z(h)-[z_0+hf(z_0)]=O(h^2)$ so that the starting procedure is of order 2. Furthermore, the 2-step midpoint scheme is of order 2 and the 2-step scheme used at a step change is of order 1 for $\tilde{r}=(h_{\nu-1}/h, h_\nu/h)$, see the example at the end of Section 4.1.2. Thus, our linear 2-step method is consistent of order 2 with any IVP$^{(2)}$ 1.

From Section 4.1.3 we know that there are linear k-step schemes of high orders for arbitrary step ratio vectors, thus the assumptions (ii) and (iii) of Theorem 4.2.2 present no difficulties if k is taken large enough. But what about starting procedures of high order of consistency?

Theorem 4.2.3. *Let ψ_s be a RK-scheme of order $p-1$ and Ψ_s the associated increment function for a given* IVP$^{(p)}$1. *Then the starting procedure*

$$\text{(4.2.4)}\qquad \begin{aligned} s_{n0}(z_0) &= z_0, \\ s_{n\nu}(z_0) &= s_{n,\nu-1}(z_0)+h_\nu \Psi_s(h_\nu, s_{n,\nu-1}(z_0)), \qquad \nu=1(1)k-1, \end{aligned}$$

is consistent of order p with the IVP$^{(p)}$ 1.

Proof. (4.2.4) may be interpreted as the generation of the first k values of the solution of a discretization of the IVP$^{(p)}$1 by a uniform RK-method based on the RK-scheme ψ_s. Therefore, we have from (3.4.15)

$$\text{(4.2.5)}\qquad s_{n\nu}(z_0)=z(t_\nu)+\frac{1}{n^{p-1}}e_{p-1}(t_\nu)+O(n^{-p}), \qquad \nu=1(1)k-1,$$

where e_{p-1} is the solution of the differential equation (3.4.16) with initial value 0. Hence $e_{p-1}(t_\nu)=O(n^{-1})$ for $\nu\le k-1$ which proves the assertion. □

Remark. While it is sufficient for the achievement of consistency of order p to start with a RK-method of order $p-1$, it may be advisable to use a starting procedure of higher order in many cases (see also Section 4.4.1).

Example. As above.

Our starting procedure is one step of the Euler-method which is of order 1 only.

Another possibility for a high order starting procedure is the use of a truncated Taylor-expansion of the true solution z about 0. The needed values of the derivatives of z at 0 may be found by a recursive differentiation of the differential equation at 0:

$$z'(0)=f(z_0),$$
$$z''(0)=f'(z_0)\,f(z_0),$$
$$z'''(0)=f''(z_0)\,f(z_0)^2+f'(z_0)\,f'(z_0)\,f(z_0),$$
etc.

(The elaboration of the right-hand sides may be left to a computer in many cases; see Section 6.1.1.)

(4.2.6) $$s_{n\nu}(z_0)=z_0+\sum_{\mu=1}^{p-1}\frac{t_\nu^\mu}{\mu!}z^{(\mu)}(0), \qquad \nu=0(1)k-1,$$

trivially satisfies (2.2.6) for f from an IVP$^{(p)}$1.

RK-methods and Taylor-expansions may also be used to "restart" the computation after the step has been changed. The reader will be able to adapt Theorems 4.2.1 and 4.2.2 to this situation.

4.2.2 Auxiliary Results

Consider the scalar linear difference equation $(\alpha_k\neq 0)$

(4.2.7) $$\rho(T)\varepsilon_{\nu-k}:=\sum_{\kappa=0}^{k}\alpha_\kappa\varepsilon_{\nu-k+\kappa}=h\delta_\nu, \qquad \nu=k,k+1,\ldots,$$

with given initial values $\varepsilon_\kappa=\delta_\kappa$, $\kappa=0(1)k-1$, and given inhomogeneities δ_ν, $\nu\geq k$; $h>0$ is an arbitrary fixed parameter. As previously, $\rho(x):=\sum_{\kappa=0}^{k}\alpha_\kappa x^\kappa$. Although most of the following remains true for complex coefficients α_κ we will assume $\alpha_\kappa\in\mathbb{R}$ throughout in view of our applications (cf. remark below).

Denote the *different* zeros of ρ by $x_{i'}\in\mathbb{C}$ and their multiplicities by $m_{i'}$, $i'=1(1)l\leq k$. As is well-known (see, e.g., Henrici [1], Section 5.2−1) the k different sequences

(4.2.8) $$\{\nu^\mu x_{i'}^\nu\}, \qquad \mu=0(1)m_{i'}-1,\ i'=1(1)l,$$

form a *fundamental system* of (4.2.7) for $\nu\geq 0$, i.e. an arbitrary solution of the homogeneous equation (4.2.7) may be represented as a linear combination of the solutions (4.2.8). We number the fundamental solutions (4.2.8) from 1 to k and denote their elements by $\xi_{i\nu}$, $i=1(1)k$, $\nu\geq 0$.

Remark. If some $x_{i'}=r_{i'}e^{i\varphi_{i'}}$ is not real the conjugate-complex $x_{i'}^*=r_{i'}e^{-i\varphi_{i'}}$ is also a zero of ρ of the same multiplicity. The corresponding pairs of fundamental solutions

$$\{\nu^\mu x_{i'}^\nu\} \quad\text{and}\quad \{\nu^\mu (x_{i'}^*)^\nu\}$$

may be replaced by the pairs

$$\{\nu^\mu r_{i'}^\nu \cos \nu \varphi_{i'}\} \quad \text{and} \quad \{\nu^\mu r_{i'}^\nu \sin \nu \varphi_{i'}\}.$$

To have a simpler notation we will generally use the form (4.2.8) without considering the reality of the $x_{i'}$. Thus various quantities may be complex in the following analysis; but any expression which refers to a discretization of a real IVP1 can always be written in a real form, at the expense of a more cumbersome notation.

For $\kappa=0(1)k-1$, $i=1(1)k$, we define numbers $c_{\kappa i}$ by

$$\text{(4.2.9)} \qquad \sum_{i=1}^{k} c_{\kappa i}\xi_{i\nu} = \begin{cases} 1, & \nu=\kappa, \\ 0, & \nu \neq \kappa, \ \nu \in \{0,1,\ldots,k-1\}. \end{cases}$$

The $c_{\kappa i}$ are uniquely defined since the $\{\xi_{i\nu}\}$, $i=1(1)k$, form a fundamental system. With

$$\text{(4.2.10)} \qquad \gamma_{\kappa\nu} := \sum_{i=1}^{k} c_{\kappa i}\xi_{i\nu}, \quad \kappa = 0(1)k-1, \qquad \nu \geq 0,$$

and

$$\text{(4.2.11)} \qquad \gamma_{\mu\nu} := \begin{cases} 0, & \nu = 0(1)\mu-1, \\ \dfrac{1}{\alpha_k}\gamma_{k-1,\nu-\mu+k-1}, & \nu \geq \mu, \end{cases} \qquad \mu \geq k,$$

the solution of (4.2.7) takes the form

$$\text{(4.2.12)} \qquad \varepsilon_\nu = \sum_{\kappa=0}^{k} \gamma_{\kappa\nu}\delta_\kappa + h\sum_{\mu=k}^{\nu} \gamma_{\mu\nu}\delta_\mu, \qquad \nu \geq 0,$$

as is easily verified.

The $\gamma_{\kappa\nu}$ may also be characterized as the coefficients in the expansion

$$\text{(4.2.13)} \qquad \frac{\alpha_k + \alpha_{k-1}x + \cdots + \alpha_{\kappa+1}x^{k-\kappa-1}}{\alpha_k + \alpha_{k-1}x + \cdots + \alpha_0 x^k} = 1 + \sum_{\nu=k}^{\infty} \gamma_{\kappa\nu}x^{\nu-\kappa}$$

for each $\kappa=0(1)k-1$. To see the equivalence of (4.2.10) and (4.2.13) we write (4.2.13) in the form

$$\alpha_\kappa x^{k-\kappa} + \cdots + \alpha_0 x^k$$
$$+ (\alpha_k + \alpha_{k-1}x + \cdots + \alpha_0 x^k)(\gamma_{\kappa k}x^{k-\kappa} + \gamma_{\kappa,k+1}x^{k-\kappa+1} + \cdots) = 0$$

which shows that the expansion coefficients satisfy (cf. (4.2.7))

$$\sum_{\lambda=0}^{k} \alpha_\lambda \gamma_{\kappa,\nu-k+\lambda} = 0, \qquad \nu = k, k+1, \ldots,$$

when we define them by (4.2.9)/(4.2.10) for $\nu=0(1)k-1$.

Lemma 4.2.4. *If and only if all zeros of* ρ *are in the closed unit disk and no multiple zeros of* ρ *are on the unit circle there exists a constant* Γ *such that the solution of* (4.2.7) *satisfies*

$$|\varepsilon_\nu| \le \Gamma\left[\max_{\kappa=0(1)k-1} |\delta_\kappa| + h\sum_{\mu=k}^{\nu} |\delta_\mu|\right] \quad \textit{for all } \nu\ge 0. \tag{4.2.14}$$

Proof. It is clear from (4.2.8) that

$$|\xi_{i\nu}| \le K < \infty \quad \text{for } i=1(1)k \quad \text{and all } \nu\ge 0 \tag{4.2.15}$$

if and only if $|x_{i'}|\le 1$, $i'=1(1)l$, and $m_{i'}=1$ if $|x_{i'}|=1$. If (4.2.15) holds we obtain (4.2.14) from (4.2.10)—(4.2.12) with

$$\Gamma := \max\left(\sum_{\kappa=0}^{k-1} c_\kappa, \frac{c_{k-1}}{|\alpha_k|}\right), \quad \text{where } c_\kappa := \max_{\nu\ge 0} |\gamma_{\kappa\nu}|, \ \kappa=0(1)k-1, \tag{4.2.16}$$

and $c_\kappa<\infty$ due to (4.2.15).

If (4.2.15) does not hold for some $i=i_0$ the solution of (4.2.7) with $\delta_\kappa=\xi_{i_0\kappa}$, $\kappa=0(1)k-1$, $\delta_\mu=0$, $\mu\ge k$, does not satisfy (4.2.14). □

Remark. The assertion of Lemma 4.2.4 can also be obtained from the representation (4.2.13) for the $\gamma_{\kappa\nu}$; see Henrici [1], Section 5.3–2.

Lemma 4.2.5. *If and only if all zeros of* ρ *are in the open unit disk except for a simple zero at* 1 *there exists a constant* Γ_0 *such that the solution of* (4.2.7) *satisfies*

$$|\varepsilon_\nu| \le \Gamma_0\left[\max_{\kappa=0(1)k-1} |\delta_\kappa| + h\max_{\mu=k(1)\nu}\left|\sum_{\mu'=k}^{\mu} \delta_{\mu'}\right|\right] \quad \textit{for all } \nu. \tag{4.2.17}$$

Proof. a) Consider at first a difference equation (4.2.7) with coefficients $\bar\alpha_\kappa$, $\kappa=0(1)\bar k$, $\bar k\ge 1$, such that *all* zeros $\bar x_{i'}$ of $\bar\rho(x):=\sum_{\kappa=0}^{\bar k}\bar\alpha_\kappa x^\kappa$ are in the open unit disk. The associated $\bar\gamma_{\bar k-1,\nu}$ of (4.2.10) satisfy

$$\sum_{\nu'=\bar k-1}^{\nu-1} |\bar\gamma_{\bar k-1,\nu'}| = \sum_{\nu'=\bar k-1}^{\nu-1}\left|\sum_{i=1}^{\bar k}\bar c_{\bar k-1,i}\xi_{i\nu'}\right| \le \sum_{i=1}^{\bar k}|\bar c_{\bar k-1,i}| \max_{i=1(1)\bar k}\sum_{\nu'=\bar k-1}^{\nu-1}|\xi_{i\nu'}|$$

which implies that the $\bar\gamma_{\mu\nu}$ of (4.2.11) satisfy

$$\sum_{\mu=\bar k}^{\nu} |\bar\gamma_{\mu\nu}| \le \bar c < \infty \quad \text{uniformly for all } \nu\ge\bar k \tag{4.2.18}$$

due to (4.2.8) and the assumption on $\bar\rho$ $\left(\sum_{\nu=0}^{\infty}\nu^m|x^\nu|\right.$ is a convergent series for fixed m and $\left.|x|<1\right)$. From (4.2.18) and the representation (4.2.12) we obtain the estimate

$$|\varepsilon_\nu| \le \bar\Gamma\left[\max_{\kappa=0(1)\bar k-1}|\delta_\kappa| + h\max_{\mu=\bar k(1)\nu}|\delta_\mu|\right], \quad \nu\ge 0, \tag{4.2.19}$$

with some $\bar\Gamma<\infty$, for the solution of (4.2.7) with $\bar\rho$.

b) Consider now an equation (4.2.7) which satisfies the hypothesis of the Lemma and define $\bar{\alpha}_\kappa$, $\kappa=0(1)k-1$, by $\bar{\rho}(x):=\rho(x)/(x-1)$. With

(4.2.20) $$h\bar{\varepsilon}_\nu := \sum_{\kappa=0}^{k-1} \bar{\alpha}_\kappa \varepsilon_{\nu-k+1+\kappa}, \qquad \nu \geq k-1,$$

the difference equation (4.2.7) turns into

(4.2.21) $$\bar{\varepsilon}_\nu - \bar{\varepsilon}_{\nu-1} = \delta_\nu, \qquad \nu = k, k+1, \ldots,$$

with the initial value

(4.2.22) $$\bar{\varepsilon}_{k-1} := \frac{1}{h} \sum_{\kappa=0}^{k-1} \bar{\alpha}_\kappa \delta_\kappa;$$

(4.2.21) trivially implies

(4.2.23) $$\bar{\varepsilon}_\nu = \bar{\varepsilon}_{k-1} + \sum_{\mu=k}^{\nu} \delta_\mu.$$

On the other hand we may consider (4.2.20), with given $\bar{\varepsilon}_\nu$, as a difference equation for the ε_ν which is now of the type regarded in part a) of this proof as all zeros of $\bar{\rho}$ are in the open unit disk. Hence (4.2.19) implies, with (4.2.22) and (4.2.23),

$$\begin{aligned} |\varepsilon_\nu| &\leq \bar{\Gamma}\left[\max_{\kappa=0(1)k-2} |\delta_\kappa| + h \max_{\mu=k-1(1)\nu} |\bar{\varepsilon}_\mu|\right] \\ &\leq \bar{\Gamma}\left[\max_{\kappa=0(1)k-2} |\delta_\kappa| + \sum_{\kappa=0}^{k-1} |\bar{\alpha}_\kappa|\,|\delta_\kappa| + h \max_{\mu=k(1)\nu}\left|\sum_{\mu'=k}^{\mu} \delta_{\mu'}\right|\right] \end{aligned}$$

which leads to (4.2.17).

c) The necessity follows from the contradiction which is obtained for $\rho(x)=x^2-1=(x-1)(x+1)$; see Example 2 in Section 2.2.4. □

4.2.3 Stability of Uniform Linear k-step Methods

Def. 4.2.2. A polynomial is said to *satisfy the root criterion* if it has no zeros outside the closed unit disk and only simple zeros on the unit circle. It is said to *satisfy the strong root criterion* if it satisfies the root criterion and 1 is its only zero on the unit circle.

Def. 4.2.3. A linear k-step scheme is called *(strongly) D-stable* if its associated polynomial[4] ρ satisfies the (strong) root criterion.

Lemma 4.2.6. *Let $\mathfrak{P}_0$ denote the trivial IVP 1*

(4.2.24) $$y(0) = z_0 \in \mathbb{R}^s, \qquad y' = 0.$$

[4] Cf. the discussion after Def. 4.1.3.

A uniform linear k-step method $\mathfrak{M}$ based upon the linear k-step scheme ψ is stable for $\mathfrak{P}_0$ if and only if ψ is D-stable. If Spijker's norm is used for the E_n^0 in place of (2.2.3) *then $\mathfrak{M}(\mathfrak{P}_0)$ is stable if and only if ψ is strongly D-stable.*

Proof. a) If $\mathfrak{M}$ has equidistant grids, with steps $h=1/n$, it is easily verified that the stability of $\mathfrak{M}(\mathfrak{P}_0)$ (w.r.t. Spijker's norm) is equivalent to the fact that all solutions of (4.2.7) admit a bound (4.2.14) (resp. (4.2.17)) uniformly for $\nu \geq 0$. Thus Lemmata 4.2.4 and 4.2.5 imply the assertion.

b) Assume that step changes from $h^{(m)}$ to $h^{(m+1)}$ occur at t_{ν_m}, $m=1(1)M$, $t_{\nu_m} < t_{\nu_{m+1}}$. For the stability analysis we have to assume that the multistep procedure used for the computation of the values at $t_{\nu_m+\mu}$, $\mu=1(1)k-1$, is perturbed by $h^{(m+1)}\delta_{\nu_m+\mu}$ so that

(4.2.25)
$$\varepsilon_{\nu_m+\mu} = \sum_{\kappa=1}^{k} a_\kappa^{(\mu)} \varepsilon_{\nu_m-k+\kappa} + h^{(m+1)} \delta_{\nu_m+\mu}, \qquad \mu=1(1)k-1,$$

the $a_\kappa^{(\mu)}$ are fixed coefficients (independent of h) specified by the method $\mathfrak{M}$ (remember that we are considering (4.2.24)!).

If an estimate

(4.2.26)
$$\|\varepsilon_\nu\| \leq \Gamma_m \left[\max_{\kappa=0(1)k-1} \|\delta_\kappa\| + \frac{1}{n} \sum_{\mu=k}^{\nu} \|\delta_\mu\| \right]$$

is valid for $\nu \leq \nu_m$, (4.2.25) implies the analogous estimate for $\nu=\nu_m+\mu$, $\mu=1(1)k-1$, with a new constant in place of Γ_m. We may now think of ε_ν, $\nu=\nu_m(1)\nu_m+k-1$, as new *starting* values and apply Lemma 4.2.4 to obtain an estimate (4.2.26), with a constant Γ_{m+1} for all $\nu \leq \nu_{m+1}$. As the process which produces Γ_{m+1} from Γ_m is independent of n we may use the above argument inductively to show that—after the *fixed* number of M step changes—we still have an estimate (4.2.26) with a constant Γ_{M+1} independent of n. This concludes the proof in the case of the norm (2.2.3).

In the case of Spijker's norm (2.2.15) we have to use the fact that

(4.2.27)
$$\max_{\mu \leq \nu} \left\| \sum_{\mu'=k}^{\mu} \delta_{\mu'} \right\| + \|\delta_{\nu+1}\| \leq 3 \max_{\mu \leq \nu+1} \left\| \sum_{\mu'=k}^{\mu} \delta_{\mu'} \right\|$$

to show that an estimate

$$\|\varepsilon_\nu\| \leq \Gamma_m \left[\max_{\kappa=0(1)k-1} \|\delta_\kappa\| + \frac{1}{n} \max_{\mu \leq \nu} \left\| \sum_{\mu'=k}^{\mu} \delta_{\mu'} \right\| \right]$$

for $\nu \leq \nu_m$ and the relation (4.2.25) imply an analogous estimate for $\nu=\nu_m+1, \ldots, \nu_{m+1}$, with a constant Γ_{m+1} which is obtained from Γ_m by a process which is independent of n; the remainder of the argument is as before. (4.2.27) is proved by the same techniques that have been used for the proof of Theorem 2.2.3; we omit the details. □

We can now very simply prove the main stability result:

Theorem 4.2.7. *A uniform linear k-step method $\mathfrak{M}$ based upon the linear k-step scheme ψ is stable (w.r.t. Spijker's norm) for an arbitrary IVP 1 if and only if ψ is (strongly) D-stable.*

Proof. We apply Corollary 2.2.7 on the stability of neighboring discretizations to $\mathfrak{D}=\mathfrak{M}(\mathfrak{P}_0)$ and $\bar{\mathfrak{D}}=\mathfrak{M}(\mathfrak{P})$ where $\mathfrak{P}$ is the given IVP 1 and $\mathfrak{P}_0$ the trivial IVP 1 (4.2.24). The mappings $G_n=\bar{F}_n-F_n$ of Corollary 2.2.7 are given by

$$(G_n\eta)(t_\nu)=\begin{cases} s_{n\nu}^{(0)}(z_0)-s_{n\nu}(z_0)\,, & \nu=0(1)k-1\,,\\ -\sum_{\kappa=0}^{k}\beta_\kappa f(\eta_{\nu-k+\kappa})\,, & \nu=k(1)n\,;\end{cases}$$

$s_{n\nu}^{(0)}$ and $s_{n\nu}$ are the starting procedures of $\mathfrak{D}$ and $\bar{\mathfrak{D}}$ resp. There may be different β_κ at step changes but they are independent of n in each case so that step changes do not influence the following argument.

By Lemma 4.2.6, $\mathfrak{D}$ is stable (w.r.t. Spijker's norm) if the hypothesis on ψ is satisfied. Furthermore the G_n satisfy the Lipschitz-condition (2.2.20): For all $\eta^{(1)}, \eta^{(2)}\in B_R(z)$,

$$\begin{aligned}\|G_n\eta^{(1)}-G_n\eta^{(2)}\|_{E_n^0}^{[\nu]} &= \frac{1}{n}\sum_{\mu=k}^{\nu}\left\|\sum_{\kappa=0}^{k}\beta_\kappa\big(f(\eta_{\mu-k+\kappa}^{(1)})-f(\eta_{\mu-k+\kappa}^{(2)})\big)\right\|\\ &\le \frac{L}{n}\left(\sum_{\kappa=0}^{k}|\beta_\kappa|\right)\sum_{\mu=0}^{\nu}\|\eta_\mu^{(1)}-\eta_\mu^{(2)}\| \quad \text{for } \nu=0(1)n\,,\end{aligned} \tag{4.2.28}$$

where L is the Lipschitz constant of f, cf. (2.1.6). For Spijker's norm in E_n^0 (4.2.28) holds, too, since trivially $\max_{\mu\le\nu}\left\|\sum_{\mu'=k}^{\mu}\delta_{\mu'}\right\|\le\sum_{\mu=k}^{\nu}\|\delta_\mu\|$ for arbitrary δ_μ.

Thus the hypotheses of Corollary 2.2.7 are satisfied for $\eta_n=\Delta_n z$ and the stability (w.r.t. Spijker's norm) of $\bar{\mathfrak{D}}=\mathfrak{M}(\mathfrak{P})$ at $\{\Delta_n z\}$ follows. The necessity of the hypothesis on ψ follows from their necessity for the stability of $\mathfrak{M}(\mathfrak{P}_0)$ by Lemma 4.2.4 and 4.2.5, resp. □

Remark. Theorem 4.2.7 shows that stability is *not* a natural property of uniform linear k-step methods. However, the stability condition is remarkably simple.

Examples. 1. All Adams-schemes are D-stable and strongly D-stable as $\rho(x)=x^k-x^{k-1}$. Thus all Adams methods, i.e. uniform linear k-step methods based on Adams schemes, are stable for arbitrary IVP 1 with the norm (2.2.3) as well as with the norm (2.2.15) in E_n^0. Cf. the remark after Theorem 2.2.4.

2. The midpoint scheme is D-stable but not strongly D-stable as $\rho(x)=x^2-1$. Hence the 2-step method of the example in Section 4.2.1 is stable for an arbitrary IVP 1 w.r.t. the norm (2.2.3) but not w.r.t. Spijker's norm.

3. The maximal order 2-step predictor

$$\frac{1}{6h}[\eta_\nu+4\eta_{\nu-1}-5\eta_{\nu-2}-h(4f(\eta_{\nu-1})+2f(\eta_{\nu-2}))]=0 \tag{4.2.29}$$

of the example after Theorem 4.1.8 is not D-stable as $\rho(x)=x^2+4x-5=(x-1)(x+5)$. Thus a uniform linear 2-step method based upon (4.2.29) is not stable and will not converge (see Theorem 4.2.9) so that its third order consistency is in vain.

Corollary 2.2.7 may also be used to obtain stability bounds S and stability thresholds r for discretizations of IVP 1 by uniform linear k-step methods in the case of equidistant grids. By part a) of the proof of Lemma 4.2.6, Γ is the stability bound for $\mathfrak{D}=\mathfrak{M}(\mathfrak{P}_0)$. Let $b:=\sum_{\kappa=0}^{k}|\beta_\kappa|$ and $0<\bar{h}\le h_0$ be such that $1/\bar{\rho}:=1-\bar{h}bL\Gamma>0$. Then Corollary 2.2.7 implies

$$\begin{aligned}&\text{for correctors: } S=\bar{\rho}\,\Gamma\, e^{\bar{\rho}bL\Gamma} \quad \text{if } h\le\bar{h},\\ &\text{for predictors: } S=\Gamma\, e^{bL\Gamma},\end{aligned} \tag{4.2.30}$$

and $r=R/S$ where R is from the IVP 1 (see (2.1.6)). According to Theorem 2.1.2 this implies that the solution of

(4.2.31)

$$\frac{1}{h}[\rho(T_h)\varepsilon_{\nu-k}-h\sigma(T_h)(f(z(t_{\nu-k}+\varepsilon_{\nu-k})-f(z(t_{\nu-k})))]=\delta_\nu, \qquad \nu>k,$$

with initial values $\varepsilon_\kappa=\delta_\kappa$, $\kappa=0(1)k-1$, satisfies

$$\|\varepsilon_\nu\|\le S\left[\max_{\kappa=0(1)k-1}\|\delta_\kappa\|+h\sum_{\mu=k}^{\nu}\|\delta_\mu\|\right] \tag{4.2.32}$$

as long as the right hand side is smaller than R and $h\le\bar{h}$ in the case of implicit methods.

By a direct approach a slightly better estimate than (4.2.32) may be obtained:

Theorem 4.2.8. *If* $\langle\rho,\sigma\rangle$ *is D-stable the solution of* (4.2.31) *satisfies, for* $0<h\le h_0/\Gamma$ (*see* (4.1.11)),

$$\|\varepsilon_\nu\|\le q\Gamma\left[e^{\nu hqbL\Gamma}\max_{\kappa=0(1)k-1}\|\delta_\kappa\|+h\sum_{\mu=k}^{\nu}e^{(\nu-\mu)hqbL\Gamma}\|\delta_\mu\|\right] \tag{4.2.33}$$

where $q:=(1-h_0L\Gamma|\beta_k/\alpha_k|)^{-1}$, *as long as the right hand side is smaller than* R.

Proof. By (4.2.12), the solution of (4.2.31) may be written as

$$\varepsilon_\nu = \sum_{\kappa=0}^{k-1} \gamma_{\kappa\nu}\delta_\kappa + h\sum_{\mu=k}^{\nu} \gamma_{\mu\nu}\delta_\mu$$

$$+h\sum_{\mu=k}^{\nu}\gamma_{\mu\nu}\sum_{\kappa=0}^{k}\beta_\kappa\left[f(z(t_{\nu-k+\kappa})+\varepsilon_{\nu-k+\kappa})-f(z(t_{\nu-k+\kappa}))\right].$$

By Lemma 4.2.4, this implies (as long as $\|\varepsilon_\nu\|<R$)

$$\|\varepsilon_\nu\| \le \Gamma\left[\max_{\kappa=0(1)k-1}\|\delta_\kappa\| + h\sum_{\mu=k}^{\nu}\|\delta_\mu\| + hbL\sum_{\mu=k}^{\nu-1}\max_{\mu'\le\mu}\|\varepsilon_{\mu'}\| + h\left|\frac{\beta_k}{\alpha_k}\right| L\|\varepsilon_\nu\|\right]$$

or, for $h\le h_0/\Gamma$,

$$\|\varepsilon_\nu\| \le q\Gamma\left[\max_{\kappa=0(1)k-1}\|\delta_\kappa\| + h\sum_{\mu=k}^{\nu}\|\delta_\mu\| + hbL\sum_{\mu=k}^{\nu-1}\max_{\mu'\le\mu}\|\varepsilon_{\mu'}\|\right].$$

Thus we have the situation of Corollary 2.1.4 and the estimate (2.1.28) yields, with $S_1=q\Gamma$ and $S_2=qbL\Gamma$,

$$\|\varepsilon_\nu\| \le q\Gamma\left[(1+qbL\Gamma h)^\nu \max_{\kappa=0(1)k-1}\|\delta_\kappa\| + h\sum_{\mu=k}^{\nu}(1+qbL\Gamma h)^{\nu-\mu}\|\delta_\mu\|\right]$$

which implies the assertion. □

Remarks. 1. For explicit linear k-step schemes $\langle\rho,\sigma\rangle$ we have $q=1$ and the restriction $h\le h_0$ disappears. For implicit $\langle\rho,\sigma\rangle$, q may be chosen arbitrarily close to 1 if one restricts oneself to sufficiently small steps h.

2. In the case of non-equidistant but coherent grid sequences Theorem 4.2.8 may be applied to each of the subintervals on which the grids have constant steps and an overall estimate may be gained recursively.

4.2.4 Convergence

By Theorem 1.2.4, a uniform linear k-step method is convergent for an arbitrary IVP1 if it is consistent and stable; it is convergent of order p for an IVP$^{(p)}$1 if it is consistent of order p and stable. The conditions for consistency (of order p) and stability of linear multistep methods have been displayed in Theorem 4.2.1 (Theorem 4.2.2) and Theorem 4.2.7.

Since consistency and stability are not trivial for uniform linear k-step methods the question regarding their *necessity* becomes interesting.

Theorem 4.2.9. *A uniform linear k-step method* $\mathfrak{M}$ *based upon the linear k-step scheme* $\langle\rho,\sigma\rangle$ *is convergent for an arbitrary* IVP1 *only if it is* *consistent with an arbitrary* IVP 1 *and only if* ρ *possesses no zeros outside the closed unit disk. If* $\langle\rho,\sigma\rangle$ *is of order* 1, $\mathfrak{M}$ *is convergent for an arbitrary* IVP 1 *only if it is stable.*

Proof. a) To establish the necessity of consistency we have to show that convergence implies the three conditions for consistency of Theorem 4.2.1:

(i) If $\mathfrak{M}$ is convergent the solution ζ_n of $\mathfrak{M}$(IVP 1) satisfies

$$\lim_{n\to\infty}\|\zeta_n(t_\nu)-z(t_\nu)\| = \lim_{n\to\infty}\|s_{n\nu}(z_0)-z(t_\nu)\| = 0, \qquad \nu=0(1)k-1.$$

(ii) Convergence of $\mathfrak{M}$ for $y'=1$, $y(0)=0$, implies $\zeta_n(t_\nu)=t_\nu+\varepsilon_\nu$, with $|\varepsilon_\nu|\to 0$ as $n\to\infty$. Introduction into (4.2.1), with $f\equiv 1$, gives

$$\frac{1}{h_\nu}\rho(T_h)(t_{\nu-k}+\varepsilon_{\nu-k}) = \sigma(1) = 1.$$

If we are not in the neighborhood of a step change we have $t_{\nu-k+\kappa} = t_{\nu-k}+\kappa h_\nu$ so that we obtain, by (4.1.3),

$$\frac{1}{h_\nu}\rho(T_h)\varepsilon_{\nu-k} = -(\rho'(1)-1) =: \delta. \tag{4.2.34}$$

If $\delta=0$ we have Condition (ii), see (4.1.27); otherwise (4.2.34) is a discretization by $\mathfrak{M}$ of $y'=\delta$ and the assumed convergence of $\mathfrak{M}$ implies $\varepsilon_\nu=\delta t_\nu+o(1)$ which is contradictory to $|\varepsilon_\nu|=o(1)$.

(iii) With a step change at t_ν, assume that $\zeta_{\nu+\mu}$, $\mu=1(1)k-1$, is computed from

$$a_k^{(\mu)}\zeta_{\nu+\mu} = -\sum_{\kappa=0}^{k-1} a_\kappa^{(\mu)}\zeta_{\nu+\mu-k+\kappa} + h\sum_{\kappa=0}^{k} b_\kappa^{(\mu)} f(\zeta_{\nu+\mu-k+\kappa}), \quad a_k^{(\mu)}\neq 0. \tag{4.2.35}$$

Convergence of $\mathfrak{M}$ for $y'=0$, $y(0)=1$, implies $\zeta_\nu=1+\varepsilon_\nu$, with $|\varepsilon_\nu|\to 0$ as $n\to\infty$. Substitution into (4.2.35) gives $\sum_\kappa a_\kappa^{(\mu)} = -\sum_\kappa a_\kappa^{(\mu)}\varepsilon_{\nu+\mu-k+\kappa}$ which implies $\sum_{\kappa=0}^{k} a_\kappa^{(\mu)}=0$ as $|\varepsilon_\nu|\to 0$.

b) To establish the necessity of stability in the stated sense we consider an IVP 1 which is a mere quadrature, i.e.,

$$\begin{pmatrix} t \\ y \end{pmatrix}' = \begin{pmatrix} 1 \\ f(t) \end{pmatrix}, \qquad \text{cf. (2.1.2).}$$

Since there is no local discretization error in the first component for a convergent and hence consistent discretization, the second component ε_ν of the global discretization error simply satisfies an equation of the type

$$\frac{1}{h}\rho(T_h)\varepsilon_{\nu-k+\kappa} = -l(t_\nu), \qquad \nu \geq k,$$

where l is the second component of the local discretization error. Assume that f has been chosen such that $l(t_\nu) = \text{const.}\, h^p$; then, by (4.2.12),

$$\varepsilon_\nu = \sum_{\kappa=0}^{k-1} \gamma_{\kappa\nu}\varepsilon_\kappa - \text{const.}\, h^{p+1} \sum_{\mu=k}^{\nu} \gamma_{\mu\nu}. \tag{4.2.36}$$

If ρ possesses zeros outside the closed unit disk the $\gamma_{\mu\nu}$ grow exponentially with ν (see the discussion in Section 4.2.2) which produces a contradiction to $\|\varepsilon\|_{E_n} \to 0$ as $n \to \infty$.

If there are no zeros of ρ outside the unit disk but multiple zeros on the unit circle the growth of the $\gamma_{\nu\mu}$ is like ν^{m-1} where m is the highest occuring multiplicity. Thus we obtain a contradiction for $m \geq 2$ if $|l(t_\mu)| = ch$ for $\nu \geq k$ and/or $\max_\kappa |\varepsilon_\kappa| = O(h)$. For $\langle\rho,\sigma\rangle$ of exact order 1 it is easy to find functions f such that the local discretization error shows this behavior. □

From the above proof it would seem that the necessity of $m=1$ for zeros of ρ on the unit circle is dubious if the method $\mathfrak{M}$ is consistent of an order greater than 1: Here $l(t_\mu)$ and the ε_κ, $\kappa = 0(1)k-1$, in (4.2.36) may decrease sufficiently rapidly as $n \to \infty$ to overcome the growth of the $\gamma_{\mu\nu}$ if the multiplicity m is not too high.

However, the global discretization error has the form (4.2.36) only for *trivial* IVP 1 like the ones used in the proof above. For such IVP 1, a method with multiple zeros of ρ on the unit circle[5] is indeed stable w.r.t. the following norm in E_n^0:

$$\|\delta\|_{E_n^0} = n^{m-1} \max_{\kappa=0(1)k-1} \|\delta_\kappa\| + n^{m-2} \sum_{\nu=k}^{n} \|\delta_\nu\|, \tag{4.2.37}$$

where m is the highest multiplicity of an essential zero.

In the proof of Theorem 4.2.7 we used Corollary 2.2.7 to derive stability for an arbitrary IVP 1 from stability for a trivial IVP 1. Now this step is no longer possible since the mappings G_n defined in that proof satisfy the Lipschitz conditions (2.2.20) only w.r.t. a norm (2.2.3) but not w.r.t. the norm (4.2.37), cf. (4.2.28). Thus the error propagation

[5] The multiple zero of ρ on the unit circle cannot be at 1 for a consistent method; see Theorem 4.2.1 and (4.1.27).

in the discretization obtained for an arbitrary IVP 1 is generally *not* restricted by ν^{m-1} with the method considered above; hence, we cannot bound the global discretization error uniformly in n even for an arbitrarily smooth IVP 1 and for a high order of consistency.

The details are best understood from the following analysis of the linear case: The solution of the discretization of the scalar IVP 1 $y'-gy=0$, $g=\text{const}\in\mathbb{C}$, is computed from

$$[\rho(T_h)-hg\sigma(T_h)]\eta_{\nu-k}=0, \qquad \nu=k(1)n. \tag{4.2.38}$$

A fundamental system of solutions for this linear difference equation with constant coefficients arises from the zeros of the polynomial $\rho(x)-hg\sigma(x)$ by (4.2.8). Naturally these zeros tend to the zeros of ρ as $h\to 0$.

Consider those m_i zeros $x_{i\mu}(h)$ of $\rho-hg\sigma$ which converge to the m_i-fold zero x_i of ρ, $|x_i|=1$. By a perturbation argument we easily find that

$$x_{i\mu}(h)=x_i+\left(\frac{m_i!}{\rho^{(m_i)}(x_i)}\sigma(x_i)hg\right)^{\frac{1}{m_i}}+o\left(h^{\frac{1}{m_i}}\right); \tag{4.2.39}$$

the m_i different values $x_{i\mu}$ arise from the m_i different (complex) values of the m_i-th root. ($\sigma(x_i)$ cannot vanish for a "minimal k representative" of the equivalence class $\langle\rho,\sigma\rangle$; cf. our discussion in Section 4.4.1.) (4.2.39) implies that

$$\left|[x_{i\mu}(h)]^{\frac{1}{h}}\right|=\left|\left[1+\left(\frac{m_i!\,\sigma(x_i)}{x_i^{m_i}\rho^{(m_i)}(x_i)}hg\right)^{\frac{1}{m_i}}\right]^{\frac{1}{h}}\right|\to\infty \quad \text{as } h\to 0, \tag{4.2.40}$$

except when $m_i=2$ and $(\sigma(x_i)/x_i^2\rho''(x_i))g$ is real and negative.

In all other cases the divergence of (4.2.40) is of an *exponential* character so that $h^r[x_{i\mu}(h)]^{1/h}$, $r>0$, diverges as well. It is thus obvious that for a given linear multistep method based on a multistep scheme which is not D-stable, we will be able to find an IVP 1 such that our method does not converge for that IVP 1. Therefore we have actually arrived at the following result:

Theorem 4.2.10. *A uniform linear k-step method based upon the linear k-step scheme $\langle\rho,\sigma\rangle$ is convergent for an arbitrary IVP 1 only if $\langle\rho,\sigma\rangle$ is D-stable.*

Remark. Henrici, who obtains the necessity of full D-stability quite elegantly in [1], Section 5.2–4, uses a different convergence definition: He considers convergence of the forward step procedure only and requires that convergence should occur with an arbitrary consistent starting procedure. Thus he is able to use a pathological starting procedure to show the necessity of D-stability.

Example. A uniform linear 3-step method $\mathfrak{M}$ based upon the predictor

$$\frac{1}{4h}\left[\eta_\nu+\eta_{\nu-1}-\eta_{\nu-2}-\eta_{\nu-3}-\frac{h}{3}(8f(\eta_{\nu-1})+2f(\eta_{\nu-2})+2f(\eta_{\nu-3}))\right]=0$$

is consistent of order 3 if used with a suitable 3rd order consistent starting procedure. The associated polynomial ρ possesses the zeros $1, -1, -1$ so that $m=2$ and $\langle\rho,\sigma\rangle$ *is not D-stable.*

Since $\rho''(-1)=-1$ and $\sigma(-1)=\frac{2}{3}$ so that $\sigma(-1)/(-1)^2\rho''(-1)<0$, the exceptional case in the above argumentation arises if and only if $g\geq 0$. It is easily checked that $\mathfrak{M}$ is *convergent of order 3* for $y'=y$; practical computations establish the same behavior even for a general scalar equation $y'=f(y)$ with $f'(z(t))>0$. On the other hand, $\mathfrak{M}$ is not convergent at all for $y'=-y$ (except theoretically with the starting procedure $\zeta_\kappa=[x_1(h)]^\kappa$, $\kappa=0,1,2$, where $x_1(0)=1$) and for "almost" every other IVP 1.

4.2.5 Highest Obtainable Orders of Convergence

Although the requirement of stability affects only the coefficients α_κ of the linear k-step scheme which is to serve as the basis of a uniform linear k-step method (see Theorem 4.2.7), it causes a severe restriction on the attainable orders of convergence. The following famous result is due to Dahlquist [1].

Theorem 4.2.11. *If the k-th degree polynomial ρ possesses no zeros outside the closed unit disk and only a simple zero at 1, the order of the linear k-step scheme $\langle\rho,\sigma\rangle$ cannot be greater than $2[k/2]+2$.*

Proof. By substituting $x=(1+u)/(1-u)$ into the order condition (4.1.24) we obtain

$$\frac{\rho\left(\dfrac{1+u}{1-u}\right)}{\log\left(\dfrac{1+u}{1-u}\right)}-\sigma\left(\frac{1+u}{1-u}\right)=O(u^p)\quad\text{as } u\to 0$$

or, with the polynomials of degree k

$$R(u):=(1-u)^k\rho\left(\frac{1+u}{1-u}\right),\qquad S(u):=(1-u)^k\sigma\left(\frac{1+u}{1-u}\right),$$

$$\frac{R(u)}{\log\left(\dfrac{1+u}{1-u}\right)}-S(u)=O(u^p)\quad\text{as } u\to 0, \tag{4.2.41}$$

which implies $R(0)=\rho(1)=0$ and $R'(0)=2\rho'(1)=2$ as necessary conditions for $p\geq 1$. Furthermore, since

$$R(u)=\alpha_k \prod_{i=1}^{k} ((1+u)-x_i(1-u))$$

where $x_i, i=1(1)k$, are the k zeros of ρ (not necessarily distinct), the hypothesis on ρ implies that R has no zeros u_i outside the closed left half plane:

$$\text{(4.2.42)} \qquad \operatorname{Re} u_i = \operatorname{Re}\left(-\frac{1-x_i}{1+x_i}\right) \leq 0, \qquad i=1(1)k.$$

As is seen from the factorization

$$R(u)=\sum_{\kappa=0}^{k} a_\kappa u^\kappa = a_k u \prod_{i=2}^{k} (u-u_i)$$

and $a_1=R'(0)=2>0$, (4.2.42) requires

$$\text{(4.2.43)} \qquad a_\kappa \geq 0, \qquad \kappa=1(1)k.$$

(If ρ has zeros at -1, R is of a lower degree than k but this does not affect the argument.)

From (4.2.41) we see that we can at best achieve $p=k+1$ through a reasonable choice of S or σ, any higher order p must come about through the vanishing of the coefficients in the expansion of $R(u)/(\log(1+u)/(1-u))$ in powers of u. The assertion of the theorem hinges on the fact that the coefficients $c_{2\mu}$ in

$$\frac{u}{\log\left(\dfrac{1+u}{1-u}\right)}=\frac{1}{2}+\sum_{\mu=1}^{\infty} c_{2\mu} u^{2\mu}$$

satisfy

$$\text{(4.2.44)} \qquad c_{2\mu}<0, \qquad \mu \geq 1;$$

for the proof of (4.2.44) we refer to Henrici [1], Section 5.2–8.

For the coefficients r_ν in

$$\frac{R(u)}{u}\,\frac{u}{\log\left(\dfrac{1+u}{1-u}\right)}=\sum_{\nu=0}^{\infty} r_\nu u^\nu$$

we now have

a) for odd k: $r_{k+1}=c_2 a_k+\cdots+c_{k+1}a_1<0$ due to (4.2.43), (4.2.44) and $a_1=2$;

b) for even k: $r_{k+1}=c_2 a_k+\cdots+c_k a_2\leq 0$. If the equality sign should hold due to $a_{2\kappa}=0$, $\kappa=1(1)k/2$, we have in any case $r_{k+2}=c_4 a_{k-1}+\cdots+c_{k+2}a_1<0$.

Thus $r_v\neq 0$ for $v=2[k/2]+2$ which proves the assertion; see (4.2.41). □

Corollary 4.2.12. *There exists no uniform linear k-step method which is convergent of an order greater than* $2[k/2]+2$ *for an arbitrary (sufficiently smooth) IVP 1.*

Proof. Consequence of Theorems 4.2.10, 4.2.11, and 4.2.2; for *uniform* linear k-step methods, the order condition (ii) in Theorem 4.2.2 is also necessary as is easily established by counter-examples. □

Example. A comparison between Theorem 4.1.8 and Theorem 4.2.11 shows that the optimal order linear k-step schemes do not lead to convergent uniform linear k-step methods for $k>2$ in the case of correctors and $k>1$ in the case of predictors.

The content of Corollary 4.2.12 has initiated a search for k-step methods which are not subject to this restriction in the order of convergence. It has turned out that the restriction is not valid for multistage (or "non-linear") k-step methods as well as for non-uniform linear k-step methods. Particularly the latter ones are hardly more complicated than uniform linear k-step methods; see Section 4.3.

4.3 Cyclic Linear k-step Methods

Quite recently, Donelson and Hansen ([1]) have investigated the possibility of basing a linear multistep method on several different k-step schemes which are used cyclically. It turns out that such a cyclic linear multistep method may be stable although the k-step schemes which it employs are not D-stable; thus the restrictions of Corollary 4.2.12 are removed. In this section we present the theory of Donelson and Hansen (in our notation). In Section 5.4, cyclic linear k-step methods will be analyzed as a particular class of cyclic forward step methods.

4.3.1 Stability of Cyclic Linear k-step Methods

Consider linear k-step schemes ψ_μ with coefficients α_κ^μ, β_κ^μ, $\kappa=0(1)k$, and associated polynomials ρ_μ, σ_μ, $\mu=1(1)m$. We assume that $\alpha_k^\mu\neq 0$, $\mu=1(1)m$; however, we do not exclude $|\alpha_0^\mu|+|\beta_0^\mu|=0$ so that some of the k-step schemes may actually be written as k'-step schemes, $k'<k$.

Def. 4.3.1. A one-stage k-step method applicable to IVP 1, with an equidistant grid sequence and E_n, E_n^0, Δ_n, Δ_n^0 (for $\nu<k$) as in Def. 4.2.1 but

$$\Delta_n^0\begin{pmatrix} d_0 \\ d(t)\end{pmatrix}(t_\nu)=\sigma_{\mu\langle\nu\rangle}(T_h)\,d(t_{\nu-k}), \qquad \nu=k(1)n,$$

(4.3.1) $$[\varphi_n(F)\eta](t_\nu)=\begin{cases}\eta_\nu-s_{n\nu}(z_0), & \nu=0(1)k-1,\\ \psi_{\mu\langle\nu\rangle}[f](h;\eta_\nu,\dots,\eta_{\nu-k}), & \nu=k(1)n,\end{cases}$$

where

(4.3.2) $$\mu\langle\nu\rangle:=\nu-k+1 \mod m, \qquad 1\le\mu\langle\nu\rangle\le m,$$

is called an *m-cyclic linear k-step method based upon the linear k-step schemes* $\psi_1,\dots,\psi_m$.

Remark. A restriction to equidistant grid sequences may not be essential; we have introduced it here for the sake of simplicity. It seems that the theory of m-cyclic linear k-step methods has to be more fully understood before step change procedures can be devised which do not destroy the desirable properties of these methods.

Example. The one-stage 2-step method with the forward step procedures

$$\frac{\eta_{2\nu}-\eta_{2\nu-2}}{2h}-\frac{1}{6}[f(\eta_{2\nu})+4f(\eta_{2\nu-1})+f(\eta_{2\nu-2})]=0,$$

$$\frac{\eta_{2\nu+1}-\eta_{2\nu}}{h}-\frac{1}{12}(5f(\eta_{2\nu+1})+8f(\eta_{2\nu})-f(\eta_{2\nu-1}))]=0,$$

for $\nu\ge1$ is a 2-cyclic linear 2-step method based upon the Simpson scheme and the 2-step Adams corrector.

In analogy with our analysis in Section 4.2.2 we regard at first the difference equation with periodic coefficients

(4.3.3) $$\sum_{\kappa=0}^{k}\alpha_{\kappa\nu}\,\varepsilon_{\nu-k+\kappa}=0, \qquad \nu\ge k,$$

with given initial values $\varepsilon_\kappa=\delta_\kappa$, $\kappa=0(1)k-1$, where

(4.3.4) $$\alpha_{\kappa\nu}=\alpha_\kappa^{\mu\langle\nu\rangle} \quad\text{so that}\quad \alpha_{\kappa,\nu+m}=\alpha_{\kappa\nu}.$$

By Q_μ we denote the Frobenius matrices of the ρ_μ

(4.3.5) $$Q_\mu:=\begin{pmatrix} 0 & 1 & & 0\\ \vdots & & \ddots & \\ 0 & \cdots & 0 & 1\\ -\dfrac{\alpha_0^\mu}{\alpha_k^\mu} & -\dfrac{\alpha_1^\mu}{\alpha_k^\mu} & \cdots & -\dfrac{\alpha_{k-1}^\mu}{\alpha_k^\mu}\end{pmatrix}$$

Lemma 4.3.1. *If the* $k \times k$ *matrix*

$$Q := Q_m Q_{m-1} \cdots Q_1 \tag{4.3.6}$$

possesses no eigenvalues outside the closed unit disk and only simple eigenvalues on the unit circle then there exists a constant C_s *such that the solution of* (4.3.3)/(4.3.4) *satisfies*

$$|\varepsilon_\nu| \le C_s \max_{\kappa=0(1)k-1} |\delta_\kappa|, \quad \textit{uniformly for } \nu \ge 0. \tag{4.3.7}$$

Proof. (4.3.3)/(4.3.4) is equivalent to

$$\begin{pmatrix} \varepsilon_{\nu-k+1} \\ \vdots \\ \varepsilon_\nu \end{pmatrix} = Q_{\mu\langle\nu\rangle} \begin{pmatrix} \varepsilon_{\nu-k} \\ \vdots \\ \varepsilon_{\nu-1} \end{pmatrix}, \qquad \nu \ge k,$$

so that

$$\begin{pmatrix} \varepsilon_{jm} \\ \vdots \\ \varepsilon_{jm+k-1} \end{pmatrix} = Q \begin{pmatrix} \varepsilon_{(j-1)m} \\ \vdots \\ \varepsilon_{(j-1)m+k-1} \end{pmatrix} = Q^j \begin{pmatrix} \varepsilon_0 \\ \vdots \\ \varepsilon_{k-1} \end{pmatrix}, \qquad j = 1, 2, \ldots$$

But $\|Q^j\|$ is bounded uniformly under our assumptions. □

Remark. We may cyclically rearrange the product (4.3.6) without changing the conclusion of Lemma 4.3.1:

$$(Q_\mu \cdots Q_1 Q_m \cdots Q_{\mu+1})^j = Q_\mu \cdots Q_1 Q^{j-1} Q_m \cdots Q_{\mu+1}$$

so that $\|(Q_\mu \cdots Q_1 Q_m \cdots Q_{\mu+1})^j\|$ is bounded uniformly in j for each $\mu = 1(1)m$. This means that the m matrices $Q_\mu \cdots Q_1 Q_m \cdots Q_{\mu+1}$ all satisfy the assumptions of Lemma 4.3.1 if one of them does. In fact all these matrices have the same eigenvalues.

We now consider the $m \times (k+m)$-matrix

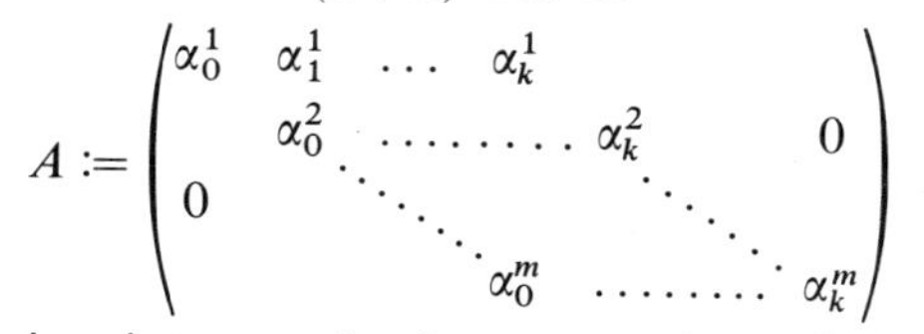

and its partitioning into quadratic $m \times m$-submatrices $A_0, A_1, \ldots, A_{j_0}$, starting from the right hand side and adding zero columns on the left hand end if necessary, e.g.

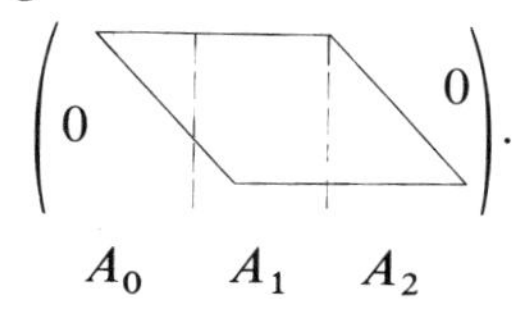

There are $j_0 + 1 = 2 + \left[\frac{k-1}{m}\right]$ submatrices.

Lemma 4.3.2. *With Q and $A_j, j=0(1)j_0$, as defined above*

$$(4.3.8) \qquad \rho(x) := x^{k-j_0 m} \det\left(\sum_{j=0}^{j_0} A_j x^j\right) = \left(\prod_{\mu=1}^{m} \alpha_k^\mu\right) \det(xI - Q).$$

Proof. Assume at first that Q has k distinct eigenvalues $x_i \neq 0$, $i=1(1)k$, and let $\xi^i = (\xi_0^i, \ldots, \xi_{k-1}^i)^T$ be the eigenvector of Q for x_i. Then the solution $\{\xi_\nu^i\}_{\nu \geq 0}$ of (4.3.3) starting with ξ_κ^i, $\kappa=0(1)k-1$, satisfies

$$(4.3.9) \qquad \begin{pmatrix} \xi_m^i \\ \vdots \\ \xi_{m+k-1}^i \end{pmatrix} = Q \begin{pmatrix} \xi_0^i \\ \vdots \\ \xi_{k-1}^i \end{pmatrix} = x_i \begin{pmatrix} \xi_0^i \\ \vdots \\ \xi_{k-1}^i \end{pmatrix},$$

i.e. it has the form

$$(4.3.10) \quad \xi_0^i, \ldots, \xi_{m-1}^i, x_i \xi_0^i, \ldots, x_i \xi_{m-1}^i, x_i^2 \xi_0^i, \ldots \quad \text{or} \quad \xi_{\nu+m}^i = x_i \xi_\nu^i, \qquad \nu \geq 0.$$

Therefore, the system of m linear equations

$$\left(\sum_{j=0}^{j_0} A_j x_i^j\right) \xi = 0$$

has the non-trivial solution $\xi = (\xi_{m-l}^i, \ldots, \xi_{m-1}^i, x_i \xi_0^i, \ldots, x_i \xi_{m-l-1}^i)^T \in \mathbb{R}^m$ (for l zero columns in A_0) which implies $\det\left(\sum_{j=0}^{j_0} A_j x_i^j\right) = 0$, for $i=1(1)k$.

It is not hard to check that the lowest power of x occuring in $\det(\sum A_j x^j)$ is $x^{j_0 m - k}$ while the highest is clearly $x^{j_0 m}$. Thus both sides of (4.3.8) are polynomials of exact degree k, with leading coefficients $\prod_{\mu=1}^{m} \alpha_k^\mu$; having the same zeros they must be identical.

Now both the coefficients of $\rho(x)$ and of $\left(\prod_{\mu=1}^{m} \alpha_k^\mu\right) \det(xI - Q)$ are, for given k and m, unique polynomials of degree m in the α_κ^μ, $\kappa=0(1)k$, $\mu=1(1)m$; hence the above result implies that these coefficient polynomials are identical. Thus (4.3.8) holds also in the case of multiple or vanishing eigenvalues. □

Def. 4.3.2. The polynomial ρ of (4.3.8) is called *associated polynomial* of the m-cyclic linear k-step method based upon the linear k-step schemes $\psi_1, \ldots, \psi_m$.

Theorem 4.3.3. *An m-cyclic linear k-step method $\mathfrak{M}$ is stable for an arbitrary IVP 1 if its associated polynomial ρ satisfies the root criterion.*

Proof. We proceed as in the proof of Theorem 4.2.7 and form the discretizations $\mathfrak{D} = \mathfrak{M}(\mathfrak{P}_0) = \{E_n, E_n^0, F_n\}_{n \in \mathbb{N}'}$ and $\bar{\mathfrak{D}} = \mathfrak{M}(\mathfrak{P}) = \{E_n, E_n^0, F_n + G_n\}_{n \in \mathbb{N}'}$, with $\mathfrak{P}_0$ and $\mathfrak{P}$ as previously.

The stability of $\mathfrak{D}$ requires an estimate

$$|\varepsilon_\nu| \le \Gamma \left[\max_{\kappa=0(1)k-1} |\delta_\kappa| + h \sum_{\mu=k}^{\nu} |\delta_\mu| \right] \tag{4.3.11}$$

for the solutions of the inhomogeneous scalar equation

$$\sum_{\kappa=0}^{k} \alpha_\kappa^{\mu\langle\nu\rangle} \varepsilon_{\nu-k+\kappa} = h\delta_\nu, \qquad \nu \ge k, \tag{4.3.12}$$

with initial values $\varepsilon_\kappa = \delta_\kappa$, $\kappa = 0(1)k-1$. The existence of an estimate (4.3.11) for (4.3.12) follows from the existence of the estimate (4.3.7) for (4.3.3) by a construction analogous to (4.2.10)—(4.2.12); cf. proof of Lemma 4.2.4. As the root criterion on ρ implies the estimate (4.3.7) due to Lemma 4.3.1 and Lemma 4.3.2, $\mathfrak{D}$ is stable under our assumptions.

The uniform Lipschitz condition (2.2.20) on the G_n is obtained as in the proof of Theorem 4.2.7; see (4.2.28). Therefore the assertion follows via Corollary 2.2.7 on the stability of neighboring discretizations. □

If $m > 1$, the root criterion on the associated polynomial ρ is not a necessary condition for the existence of an estimate (4.3.11) as far as multiple zeros on the unit circle are concerned. This is due to the fact that the powers of Q are also bounded uniformly if there are multiple eigenvalues on the unit circle but the associated elementary divisors are linear.

To see how such a situation may actually occur we assume that $\rho_1 = \rho_2 = \cdots = \rho_m =: \rho_0$ which may happen even if the ψ_μ are distinct. Then the eigenvalues of Q and the zeros of ρ are the m-th powers of the zeros of ρ_0 and separated zeros of ρ_0 on the unit circle may coincide for ρ. (Take $m=2$, $x_1=1$, $x_2=-1$.) On the other hand, if ρ_0 satisfies the root criterion we have the situation of Section 4.2.2 and (4.3.11) certainly holds. At this point we will be satisfied with Theorem 4.3.3; a sharper formulation will be made in Section 5.4.2.

Corollary 4.3.4. *An m-cyclic linear k-step method* $\mathfrak{M}$ *based upon the linear k-step schemes* ψ_μ *of orders* $p_\mu \ge p$, $\mu = 1(1)m$, *is convergent of order p for an* $IVP^{(p)}$ *if its starting procedure is of order p and if its associated polynomial* ρ *satisfies the root criterion.*

Proof. The consistency of order p of $\mathfrak{M}$ with an $\text{IVP}^{(p)}\,1$ is shown as in the proof of Theorem 4.2.2 under our assumptions. The assertion thus follows from Theorem 4.3.3 and Theorem 1.2.4. □

Example. As above (after Def. 4.3.1). We have $Q_1 = \begin{pmatrix} 0 & 1 \\ 1 & 0 \end{pmatrix}$, $Q_2 = \begin{pmatrix} 0 & 1 \\ 0 & 1 \end{pmatrix}$, $Q = Q_2 Q_1 = \begin{pmatrix} 1 & 0 \\ 1 & 0 \end{pmatrix}$, with eigenvalues $x_1 = 1$, $x_2 = 0$.

$$A=\begin{pmatrix} -\frac{1}{2} & 0 & \vline & \frac{1}{2} & 0 \\ 0 & 0 & \vline & -1 & 1 \end{pmatrix}, \quad A_0=\begin{pmatrix} -\frac{1}{2} & 0 \\ 0 & 0 \end{pmatrix}, \quad A_1=\begin{pmatrix} \frac{1}{2} & 0 \\ -1 & 1 \end{pmatrix}, \qquad k=m=2,\, j_0=1.$$

$$\rho(x)=\det(A_0+A_1 x)=\det\begin{pmatrix} \dfrac{x-1}{2} & 0 \\ -x & x \end{pmatrix}=\frac{1}{2}x(x-1)=\alpha_2^1\alpha_2^2\det\begin{pmatrix} x-1 & 0 \\ -1 & x \end{pmatrix}.$$

Note that the zeros of ρ are 1 and 0 as for the polynomial ρ_2 of the Adams corrector, the zero -1 of the polynomial ρ_1 of the Simpson scheme no longer appears. Thus ρ satisfies the strong root criterion.

Corollary 4.3.4 implies that the method is convergent of order 3 (with proper starting); actually it will turn out to be convergent of order 4, see the example at the end of Section 4.3.2.

4.3.2 The Auxiliary Method

The order of convergence of a cyclic linear k-step method may be greater than the minimal order of the linear k-step schemes employed. One possible approach to the analysis of this question is to regard m steps, i. e. one cycle of the method as one step of an m-stage j_0-step method. This is particularly appropriate if $m\geq k$ so that $j_0=1$; see the partitioning of A preceding Lemma 4.3.2. We will follow this approach in Section 5.4.

The approach used by Donelson and Hansen [1] consists of constructing an equivalent m-cyclic linear $\bar{k}$-step method $\overline{\mathfrak{M}}$ whose order of convergence is easier to analyze. The construction of this *auxiliary method* proceeds as follows:

We enlarge the matrix A of Section 4.3.1 by adding more rows of the type $\alpha_0^\mu \ldots \alpha_k^\mu$ in a cyclic order until we have arrived at a total of

(4.3.13) $$\bar{m}:=(m-1)k+1=m+(m-1)(k-1) \text{ rows}.$$

Obviously, $\bar{m}>m$ in each non-trivial case. The non-vanishing elements in each row are shifted towards the right so that the total matrix has $\bar{m}$ rows and $\bar{m}+k$ columns.

Altogether we form m different such $\bar{m}\times(\bar{m}+k)$-matrices $\bar{A}^{(\mu)}$, $\mu=1(1)m$, where $\bar{A}^{(\mu)}$ has the coefficients α_κ^μ in the first row, with α_0^μ in the upper left corner, the coefficients $\alpha_\kappa^{\mu+1}$ in the second row translated towards the right by one position, etc. continuing cyclically. In exactly the same fashion we form m different $\bar{m}\times(\bar{m}+k)$-matrices $\bar{B}^{(\mu)}$ from the coefficients β_k^μ of the linear k-step schemes ψ_μ.

Then for each $\mu=1(1)m$, we determine the vector $d^\mu\in\mathbb{R}^{\bar{m}}$, $d^\mu\neq 0$, with components d_λ^μ, $\lambda=1(1)\bar{m}$, such that the vector

(4.3.14) $$(a^\mu)^T:=(d^\mu)^T\bar{A}^{(\mu)}\in\mathbb{R}^{\bar{m}+k}=\mathbb{R}^{mk+1},$$

with components $a_j^\mu, j=0(1)mk$, satisfies

(4.3.15) $$a_j^\mu = 0,\ j \neq 0 \bmod m; \qquad a_{mk}^\mu = \prod_{\mu=1}^{m} \alpha_k^\mu .$$

For this purpose, d^μ has to annihilate $k(m-1)=\bar{m}-1$ columns of $\bar{A}^{(\mu)}$ which are of length $\bar{m}$ and thus linearly dependent so that $d^\mu \neq 0$ exists. The required normalization is possible since a_{mk}^μ cannot vanish due to the form of $\bar{A}^{(\mu)}$. With the same vectors d^μ we also form, for $\mu=1(1)m$,

(4.3.16) $(b^\mu)^T := (d^\mu)^T \bar{B}^{(\mu)}$, with components $b_j^\mu, j=0(1)mk$.

For the remainder of Section 4.3 we will use the following convention: If a quantity depends on the coefficients of the ψ_μ, $\mu=1(1)m$, $m>1$:

$$q = q(\alpha_0^1, \ldots, \alpha_k^1; \alpha_0^2, \ldots, \alpha_k^2; \ldots; \alpha_0^m, \ldots, \alpha_k^m)$$

then we will denote by πq the quantity

$$\pi q := q(\alpha_0^2, \ldots, \alpha_k^2; \alpha_0^3, \ldots, \alpha_k^3; \ldots; \alpha_0^m, \ldots, \alpha_k^m; \alpha_0^1, \ldots, \alpha_k^1)$$

so that

$$\pi^{\mu-1} q := q(\alpha_0^\mu, \ldots, \alpha_k^\mu; \ldots; \alpha_0^{\mu-1}, \ldots, \alpha_k^{\mu-1}), \qquad \mu=2(1)m,$$

and $\pi^m q = q$. If $\pi q = q$ although $m>1$ we will call q *π-invariant.*

Obviously, $\bar{A}^{(\mu)} = \pi^{\mu-1} \bar{A}^{(1)}$ and $d^\mu = \pi^{\mu-1} d^1$; by the second relation it is sufficient to compute d^1 and then to exchange cyclically the α_κ^μ to obtain the d^μ, $\mu>1$.

Lemma 4.3.5. *The vectors a^μ of* (4.3.14)—(4.3.15) *are identical, i.e.,*

(4.3.17) $$a_{\kappa m}^\mu = a_\kappa,\ \kappa=0(1)k, \quad \textit{for each } \mu=1(1)m.$$

Furthermore $\sum_{\kappa=0}^{k} a_\kappa x^\kappa = \rho(x)$, *the associated polynomial of* $\mathfrak{M}$.

Proof. As in the proof of Lemma 4.3.2, we assume at first that Q has k distinct eigenvalues x_i and we consider the solution sequences $\{\xi_\nu^i\}$ of (4.3.3) generated by the eigenvectors ξ^i of Q; see (4.3.9)/(4.3.10). We also assume that $\xi_\nu^i \neq 0$, $\nu=0(1)m-1$.

Let $\xi^{(\mu)} := (\xi_{\mu-1}^i, \ldots, \xi_{\mu+mk-1}^i)^T \in \mathbb{R}^{\bar{m}+k}$ be the indicated sections of $\bar{m}+k=mk+1$ consecutive elements of $\{\xi_\nu^i\}$. From the construction of $\bar{A}^{(\mu)}$ it follows that $\bar{A}^{(\mu)} \xi^{(\mu)} = 0$ which implies

(4.3.18) $$(a^\mu)^T \xi^{(\mu)} = (d^\mu)^T \bar{A}^{(\mu)} \xi^{(\mu)} = 0, \qquad \mu=1(1)m.$$

But by (4.3.15) and (4.3.10),

(4.3.19) $$(a^\mu)^T \xi^{(\mu)} = \sum_{\kappa=0}^{k} a_{\kappa m}^\mu \xi_{\mu-1+\kappa m}^i = \left(\sum_{\kappa=0}^{k} a_{\kappa m}^\mu x_i^\kappa \right) \xi_{\mu-1}^i .$$

Since $\xi_{\mu-1}^{i} \neq 0$ by our preliminary assumption we find from (4.3.18)/(4.3.19) that each of the k eigenvalues x_i of Q is a zero of each of the k-th degree polynomials $\bar{\rho}_\mu(x) = \sum_{\kappa=0}^{k} \alpha_{\kappa m}^{\mu} x^\kappa$ which all have the same leading coefficient as ρ. According to Lemma 4.3.2 this implies the identity of the $\bar{\rho}_\mu$ with the associated polynomial ρ of $\mathfrak{M}$.

Again, as in the proof of Lemma 4.3.2, all the $a_{\kappa m}^{\mu}$ as well as the coefficients a_κ of ρ are m-th degree polynomials in the α_κ^μ so that the identity which we have derived is actually an identity in the α_κ^μ. Thus, it must also hold if Q has multiple eigenvalues or if some of the ξ_ν^i, $\nu = 0(1)m-1$, vanish. □

Remark. If the $\bar{m}-1$ columns of $\bar{A}^{(\mu)}$ which define d^μ form a matrix of rank less than $\bar{m}-1$ the d^μ are not unique. We can circumvent this difficulty by requiring that we take d^μ to be that unique function of the α_κ^μ which arises in the case of rank $\bar{m}-1$.

We now write down the forward step procedure of our m-cyclic linear k-step method $\mathfrak{M}$ for $\bar{m}$ successive values of ν

$$\frac{1}{h} \sum_{\kappa=0}^{k} [\alpha_\kappa^{\mu\langle\nu\rangle} \eta_{\nu-k+\kappa} - h \beta_\kappa^{\mu\langle\nu\rangle} f(\eta_{\nu-k+\kappa})] = 0, \qquad \nu = \nu_0(1)\nu_0 + \bar{m} - 1,$$

and form their linear combination with the components of $d^{\mu\langle\nu_0\rangle}$ as weights. The result is a linear mk-step procedure (note that $mk = \bar{m} + k - 1$ by (4.3.13)):

$$\text{(4.3.20)} \quad \begin{aligned} &\frac{1}{h} \sum_{\lambda=0}^{\bar{m}-1} d_{\lambda+1}^{\mu\langle\nu_0\rangle} \left[\sum_{\kappa=0}^{k} (\alpha_\kappa^{\mu\langle\nu_0+\lambda\rangle} \eta_{\nu_0+\lambda-k+\kappa} - h \beta_\kappa^{\mu\langle\nu_0+\lambda\rangle} f(\eta_{\nu_0+\lambda-k+\kappa})) \right] \\ &= \frac{1}{h} \sum_{\kappa=0}^{mk} [a_\kappa^{\mu\langle\nu_0\rangle} \eta_{\nu_0-k+\kappa} - h b_\kappa^{\mu\langle\nu_0\rangle} f(\eta_{\nu_0-k+\kappa})] = 0, \end{aligned}$$

with a_κ^μ from (4.3.14) and b_κ^μ from (4.3.16).

For each of m successive values of ν_0 we obtain a different mk-step procedure (4.3.20), with the value of the superscript of the a_κ^μ and b_κ^μ taking the values $1(1)m$ in some cyclical order. Thus (4.3.20), with $\nu_0 = mk(1)n$, represents the forward step procedure of an m-cyclic linear mk-step method $\bar{\mathfrak{M}}$ based upon linear mk-step schemes $\bar{\psi}_1, \ldots, \bar{\psi}_m$ with coefficients from (4.3.14)/(4.3.16). Due to (4.3.17) the associated polynomials $\bar{\rho}_\mu$ of the $\bar{\psi}_\mu$ are all equal and

$$\text{(4.3.21)} \qquad \bar{\rho}_\mu(x) = \bar{\rho}(x) = \sum_{\kappa=0}^{k} a_\kappa x^{m\kappa} = \rho(x^m)$$

while the coefficients b_κ^μ are normally different for each μ.

Def. 4.3.3. For a given m-cyclic linear k-step method $\mathfrak{M}$, the m-cyclic linear mk-step method $\overline{\mathfrak{M}}$ based upon the linear mk-step schemes $\overline{\psi}_\mu$, $\mu=1(1)m$, with the starting procedure

$$\eta_\nu - \zeta_n(\nu h) = 0, \qquad \nu = 0(1)mk-1, \tag{4.3.22}$$

where ζ_n is the solution of $\mathfrak{M}$(IVP 1), is called the *auxiliary method* of $\mathfrak{M}$.

Theorem 4.3.6. *Let $\mathfrak{M}$ be some m-cyclic linear k-step method which satisfies the assumptions of Corollary 4.3.4. Then the auxiliary method $\overline{\mathfrak{M}}$ of $\mathfrak{M}$ is stable, its linear mk-step schemes $\overline{\psi}_\mu$ are of orders $\overline{p}_\mu \geq p$, $\mu=1(1)m$, and its starting procedure (4.3.22) is of order p. Furthermore, the solutions ζ_n of $\mathfrak{M}$(IVP 1) and $\overline{\zeta}_n$ of $\overline{\mathfrak{M}}$(IVP 1) are identical for an arbitrary IVP 1, at least for sufficiently small h.*

Proof. Since the associated polynomials $\overline{\rho}_\mu$ of the $\overline{\psi}_\mu$ are all equal to $\overline{\rho}$ it follows from a trivial modification in the proof of Theorem 4.2.7 that $\overline{\mathfrak{M}}$ is stable if $\overline{\rho}$ satisfies the root criterion, which is the case according to (4.3.21) and our assumption on ρ.

The associated difference operators of the $\overline{\psi}_\mu$ are finite linear combinations (with weights independent of h) of the associated difference operators of the ψ_μ, with appropriate shifts. Therefore, $\overline{p}_\mu \geq \min_\mu p_\mu \geq p$, $\mu=1(1)m$.

By Corollary 4.3.4, $\mathfrak{M}$ is convergent of order p for an $\mathrm{IVP}^{(p)}1$ so that $\zeta_n(\nu h) - z(\nu h) = O(h^p)$ which implies the consistency of order p of the starting procedure (4.3.22) of $\overline{\mathfrak{M}}$. As $\overline{\mathfrak{M}}$ is thus stable and consistent of order p it is also convergent of order p for an $\mathrm{IVP}^{(p)}1$.

Finally, ζ_n is clearly a solution of $\overline{\mathfrak{M}}$(IVP 1) by the construction of $\overline{\mathfrak{M}}$. Since both ζ_n and $\overline{\zeta}_n$ are unique for sufficiently small h by Theorem 1.2.3, we have $\zeta_n = \overline{\zeta}_n$. □

Example. As in Section 4.3.1. From $k=2$, $m=2$, we have $\overline{m}=3$ and the 3×5-matrices $\overline{A}^{(1)}$ and $\overline{A}^{(2)}$ are

$$\overline{A}^{(1)} = \begin{pmatrix} -\frac{1}{2} & 0 & \frac{1}{2} & 0 & 0 \\ 0 & 0 & -1 & 1 & 0 \\ 0 & 0 & -\frac{1}{2} & 0 & \frac{1}{2} \end{pmatrix}, \qquad \overline{A}^{(2)} = \begin{pmatrix} 0 & -1 & 1 & 0 & 0 \\ 0 & -\frac{1}{2} & 0 & \frac{1}{2} & 0 \\ 0 & 0 & 0 & -1 & 1 \end{pmatrix}.$$

With $d^1 = \begin{pmatrix} 0 \\ 0 \\ 1 \end{pmatrix}$ and $d^2 = \begin{pmatrix} -\frac{1}{2} \\ 1 \\ \frac{1}{2} \end{pmatrix}$ we have

$$(d^1)^T \overline{A}^{(1)} = (d^2)^T \overline{A}^{(2)} = (0 \quad 0 \quad -\tfrac{1}{2} \quad 0 \quad \tfrac{1}{2}) = (a^1)^T = (a^2)^T$$

and

$$\sum_{\kappa=0}^{k} a_\kappa x^\kappa = \tfrac{1}{2}x^2 - \tfrac{1}{2}x = \tfrac{1}{2}x(x-1) = \rho(x);$$

see Section 4.3.1. Note that d^1 is not unique; but if we replace the vanishing coefficient α_1^1 of the Simpson scheme by a non-vanishing one and let it tend to zero we arrive at the above d^1 which satisfies Lemma 4.3.4.

For the b_κ^μ of the schemes $\bar{\psi}_\mu$ of the auxiliary method we find

$$b_\kappa^1 = \begin{cases} 0, & \kappa = 0,1, \\ \beta_{\kappa-2}^1, & \kappa = 2(1)4, \end{cases} \quad \text{due to the simplicity of } d^1$$

and

$$b_0^2 = \tfrac{1}{24}, \quad b_1^2 = -\tfrac{1}{6}, \quad b_2^2 = \tfrac{5}{12}, \quad b_3^2 = \tfrac{1}{2}, \quad b_4^2 = \tfrac{5}{24}$$

so that the auxiliary 2-cyclic linear 4-step method $\bar{\mathfrak{M}}$ is based upon the schemes $\bar{\psi}_1$ and $\bar{\psi}_2$ with coefficient matrices

$$\begin{pmatrix} 0 & 0 & -\frac{1}{2} & 0 & \frac{1}{2} \\ 0 & 0 & \frac{1}{6} & \frac{2}{3} & \frac{1}{6} \end{pmatrix} \text{ and } \begin{pmatrix} 0 & 0 & -\frac{1}{2} & 0 & \frac{1}{2} \\ \frac{1}{24} & -\frac{1}{6} & \frac{5}{12} & \frac{1}{2} & \frac{5}{24} \end{pmatrix}.$$

$\bar{\psi}_1$ and $\bar{\psi}_2$ are *both* of order 4 which proves the statement at the end of Section 4.3.1.

4.3.3 Attainable Order of Cyclic Linear Multistep Methods

As mentioned at the start of Section 4.3, cyclic linear k-step methods become interesting through the fact that ρ may satisfy the root criterion although none of the ρ_i does as is shown by examples in Donelson and Hansen [1]. However, no systematic theory of the relation between the zeros of ρ and those of the ρ_i has as yet been elaborated.

It follows immediately from (4.3.8) that $\rho(1)=0$ since the row sums of A and thus those of $\sum_{j=0}^{j_0} A_j$ vanish due to (4.1.3) so that $\det\left(\sum_{j=0}^{j_0} A_j\right)=0$. If we *prescribe* the values for the $k-1$ remaining zeros of ρ we have $k-1$ conditions on the α_κ^μ to be satisfied. We may hope to satisfy these conditions if we have $k-1$ free parameters available. (Of course, the conditions are non-linear in the α_κ^μ as we have seen.)

By arguments analogous to the ones in Section 4.1.3 one can show that there is a one-parameter family of linear k-step schemes of order $2k-1$ (according to Theorem 4.1.8 there is exactly one k-step scheme of order $2k$). Therefore, we might try to choose $k-1$ of these $(2k-1)$-order k-step schemes such that the associated polynomial of the $(k-1)$-cyclic linear k-step method $\mathfrak{M}$ based upon them has prescribed zeros inside the unit disk (besides the zero at 1). $\mathfrak{M}$ would then be convergent of order $2k-1$ for sufficiently smooth IVP 1 by Corollary 4.3.4 which is more than can be achieved by uniform linear k-step methods for $k \geq 3$.

Conjecture. *For each k there exist $(k-1)$-cyclic linear k-step methods which are convergent of order $2k-1$.*

Example. $k=3$, $m=2$. For each value of $c \neq 0$, the linear 3-step scheme

$$\begin{pmatrix} c - \dfrac{11}{30} & -c - \dfrac{8}{30} & -c + \dfrac{19}{30} & c \\[2ex] \dfrac{c}{3} - \dfrac{1}{90} & -c + \dfrac{19}{30} & c + \dfrac{8}{30} & -\dfrac{c}{3} + \dfrac{1}{9} \end{pmatrix} \tag{4.3.23}$$

is of order 5. From

$$A=\begin{pmatrix} 0 & \alpha_0^1 & | & \alpha_1^1 & \alpha_2^1 & | & \alpha_3^1 & 0 \\ 0 & 0 & | & \alpha_0^2 & \alpha_1^2 & | & \alpha_2^2 & \alpha_3^2 \end{pmatrix}, \qquad j_0=2,$$

we obtain

$$\rho(x)=x^{-1}\det\left(\sum_{j=0}^{2} A_j x^j\right)=\frac{1}{x}\det\begin{pmatrix}(\alpha_3^1 x+\alpha_1^1)x & \alpha_2^1 x+\alpha_0^1 \\ (\alpha_2^2 x+\alpha_0^2)x & (\alpha_3^2 x+\alpha_1^2)x\end{pmatrix}$$
$$=x(\alpha_3^1 x+\alpha_1^1)(\alpha_3^2 x+\alpha_1^2)-(\alpha_2^1 x+\alpha_0^1)(\alpha_2^2 x+\alpha_0^2)$$
$$=x(c_1 x-c_1-\tfrac{8}{30})(c_2 x-c_2-\tfrac{8}{30})-((\tfrac{19}{30}-c_1)x+c_1-\tfrac{11}{30})((\tfrac{19}{30}-c_2)x+c_2-\tfrac{11}{30})$$

for the 5-th order schemes (4.3.23).

If we prescribe $x_2=0$ the constant term $(c_1-\frac{11}{30})(c_2-\frac{11}{30})$ of ρ has to vanish so that $c_1=\frac{11}{30}$. For prescribed x_3, c_2 has to satisfy (by Vieta's Theorem)

$$\tfrac{19}{30}(c_2+\tfrac{8}{30})-\tfrac{8}{30}(c_2-\tfrac{11}{30})=\tfrac{11}{30}c_2 x_3 \qquad \text{or} \qquad c_2=-\frac{8}{11(1-x_3)}.$$

Thus we can arbitrarily prescribe $x_3\in[-1,+1)$ and find the appropriate c_2. Therefore there exists a set of pairs of linear 3-step schemes of order 5 such that the 2-cyclic linear 3-step method based upon them is stable and hence convergent of order 5 for an $\mathrm{IVP}^{(5)}1$ if it is started sufficiently accurately. For $x_3=-\frac{1}{11}$, Hansen gives the following schemes:

$$\psi_1:\tfrac{1}{30}\begin{pmatrix}0 & -19 & 8 & 11 \\ -\frac{1}{3} & 8 & 19 & \frac{10}{3}\end{pmatrix}, \qquad \psi_2:\tfrac{1}{30}\begin{pmatrix}31 & 12 & 39 & -20 \\ 10 & 39 & -12 & -7\end{pmatrix}.$$

Even more interesting is the question whether one could achieve a convergence of order $2k$ by combining sufficiently many k-step schemes of order $2k-1$ each. Note that we cannot choose k-step schemes which are all of order $2k$ for our cyclic linear k-step method since there is only *one* such scheme so that we would obtain a uniform linear k-step method with the restrictions of Corollary 4.2.12.

Theorem 4.3.7. *Consider an m-cyclic linear k-step method $\mathfrak{M}$ based upon linear k-step schemes of orders $p_\mu\ge p$, $\mu=1(1)m$, with an associated polynomial which satisfies the root criterion. If the starting procedure of $\mathfrak{M}$ is of order $p+1$ and if the linear mk-step schemes $\overline{\psi}_\mu$ of the auxiliary method $\overline{\mathfrak{M}}$ of $\mathfrak{M}$ are of orders $\overline{p}_\mu\ge p+1$, $\mu=1(1)m$, then $\mathfrak{M}$ is convergent of order $p+1$ for an* $\mathrm{IVP}^{(p+1)}1$.

Proof. The starting procedure (4.3.22) of $\overline{\mathfrak{M}}$ is of order $p+1$: Under our assumptions on $\mathfrak{M}$ its local discretization error $l_n\in E_n$ satisfies

$$l_n(t_\nu)=\begin{cases}O(h^{p+1}), & \nu=0(1)k-1,\\ O(h^p), & \nu\ge k.\end{cases}$$

Thus, the stability of $\mathfrak{M}$ (see Theorem 4.3.3) implies via Theorem 2.1.2 that

$$\|\zeta_n(t_\nu)-z(t_\nu)\|\le S\left[\max_{\kappa=0(1)k-1}\|l_n(t_\kappa)\|+h\sum_{\mu=k}^{\nu}\|l_n(t_\mu)\|\right]=O(h^{p+1})$$

for $\nu=0(1)mk-1$. Therefore, $\overline{\mathfrak{M}}$ is consistent of order $p+1$ and Theorem 4.3.6 establishes the convergence of order $p+1$ of $\overline{\mathfrak{M}}$ and of $\mathfrak{M}$. □

Since we had needed $k-1$ different k-step schemes for a convergence order $2k-1$ of $\mathfrak{M}$, one would expect that *at least* k different k-step schemes are necessary for convergence of order $2k$. In this attempt to construct k-cyclic linear k-step methods with $p=2k-1$ but $\overline{p}_\mu=2k$, $\mu=1(1)k$, Donelson and Hansen discovered the following phenomenon (for $k=2,3,4$):

Consider the error constants $\overline{C}_\mu$ of the linear k^2-step schemes $\overline{\psi}_\mu$ of order $2k-1$ of the auxiliary method $\overline{\mathfrak{M}}$ as a function of the parameters c_μ of the original k-step schemes ψ_μ of order $2k-1$, $\mu=1(1)k$ (cf. the example above). It is clear from the construction of the $\overline{\psi}_\mu$ that (see Section 4.3.2 above Lemma 4.3.5)

$$\overline{C}_\mu(c_1,c_2,\dots,c_k)=\pi^{\mu-1}\,\overline{C}_1(c_1,c_2,\dots,c_k)=\overline{C}_1(c_\mu,c_{\mu+1},\dots,c_{\mu-1}).$$

It turns out, however, that the functions $\overline{C}_\mu(c_1,\dots,c_k)$ are π-invariant so that $\overline{C}_1=\overline{C}_2=\dots=\overline{C}_k=:\overline{C}$. Thus, the *only* condition to be satisfied in order to have *all* $\overline{\psi}_\mu$ of order $2k$ is

$$\overline{C}(c_1,\dots,c_k)=0. \tag{4.3.24}$$

No formal proof of the π-invariance has as yet been given; in Section 5.4.4 we will return to this subject in a different context which will make it clear that one condition is sufficient.

Since we have $k-1$ conditions for the stability of $\mathfrak{M}$ as previously and (4.3.24) represents only one further condition, we may hope to satisfy these k conditions by a proper choice of the k parameters c_μ, $\mu=1(1)k$. Hansen has demonstrated the existence of k-cyclic linear k-step methods which are convergent of order $2k$ for $k=2,3,4$.

Conjecture. *For each k there exist k-cyclic linear k-step methods which are convergent of order $2k$.*

Examples. 1. $k=2$, see end of Section 4.3.2: Since it is obvious that $\overline{\psi}_1$ is of order 4, the π-invariance of $\overline{C}$ implies that $\overline{\psi}_2$ is also of order 4 which may easily be checked. Thus the alternate use of a Simpson and an implicit Adams 2-step procedure with a starting procedure of order 4 leads to a linear 2-step method which is convergent of order 4. It has the advantage over the ordinary Simpson method that its associated polynomial has no zeros of modulus 1 except $x_1=1$.

2. $k=3$, see Donelson and Hansen [1]. The 3-cyclic linear 3-step method $\mathfrak{M}$ based upon the 3-step schemes of order 5

$$\psi_1:\tfrac{1}{90}\begin{pmatrix}0 & -57 & 24 & 33\\ -1 & 24 & 57 & 10\end{pmatrix},\quad \psi_2:\tfrac{1}{30}\begin{pmatrix}-136 & 117 & 144 & -125\\ 45 & 144 & -117 & -42\end{pmatrix}$$

$$\psi_3:\tfrac{1}{930}\begin{pmatrix}-283 & -306 & 531 & 58\\ 84 & 531 & 306 & 9\end{pmatrix}$$

has the characteristic polynomial $\rho(x)=\frac{1}{83700}(x-1)x(7975x-361)$ and the 9-step schemes of the auxiliary method are all of order 6; thus, $\mathfrak{M}$ is convergent of order 6 if started with a starting procedure of order 6.

Added in proof: The two conjectures above have been proved in the meantime by Mischak [1] who has also established the missing parts of the theory.

4.4 Asymptotic Expansions

Although the discretization methods regarded in Chapter 4 are "linear", i. e., they do not use iterated substitutions into f, the structure of their discretization error is quite complicated. This is due to the fact that k-th order difference equations have been used to approximate first order differential equations.

4.4.1 The Local Discretization Error

The local discretization error $l_n=\varphi_n(F)\Delta_n z$ of a uniform linear k-step method based upon the linear k-step scheme ψ is given by (cf. Def. 4.2.1)

$$(4.4.1)\quad l_n(t_\nu)=\begin{cases} z(t_\nu)-s_{n\nu}(z_0), & \nu=0(1)k-1,\\ (L_{h_\nu}[\psi]z)(t_{\nu-k}), & \nu=k(1)n,\ \text{except near step changes,}\end{cases}$$

where $L_h[\psi]$ is the associated difference operator of ψ (Def. 4.1.6). If there is a step change at t_ν, $L_h[\psi]$ has to be replaced by the appropriate $L_h[\hat{\psi},\hat{r}]$ for $l_n(t_{\nu+\kappa})$, $\kappa=1(1)k-1$; see Section 4.1.2.

For sufficiently smooth IVP 1, the right-hand sides of (4.4.1) will normally possess asymptotic expansions in powers of $1/n$ at each t_ν. For the forward step part of (4.4.1) this is trivial since, for $z\in D^{(J+1)}$ and ψ of order p,

$$(4.4.2)\qquad (L_h[\psi]z)(t_\nu)=\sum_{j=p}^{J}\frac{h^j}{j!}\left(\sum_{\kappa=1}^{k}\frac{\kappa^{j+1}\alpha_\kappa}{j+1}-\kappa^j\beta_\kappa\right)z^{(j+1)}(t_\nu)+O(h^{J+1}),$$

and similar expansions also hold for unequally spaced t_ν, see (4.1.26).

For the starting part of (4.4.1) we will assume—without seriously restricting the generality of the discussion—that we have, for $z\in D^{(J+1)}$ and a starting procedure of order p,

$$(4.4.3)\quad s_{n\kappa}(z_0)=z(t_\kappa)+\sum_{j=p}^{J}s_\kappa^{(j)}h_1^j+O(h_1^{J+1}),\qquad \text{for } \kappa=0(1)k-1,$$

where h_1 is the constant stepsize at the beginning of the grid. (With a coherent grid sequence the first k steps have to be equal for sufficiently large n.)

But (4.4.2)/(4.4.3) is not sufficient for the existence of an asymptotic expansion of the local discretization error in the sense of Section 1.3.1 (see (1.3.4)) which requires the existence of elements $\lambda_j z \in E^0$, $j=p(1)J$, such that

$$l_n = \sum_{j=p}^{J} \frac{1}{n^j} \Delta_n^0(\lambda_j z) + O(n^{-(J+1)}). \tag{4.4.4}$$

Here a step change presents serious difficulties: The second component functions of the $\lambda_j z$ (cf. (2.1.4)) may have a finite jump at the points $\bar{t}_\mu$ where step changes occur but they must be *independent of n*. Thus they cannot respond to the fact that the local discretization error arises from $k-1$ different operators $L_h[\hat{\psi}_\kappa, \hat{r}_\kappa]$ within the intervals $(\bar{t}_\mu, \bar{t}_\mu+(k-1)h_{\mu+1}]$, where $h_{\mu+1}$ is the step in $(\bar{t}_\mu, \bar{t}_{\mu+1}]$, because the lengths $(k-1)h_{\mu+1}$ of these intervals depend on n and tend to zero as $n \to \infty$.

Even the inclusion of the asymptotic expansion (4.4.3) of the starting procedure into (4.4.4) is generally not feasible: The $\lambda_j z \in E^0$ have to be of the form

$$\begin{pmatrix} (\lambda_j z)_0 \\ (\lambda_j z)(t) \end{pmatrix}.$$

Since the functions $(\lambda_j z)(t)$ are fully specified by (4.4.2) we have only the quantities $(\lambda_j z)_0 \in \mathbb{R}^s$ left to match the expansions (4.4.3) and (4.4.4) at the k grid points $t_0, \ldots, t_{k-1}$. Thus, no matter how we choose the Δ_n^0 (cf. e.g. (4.2.1)) we can hope to achieve that matching only for very special starting procedures which depend strongly on the terms in (4.4.2).

Hence the only result for realistic uniform k-step methods along the lines of approach in Section 1.3.1 is the existence of a principal part of the local discretization error (see Def. 1.3.5):

Lemma 4.4.1. *The local discretization error l_n of a uniform linear k-step method $\mathfrak{M}$ based upon the linear k-step scheme ψ of order p possesses an asymptotic expansion to the order p if its starting procedure is of order $p+1$ and if linear multistep schemes which are of order p for the appropriate step ratio vectors are used at step changes.*

Proof. Let θ be the stepsize function of $\mathfrak{M}$ and C the error constant of ψ and set

$$\lambda_p z = \begin{pmatrix} 0 \\ C\,\theta(t)^p z^{(p+1)}(t) \end{pmatrix} \in E^0. \tag{4.4.5}$$

With Δ_n^0 from (4.2.1) we obtain, with $h := 1/n$,

$$\|l_n - h^p \Delta_n^0 \lambda_p z\|_{E_n^0} = \max_{\kappa=0(1)k-1} \left\| z(t_\kappa) - s_{n\kappa}(z_0) - h^p \int_0^{t_\kappa} C_p \theta(\tau)^p z^{(p+1)}(\tau) d\tau \right\|$$
$$+ \frac{1}{n} \sum_{v=k}^{n} \|L_{h_v}[\psi] z(t_{v-k}) - h^p C \sigma(T_{h_v}) \theta^p z^{(p+1)}(t_{v-k})\|;$$

in the wake of a step change $L_h[\psi]$ has to be replaced by the appropriate $L_h[\hat{\psi}, \hat{r}]$.

The first term on the right-hand side is $O(n^{-(p+1)})$ due to our assumption on the starting procedure and since $\int_0^{t_\kappa} \dots d\tau = O(n^{-1})$. The elements in the sum are $O(n^{-(p+1)})$ by (4.1.31) except near step changes. After step changes $(L_{h_v}[\hat{\psi}, \hat{r}] z)(t_{v-k})$ is of order $O(n^{-p})$ but generally not equal to $h^p C \sigma(T_{h_v}) \theta^p z^{(p+1)}(t_{v-k})$ so that the difference remains $O(n^{-p})$. Since there are only finitely many step changes as $n \to \infty$, the factor $1/n$ in front of the sum produces the desired order. □

Def. 4.4.1. For a given linear k-step scheme ψ of exact order p with error constant C and for a given IVP$^{(p+1)}$1, the mapping $\varphi: \mathbb{R}^s \to \mathbb{R}^s$ defined by

$$\varphi(\xi) := -C \frac{\partial^{p+1}}{\partial h^{p+1}} \bar{z}(h, \xi) \bigg|_{h=0}$$

is called the *principal error function* of ψ for that IVP 1. ($\bar{z}$ is the local solution of the IVP 1; see Def. 3.1.6.)

Remark. It is clear from (4.4.1) and (4.1.31) that

$$\text{(4.4.6)} \qquad \begin{aligned} l_n(t_v) &= -n^{-p} \theta(t_{v-k})^p \varphi(z(t_{v-k})) + O(n^{-(p+1)}) \\ &= -n^{-p} \theta(t_{v-k})^p \sigma(T_h) \varphi(z(t_{v-k})) + O(n^{-(p+1)}) \end{aligned}$$

for all $v \geq k$ which are not "close" to a step change. Compare also Def. 3.4.1 and Lemma 3.4.1 for RK-methods.

Since the requirements for the existence of a local error mapping are even more stringent than those for the existence of an asymptotic expansion (4.4.4), we cannot expect to have local error mappings with uniform linear k-step methods except in very special circumstances. While this leaves little hope for the existence of a general asymptotic expansion (1.3.7) of the global discretization error, it will turn out that for many uniform linear k-step methods $\hat{\Delta}$-reduced asymptotic expansions (1.4.3) do exist for appropriate types of $\hat{\Delta}$-reductions, cf. Section 1.4.1.

In preparation for the rather complicated derivation of these results we introduce the following quantities:

Def. 4.4.2. Consider a D-stable linear k-step scheme $\langle\rho,\sigma\rangle$ of (minimal) order 1. For each zero x_i of ρ we define the polynomials

$$\text{(4.4.7)}\qquad \begin{aligned} \rho_i(x) &:= \rho(x_i x) = \sum_{\kappa=0}^{k} (\alpha_\kappa x_i^\kappa) x^\kappa, \\ \sigma_i(x) &:= \sigma(x_i x) = \sum_{\kappa=0}^{k} (\beta_\kappa x_i^\kappa) x^\kappa. \end{aligned}$$

For each of the $q\geq 1$ "essential" zeros x_i of ρ, $|x_i|=1$, we call the quantity

$$\text{(4.4.8)}\qquad \lambda_i := \frac{\sigma_i(1)}{\rho_i'(1)} = \frac{\sigma(x_i)}{x_i\,\rho'(x_i)}, \qquad i=1(1)q,$$

the *growth parameter* of x_i. Furthermore, we define for $i=1(1)q$

$$\text{(4.4.9)}\qquad L_h^{(i)}[\langle\rho,\sigma\rangle] := \frac{1}{h}\left[\rho_i(T_h) - h\frac{1}{\lambda_i}\sigma_i(T_h)\,\partial\right]; \quad \text{cf. (4.1.20).}$$

Note that $\sigma_i(1)=\sigma(x_i)\neq 0$ since ρ and σ have no common divisor (see Def. 4.1.6 and the discussion after Def. 4.1.3) so that $\lambda_i\neq 0$ and $L_h^{(i)}$ is well-defined. Also $\rho_i'(1)=\rho'(x_i)\neq 0$ due to the D-stability of $\langle\rho,\sigma\rangle$. We will always denote the essential zero at 1 by x_1 so that $\rho_1=\rho$, $\sigma_1=\sigma$, $\lambda_1=1$, and $L_h^{(1)}[\langle\rho,\sigma\rangle]\equiv L_h[\langle\rho,\sigma\rangle]$. Any further essential zeros x_i of ρ, $i=2(1)q$, will be called *extraneous* essential zeros.

We will denote by $c_j^{(i)}$, $i=1(1)q$, $j=1(1)J$, the coefficients of the expansions ($y\in D^{(J+1)}$)

$$\text{(4.4.10)}\qquad (L_h^{(i)}[\langle\rho,\sigma\rangle]y)(t) = \sigma_i(T_h)\sum_{j=1}^{J} c_j^{(i)} h^j y^{(j+1)}(t) + O(h^{J+1})$$

for each of the essential zeros x_i of ρ. According to Lemma 4.1.5, the $c_j^{(i)}$ are determined by the generating functions

$$\frac{\rho_i(e^h) - h\lambda_i^{-1}\sigma_i(e^h)}{h\,\sigma_i(e^h)} = \sum_{j=1}^{\infty} c_j^{(i)} h^j;$$

the $c_0^{(i)}$ vanish due to the definition of the λ_i.

If $\langle\rho,\sigma\rangle$ is of order p, we have $c_1^{(1)}=\cdots=c_{p-1}^{(1)}=0$ and $c_p^{(1)}=C$, the error constant of $\langle\rho,\sigma\rangle$, according to Corollary 4.1.6. For given fixed $J\geq p$, we will denote by $\bar{L}_h[\langle\rho,\sigma\rangle]$ the linear operator defined by

$$\text{(4.4.11)}\qquad \bar{L}_h[\langle\rho,\sigma\rangle] := \sum_{j=p}^{J} c_j^{(1)} h^j \partial^{j+1}, \quad \text{cf. (4.4.10).}$$

4.4.2 Asymptotic Expansion of the Global Discretization Error; Preparations

We will at first display the relation that would have to exist between the starting and the forward-step procedures of a uniform linear k-step method in order that its global discretization error possess an asymptotic expansion to order J, at least for equidistant grids:

Theorem 4.4.2 (cf. Gragg [1]). *For a D-stable linear k-step scheme ψ of order p consider the functions $\bar{e}_j \in E$, $j=0(1)J$, generated for a given $IVP^{(J+1)}1$ through to the construction* (1.3.9)/(1.3.10) *of Theorem* 1.3.1 *by*

$$(4.4.12)\qquad \Lambda_n y = \begin{pmatrix} h^p \bar{b}_0(h) \\ \bar{L}_h[\psi]y \end{pmatrix} \in E^0, \qquad h = \frac{1}{n},$$

where $\bar{b}_0$ is some polynomial and $\bar{L}_h$ is from (4.4.11). *For that $IVP^{(J+1)}1$, the global discretization error of a uniform linear k-step method $\mathfrak{M}$ with equidistant grids based upon ψ has the asymptotic expansion*

$$(4.4.13)\qquad \bar{\zeta}_n = \Delta_n z + \sum_{j=p}^{J} h^j \Delta_n \bar{e}_j + O(h^{J+1})$$

if and only if the starting procedure of $\mathfrak{M}$ satisfies

$$(4.4.14)\quad \bar{s}_{n\kappa}(z_0) = z(t_\kappa) + \sum_{j=p}^{J} h^i \bar{e}_j(t_\kappa) + O(h^{J+1}), \qquad \kappa = 0(1)k-1.$$

Proof. The $\bar{e}_j$ are well-defined since Λ_n possesses the asymptotic expansion

$$(4.4.15)\qquad \Lambda_n y = \sum_{j=p}^{J} h^j \begin{pmatrix} \bar{b}_{j0} \\ c_j^{(1)} y^{(j+1)} \end{pmatrix} + O(h^{J+1})$$

by (4.4.11); it is J,p-smooth and the construction (1.3.9)/(1.3.10) can be carried out in a straightforward manner for F from an $\mathrm{IVP}^{(J+1)}1$.

According to (4.2.1), (4.2.3), and (4.4.10), for $y \in D^{(J+1)}$

$$(F_n \Delta_n y)(t_\nu) = [\Delta_n^0 (F + \Lambda_n) y](t_\nu) + O(h^{J+1}), \qquad \nu = k(1)n,$$

which implies

$$\left[F_n \Delta_n \left(z + \sum_{j=p}^{J} h^j \bar{e}_j\right)\right](t_\nu) = O(h^{J+1}), \qquad \nu = k(1)n,$$

according to the proof of Theorem 1.3.1. If (4.4.14) holds then also

$$\left[F_n \Delta_n \left(z + \sum_{j=p}^{J} h^j \bar{e}_j\right)\right](t_\kappa) = O(h^{J+1}), \qquad \kappa = 0(1)k-1,$$

so that $\|F_n \Delta_n(z+\sum h^j \bar{e}_j)\|_{E_n^0}=O(h^{J+1})$ which implies (4.4.13) due to $F_n\bar{\zeta}_n=0$ and the stability of $\overline{\mathfrak{M}}$.

A violation of (4.4.14) trivially violates (4.4.13) due to the norm (2.2.2) in E_n. □

Remarks. 1. Unfortunately, the immediate use of the starting procedure (4.4.14) (without the remainder term) is unrealistic except in trivial cases since the $\bar{e}_j$ depend in a rather complicated manner on the data of the IVP 1. The construction of a RK-scheme such that the starting procedure (4.2.4) satisfies (4.4.14) for $J>p$ seems not to have been investigated so far.

2. The use of $s_{n\kappa}(z_0)=z(t_\kappa)+O(h^{J+1})$ does *not* satisfy (4.4.14) for $J>p$. Thus, the use of an expansion (4.2.6) to a sufficiently high power of h (or even the use of the exact values $z(t_\kappa)$) for a starting procedure generally does not produce an asymptotic expansion for the global discretization error to an order $J>p$.

For the case $J=p$ we have a satisfactory result:

Corollary 4.4.3. *For an $IVP^{(p+1)}$ 1, the global discretization error ε_n of a uniform linear k-step method based upon the linear k-step scheme ψ of exact order $p\geq 1$, with a starting procedure of order $p+1$ and linear multistep schemes of order p at step changes, satisfies*

$$\varepsilon_n=\frac{1}{n^p}\Delta_n e_p+O(n^{-(p+1)}) \tag{4.4.16}$$

where e_p is the solution of the differential equation

$$\begin{cases} e_p(0)=0, \\ e_p'-f'(z(t))e_p=-C\theta(t)^p z^{(p+1)}(t)=\theta(t)^p\varphi(z(t)); \end{cases} \tag{4.4.17}$$

C is the error constant and φ the principal error function of ψ.

Proof. For equidistant grids, the assertion is a consequence of Theorem 4.4.2 since $h^p\bar{e}_p(t_\kappa)=O(h^{p+1})$ in (4.4.14) follows from $e_p(0)=0$. In the general case, we can apply Theorem 1.3.3. Assumption (ii) of Theorem 1.3.3 follows from Lemma 4.4.1; the other assumptions are easily verified. □

Remark. According to (4.4.16), Richardson extrapolation with $r=1$ is justified in the situation of Corollary 4.4.3.

On the basis of Theorem 4.4.2 we can now visualize the problem: The specified starting procedure of order p of a given method $\mathfrak{M}$ may be interpreted as a perturbation of the "natural" starting procedure with $\bar{s}_{n\kappa}$ from (4.4.14) (cf. (4.4.3)):

$$s_{n\kappa}(z_0) = \bar{s}_{n\kappa}(z_0) + \sum_{j=p}^{J} h^j(s_\kappa^{(j)} - \bar{e}_j(t_\kappa)) + O(h^{J+1}). \tag{4.4.18}$$

Assume initially that $f \equiv 0$: According to our analysis in Section 4.2.2, a perturbation of the initial values will generally produce a perturbation of the solution of $F_n\zeta_n = 0$ which contains components of *all* fundamental solutions $\xi_{i\nu}$ of $\rho(T_h)\varepsilon_\nu = 0$. But only the fundamental solution $\xi_{1\nu}$ arising from the principal zero $x_1 = 1$ of ρ approximates a solution of $e' = 0$, the other fundamental solutions *cannot be interpreted as discretizations of continuous functions* at all. Thus there are no functions $e_j \in E$ independent of n such that

$$\zeta_n - \bar{\zeta}_n = \Delta_n \sum_{j=p}^{J} h^j(e_j - \bar{e}_j) + O(h^{J+1}).$$

For the general case, this reasoning suggests the *impossibility of an asymptotic expansion* for a starting procedure different from (4.4.14).

On the other hand, in the trivial case $f \equiv 0$ we may expect the following behavior:

a) If ρ has *no extraneous essential zeros*, all fundamental solutions $\xi_{i\nu}$, $i > 1$, tend to zero exponentially as $\nu \to \infty$. Thus, at any fixed $t \in (0,1]$ the contributions from these $\xi_{i\nu}$ will vanish faster than a fixed power of h as $h \to 0$. If we select a sequence of grids $\mathbb{G}_n$ which all contain t and define $\hat{\Delta}$ to be the reduction to the value at t, then there will exist a $\hat{\Delta}$-reduced asymptotic expansion for the global discretization error (cf. Def. 1.4.1 and 1.4.2).

b) If ρ has $q > 1$ essential zeros x_i, the corresponding fundamental solutions $\xi_{i\nu}$, $i = 1(1)q$, are of the form x_i^ν since ρ is D-stable. If all the essential zeros x_i are *roots of unity*, there will be a smallest value ν_0 such that $x_i^{\nu_0} = 1$, $i = 1(1)q$. Thus if we consider a fixed $t \in (0,1]$ and a sequence of grids for which $t = t_{m\nu_0}$, $m \in \mathbb{N}$, we will again have a $\hat{\Delta}$-reduced asymptotic expansion.

The fact that both results also hold in the case of a general IVP 1 has been established by Gragg in his thesis [1] on which we will rely heavily in the following.

4.4.3 The Case of no Extraneous Essential Zeros

Before proving the main theorem on the case in which ρ satisfies the strong root criterion (Def. 4.2.2) we establish

Lemma 4.4.4. *Let ρ be a polynomial of degree $k \geq 1$ and p a polynomial of degree $m \geq -1$ (the zero polynomial is assigned the degree -1). The polynomial x which satisfies the difference equation*

$$\rho(T)x(\nu) = p(\nu) \tag{4.4.19}$$

has $\begin{cases} \text{degree } m \text{ and is uniquely determined} \\ \text{degree } m+1 \text{ and is determined except for its constant term} \end{cases}$

if $\begin{cases} \rho(1) \neq 0 \\ \rho(1) = 0 \text{ but } \rho'(1) \neq 0. \end{cases}$

Proof. a) $\rho(1) \neq 0$: Let $x(\nu) = \sum_{\mu=0}^{m} c_\mu \nu^\mu$, $p(\nu) = \sum_{\mu=0}^{m} d_\mu \nu^\mu$, then (4.4.19) implies

$$\rho(T) \sum_{\mu=0}^{m} c_\mu \nu^\mu = \sum_{\mu=0}^{m} c_\mu \sum_{\kappa=0}^{k} \alpha_\kappa (\nu+\kappa)^\mu$$

$$= \sum_{\mu=0}^{m} \left(\sum_{\lambda=\mu}^{m} c_\lambda \binom{\lambda}{\lambda-\mu} \sum_{\kappa=0}^{k} \alpha_\kappa \kappa^{\lambda-\mu} \right) \nu^\mu = \sum_{\mu=0}^{m} d_\mu \nu^\mu .$$

Comparison of coefficients yields an upper triangular linear system for the c_λ, $\lambda = 0(1)m$, with diagonal coefficients $\rho(1)$ so that the c_λ are uniquely determined. For $m=-1$, $p \equiv 0$ implies $x \equiv 0$.

b) $\rho(1)=0$, $\rho'(1) \neq 0$: With $x(\nu) = \sum_{\mu=0}^{m+1} c_\mu \nu^\mu$ we have

$$\rho(T) \sum_{\mu=0}^{m+1} c_\mu \nu^\mu = \sum_{\mu=0}^{m+1} \left(\sum_{\lambda=\mu}^{m+1} c_\lambda \binom{\lambda}{\lambda-\mu} \sum_{\kappa=0}^{k} \alpha_\kappa \kappa^{\lambda-\mu} \right) \nu^\mu$$

$$= \sum_{\mu=0}^{m} \left(\sum_{\lambda=\mu+1}^{m+1} c_\lambda \binom{\lambda}{\lambda-\mu} \sum_{\kappa=0}^{k} \alpha_\kappa \kappa^{\lambda-\mu} \right) \nu^\mu = \sum_{\mu=0}^{m} d_\mu \nu^\mu$$

which now yields an upper triangular system for c_λ, $\lambda = 1(1)m+1$, with diagonal coefficients $(\mu+1)\rho'(1)$. For $p \equiv 0$ we have $x = c_0$. c_0 remains undetermined since $\rho(T)c_0 = 0$. □

Theorem 4.4.5. *Consider a uniform linear k-step method $\mathfrak{M}$ with equidistant grids based upon the linear k-step scheme $\langle \rho, \sigma \rangle$ of order p, with a starting procedure of order p satisfying* (4.4.3). *If $\langle \rho, \sigma \rangle$ is strongly D-stable there exist functions $e_j \in D^{(J-j+1)}$, $j = p(1)J$, such that for an $IVP^{(J+1)}$ 1 the solution ζ_n of its discretization by $\mathfrak{M}$ satisfies* $(h = 1/n)$

$$\zeta_n(t) = z(t) + \sum_{j=p}^{J} h^j e_j(t) + O(h^{J+1}), \quad n \in \overline{\mathbb{N}}, \tag{4.4.20}$$

at each fixed $t>0$ *which is a common gridpoint of the infinitely many grids* $\mathbb{G}_n, n\in\bar{\mathbb{N}}\subset\mathbb{N}'$ (*cf. the example after Def.* 1.4.1).

Proof. a) The main idea of the proof is to show that the global discretization error $\varepsilon_n=\zeta_n-\Delta_n z\in E_n$ of $\mathfrak{M}$ for an IVP$^{(J+1)}$1 can be represented in the form

$$\varepsilon_n=\Delta_n\sum_{j=p}^{J}h^j e_j+\sum_{j=p}^{J}h^j\omega_j+O(h^{J+1}) \tag{4.4.21}$$

where the $e_j\in E$ are solutions of

$$F'(z)e_j=\begin{pmatrix} e_j(0)\\ e_j'-f'(z(t))e_j\end{pmatrix}=\begin{pmatrix} b_{j0}\\ b_j(t)\end{pmatrix}=b_j\in E^0,\qquad j=p(1)J\,, \tag{4.4.22}$$

with appropriate $b_j\in E^0$, and the $\omega_j\in E_n$ are of the form

$$(\omega_j)_\nu=\sum_i x_i^\nu\, w_{ji}(\nu),\qquad j=p(1)J\,. \tag{4.4.23}$$

Here the x_i are the extraneous zeros of ρ (for $i=2(1)k$) and certain of their products (for $i>k$, see below after (4.4.32)) while the w_{ji} are appropriate $\mathbb{R}^s$-valued polynomials in ν of a known maximal degree which is independent of h. If 0 is a zero of ρ, a slight modification of (4.4.23) is needed.

Let $\max\limits_{i=2(1)k}|x_i|<\hat{x}<1$ ($\hat{x}$ exists by our assumption on ρ); then there exist constants M_j such that

$$\|(\omega_j)_\nu\|=\hat{x}^\nu\left\|\sum_i\left(\frac{x_i}{\hat{x}}\right)^\nu w_{ji}(\nu)\right\|\le M_j\hat{x}^\nu\quad\text{for all }\nu\ge 0\,.$$

But for a fixed $t\in(0,1]$, $\hat{x}^{t/h}$ tends to zero faster than any given power of h with $h\to 0$ so that $\sum\limits_j h^j\omega_j(t)=O(h^{J+1})$ for any J and (4.4.20) holds.

b) To establish (4.4.21) we set

$$\zeta_n=\Delta_n z+\Delta_n\sum_{j=p}^{J}h^j e_j+\sum_{j=p}^{J+1}h^j\omega_j+\bar{\varepsilon}_n\,, \tag{4.4.24}$$

with e_j and ω_j from (4.4.22) and (4.4.23) and $\bar{\varepsilon}_n\in E_n$; the second sum runs to $J+1$ for technical reasons. Since $\mathfrak{M}$ converges of order p under our assumptions we have

$$\|\bar{\varepsilon}_n\|=O(h^p)\,. \tag{4.4.25}$$

By (4.4.24), with F_n from $\mathfrak{M}(\mathrm{IVP}^{(J+1)}1)$,

$$
\begin{aligned}
0=F_n\zeta_n=F_n\Delta_n z+F_n'(\Delta_n z)\Delta_n\sum_j h^j e_j+F_n'(\Delta_n z)\sum_j h_j\omega_j \\
+\sum_{m=2}^{[J/p]}\frac{1}{m!}F_n^{(m)}(\Delta_n z)\left(\Delta_n\sum_j h^j e_j\right)^m \\
+\sum_{m=2}^{[J/p]}\frac{1}{m!}\sum_{\mu=1}^{m}F_n^{(m)}(\Delta_n z)\binom{m}{\mu}\left(\Delta_n\sum_j h^j e_j\right)^{m-\mu}\left(\sum_j h^j\omega_j\right)^{\mu} \\
+F_n'(\Delta_n z)\overline{\varepsilon}_n+O(h^p)O(\|\overline{\varepsilon}_n\|)+O(h^{J+1}).
\end{aligned}
\tag{4.4.26}
$$

For $m\geq 2$ and $\varepsilon\in E_n$,

$$
(F_n^{(m)}(\Delta_n z)\varepsilon^m)(t_\nu)=\begin{cases}0, & \nu=0(1)k-1,\\ -\sigma(T_h)f^{(m)}(z(t_{\nu-k}))\varepsilon^m_{\nu-k}, & \nu=k(1)n;\end{cases}
\tag{4.4.27}
$$

the needed derivatives of f exist and are Lipschitz-continuous and $\|F_n^{(m)}(\Delta_n z)\|$ is bounded uniformly in n which permits the replacement of the multilinear terms involving $\overline{\varepsilon}_n$ by $O(h^p)O(\|\overline{\varepsilon}_n\|)$ (see (4.4.25)) and the remainder estimate $O(h^{J+1})$.

It will be shown below that the $b_j\in E^0$ in (4.4.22) and the polynomials w_{ji} in (4.4.23) can be chosen such that all terms not involving $\overline{\varepsilon}_n$ on the right hand side of (4.4.26) cancel which leaves

$$
F_n'(\Delta_n z)\overline{\varepsilon}_n=O(h^p)O(\|\overline{\varepsilon}_n\|)+O(h^{J+1}).
\tag{4.4.28}
$$

The discretization $\{E_n, E_n^0, F_n'(\Delta_n z)\}$ is stable by Theorem 4.2.7 and the D-stability of ρ so that (4.4.28) implies

$$
\|\overline{\varepsilon}_n\|=O(h^p)O(\|\overline{\varepsilon}_n\|)+O(h^{J+1});
\tag{4.4.29}
$$

from (4.4.25) and (4.4.29) we can conclude recursively that $\|\overline{\varepsilon}_n\|=O(h^{J+1})$ which reduces (4.4.24) to (4.4.21).

c) The technical part of the proof consists in effecting the above-mentioned cancellation of the terms in (4.4.26). To simplify slightly the argument and the notation we will assume that all zeros of ρ are simple and different from 0, the modifications needed otherwise may be found in Gragg's thesis [1].

First consider the various parts of (4.4.26) at a point $t_\nu \in \mathbb{G}_n$, $\nu \geq k$:

$$\left[F_n \Delta_n z + F_n'(\Delta_n z)\Delta_n \sum_j h^j e_j + \sum_{m=2}^{[J/p]} \frac{1}{m!} F_n^{(m)}(\Delta_n z)\left(\Delta_n \sum_j h^j e_j\right)^m\right](t_\nu)$$

$$\begin{aligned}
&= (L_h[\langle\rho,\sigma\rangle]z)(t_{\nu-k}) + \frac{1}{h}[\rho(T_h) - h\sigma(T_h) f'(z(t_{\nu-k}))] \sum_j h^j e_j(t_{\nu-k}) \\
&\quad - \sigma(T_h) \sum_{m=2}^{[J/p]} f^{(m)}(z(t_{\nu-k}))\left(\sum_j h^j e_j(t_{\nu-k})\right)^m \\
&= \sigma(T_h) \sum_{j=p}^{J} h^j [c_j^{(1)} z^{(j+1)}(t_{\nu-k}) + \sum_{l=p}^{J-j} h^l c_l^{(1)} e_j^{(l+1)}(t_{\nu-k}) + b_j(t_{\nu-k}) \\
&\quad + g_j(t_{\nu-k}; e_p(t_{\nu-k}), \ldots, e_{j-p}(t_{\nu-k}))] + O(h^{J+1})
\end{aligned} \tag{4.4.30}$$

by (4.4.10) and (4.4.22); the g_j are constructed as in (1.3.9). Thus, (4.4.30) vanishes with

$$b_j(t) = -\left[c_j^{(1)} z^{(j+1)}(t) + \sum_{l=p}^{j-p} c_l^{(1)} e_{j-l}^{(l+1)}(t) + g_j(t; e_p(t), \ldots, e_{j-p}(t))\right] \tag{4.4.31}$$

for $\nu = k(1)n$ except for a remainder term of order $O(h^{J+1})$. This is, of course, the same recursive construction as in Theorem 4.4.2; the initial values $b_{j0}, j = p(1)J$, remain undetermined at this stage. Furthermore, it can be established recursively from $z \in D^{(J+1)}$ that $b_j \in D^{(J-j)}$ and $e_j \in D^{(J-j+1)}$, $j = p(1)J$, in agreement with the smoothness assertions on the e_j.

Now we take, at some $t_\nu \in \mathbb{G}_n$, $\nu \geq k$, the terms

(4.4.32)

$$\left[F_n'(\Delta_n z)\sum_j h^j \omega_j + \sum_{m=2}^{[J/p]} \frac{1}{m!} F_n^{(m)}(\Delta_n z) \sum_{\mu=1}^{m} \binom{m}{\mu}\left(\Delta_n \sum_j h^j e_j\right)^{m-\mu}\left(\sum_j h^j \omega_j\right)^{\mu}\right](t_\nu)$$

$$= \frac{1}{h}[\rho(T_h) - h\sigma(T_h) f'(z(t_{\nu-k}))] \sum_j h^j \sum_i x_i^{\nu-k} w_{ji}(\nu-k)$$

$$-\sigma(T_h)\sum_{m=2}^{[J/p]} \frac{1}{m!} f^{(m)}(z(t_{\nu-k})) \sum_{\mu=1}^{m} \binom{m}{\mu}\left(\sum_j h^j e_j(t_{\nu-k})\right)^{m-\mu}\left(\sum_j h^j \sum_i x_i^{\nu-k} w_{ji}(\nu-k)\right)^{\mu}.$$

The range of the i-sums in (4.4.32) is specified as follows: For $j = p(1)2p$, the sum ranges over the extraneous zeros x_i of ρ, $i = 2(1)k$; for $j = 2p+1(1)3p$, we include also the products $x_{i_1} x_{i_2}$, $i_1, i_2 = 2(1)k$, which are different from any of the x_i; etc.

By expanding the functions of t on the right hand side of (4.4.32) in powers of $t_\nu = \nu h$ about zero with a remainder term of $O(h^{J+1})$ which is possible by the established smoothness of f, z and the e_j, and by using

$$\rho(T_h)\, x_i^\nu\, \varepsilon_\nu = x_i^\nu\, \rho_i(T_h)\, \varepsilon_\nu\,, \qquad \text{see (4.4.7)}, \tag{4.4.33}$$

we obtain for (4.4.32) the expression

$$\sum_{j=p-1}^{J} h^j \sum_i x_i^{\nu-k} [\rho_i(T_h)\, w_{j+1,i}(\nu-k) - W_{ji}(\nu-k)] + O(h^{J+1}),$$

where the W_{ji} are polynomials in ν which depend on the functions f, z, and e_j, and on the $w_{j'i}$, $j' \leq j$. We can now resursively choose the w_{ji}, $j = p(1)J+1$, such that this expression vanishes except for a remainder $O(h^{J+1})$:

For $j = p(1)J+1$ and $i = 2(1)k$, the $W_{j-1,i}$ are of degree $j-1-p$ so that the w_{ji} of degree $j-p$ are determined except for their constant terms w_{ji0} according to Lemma 4.4.4 since $\rho_i(1) = \rho(x_i) = 0$. (I. e., we obtain the polynomials $w_{ji}(\nu) - w_{ji0}$.)

If r-fold products of zeros have to be taken care of for $j > rp$, $r = 2, 3, \ldots$, the corresponding $W_{j-1,i}$ are of degree $j-1-rp$. Since now $\rho_i(1) \neq 0$, $i > k$, these w_{ji} are of degree $j-1-rp$ and fully determined according to Lemma 4.4.4.

We now consider the terms (4.4.30) and (4.4.32) at t_κ, $\kappa = 0(1)k-1$; with (4.4.27) we obtain

$$\begin{aligned} z(t_\kappa) + \sum_{j=p}^{J} h^j e_j(t_\kappa) + \sum_{j=p}^{J+1} h^j \omega_j(t_\kappa) - s_{n\kappa}(z_0) = \sum_{j=p}^{J} h^j \Bigg[\Bigg(b_{j0} + \sum_{i=2}^{k} x_i^\kappa w_{ji0} \Bigg) \\ + \Bigg(-s_\kappa^{(j)} + \sum_{l=1}^{j-p} \frac{\kappa^l}{l!} e_{j-l}^{(l)}(0) + \sum_{i=2}^{k} x_i^\kappa (w_{ji}(\kappa) - w_{ji0}) + \sum_{i>k} x_i^\kappa w_{ji}(\kappa) \Bigg) \Bigg] + O(h^{J+1}). \end{aligned} \tag{4.4.34}$$

We can now recursively determine the k constants $b_{j0} \in \mathbb{R}^s$, $w_{ji0} \in \mathbb{R}^s$, $i = 2(1)k$, for each $j = p(1)J$, to make the sum in (4.4.34) vanish: The terms in the second parentheses are known at the time they are needed so that we have linear systems of the form

$$\sum_{i=1}^{k} x_i^\kappa c_i = d_\kappa\,, \qquad \kappa = 0(1)k-1,$$

which are non-singular because of their Vandermonde matrix and the assumed disjointness of the zeros of ρ. The $w_{J+1,i0}$, $i = 2(1)k$, which have remained undetermined may be set 0 without loss of generality since $\rho_i(T_h)\, w_{J+1,i0} = 0$ and $h^{J+1} w_{J+1,i0} = O(h^{J+1})$.

We have thus proved the statement in part b) which led to (4.4.28) so that the proof of Theorem 4.4.5 is complete. At the same time we have displayed the recursive construction of the elements $b_j \in E^0$ in (4.4.22) and of the $\mathbb{R}^s$-valued polynomials w_{ji} in (4.4.23). □

Remarks. 1. For each $t \in (0,1]$ of the sort described in Theorem 4.4.5 the uniqueness of the *values* $e_j(t)$ in (4.4.20) is trivial. If these t are dense in $[0,1]$, which is the case for usual choices of $\mathbb{N}'$, then the continuity of the e_j implies the uniqueness of the *functions* e_j.
2. From part a) of the proof it follows that the remainder term in (4.4.20) may be considered as uniform in each fixed interval $[\bar{t},1]$, $\bar{t}>0$, i.e., bounded by $M h^{J+1}$ with M *independent* of t.

Corollary 4.4.6. *If we permit a coherent grid sequence for* $\mathfrak{M}$ *in the situation of Theorem* 4.4.5, *with linear multistep schemes of order* $p-1$ *for the step changes, the assertion remains valid except that the definition of the* e_j *changes in each interval of constant steps.*

Proof. There are only a fixed finite number of step changes at points $\bar{t}_\mu$ which are common gridpoints of all $\mathbb{G}_n$, $n \in \mathbb{N}'$; see Def. 2.1.11. Assume inductively that an expansion (4.4.20) holds for $t \le \bar{t}_\mu$ (this is true for $t \le \bar{t}_1$ by Theorem 4.4.5). Then the values of ζ_n at the first $k-1$ gridpoints after $\bar{t}_\mu$ satisfy ($h_{\mu+1}$ is the step in $(\bar{t}_\mu, \bar{t}_{\mu+1}]$)

$$\zeta_n(\bar{t}_\mu + \kappa h_{\mu+1}) = z(\bar{t}_\mu + \kappa h_{\mu+1}) + \sum_{j=p}^{J} h_{\mu+1}^j \bar{s}_{\mu\kappa}^{(j)} + O(h_{\mu+1}^{J+1}), \qquad \kappa = 1(1)k-1,$$

with well-defined quantities $\bar{s}_{\mu\kappa}^{(j)} \in \mathbb{R}^s$. If we consider these $\zeta_n(\bar{t}_\mu + \kappa h_{\mu+1})$, $\kappa = 0(1)k-1$, as new starting values and apply Theorem 4.4.5 to the interval $[\bar{t}_\mu, \bar{t}_{\mu+1}]$ we find that an expansion (4.4.20) holds also for suitable $t \in (\bar{t}_\mu, \bar{t}_{\mu+1}]$. □

In all practical applications Richardson extrapolation is used only on the values at a fixed point t. Thus Theorem 4.4.5 and Corollary 4.4.6 have established effectively that Richardson extrapolation is usable for uniform linear k-step methods based on strongly D-stable linear k-step schemes, independently of the (sufficiently accurate) starting procedure.

4.4.4 The Case of Extraneous Essential Zeros

It remains to analyze the case where ρ possesses essential zeros besides $x_1 = 1$. From the considerations in Section 4.4.2 we may not expect that the effect of the starting perturbations in (4.4.18) tends to zero with increasing v. Instead it will turn out that it may be represented by terms of the form $x_i^v e_{ji}(t_v)$ where the x_i are the extraneous essential zeros of ρ or certain of their products and the e_{ji} are solutions of differential equations which are, however, normally *not* variational problems of the given IVP 1. The analysis of this situation is again due to Gragg [1]; the case $J=p$ had already been treated by Henrici [1], Section 5.3–5. We consider first the case where *all* zeros of ρ have modulus 1, they are necessarily simple for a D-stable $\langle\rho,\sigma\rangle$.

To facilitate the formulation of the results, let $x_1=1, x_2,\dots,x_k$ be the zeros of ρ, denote by $x_{k+1},\dots,x_{k_2}$ the values of the products $x_{i_1}x_{i_2}$ different from any of the x_i, $i=1(1)k$, and generally by $x_{k_{m-1}+1},\dots,x_{k_m}$ the values of m-fold products $\prod_{\mu=1}^{m} x_{i_\mu}$, $i_\mu\in\{1,\dots,k\}$ which are different from all x_i, $i\le k_{m-1}$. Then, with $k_1:=k$, set

$$I_j := \begin{cases}\{1,\dots,k\}, & j=p,\\ \{1,\dots,k_{\lceil\frac{j-1}{p}\rceil}\}, & j>p.\end{cases} \tag{4.4.35}$$

Note that $k_m=k_{m+1}=\cdots$ for some $m\ge 1$ if and only if the x_i, $i=1(1)k$, are all roots of unity.

Theorem 4.4.7. *In the situation of Theorem 4.4.5, if $\langle\rho,\sigma\rangle$ is D-stable and ρ possesses only essential zeros then there exist unique functions $e_{ji}\in D^{(J-j+1)}$, $j=p(1)J$, $i\in I_j$, such that for an $IVP^{(J+1)}$ 1 the solution ζ_n of its discretization by $\mathfrak{M}$ satisfies $(h=1/n)$*

$$\zeta_n = \Delta_n z + \sum_{j=p}^{J} h^j\left[\Delta_n e_{j1} + \sum_{\substack{i\in I_j\\ i\neq 1}} \varepsilon_{ji}\right] + O(h^{J+1}) \tag{4.4.36}$$

where the $\varepsilon_{ji}\in E_n$ are defined by

$$(\varepsilon_{ji})_\nu := x_i^\nu e_{ji}(t_\nu), \qquad i\in I_j. \tag{4.4.37}$$

Proof. a) We will first establish the uniqueness of (4.4.36)/(4.4.37): Assume the existence of two sets of functions e_{ji} and $\hat e_{ji}\in E$ satisfying (4.4.36)/(4.4.37). Assume that we have shown $e_{j'i}=\hat e_{j'i}$ for $j'<j$; then we obtain by subtraction of the two expansions and division by h^j

$$\sum_{i\in I_j} x_i^\nu\,[e_{ji}(t_\nu)-\hat e_{ji}(t_\nu)] = O(h) \quad \text{for } \nu=0(1)n. \tag{4.4.38}$$

Let $q_j:=|I_j|$ and consider (4.4.38) for q_j consecutive values $\nu=\nu_0(1)\nu_0+q_j-1$ but replace the arguments of e_{ji} and $\hat e_{ji}$ by t_{ν_0} which introduces only an error of $O(h)$. The linear system

$$\sum_{i=1}^{q_j} x_i^\nu\,[x_i^{\nu_0}(e_{ji}(t_{\nu_0})-\hat e_{ji}(t_{\nu_0}))] = \delta_\nu, \qquad \nu=0(1)q_j-1, \tag{4.4.39}$$

with $\|\delta_\nu\|=O(h)$ has a non-singular Vandermonde matrix (see the definition of the x_i). (4.4.39) holds for any ν_0 and implies $\Delta_n(e_{ji}-\hat e_{ji})=O(h)$ so that (see Def. 1.1.2)

$$\|e_{ji}-\hat e_{ji}\| = \lim_{n\to\infty}\|\Delta_n(e_{ji}-\hat e_{ji})\| = 0.$$

b) As in part b) of the proof of Theorem 4.4.5 we set

(4.4.40)
$$\zeta_n = \Delta_n z + \sum_{j=p}^{J} h^j \sum_{i\in I_j} \varepsilon_{ji} + \bar{\varepsilon}_n,$$

with ε_{ji} from (4.4.37) and $\bar{\varepsilon}_n \in E_n$, where the e_{ji}, $i=1(1)k$, are solutions of

(4.4.41)
$$\begin{cases} e_{ji}(0) = b_{ji0}, \\ e'_{ji} - \lambda_i f'(z(t)) e_{ji} = b_{ji}(t), \end{cases}$$

with λ_i the growth parameters of $\langle\rho,\sigma\rangle$, see Def. 4.4.2.

We then show that we can choose the $b_{ji} = \begin{pmatrix} b_{ji0} \\ b_{ji}(t) \end{pmatrix} \in E^0$, $i=1(1)k$, and the $e_{ji} \in E$, $i \in I_j - I_1$, such that (4.4.28) holds which implies $\|\bar{\varepsilon}_n\| = O(h^{J+1})$ and the validity of (4.4.36) as in the proof of Theorem 4.4.5.

c) For this purpose we form (cf. (4.4.26), (4.4.30), and (4.4.40)) for some $t_\nu \in \mathbb{G}_n$, $\nu \geq k$,

(4.4.42)
$$\begin{aligned} 0 = F_n \zeta_n(t_\nu) &= L_h[\langle\rho,\sigma\rangle] z)(t_{\nu-k}) \\ &+ \frac{1}{h}[\rho(T_h) - h\sigma(T_h) f'(z(t_{\nu-k}))] \sum_{j=p}^{J} h^j \sum_{i=1}^{k} x_i^{\nu-k} e_{ji}(t_{\nu-k}) \\ &+ \frac{1}{h}[\rho(T_h) - h\sigma(T_h) f'(z(t_{\nu-k}))] \sum_{j=2p+1}^{J+1} h^j \sum_{i\in I_j - I_1} x_i^{\nu-k} e_{ji}(t_{\nu-k}) \\ &- \sigma(T_h) \sum_{m=2}^{[J/p]} \frac{1}{m!} f^{(m)}(z(t_{\nu-k})) \left(\sum_{j=p}^{J} h^j \sum_{i\in I_j} x_i^{\nu-k} e_{ji}(t_{\nu-k}) \right)^m \\ &+ (F'_n(\Delta_n z)\bar{\varepsilon}_n)(t_\nu) + O(h^p) O(\|\bar{\varepsilon}_n\|) + O(h^{J+1}); \end{aligned}$$

the $e_{J+1,i}$, $i \in I_{J+1} - I_1$, have been introduced to simplify slightly the proof.

The first term in (4.4.42) is $\sigma(T_h) \sum_j h^j c_j^{(1)} z^{(j+1)}(t_{\nu-k})$ by (4.4.10). Into the second term we substitute (4.4.41) to obtain, with (4.4.9) and (4.4.10),

$$\begin{aligned} &\sum_j h^j \sum_{i=1}^{k} x_i^{\nu-k} \left[(L_h^{(i)}[\langle\rho,\sigma\rangle] e_{ji})(t_{\nu-k}) + \frac{\sigma_i(T_h)}{\lambda_i} b_{ji}(t_{\nu-k}) \right] \\ &= \sum_j h^j \sum_{i=1}^{k} x_i^{\nu-k} \sigma_i(T_h) \left[\sum_{l=1}^{J-j} c_l^{(i)} h^l e_{ji}^{(l+1)}(t_{\nu-k}) + \frac{1}{\lambda_i} b_{ji}(t_{\nu-k}) \right] + O(h^{J+1}) \\ &= \sum_j h^j \sum_{i=1}^{k} x_i^{\nu-k} \sigma_i(T_h) \left[\sum_{l=1}^{j-p} c_l^{(i)} e_{j-l,i}^{(l+1)}(t_{\nu-k}) + \frac{1}{\lambda_i} b_{ji}(t_{\nu-k}) \right] + O(h^{J+1}). \end{aligned}$$

In the third term we expand in powers of h about $t_{\nu-k}$, assuming $e_{ji}\in D^{(J-j+1)}$, with ρ_i and σ_i defined for x_i, $i>k$, as in (4.4.7):

$$\frac{1}{h}\left[\rho_i(T_h)-h\sigma_i(T_h)f'(z(t_{\nu-k}))\right]e_{ji}(t_{\nu-k})$$
$$=\frac{1}{h}\rho_i(1)e_{ji}(t_{\nu-k})+\sum_{l=0}^{J-j}\bar{c}_l^{(i)}h^l E_{ji}^{(l)}(t_{\nu-k})+O(h^{J-j+1})$$

where $\rho_i(1)=\rho(x_i)\neq 0$ since $i>k$. Thus we obtain for the third term in (4.4.42)

$$\sum_{j=2p+1}^{J+1}h^{j-1}\sum_{i\in I_j-I_1}x_i^{\nu-k}\left[\rho_i(1)e_{ji}(t_{\nu-k})+\sum_{l=0}^{j-2p-2}\bar{c}_l^{(i)}E_{j-l-1,i}^{(l)}(t_{\nu-k})\right]+O(h^{J+1}).$$

The fourth term we rearrange in powers of h:

$$\sum_{j=2p}^{J}h^j\left[\sum_{i=1}^{k}x_i^{\nu-k}\sigma_i(T_h)g_{ji}(t_{\nu-k})+\sum_{i\in I_{j+1}-I_1}x_i^{\nu-k}G_{ji}(t_{\nu-k})\right]+O(h^{J+1})$$

where the g_{ji} and G_{ji} depend only on $e_{j'i}$ with $j'\leq j-p$, the G_{ji} involve an expansion of $\sigma_i(T_h)g_{ji}$ in powers of h about $t_{\nu-k}$.

We can now recursively choose

$$(4.4.43)\quad\begin{cases}\left.\begin{aligned}b_{j1}(t)&=-\left[c_j^{(1)}z^{(j+1)}(t)+\sum_{l=p}^{j-p}c_l^{(1)}e_{j-l,1}^{(l+1)}(t)+g_{j1}(t)\right],\\ b_{ji}(t)&=-\lambda_i\left[\sum_{l=1}^{j-p}c_l^{(i)}e_{j-l,i}^{(l+1)}(t)+g_{ji}(t)\right],\qquad i=2(1)k,\end{aligned}\right\}\quad j=p(1)J,\\ e_{ji}(t)=-\dfrac{1}{\rho_i(1)}\left[\sum_{l=0}^{j-2p-2}\bar{c}_l^{(i)}E_{j-l-1,i}^{(l)}(t)+G_{j-1,i}(t)\right],\\ \qquad\qquad j=2p+1(1)J+1,\quad i\in I_j-I_1,\end{cases}$$

to achieve a cancellation of all terms in (4.4.42) not involving $\bar{\varepsilon}_n$ for an arbitrary $t_\nu\in\mathbb{G}_n$, $\nu\geq k$, with a remainder of $O(h^{J+1})$. The initial values b_{ji0}, $i=1(1)k, j=p(1)J$, are as yet undetermined. It is easily checked that the recursive process attributes the asserted smoothness to the e_{ji}.

The consideration of $F_n\zeta_n$ at t_κ, $\kappa=0(1)k-1$, yields with (4.4.40)

$$(4.4.44)\quad\begin{aligned}0&=z(t_\kappa)+\sum_{j=p}^{J}h^j\sum_{i\in I_j}x_i^\kappa e_{ji}(t_\kappa)+\bar{\varepsilon}_n(t_\kappa)-s_{n\kappa}(z_0)\\&=\sum_{j=p}^{J}h^j\left[\sum_{i=1}^{k}x_i^\kappa b_{ji0}+\left(-s_\kappa^{(j)}+\sum_{i\in I_j}x_i^\kappa S_{\kappa i}^{(j)}\right)\right]+\bar{\varepsilon}_n(t_\kappa)+O(h^{J+1}),\end{aligned}$$

where the $S_{\kappa i}^{(j)}$ arise from expanding $e_{ji}(t_\kappa)-b_{ji0}$ for $i\leq k$ and $e_{ji}(t_\kappa)$ for $i>k$ in powers of $t_\kappa=\kappa h$ about 0 and rearranging by powers of h; they depend only on $e_{j'i}$, $j'<j$, for $i\leq k$. Since the linear system with the coefficients x_i^κ, $i=1(1)k$, $\kappa=0(1)k-1$, is non-singular, the b_{ji0} may

be recursively determined (after the $e_{j-1,i}$, $i \leq k$, and e_{ji}, $i > k$, have been found) as to annihilate the brackets in (4.4.44).

We have thus established (4.4.28) as claimed in b) above and the proof is complete. □

Remark. We have tacitly admitted complex-valued functions e_{ji}, $i > 1$, in accordance with the remark after the introduction of the x_i in Section 4.2.2. It is clear that everything may be written in terms of real functions and real coefficients by a suitable combination of complex-conjugate quantities.

These case where all x_i are *roots of unity* merits special attention: Let, for $i = 2(1)k$,

$$x_i^{d_i} = 1 \quad \text{while } x_i^d \neq 1 \quad \text{for } d < d_i, \quad d, d_i \in \mathbb{N}, \tag{4.4.45}$$

and let $\bar{d}$ be the smallest common multiple of the d_i. Take a fixed set $\mathbb{G} := \{\bar{t}_\mu, \mu = 0(1)m \colon 0 = \bar{t}_0 < \bar{t}_1 < \cdots < \bar{t}_m = 1\}$, $m \geq 1$; the grids $\mathbb{G}_n^{(\bar{d})}$ which arise from subdividing each interval $[\bar{t}_\mu, \bar{t}_{\mu+1}]$ into $n\bar{d}$ equal steps form a coherent grid sequence $\{\mathbb{G}_n^{(\bar{d})}\}_{n \in \bar{\mathbb{N}}}$, $\bar{\mathbb{N}} \subset \mathbb{N}$. Let $\hat{E}$ be the space of functions $\mathbb{G} \to \mathbb{R}^s$ and $\hat{\Delta} \colon E \to \hat{E}$ the restriction to $\mathbb{G}$ (cf. Def. 1.4.2).

Corollary 4.4.8. *In the situation of Theorem 4.4.7, assume that (4.4.45) holds for all zeros x_i of ρ. If $\mathfrak{M}$ uses a coherent grid sequence $\{\mathbb{G}_n^{(\bar{d})}\}_{n \in \bar{\mathbb{N}}}$ of the structure described above and linear k-step schemes of order $p-1$ at step changes then the global discretization error of $\mathfrak{M}$ for an $IVP^{(J+1)}1$ possesses a $\hat{\Delta}$-reduced asymptotic expansion to order J, with $\hat{E}$ defined above.*

Proof. If the $\bar{t}_\mu$ are equidistant, $\bar{t}_{\mu+1} - \bar{t}_\mu = 1/m$, it follows from Theorem 4.4.7 and the structure of the $\mathbb{G}_n^{(\bar{d})}$ that

$$\zeta_n(\bar{t}_\mu) = z(\bar{t}_\mu) + \sum_{j=p}^{J} h^j \left(\sum_{i \in I_j} e_{ji}(\bar{t}_\mu) \right) + O(h^{J+1}), \qquad \bar{t}_\mu \in \mathbb{G}, \tag{4.4.46}$$

where h is the step of $\mathbb{G}_n^{(\bar{d})}$.

If the $\bar{t}_\mu$ are not equidistant, so that there are step changes at the $\bar{t}_\mu$, we assume inductively that we have shown that an expansion (4.4.36)/(4.4.37) holds in $[\bar{t}_{\mu-1}, \bar{t}_\mu]$. At the gridpoints $\bar{t}_\mu - \kappa h_\mu$, $\kappa = 0(1)k-1$, with $h_\mu = (\bar{t}_\mu - \bar{t}_{\mu-1})/(n\bar{d})$ we have

$$\begin{aligned}\zeta_n(\bar{t}_\mu - \kappa h_\mu) &= z(\bar{t}_\mu - \kappa h_\mu) + \sum_{j=p}^{J} h_\mu^j \sum_{i \in I_j} x_i^{-\kappa} e_{ji}(\bar{t}_\mu - \kappa h_\mu) + O(h_\mu^{J+1}) \\ &= z(\bar{t}_\mu - \kappa h_\mu) + \sum_{j=p}^{J} h_\mu^j s_{-\kappa}^{(j)} + O(h_\mu^{J+1}),\end{aligned}$$

where the $s_{-\kappa}^{(j)}$ are obtained by expanding the e_{ji} about $\bar{t}_\mu$, and are independent of n. Hence, after the execution of the step change by linear k-step procedures of order $p-1$ we will have

(4.4.47) $$\zeta_n(\bar{t}_\mu+\kappa h_{\mu+1})=z(\bar{t}_\mu+\kappa h_{\mu+1})+\sum_{j=p}^{J} h_{\mu+1}^j s_\kappa^{(j)}+O(h_{\mu+1}^{J+1})$$

with quantities $s_\kappa^{(j)}$, $\kappa=0(1)k-1$, which depend on the $s_{-\kappa}^{(j)}$ and the multistep procedures used but which are again independent of n. Thus we may consider (4.4.47) as a new starting procedure (4.4.3) and obtain from Theorem 4.4.7 the existence of an expansion (4.4.36)/(4.4.37) in $[t_\mu, t_{\mu+1}]$, with new e_{ji}, of course. Since the induction assumption was satisfied in $[0, \bar{t}_1]$ by Theorem 4.4.7, the assertion follows. □

Corollary 4.4.8 shows that in the case (4.4.45) the situation is very similar to that in the case of a strongly D-stable $\langle\rho,\sigma\rangle$. If there is but one zero x_i of ρ which is not a root of unity there exists no ($\hat{\Delta}$-reduced) asymptotic expansion of the global discretization error.

Example. Take the method of the example in Section 4.2.1 based upon the 2-step midpoint scheme. We have $x_1=1$, $x_2=-1$, so that (4.4.45) holds with $\bar{d}=2$. If we restrict ourselves to a coherent grid sequence such that the points of step change are an even number of steps from the origin in each grid, an asymptotic expansion exists at each "even" gridpoint t:

$$\zeta(t)=z(t)+\sum_{j=p}^{J} h^j e_j(t)+O(h^{J+1}).$$

For more details see Section 6.3.2.

Finally we state the general theorem on asymptotic expansions for uniform linear multistep methods the proof of which combines the ideas of the proofs of Theorems 4.4.5 and 4.4.7 (see Gragg [1]):

Theorem 4.4.9. *In the situation of Theorem 4.4.5, if $\langle\rho,\sigma\rangle$ is D-stable and ρ possesses $q\le k$ essential zeros there exist functions $e_{ji}\in D^{(J-j+1)}$, $i\in I_j$, $j=p(1)J$, for each IVP$^{(J+1)}$ 1 such that the solution ζ_n of $\mathfrak{M}$(IVP$^{(J+1)}$1) satisfies ($h=1/n$)*

(4.4.48) $$\zeta_n(t)=z(t)+\sum_{j=p}^{J} h^j \sum_{i\in I_j} x_i^{\frac{t}{h}} e_{ji}(t)+O(h^{J+1}),\qquad n\in\bar{\mathbb{N}},$$

at each fixed $t>0$ which is a common gridpoint of the infinitely many equidistant grids $\mathbb{G}_n$, $n\in\bar{\mathbb{N}}\subset\mathbb{N}'$. (The definition of the x_i, $i>q$, and of the sets I_j is analogous to the one at the beginning of this section.)

4.5 Further Analysis of the Discretization Error

4.5.1 Weak Stability

A very important feature of Theorems 4.4.7 and 4.4.9 is the fact that the functions e_{ji}, $i>1$, are solutions to differential equations of the form (see (4.4.41))

(4.5.1) $$e_{ji}'-\lambda_i f'(z(t))e_{ji}=b_{ji}(t),\qquad i=2(1)q,$$

where the λ_i are the growth parameters of the underlying linear k-step scheme (see Def. 4.4.2).

Only for $\lambda_i = 1$, (4.5.1) is a variational equation of the given IVP1 so that e_{ji} may be regarded as the effect of a perturbation of the IVP1. If $\lambda_i \neq 1$ there are always IVP1 such that, with increasing t, solutions of (4.5.1) grow exponentially relative to small perturbations of the IVP 1; this explains the term growth parameter for the λ_i.

To see clearly the implications, we consider the scalar IVP1

$$y' - gy = 0, \; y(0) = 1, \quad g \in \mathbb{C}, \tag{4.5.2}$$

as a *model problem*. For (4.5.2), (4.5.1) becomes $e'_{ji} - \lambda_i g e_{ji} = b_{ji}(t)$ so that

$$e_{ji}(t) = b_{ji0} e^{\lambda_i g t} + \int_0^t e^{\lambda_i g(t-\tau)} b_{ji}(\tau) d\tau. \tag{4.5.3}$$

If λ_i is *real*—which is the most frequent case, see Theorems 4.5.2 and 4.5.4—we can distinguish 3 cases:

(i) $\lambda_i < 0$: If $\operatorname{Re} g > 0$ in (4.5.2) then the effects of the starting perturbation b_{ji0} and of the perturbation $b_{ji}(t)$ of the forcing function decrease although they would actually grow with the IVP1. However, for $\operatorname{Re} g < 0$ the opposite situation arises and the e_{ji} grow very rapidly both absolutely and relative to the solution $z = e^{gt}$ of (4.5.2).

(ii) $0 < \lambda_i < 1$: Again this case is harmless for $\operatorname{Re} g > 0$. For $\operatorname{Re} g < 0$ the modulus of the e_{ji} grows relatively to z although $\operatorname{Re} \lambda_i g < 0$.

(iii) $1 < \lambda_i$: Now the harmless situation arises for $\operatorname{Re} g < 0$ while for $\operatorname{Re} g > 0$ the exponential growth of the e_{ji} is stronger than that of z.

For complex λ_i we obtain similar distinctions according to the relative signs and sizes of $\operatorname{Re} g$ and $\operatorname{Re} \lambda_i g$.

The most undesirable case is clearly (i) because here an exponential growth occurs precisely when the IVP1 is exponentially stable in the sense of Def. 2.3.2 and its integration over long time intervals is meaningful even under persistent perturbations. The phenomenon that the discretization of an exponentially stable IVP1 is never exponentially stable (not even for arbitrarily small steps h) has been called *weak stability* in our discussion in Section 2.3.5. It is clear that a discretization method which is weakly stable w.r.t. the class of model problems (4.5.2) will be weakly stable w.r.t. any meaningful class of IVP1. Thus we have:

Theorem 4.5.1. *A uniform linear k-step method based upon a D-stable linear k-step scheme $\langle \rho, \sigma \rangle$ is weakly stable w.r.t. any class $J \supset J_c$ of IVP1 if ρ possesses at least one essential zero x_i with a real growth parameter $\lambda_i < 0$.*

We omit the proof here since we will analyze the strong stability behavior of linear k-step methods in more detail in Section 4.6.2; see

Theorem 4.6.4. It should be emphasized, however, that weak stability does not concern the stability of the method in the sense of Def. 1.1.10: Over a *fixed* interval of integration the effect of a sufficiently small perturbation of the discretization remains bounded relative to the size of the perturbation as $h \to 0$. But the stability bound is very large (for longer intervals) in comparison with the natural stability bound 1 for exponentially stable problems (4.5.2) and it grows without bound as the length of the interval increases.

While weak stability in the sense of Def. 2.3.11 occurs only if some of the growth parameters are negative, the term is often used in the literature whenever ρ possesses extraneous essential zeros. This is made plausible by the unfortunate fact that the case (i) in our above discussion is the rule rather than the exception:

Theorem 4.5.2. *In a linear k-step scheme $\langle \rho, \sigma \rangle$, if ρ has only essential zeros and σ is determined according to Theorem* 4.1.9 *(to achieve maximum order) then*

(i) *all growth parameters are real,*

(ii) $\sum_{i=2}^{k} \lambda_i < 0$,

(iii) *for even k, the growth parameter λ_2 associated with $x_2 = -1$ satisfies*

$$\lambda_2 \leq \begin{cases} -1 & \text{for a predictor,} \\ -\frac{1}{3} & \text{for a corrector.} \end{cases}$$

Proof. See Dahlquist [2], p. 39—43, and Henrici [1], Section 5.4–6. □

On the other hand, for a given polynomial ρ with $q > 1$ essential zeros the conditions

$$\lambda_i = \frac{\sigma(x_i)}{x_i \rho'(x_i)} = 1, \qquad i = 2(1)q, \tag{4.5.4}$$

are linear in the coefficients of σ. If they are imposed together with the order conditions (4.1.22) in the determination of σ, the attainable order will normally decrease by $q-1$, but the resulting method will show none of the effects discussed above.

Examples. 1. For $\rho(x) = \frac{1}{2}(x^2 - 1)$ the requirement $\lambda_2 = 1$ for $x_2 = -1$ leads to $\sigma(x) = \frac{1}{2}(x^2 + 1)$. We have thus arrived at the trapezoidal RK-method, with steps of size $2h$, proceeding separately on even and odd numbered gridpoints.

2. To admit a polynomial σ of degree 3 we take $\rho(x) = \frac{1}{2}(x^3 - x)$ and supplement the conditions (4.1.22) by $\sigma(-1) = (-1)\rho'(-1) = -2$. We obtain a 3-step corrector and predictor with generating matrices

$$\frac{1}{2}\begin{pmatrix} 0 & -1 & 0 & 1 \\ -\frac{1}{3} & \frac{4}{3} & \frac{1}{3} & \frac{2}{3} \end{pmatrix} \quad \text{and} \quad \frac{1}{2}\begin{pmatrix} 0 & -1 & 0 & 1 \\ -\frac{3}{2} & 3 & \frac{1}{2} & 0 \end{pmatrix}$$

and orders 3 and 2 resp. In the asymptotic expansion of the global discretization error the functions e_{j1} and e_{j2} both satisfy differential equations (4.5.1) with $\lambda_i = 1$ so that there is no undue growth for an IVP 1 (4.5.2) with $\operatorname{Re} g < 0$.

The phenomenon of a strong growth of the global discretization error with (4.5.2) and $\operatorname{Re} g < 0$ may also occur for uniform linear k-step methods based on a *strongly* D-stable linear k-step scheme $\langle \rho, \sigma \rangle$ if the steps h are not sufficiently small. We will return to this phenomenon in Section 4.6.

4.5.2 Smoothing

Consider a linear k-step method based on a k-step scheme with extraneous essential zeros: Even if the special starting procedures of Theorem 4.4.2 are used which introduce no "parasitic" terms into the expansion of the global discretization error up to some order J, such terms will be introduced into the total error of the *computational* solution of the discretization through the local computing errors which can never be completely excluded; see also Henrici [1], Sections 5.4–2 and 3, and [2], Section 5.5.

Therefore, if one wishes to use a weakly stable linear k-step method on an exponentially stable IVP1 one has to attempt to eliminate the parasitic terms from time to time as the integration proceeds. Fortunately, the oscillatory character of these terms permits the use of *smoothing operators* for this purpose. The following general approach to the construction of such smoothing operators has been suggested by Iri [1].

Theorem 4.5.3. *Consider $q-1$ disjoint complex numbers x_i, $x_i \neq 1$ and $\neq 0$, $i=2(1)q$, and integers $r>0$ and $s_i>0$, $i=2(1)q$. For an arbitrary integer v_0 let v be the polynomial of degree $r-1$ such that*

$$\frac{x^{v_0}}{\prod_{i=2}^{q} (x-x_i)^{s_i}} - v(x) = O((x-1)^r). \tag{4.5.5}$$

Then

$$\pi(x) := \frac{\prod_{i=2}^{q} (x-x_i)^{s_i}}{x^{v_0}} v(x) \tag{4.5.6}$$

satisfies

$$\pi(T_h) y(t) = y(t) + O(h^r) \qquad \text{for } y \in D^{(r-1)}, \tag{4.5.7}$$

$$\pi(T_h) x_i^{\frac{t}{h}} y(t) = h^{s_i} r_i x_i^{\frac{t}{h}} y^{(s_i)}(t) + O(h^{s_i+1}) \quad \text{for } y \in D^{(s_i)}, \tag{4.5.8}$$

$i=2(1)q$, where the r_i are constants independent of h.

Proof. With our assumptions $x^{\nu_0}\left[\prod_i (x-x_i)^{s_i}\right]^{-1}$ possesses a unique expansion in powers of $(x-1)$ the first $r-1$ terms of which define v. By (4.5.5) and (4.5.6)

$$\pi(x) = 1 + x^{-\nu_0}(x-1)^r \bar{\pi}(x) \tag{4.5.9}$$

with some polynomial $\bar{\pi}$ depending on the x_i and s_i. As $(T_h-1)^r$ is the operator of the r-th difference which annihilates polynomials of degree not larger than $r-1$, $\pi(T_h)$ reproduces such polynomials according to (4.5.9). This implies that in the Taylor-expansion

$$\pi(T_h)y(t) = y(t) + \sum_{j=1}^{r-1} c_j h^j y^{(j)}(t-\nu_0 h) + O(h^r)$$

all coefficients c_j, $j=1(1)r-1$, vanish which proves (4.5.7).

To establish (4.5.8) we define $v_i(x) := v(x) \prod_{l \neq i}(x-x_l)^{s_l}$ so that

$$\pi(x) = v_i(x)\,(x-x_i)^{s_i}\,x^{-\nu_0} \quad \text{for each } i=2(1)q.$$

Using

$$(T_h-1)^s y(t-\nu_0 h) = h^s y^{(s)}(t) + O(h^{s+1}) \quad \text{for } y \in D^{(s)}$$

we obtain, with an integer $\nu := t/h$,

$$\begin{aligned} \pi(T_h)x_i^\nu y(t) &= x_i^{\nu-\nu_0} v_i(x_i T_h)\,(x_i T_h - x_i)^{s_i} y(t-\nu_0 h) \qquad \text{(cf. (4.4.33))} \\ &= x_i^{\nu-\nu_0+s_i} v_i(x_i T_h)\,(T_h-1)^{s_i} y(t-\nu_0 h) \\ &= x_i^\nu x_i^{s_i-\nu_0} v_i(x_i)\,[h^{s_i} y^{(s_i)}(t) + O(h^{s_i+1})] \end{aligned}$$

which is (4.5.8) with $r_i = x_i^{s_i-\nu_0} v_i(x_i)$. □

Remark. It has not been assumed that $|x_i|=1$; thus the operator π may also be constructed to act on components $x_i^\nu e_i(t_\nu)$ with $|x_i|<1$.

For the application of Theorem 4.5.3 to the solution ζ_n of a discretization generated by a weakly stable uniform linear k-step method of order p, one will usually take $r \geq p$ (to preserve the order of convergence) and $s_i = s \geq 2$ for each of the $q-1$ extraneous essential zeros x_i, and assume that

$$\zeta_n(t_\nu) = z(t_\nu) + e(t_\nu; h) + \sum_{i=2}^{q} x_i^\nu e_i(t_\nu; h) \tag{4.5.10}$$

at the $r+(q-1)s$ consecutive gridpoints of $\mathbb{G}_n$ which enter the formation of $\pi(T_h)\zeta_n(t_\nu)$. ν_0 may be chosen to make $\pi(T_h)$ a backward difference operator; but other choices may be preferable in certain cases (see Theorem 4.5.9).

If ζ_n of (4.5.10) represents the exact solution of the discretization the smoothness assumptions of Theorem 4.5.3 will hold at least for the lower order parts of the e_i with a sufficiently smooth IVP 1; see Theorem 4.4.7.

If ζ_n of (4.5.10) includes the effects of computing errors which are small enough for second order effects to be negligible, we may also expect sufficient smoothness for the e_i-components in (4.5.10) due to computing errors: For we may regard the local computing error at each gridpoint as starting value for an individual perturbation, with values 0 at the $k-1$ preceding gridpoints; the further development of these individual perturbations may then be approximated by terms of the form $\sum_i x_i^\nu e_i(t_\nu, h)$ which are constructed as in Theorem 4.4.7. Thus the application of the smoothing operator π should also be effective in reducing the parasitic terms generated by small computing errors.

Example. $q=2, x_2=-1$ (e.g. for the 2-step midpoint method). For $r=2, s_2=2, v_0=1$, we obtain

$$v(x) = \frac{x}{(x+1)^2} + O((x-1)^2) = \frac{1}{4}, \quad \pi(x) = x^{-1}(x+1)^2 \frac{1}{4} = \frac{x^{-1}+2+x}{4},$$

so that

$$\pi(T_h)y(t_\nu) = \frac{1}{2}\left[y(t_\nu) + \frac{y(t_{\nu-1})+y(t_{\nu+1})}{2}\right].$$

This is the well-known symmetric smoothing operator of Gragg, see also Section 6.3.1.

4.5.3 Symmetric Linear k-step Schemes

In our treatment of RK-methods, we placed special emphasis on those RK-schemes which generate RK-methods for which the asymptotic expansion of the global discretization error contains only even powers of h. The corresponding considerations for linear k-step methods are more complicated and lead to the following class of linear k-step schemes:

Def. 4.5.1. A linear k-step scheme $\langle \rho, \sigma \rangle$ is called *symmetric* if

$$\rho(x) = -x^k \rho\left(\frac{1}{x}\right), \quad \sigma(x) = x^k \sigma\left(\frac{1}{x}\right), \tag{4.5.11}$$

or, equivalently,

$$\alpha_\kappa = -\alpha_{k-\kappa}, \quad \beta_\kappa = \beta_{k-\kappa}, \quad \kappa = 0(1)k. \tag{4.5.12}$$

Theorem 4.5.4. *For a D-stable symmetric linear k-step scheme* $\langle\rho,\sigma\rangle$ *the following hold:*

(i) *All zeros of* ρ *are essential; for even* k, -1 *is a zero of* ρ.

(ii) *All growth parameters* λ_i *are real,* $i=2(1)k$.

(iii) *In the expansion* (4.4.10),

$$(4.5.13)\qquad \begin{cases} \qquad\qquad\quad c_j^{(1)}=0 \quad \text{for odd } j, \quad j\geq p; \\ \text{for } i=2(1)k:\ \operatorname{Re} c_j^{(i)}=0 \quad \text{for odd } j \\ \qquad\qquad\quad \operatorname{Im} c_j^{(i)}=0 \quad \text{for even } j \end{cases} \; j\geq 1; \\ \text{if } x_2=-1, \qquad c_j^{(2)}=0 \quad \text{for odd } j.$$

(iv) $\langle\rho,\sigma\rangle$ *is of even order.*

Proof. (i) By (4.5.11), $1/x_i$ has to be a zero of ρ together with x_i but both have to be of modulus not larger than 1 for D-stability. $\rho(-1)=-(-1)^k\rho(-1)$ implies $\rho(-1)=0$ for even k.

(ii) $|x_i|=1$ implies $x_i^{-1}=x_i^*$. From (4.5.11), $\rho'(x)=-k\,x^{k-1}\rho(1/x)+x^{k-2}\rho'(1/x)$ so that $\rho'(x_i)=x_i^{k-2}\rho'(x_i^*)$. Hence,

$$\lambda_i=\frac{\sigma(x_i)}{x_i\,\rho'(x_i)}=\frac{x_i^k\,\sigma(x_i^*)}{x_i^{k-1}\,\rho'(x_i^*)}=\frac{\sigma(x_i^*)}{x_i^*\,\rho'(x_i^*)}=\lambda_i^*$$

since ρ and σ are real.

(iii) Consider the generating function of the $c_j^{(i)}$ (see Lemma 4.1.5):

$$\frac{\rho_i(e^h)-h\lambda_i^{-1}\sigma_i(e^h)}{h\sigma_i(e^h)}=\frac{\rho(x_i e^h)-h\lambda_i^{-1}\sigma(x_i e^h)}{h\sigma(x_i e^h)}=\frac{-\rho(x_i^* e^{-h})-h\lambda_i^{-1}\sigma(x_i^* e^{-h})}{h\sigma(x_i^* e^{-h})}$$

$$=\left[\frac{\rho_i(e^{-h})+h\lambda_i^{-1}\sigma_i(e^{-h})}{-h\sigma_i(e^{-h})}\right]^*$$

due to the reality of ρ,σ and λ_i so that

$$\sum_{j=1}^{\infty} c_j^{(i)}h^j=\left[\sum_{j=1}^{\infty} c_j^{(i)}(-h)^j\right]^*=\sum_{j=1}^{\infty} c_j^{(i)*}(-h)^j$$

which implies (4.5.13) (note that the $c_j^{(i)}$ are real for $x_1=1$ and $x_2=-1$).

(iv) Follows from (iii) and Def. 4.1.7. □

(4.5.13) implies that the local discretization error at t_ν, $\nu\geq k$, is an even function of h and its expansion (4.4.10) at such points contains only even powers of h (cf. (4.4.1)): For $z\in D^{(2J+2)}$

$$(4.5.14)\quad l_n(t_\nu)=(L_h[\psi]z)(t_{\nu-k})=\sigma(T_h)\sum_{j=\frac{p}{2}}^{J} c_{2j}^{(1)}h^{2j}z^{(2j+1)}(t_{\nu-k})+O(h^{2J+2}).$$

Corollary 4.5.5. *In the situation of Theorem* 4.4.2, *if* ψ *is symmetric and* $\bar{b}_0$ *in* (4.4.12) *is an even polynomial, the* $\bar{e}_j$ *vanish for odd* j *and the expansion*

$$\bar{\zeta}_n = \Delta_n z + \sum_{j=\frac{p}{2}}^{J} h^{2j} \Delta_n \bar{e}_{2j} + O(h^{2J+2}) \tag{4.5.15}$$

holds for an $IVP^{(2J+2)}1$ *if and only if the starting procedure of* $\mathfrak{M}$ *satisfies*

$$\bar{s}_{n\kappa}(z_0) = z(t_\kappa) + \sum_{j=\frac{p}{2}}^{J} h^{2j} \bar{e}_{2j}(t_\kappa) + O(h^{2J+2}), \qquad \kappa = 0(1)k-1. \tag{4.5.16}$$

Proof. The $\bar{e}_j$ vanish for odd j by (4.5.14) and Theorem 1.3.2. Thus we may replace h by h^2 in Theorem 4.4.2. □

A first conclusion that can be drawn from Corollary 4.5.5 is the fact that the expansion (4.4.3) of a starting procedure which produces an even asymptotic expansion (4.5.15) for a uniform k-step method based on a symmetric scheme ψ is *not* itself in even powers of h, as is easily seen from (4.5.16):

$$\begin{aligned}
\bar{s}_{n\kappa}(z_0) - z(\kappa h) &= \sum_{j=p}^{2J} \bar{s}_\kappa^{(j)} h^j + O(h^{2J+1}) = \sum_{j=\frac{p}{2}}^{J} h^{2j} \bar{e}_{2j}(\kappa h) + O(h^{2J+2}) \\
&= \sum_{j=\frac{p}{2}}^{J} h^{2j} \sum_{l=0}^{J-2j} \frac{(\kappa h)^l}{l!} \bar{e}_{2j}^{(l)}(0) + O(h^{2J+1}) \\
&= \sum_{j=\frac{p}{2}}^{J} h^{2j} \sum_{l=0}^{j-\frac{p}{2}} \frac{\kappa^{2l}}{(2l)!} \bar{e}_{2j-2l}^{(2l)}(0) \\
&\quad + \sum_{j=\frac{p}{2}}^{J-1} h^{2j+1} \sum_{l=0}^{j-\frac{p}{2}} \frac{\kappa^{2l+1}}{(2l+1)!} \bar{e}_{2j-2l}^{(2l+1)}(0) + O(h^{2J+1}).
\end{aligned}$$

so that

$$\bar{s}_\kappa^{(j)} = \begin{cases} \displaystyle\sum_{l=0}^{\frac{j-p}{2}} \frac{\kappa^{2l}}{(2l)!} \bar{e}_{j-2l}^{(2l)}(0), & \text{for } j \text{ even}, \\ \displaystyle\sum_{l=0}^{\frac{j-p-1}{2}} \frac{\kappa^{2l+1}}{(2l+1)!} \bar{e}_{j-1-2l}^{(2l+1)}(0), & \text{for } j \text{ odd}. \end{cases} \tag{4.5.17}$$

For $\bar{s}_{n\kappa}(z_0) = \bar{s}_\kappa(h)$ define

$$\bar{s}_{n,-\kappa}(z_0) := \bar{s}_\kappa(-h);$$

(4.4.3) implies for sufficiently smooth z that

(4.5.18) $$\bar{s}_{n,-\kappa}(z_0) = z(-\kappa h) + \sum_{j=p}^{2J} \bar{s}_\kappa^{(j)}(-h)^j + O(h^{2J+1}),$$

where $z(-\kappa h)$ is defined by expansion about the origin except for a remainder term $O(h^{2J+1})$. Then the calculation above (4.5.17) implies

(4.5.19) $$\bar{s}_{n,-\kappa}(z_0) = z(-t_\kappa) + \sum_{j=\frac{p}{2}}^{J} h^{2j}\bar{e}_{2j}(-t_\kappa) + O(h^{2J+2}).$$

The following observation is the key to the analysis of k-step methods based on symmetric linear multistep schemes:

Lemma 4.5.6. *For the starting procedure of Corollary* 4.5.5, *the sequence* $\{\bar{s}_{n,-k+1}, \bar{s}_{n,-k+2}, \ldots, \bar{s}_{n0}, \ldots, \bar{s}_{n,k-1}\}$, *with* $\bar{s}_{n\kappa}$ *defined by* (4.5.19) *for* $\kappa<0$, *satisfies*

(4.5.20) $$\begin{aligned} &\psi[f](h; \bar{s}_{n,-\kappa+k}, \ldots, \bar{s}_{n,-\kappa}) \\ &= \frac{1}{h}[\rho(T_h)\bar{s}_{n,-\kappa} - h\sigma(T_h) f(\bar{s}_{n,-\kappa})] = O(h^{2J+1}) \end{aligned}$$

for $\kappa = 1(1)k-1$.

Proof. We can define ζ_n at $-\kappa h$, $\kappa=1(1)k-1$, by the requirement

(4.5.21) $$\frac{1}{h}[\rho(T_h)\zeta_n(-\kappa h) - h\sigma(T_h) f(\zeta_n(-\kappa h))] = 0, \qquad \kappa = 1(1)k-1,$$

since the linear k-step procedures for a symmetric scheme are solvable "in both directions" due to (4.5.12). On the other hand, if we define z and the $\bar{e}_{2j}$ for $-\kappa h$, $\kappa=1(1)k-1$, by their differential equations (or the power expansions about 0 defined by these, with a remainder term of suitably high order), (4.5.15) holds also at the supplemented grid-points $t_{-\kappa} = -\kappa h$ as is clear from its derivation (see Theorems 4.4.2 and 1.3.1). Thus (4.5.16) and (4.5.19) imply

(4.5.22) $$\bar{s}_{n\kappa} = \zeta_n(t_\kappa) + O(h^{2J+2}), \qquad \kappa = -k+1(1)k-1,$$

or

$$\begin{aligned} &\frac{1}{h}[\rho(T_h)\bar{s}_{n,-\kappa} - h\sigma(T_h) f(\bar{s}_{n,-\kappa})] \\ &\qquad = \frac{1}{h}[\rho(T_h)\zeta_n(-\kappa h) - h\sigma(T_h) f(\zeta_n(-\kappa h))] + O(h^{2J+1}). \end{aligned}$$

Now (4.5.20) follows from (4.5.21). $\square$

(4.5.20) has a simple intuitive meaning: If the starting procedure is applied with the negative step $-h$ the values $\zeta_{-\kappa}$, $\kappa=1(1)k-1$, created in this fashion have to satisfy the difference equation together with the ζ_κ, $\kappa=0(1)k-1$, except perhaps for a remainder of order $O(h^{2J+1})$. Thus, (4.5.20) extends the concept of symmetry to the starting procedure.

Example. The starting procedure of the 2-step method in the example in Section 4.2.1 produces

$$s_{n,-1}=z_0-hf(z_0), \quad s_{n0}=z_0, \quad s_{n1}=z_0+hf(z_0)$$

and

$$\psi[f](h;s_{n1},s_{n0},s_{n,-1}) = \frac{s_{n1}-s_{n,-1}}{2h} - f(s_{n0})=0$$

so that (4.5.20) holds for any J.

If we now assume only the symmetry condition (4.5.20) but not (4.5.16) for the starting procedure of a k-step method with a symmetric scheme $\langle\rho,\sigma\rangle$, we obtain

Theorem 4.5.7. *Consider the discretization generated for an* $IVP^{(2J+2)}\,1$ *by a uniform linear k-step method* $\mathfrak{M}$ *with equidistant grids based on a D-stable symmetric linear k-step scheme* $\langle\rho,\sigma\rangle$ *of order p. If and only if the starting procedure of* $\mathfrak{M}$ *satisfies* (4.5.20) *do there exist* $\mathbb{R}^s$*-valued*[7] *functions* $e_{2j,i}\in D^{(2J-2j+1)}$, $i\in I_{2j}^*\cup\{1\}$, *and* $e_{2j+1,i}\in D^{(2J-2j)}$, $i\in I_{2j+1}^*$, *such that the solution* ζ_n *of the discretization satisfies at each* $t_\nu\in\mathbb{G}_n$

$$(4.5.23)\quad \begin{aligned}\zeta_n(t_\nu) &= z(t_\nu) + \sum_{j=\frac{p}{2}}^{J} h^{2j}\Big[e_{2j,1}(t_\nu) + \sum_{i\in I_{2j}^*} e_{2j,i}(t_\nu)\cos\nu\varphi_i\Big]\\ &\quad + \sum_{j=\frac{p}{2}}^{J-1} h^{2j+1} \sum_{i\in I_{2j+1}^*} e_{2j+1,i}(t_\nu)\sin\nu\varphi_i + O(n^{-(2J+1)}).\end{aligned}$$

Here $x_i=e^{i\varphi_i}$, $0\le\varphi_i<2\pi$, *and* $I_j^*:=\{x_i\in I_j: 0<\varphi_i\le\pi\}$, *with* I_j *from* (4.4.35).

Proof. a) That (4.5.23) implies (4.5.20) is simply an extension of Lemma 4.5.6 to expansions of the structure (4.5.23) instead of (4.5.15). We only have to show that the analogues of the right hand sides of (4.5.17) are even in κ for even j and odd for odd j:

The terms arising from the $e_{2j,1}$ are as in (4.5.17). The terms from the $e_{2j,i}$, $i>1$, carry factors $\cos\kappa\varphi_i$ which are even in κ and hence do not change the parity. The terms from the $e_{2j+1,i}$, $i>1$, are odd in κ for $\bar{s}_\kappa^{(j)}$ with an even superscript and vice versa so that the odd factors $\sin\kappa\varphi_i$ give them the right behavior.

---

[7] The emphasis is on the fact that the e_{ji} for *complex* x_i are *real*-valued.

b) We now have to show that the expansion (4.4.36)/(4.4.37) is of the form (4.5.23) if (4.5.20) holds. We note first that (4.5.23) can be valid only if the e_{ji} in (4.4.36)/(4.4.37) satisfy (cf. remark after Theorem 4.4.7)

$$(4.5.24)\quad \begin{cases} e_{2j+1,i}\equiv 0 \text{ for } x_1=1 \text{ and, in the case of even } k, \text{ for } x_2=-1; \\ e_{ji} \text{ is } \begin{cases}\text{real} \\ \text{purely imaginary}\end{cases} \text{ for } j \begin{cases}\text{even} \\ \text{odd}\end{cases}, \quad i>1. \end{cases}$$

That this is in fact equivalent to (4.5.23) follows from the fact that the x_i occur in conjugate-complex pairs x_i, $x_{i^*}:=x_i^*$ which implies $e_{ji^*}=e_{ji}^*$ due to the reality of the k-step scheme and the IVP 1. Thus for real e_{ji} we have $e_{ji^*}=e_{ji}$ and

$$x_i^\nu e_{ji}+x_{i^*}^\nu e_{ji^*}=(e^{i\varphi_i\nu}+e^{-i\varphi_i\nu})\,e_{ji}=2\cos\nu\varphi_i\, e_{ji};$$

similarly we obtain $2\sin\nu\varphi_i \operatorname{Im} e_{ji}$ for the sum of the two corresponding terms in the case of a purely imaginary $e_{ji}=-e_{ji^*}$.

We now observe that (4.5.20) and (4.5.21) imply (4.5.22). This follows from the stability of the discretization and Theorem 2.1.2 applied to the discretization on the grids $\mathbb{G}_n\cup\{-\kappa h, \kappa=1(1)k-1\}$, with the starting values given at $t_{-\kappa}$, $\kappa=0(1)k-1$. According to the construction (4.5.18) of the $\bar{s}_{n,-\kappa}$, (4.5.22) implies that the terms of the expansion

$$\zeta_n(\kappa h)=z(\kappa h)+\sum_{j=p}^{2J}\bar{\zeta}_\kappa^{(j)}h^j+O(h^{2J+1}), \qquad \kappa=-k+1(1)k-1,$$

satisfy

$$(4.5.25)\qquad \bar{\zeta}_{-\kappa}^{(j)}=(-1)^j\bar{\zeta}_\kappa^{(j)}, \qquad \kappa=0(1)k-1.$$

By introducing (4.4.36)/(4.4.37) into (4.5.25) and using the relations (4.4.41), (4.4.43), and (4.5.13) we can show recursively that in (4.4.41)

$$(4.5.26)\quad \begin{cases} e_{2j+1,i}(0) \text{ and } b_{2j+1,i}(t) \text{ vanish for } i=1, \text{ and for } i=2 \text{ if } x_2=-1, \\ e_{ji}(0) \text{ and } b_{ji}(t) \text{ are} \begin{cases}\text{real} \\ \text{purely imaginary}\end{cases} \text{ for } j\begin{cases}\text{even} \\ \text{odd}\end{cases}, \quad i\in I_j^*. \end{cases}$$

This implies (4.5.24), which is equivalent to (4.5.23) as we have seen above. The lengthy technical details of the recursive establishment of (4.5.26) are omitted. □

Example. $$k=2, \quad \langle\rho,\sigma\rangle=\begin{pmatrix}-\frac{1}{2} & 0 & \frac{1}{2} \\ \beta_0 & 1-2\beta_0 & \beta_0\end{pmatrix}.$$

Here (4.5.20) poses the single condition

$$\frac{s_{n1}-s_{n,-1}}{2h}-[\beta_0(f(s_{n1})+f(s_{n,-1}))+(1-2\beta_0)f(s_{n0})]=O(h^{2J+1})$$

for the reduction of the general expansion (4.4.36)/(4.4.37) to

$$\zeta_n(t_\nu)=z(t_\nu)+\sum_{j=1}^{J} h^{2j}[e_{2j,1}(t_\nu)+(-1)^\nu e_{2j,2}(t_\nu)]+O(h^{2J+2}). \tag{4.5.27}$$

4.5.4 Asymptotic Expansions in Powers of h²

At first sight, the result of Theorem 4.5.7 looks disappointing since (4.5.23) is not an expansion in even powers of h; but we will see below that it is the natural result under the circumstances.

Def. 4.5.2. A linear k-step method with equidistant grids based upon a symmetric D-stable linear k-step scheme $\langle\rho,\sigma\rangle$ is called *symmetric* if its starting procedure satisfies (4.5.20) for arbitrary J (for a sufficiently smooth f); it is called $2J$-*symmetric* if its starting procedure satisfies (4.5.20). The starting procedure itself is also called ($2J$-)symmetric under these circumstances.

Remark. For an implicit symmetric scheme $\langle\rho,\sigma\rangle$ and nonlinear f it is normally not possible to satisfy (4.5.20) for an arbitrary J except when the starting procedure is *defined* implicitly by (4.5.20) and $s_{n,-\kappa}(z_0)=s_{-n,\kappa}(z_0)$.

Corollary 4.5.8. *In the situation of Theorem* 4.5.7, *if* $\mathfrak{M}$ *is* $2J$-*symmetric and* (4.4.45) *holds for all zeros of* ρ *(i.e., all zeros of* ρ *are roots of unity) the global discretization error of* $\mathfrak{M}$ *for an* $IVP^{(2J+2)}1$ *possesses a* $\hat{\Delta}$-*reduced asymptotic expansion to order* $2J$ *in even powers of* h, *where* $\hat{\Delta}$ *denotes the reduction described above Corollary* 4.4.8.

Proof. If ν is a multiple of $\bar{d}$ we have $\cos\nu\varphi_i=1$ and $\sin\nu\varphi_i=0$ for each $i>1$ so that (4.5.23) is in even powers of h. Thus if we choose $\mathbb{G}$ and $\{\mathbb{G}_n^{(d)}\}_{n\in\bar{\mathbb{N}}}$ as above Corollary 4.4.8, with $\bar{t}_\mu-\bar{t}_{\mu-1}=1/m$, then (4.5.23) becomes, for $\bar{t}_\mu\in\mathbb{G}$,

$$\zeta_n(\bar{t}_\mu)=z(\bar{t}_\mu)+\sum_{j=\frac{p}{2}}^{J} h^{2j}\left(\sum_{i\in I_{2j}^*\cup\{1\}} e_{2j,i}(\bar{t}_\mu)\right)+O(h^{2J+2}) \tag{4.5.28}$$

for all grids from the sequence $\{\mathbb{G}_n^{(d)}\}$. □

Corollary 4.5.8 shows that (4.5.23) is the proper generalization of (4.5.27) to the case $k>2$: The only case where we can reasonably expect a $\hat{\Delta}$-reduced asymptotic expansion of the type (1.4.2) to exist at all, occurs when all x_i are roots of unity; see Corollary 4.4.8. In this case we obtain an asymptotic expansion proper in powers of h^2 if and only if (4.5.20) holds. Thus, (4.5.20) is the appropriate symmetry condition on the starting procedure and Def. 4.5.2 is meaningful.

Example. $k=3$, $\rho=\frac{1}{3}(x^3-1)$, $\sigma=\frac{1}{2}(x^2+x)$, $p=2$. $\langle\rho,\sigma\rangle$ is obviously symmetric, the symmetry condition on the starting procedure is (with $s_{n\kappa}(z_0)=: s_\kappa$):

$$\left.\begin{array}{l} \dfrac{s_1-s_{-2}}{3h}-\dfrac{1}{2}[f(s_0)+f(s_{-1})] \\ \\ \dfrac{s_2-s_{-1}}{3h}-\dfrac{1}{2}[f(s_1)+f(s_0)] \end{array}\right\} = 0 \quad \text{or} \quad O(h^{2J+1}).$$

If we choose

$$s_0=z_0\,, \qquad s_1=z_0+hf(z_0)\,, \qquad s_2=s_1+\frac{h}{2}[3f(s_1)-f(s_0)]\,, \tag{4.5.29}$$

we find that we have achieved full symmetry. Thus the uniform linear 3-step method based upon $\langle\rho,\sigma\rangle$ with the starting procedure (4.5.29) is symmetric; for an $\mathrm{IVP}^{(2J+2)}\,1$ it produces an expansion

$$\zeta_n(t)=z(t)+\sum_{j=1}^{J} h^{2j}e_{2j}(t)+O(h^{2J+2})$$

at each t which is a multiple of $3h$ in infinitely many grids. Hence, "quadratic" Richardson extrapolation may be applied at all such points.

Symmetric linear k-step methods will normally suffer from weak stability since all zeros of ρ are essential; see Theorem 4.5.4. (In the example above, the e_{2j} are really the sum of a solution $e_{2j,1}$ of $e'_{2j,1}-f'(z(t))e_{2j,1}=b_{2j,1}$ and of a solution $e_{2j,2}$ of $e'_{2j,2}+\frac{1}{2}f'(z(t))e_{2j,2}=b_{2j,2}$, since $\lambda_2=\lambda_3=-\frac{1}{2}$.) On the other hand, linear smoothing procedures as constructed in Section 4.5.2 will generally destroy the special structure of (4.5.23) which made the $\hat{\Delta}$-reduced asymptotic expansion in powers of h^2 possible. Fortunately, under the conditions of Corollary 4.5.8 we can choose the smoothing operator $\pi(T_h)$ such that $\pi(T_h)\zeta_n(\bar{t}_\mu)$ also possesses an asymptotic expansion (4.5.28):

Theorem 4.5.9. *In the construction of Theorem 4.5.3 of the linear smoothing operator $\pi(T_h)$ for a symmetric linear k-step scheme $\langle\rho,\sigma\rangle$, choose $r=2\bar{r}-1$, $s_i=2\bar{s}$, $i=2(1)k$, and $v_0=(k-1)\bar{s}+\bar{r}-1$, with $\bar{r},\bar{s}\in\mathbb{N}$. The operator thus generated is symmetric, i. e.,*

$$\pi(x)=\pi\left(\frac{1}{x}\right); \tag{4.5.30}$$

furthermore, (4.5.7) holds for $r=2\bar{r}$.

Proof. For the zeros $x_i\neq 1$ of ρ, $i=2(1)k$, consider

$$\begin{aligned}\delta(x) &:= x^{-(k-1)\bar{s}}\prod_{i=2}^{k}(x-x_i)^{2\bar{s}} = x^{(k-1)\bar{s}}\prod_{i=2}^{k}\left(1-\frac{1}{xx_i}\right)^{2\bar{s}} \\ &= \left(\frac{1}{x}\right)^{-(k-1)\bar{s}}\prod_{i=2}^{k}\left(\frac{1}{x}-x_i\right)^{2\bar{s}} = \delta\left(\frac{1}{x}\right)\end{aligned}$$

where we have used the fact that x_i^{-1} is a zero of ρ with x_i and that $\left|\prod_{i=2}^{k} x_i\right| = \left|\prod_{i=1}^{k} x_i\right| = |\alpha_0/\alpha_k| = 1$ due to (4.5.12). Next we consider

$$\frac{x^{\bar{r}-1}}{\delta(x)} = \sum_{j=0}^{2\bar{r}-2} c_j(x-1)^j + O((x-1)^{2\bar{r}-1})$$

$$= x^{2(\bar{r}-1)} \frac{\left(\frac{1}{x}\right)^{\bar{r}-1}}{\delta\left(\frac{1}{x}\right)} = x^{2(\bar{r}-1)} \sum_{j=0}^{2\bar{r}-2} c_j \left(\frac{1}{x} - 1\right)^j + O((x-1)^{2\bar{r}-1})$$

due to $\delta(x) = \delta(1/x)$ and $(1/x) - 1 = O(x-1)$ as $x \to 1$.

As $v(x) = \sum_{j=0}^{2\bar{r}-2} c_j(x-1)^j$ this implies $v(x) = x^{2(\bar{r}-1)} v(1/x)$. Thus

$$\pi(x) = x^{-(\bar{r}-1)} \delta(x) v(x) = \left(\frac{1}{x}\right)^{-(\bar{r}-1)} \delta\left(\frac{1}{x}\right) v\left(\frac{1}{x}\right) = \pi\left(\frac{1}{x}\right).$$

By (4.5.30), the expansion of the left-hand side of (4.5.7) can contain only even powers of h so that (4.5.7) has to hold for $r = 2\bar{r}$. □

Corollary 4.5.10. *Under the assumptions of Corollary* 4.5.8, *if* $\pi(T_h)$ *is the operator of Theorem* 4.5.9 *with* $\bar{r} \geq p/2$, *then*

$$\pi(T_h)\zeta_n(\bar{t}_\mu) = z(\bar{t}_\mu) + \sum_{j=\frac{p}{2}}^{J} h^{2j} \tilde{e}_{2j}(\bar{t}_\mu) + O(h^{2J+2}), \qquad \bar{t}_\mu \in \mathbb{G}, \tag{4.5.31}$$

for grids from the sequence $\{\mathbb{G}_n^{(d)}\}_{n \in \bar{\mathbb{N}}}$.

Proof. We apply $\pi(T_h)$ to the right hand side of (4.5.23) and consider the various terms separately:

$$\pi(T_h)\left[z(\bar{t}_\mu) + \sum_{j=\frac{p}{2}}^{J} h^{2j} e_{2j,1}(\bar{t}_\mu)\right] = z(\bar{t}_\mu) + \sum_{j=\frac{p}{2}}^{J} h^{2j} \tilde{e}_{2j,1}(\bar{t}_\mu) + O(h^{2J+2})$$

due to the symmetry (4.5.30) of π and (4.5.7) with $r = 2\bar{r} \geq p$. Now let $\bar{t}_\mu := t_{\bar{v}}$; then

$$\pi(T_h)\left(\cos \bar{v}\varphi_i \sum_{j=\frac{p}{2}}^{J} h^{2j} e_{2j,i}(t_{\bar{v}})\right) = \pi_i(T_h) \sum_{j=\frac{p}{2}}^{J} h^{2j} e_{2j,i}(\bar{t}_\mu)$$

$$= \sum_{j=\frac{p}{2}+\bar{s}}^{J} h^{2j} \tilde{e}_{2j,i}(\bar{t}_\mu) + O(h^{2J+2}) \quad \text{for } i \in I_p^*,$$

where $\pi_i(x) := \sum_{\kappa=-\nu_0}^{\nu_0} (\cos\kappa\varphi_i) p_\kappa x^\kappa$ for $\pi(x) = \sum_{\kappa=-\nu_0}^{\nu_0} p_\kappa x^\kappa$ so that π_i also satisfies (4.5.30). The first $2\bar{s}$ powers in the expansion of $\pi_i(e^h)$ in powers of h have to vanish by (4.5.8) and the construction of π. Similarly,

$$\pi(T_h)\left(\sin\bar{v}\varphi_i \sum_{j=\frac{p}{2}}^{J-1} h^{2j+1} e_{2j+1,i}(t_{\bar{v}})\right) = \pi_i^*(T_h) \sum_{j=\frac{p}{2}}^{J-1} h^{2j+1} e_{2j+1,i}(\bar{t}_\mu)$$

$$= \sum_{j=\frac{p}{2}+\bar{s}+1}^{J} h^{2j} \tilde{e}_{2j+1,i}(\bar{t}_\mu) + O(h^{2J+2}) \quad \text{for } i \in I_p^*;$$

here $\pi_i^* := \sum_{\kappa=-\nu_0}^{\nu_0} (\sin\kappa\varphi_i) p_\kappa x^\kappa$ satisfies $\pi_i^*(x) = -\pi_i^*(1/x)$ so that $\pi_i^*(e^h)$ possesses an expansion in odd powers of h which begins with $h^{2\bar{s}+1}$.

If x_i, $i > k$, appear in (4.5.23), the symmetry and antisymmetry of the corresponding operators π_i and π_i^* also produces expansions in even powers of h although no extinction of leading terms occurs. □

Remark. The x_i for $i > k$—if any—could also have been included in the construction of the symmetric smoothing operator $\pi(T_h)$ since they also occur in complex-conjugate pairs. However, the damping of the corresponding terms is not so important since they carry powers of h greater than $2p$.

Example (see also the example at the end of Section 4.5.2): For the example following Corollary 4.5.8 with $x_{2,3} = e^{\pm i(2\pi/3)}$ take $\bar{r} = p/2 = 1$, $\bar{s} = 1$, $\nu_0 = \bar{s}(k-1) + \bar{r} - 1 = 2$, to obtain

$$\frac{x^2}{(x - e^{i(2\pi/3)})^2 (x - e^{-i(2\pi/3)})^2} = \frac{1}{9} + O((x-1)) \quad (\text{actually } O((x-1)^2))$$

and

$$\pi(x) = \frac{(x - e^{i(2\pi/3)})^2 (x - e^{-i(2\pi/3)})^2}{9x^2} = \frac{1}{9}(x^{-2} + 2x^{-1} + 3 + 2x + x^2)$$

with $\pi(e^h) = 1 + O(h^2)$. Furthermore,

$$\pi_2(x) = \tfrac{1}{18}(-x^{-2} - 2x^{-1} + 6 - 2x - x^2) \quad \text{satisfies } \pi_2(e^h) = O(h^2),$$

$$\pi_2^*(x) = \frac{1}{6\sqrt{3}}(x^{-2} - 2x^{-1} + 2x - x^2) \quad \text{satisfies } \pi_2^*(e^h) = O(h^3),$$

and the symmetry properties are evident.

4.5.5 Estimation of the Discretization Error

For reasons explained in Section 3.4.4, it is desirable to be able to obtain approximate values of

$$(4.5.32)\qquad \begin{aligned} -l_n(t_\nu) &= h_\nu^p\,\sigma(T_h)\,\varphi(z(t_{\nu-k}))+O(h_\nu^{p+1}) \\ &= -C h_\nu^p\,\sigma(T_{h_\nu})\,z^{(p+1)}(t_{\nu-k})+O(h_\nu^{p+1}); \end{aligned}$$

see Def. 4.4.1 and (4.4.6).

Basically, we can use the same two approaches as in Section 3.4.4 for the estimation of the local discretization error:

a) We take $\zeta(t_{\nu-\bar{k}}),\dots,\zeta(t_\nu)$ as computed by our linear k-step method and insert these values into the forward step procedure associated with *another* linear $\bar{k}$-step scheme $\bar{\psi}$ to estimate (4.5.32) from the residual obtained, cf. Theorem 3.4.9.

b) From $\zeta(t_{\nu-\bar{k}}),\dots,\zeta(t_{\nu-1})$ we compute a value $\bar{\zeta}(t_\nu)$ by means of a linear $\bar{k}$-step procedure associated with $\bar{\psi}$ and estimate (4.5.32) from the value of $\bar{\zeta}(t_\nu)-\zeta(t_\nu)$, cf. (3.4.32).

Due to the linear nature of our k-step methods, the two approaches become almost identical and b) is much simpler than with RK-methods.

Assume that we have used a uniform linear k-step method $\mathfrak{M}$ based on the linear k-step scheme $\langle\rho,\sigma\rangle$ of order $p\geq 2$ to discretize an $\mathrm{IVP}^{(p+2)}1$ and that the step has been constant in $[t_{\nu-\bar{k}},t_\nu]$; we will denote it by h and assume $h=1/n$ without loss of generality. Furthermore, we will assume that the solution ζ_n of $\mathfrak{M}(\mathrm{IVP}^{(p+2)}1)$ satisfies

$$(4.5.33)\qquad \zeta_n(t_\mu)=z(t_\mu)+h^p e_p(t_\mu)+h^{p+1}\sum_{i=1}^{q} x_i^\mu\, e_{p+1,i}(t_\mu)+O(h^{p+2})$$

for $\mu=\nu-\bar{k}(1)\nu$, where the x_i, $i=1(1)q$, are the essential zeros of ρ, $e_p\in D^{(2)}$, and the $e_{p+1,i}\in D^{(1)}$.

(Corollary 4.4.3 and Theorems 4.4.5—4.4.9 give some insight into the conditions under which (4.5.33) will hold: E. g., if $q=1$ the starting procedure can be of order p; if $q>1$ we need a starting procedure of order $p+1$. Since we are considering only one specific value of h resp. n it is not important that the $e_{p+1,i}$ are independent of n and step changes may also have occurred if the step change procedure was of order p.)

Finally, we will say that the linear $\bar{k}$-step scheme $\langle\bar{\rho},\bar{\sigma}\rangle$ is *compatible* with $\langle\rho,\sigma\rangle$ if

(i) the order $\bar{p}$ of $\langle\bar{\rho},\bar{\sigma}\rangle$ is not smaller than the order p of $\langle\rho,\sigma\rangle$;

(ii) $\bar{\rho}(x_i)=0$, $i=2(1)q$, for the extraneous essential zeros of ρ (if any);

(iii) $\bar{C}$ is different from the error constant C of $\langle\rho,\sigma\rangle$, where

$$(4.5.34)\qquad \bar{C} := \begin{cases} \text{error constant of } \langle\bar{\rho},\bar{\sigma}\rangle & \text{if } \bar{p}=p, \\ 0 & \text{if } \bar{p}>p. \end{cases}$$

Theorem 4.5.11. *If* (4.5.33) *holds for the solution* ζ_n *of the discretization* $\mathfrak{M}(IVP^{(p+2)}\,1)$ *described above then*

$$(4.5.35)\qquad h^{p+1}\sigma(T_h)\varphi(z(t_{\nu-k})) = \frac{C}{C-\bar{C}}\,[\bar{\rho}(T_h)\zeta_n(t_{\nu-\bar{k}}) - h\bar{\sigma}(T_h)f(\zeta_n(t_{\nu-\bar{k}}))] + O(h^{p+2})$$

for any linear $\bar{k}$*-step scheme* $\langle\bar{\rho},\bar{\sigma}\rangle$ *compatible with* $\langle\rho,\sigma\rangle$.

Proof. With (4.5.33) and (4.5.34) we have

$$(4.5.36)\qquad [\bar{\rho}(T_h)\zeta_n(t_{\nu-\bar{k}}) - h\bar{\sigma}(T_h)f(\zeta_n(t_{\nu-\bar{k}}))]$$

$$\begin{aligned}
&= h\,L_h[\langle\bar{\rho},\bar{\sigma}\rangle]z(t_{\nu-\bar{k}}) + h^p\bar{\rho}(T_h)e_p(t_{\nu-\bar{k}}) - h^{p+1}\bar{\sigma}(T_h)f'(z(t_{\nu-k}))e_p(t_{\nu-\bar{k}}) \\
&\quad + h^{p+1}\bar{\rho}(T_h)\sum_{i=1}^{q} x_i^{\nu-\bar{k}} e_{p+1,i}(t_{\nu-\bar{k}}) + O(h^{p+2}) \\
&= \bar{C}h^{p+1}\bar{\sigma}(T_h)z^{(p+1)}(t_{\nu-\bar{k}}) + h^{p+1}L_h[\langle\bar{\rho},\bar{\sigma}\rangle]e_p(t_{\nu-\bar{k}}) - C h^{p+1}\bar{\sigma}(T_h)z^{(p+1)}(t_{\nu-\bar{k}}) \\
&\quad + h^{p+1}\sum_{i=1}^{q} x_i^{\nu-\bar{k}}\bar{\rho}_i(T_h)e_{p+1,i}(t_{\nu-\bar{k}}) + O(h^{p+2}) \\
&= (\bar{C}-C)h^{p+1}\sigma(T_h)z^{(p+1)}(t_{\nu-k}) + O(h^{p+2}).
\end{aligned}$$

Here we have used the fact that $e_p'(t) - f'(z(t))e_p(t) = -C z^{(p+1)}(t)$, see (4.4.17), and $\bar{\rho}_i(1) = \bar{\rho}(x_i) = 0$ according to the compatibility assumption (ii) so that $\bar{\rho}_i(T_h)e_{p+1,i} = O(h)$; furthermore, we have used $\bar{\sigma}(1) = \sigma(1) = 1$ to switch from $\bar{\sigma}(T_h)$ to $\sigma(T_h)$ with an error of $O(h)$. Multiplication of (4.5.36) by $C/(C-\bar{C})$ yields (4.5.35), see (4.5.32). □

Remarks. 1. The evaluation of the right hand side of (4.5.35) requires only arithmetic operations since $\zeta_n(t_{\nu-\kappa})$ and $f(\zeta_n(t_{\nu-\kappa}))$, $\kappa = 0(1)\bar{k}$, are known (or needed in the further computation) anyway. There is no stability requirement on $\langle\bar{\rho},\bar{\sigma}\rangle$!

2. In the analogous Theorem 3.4.9 for RK-methods we had to choose $\bar{p}>p$ due to the more complicated structure of φ for RK-methods.

Corollary 4.5.12. *Under the conditions of Theorem* 4.5.11, *if* $\bar{\zeta}_\nu$ *is computed from* $\zeta_n(t_{\nu-\kappa})$, $\kappa = 1(1)\bar{k}$, *by the linear* $\bar{k}$*-step procedure associated with some linear* $\bar{k}$*-step scheme* $\langle\bar{\rho},\bar{\sigma}\rangle$ *compatible with* $\langle\rho,\sigma\rangle$, *then*

$$(4.5.37)\qquad h^{p+1}\sigma(T_h)\varphi(z(t_{\nu-k})) = \frac{C}{\bar{C}-C}\,\bar{\alpha}_{\bar{k}}(\bar{\zeta}_\nu - \zeta_n(t_\nu)) + O(h^{p+2}).$$

Proof. Let
$$\overline{\zeta}_n(t_{\nu-\kappa}) := \begin{cases} \zeta_n(t_{\nu-\kappa}), & \kappa = 1(1)\overline{k}, \\ \overline{\zeta}_\nu, & \kappa = 0, \end{cases} \quad \text{then}$$

$$\begin{aligned} 0 &= \overline{\rho}(T_h)\overline{\zeta}_n(t_{\nu-\overline{k}}) - h\overline{\sigma}(T_h)\, f(\overline{\zeta}_n(t_{\nu-\overline{k}})) \qquad \text{(according to the definition of } \overline{\zeta}_\nu) \\ &= \overline{\rho}(T_h)\zeta_n(t_{\nu-\overline{k}}) - h\overline{\sigma}(T_h)\, f(\zeta_n(t_{\nu-\overline{k}})) + \overline{\alpha}_k(\overline{\zeta}_\nu - \zeta_n(t_\nu)) - h\overline{\beta}_k(f(\overline{\zeta}_\nu) - f(\zeta_n(t_\nu))) \\ &= (\overline{C} - C)h^{p+1}\sigma(T_h) z^{(p+1)}(t_{\nu-k}) + \overline{\alpha}_k(\overline{\zeta}_\nu - \zeta_n(t_\nu)) + O(h^{p+2}) \end{aligned}$$

by (4.5.36); $f(\overline{\zeta}_\nu) - f(\zeta_n(t_\nu)) = O(h^{p+1})$ follows from the Lipschitz continuity of f and $\overline{\zeta}_\nu - \zeta_n(t_\nu) = O(h^{p+1})$. □

Remarks. 1. (4.5.37) with a *predictor* scheme $\langle\overline{\rho}, \overline{\sigma}\rangle$ is the device normally used in practical computations. If $\overline{\beta}_k \neq 0$, we can replace $f(\overline{\zeta}_\nu)$ by $f(\zeta_n(t_\nu))$ in the computation of $\overline{\zeta}_\nu$. In both cases, (4.5.37) is completely equivalent to (4.5.35); this shows that the approach a) is really the more fundamental one. Theorem 5.13 of Henrici [1] is equivalent to our Corollary 4.5.12 but uses more restrictive assumptions.

2. The remarks after Theorem 4.5.11 apply again; the analogous procedure for RK-methods is now that of Fehlberg where we need again $\overline{p} > p$.

If (4.5.33) holds, so that the principal term of the *global discretization error* consists only of $n^{-p}\Delta_n e_p$, with e_p from (4.4.17), then one can use the estimates of the principal error function for an approximate integration of the differential equation (4.4.17) to obtain an estimate of the global discretization error. A simple discretization method and a wide grid (the principal error function will be estimated only at selected gridpoints normally) will generally suffice.

Unfortunately, the interesting approach to the estimation of the global error based upon Lemma 3.4.7 cannot be applied to uniform linear k-step methods: Due to the form of the principal error function φ, (3.4.21) can never be fulfilled so that the differential equation (4.4.17) for e_p is never exact.

Lemma 4.5.13. *If the principal error function φ is a sum of elementary differentials of f of order $p+1$ and satisfies* (3.4.21) *then φ vanishes for a linear IVP1 with constant coefficients.*

Proof. For $f(y) = gy$, g a constant matrix, all elementary differentials of order $p+1$ vanish except $\{_p f\}_p(\xi) = g^{p+1}\xi$. But $\{_p f\}_p$ cannot be generated from the right hand side of (3.4.21):

For a term $\gamma\{_{p-1} f\}_{p-1}$ in Φ we have $\Phi' f - f'\Phi = \gamma\{_p f\}_p +$ (terms with second or higher derivatives of f) $- \gamma\{_p f\}_p$; with all terms in Φ containing second or higher derivatives of f these derivatives are retained in $\Phi' f - f'\Phi$. Thus a φ satisfying (3.4.21) cannot contain $\{_p f\}_p$ and has to vanish for $f(y) = gy$. □

The principal error function of a linear k-step method of order p (see Def. 4.4.1) is $\varphi(\xi) = -Cg^{p+1}\xi$ for $f(y) = gy$, which obviously does not vanish. However, the principal error functions of multistage k-step methods, e.g. predictor-corrector methods, can very well be of the form (3.4.21); see Section 5.2.5.

4.6 Strong Stability of Linear Multistep Methods

When we consider the discretizations of an $\{\mathrm{IVP}\,1\}_{\mathbb{T}}$ by a linear multistep method $\mathfrak{M}_{\mathbb{T}}$, the introductory remarks of Section 3.5 apply once more. Again, we have to distinguish the role of t as an independent variable; all formulas in the preceding sections of Chapter 4 may be accordingly rewritten by the use of the interpretation (2.1.2).

4.6.1 Strong Stability for Sufficiently Large n

In Section 3.5.1 we were able to show that RK-methods are very suitable for the numerical integration of $\{\mathrm{IVP}\,1\}_{\mathbb{T}}$ if a sufficiently large n is used. The results of Section 3.5.1, however, cannot be extended to all stable uniform linear multistep methods but only to those which are stable w.r.t. Spijker's norm, see Def. 4.2.3 and Theorem 4.2.7. This fact has already been pointed out by Babuška, Práger, Vitásek [1] where a slightly less general version of the following theorem has been proved.

Theorem 4.6.1. *Consider a totally stable $\{\mathrm{IVP}\,1\}_{\mathbb{T}}$ and a uniform linear k-step method $\mathfrak{M}_{\mathbb{T}}$, with equidistant grids, based upon a strongly D-stable k-step scheme of (minimal) order* 1. *Let ζ_n be the solution of $\mathfrak{M}_{\mathbb{T}}(\{\mathrm{IVP}\,1\}_{\mathbb{T}})$ for $n \in \mathbb{N}'$. For each given $\rho < R/2a$ (see Lemma* 4.1.2), *$\rho > 0$, there exists an $n_0(\rho)$ such that ζ_n exists and*

$$\|\zeta_n(t_\nu) - z(t_\nu)\| < \rho \quad \text{for all } t_\nu \in \mathbb{G}_n \tag{4.6.1}$$

for each $n \geq n_0$, $n \in \mathbb{N}'$.

Proof. a) For the associated polynomial ρ of our linear k-step scheme ψ let

$$\bar{\rho}(x) := \frac{\rho(x)}{x-1} = \sum_{\kappa=0}^{k-1} \bar{\alpha}_\kappa x^\kappa;$$

$\sum_\kappa \bar{\alpha}_\kappa = \rho'(1) = 1$ due to the assumed consistency and $\bar{\alpha}_{k-1} = \alpha_k \neq 0$. By the strong D-stability of ψ, $\bar{\rho}$ has all its zeros inside the open unit disk. Thus, the corresponding $\bar{\gamma}_{\mu\nu}$ of (4.2.11) satisfy (4.2.18), see Section 4.2.2.

The quantities $\overline{\zeta}_n(t_\nu)$ defined by

$$(4.6.2) \qquad \begin{cases} \overline{\zeta}_n(t_\nu) := \zeta_n(t_\nu), & \nu = 0(1)k-2, \\ \overline{\zeta}_n(t_\nu) := \sum_{\kappa=0}^{k-1} \overline{\alpha}_\kappa \zeta_n(t_{\nu-k+1+\kappa}), & \nu \geq k-1, \end{cases}$$

satisfy

$$(4.6.3) \qquad \overline{\zeta}_n(t_\nu) = \overline{\zeta}_n(t_{\nu-1}) + h \sum_{\kappa=0}^{k} \beta_\kappa f(t_{\nu-k+\kappa}, \zeta_n(t_{\nu-k+\kappa})), \qquad \nu \geq k.$$

Hence, by (4.6.2) and $\sum \overline{\alpha}_\kappa = 1$, for $\nu \geq k-1$

$$(4.6.4) \qquad \begin{aligned} &\sum_{\kappa=0}^{k-1} \overline{\alpha}_\kappa (\zeta_n(t_{\nu-k+1+\kappa}) - \overline{\zeta}_n(t_{\nu-k+1+\kappa})) \\ &\qquad = \sum_{\kappa=0}^{k-2} \overline{\alpha}_\kappa (\overline{\zeta}_n(t_\nu) - \overline{\zeta}_n(t_{\nu-k+1+\kappa})) =: \delta_\nu \end{aligned}$$

where, for $\nu \geq 2(k-1)$,

$$(4.6.5) \qquad \|\delta_\nu\| \leq \sum_{\kappa=0}^{k-2} |\overline{\alpha}_\kappa| (k-1-\kappa) \left(\sum_{\kappa=0}^{k} |\beta_\kappa| \right) h M_f =: \overline{M} h$$

if the values of ζ_n are in $B_R(z)$. Furthermore, an estimate (4.6.5)—possibly with a larger value of $\overline{M}$, which would then be used throughout—trivially holds also for $k-1 \leq \nu \leq 2k-3$. Thus

$$(4.6.6) \qquad \|\zeta_n(t_\nu) - \overline{\zeta}_n(t_\nu)\| \leq \left\| \sum_{\mu=k-1}^{\nu} \overline{\gamma}_{\mu\nu} \delta_\mu \right\| \leq \overline{c} \overline{M} h.$$

b) For $t \geq t_{k-2}$, let $\overline{z}_n$ be defined by

$$(4.6.7) \qquad \overline{z}_n(t) := \overline{\zeta}_n(t_{\nu-1}) + (t - t_{\nu-1}) \sum_{\kappa=0}^{k} \beta_\kappa f(t_{\nu-k+\kappa}, \zeta(t_{\nu-k+\kappa})), \qquad t \in [t_{\nu-1}, t_\nu],$$

so that $\overline{z}_n(t_\nu) = \overline{\zeta}_n(t_\nu)$ for $\nu \geq k-2$. Then

$$\overline{z}_n'(t) = f(t, z_n(t)) + \delta_n(t)$$

where

$$(4.6.8) \qquad \begin{aligned} \delta_n(t) = &\sum_{\kappa=0}^{k} \beta_\kappa [f(t_{\nu-k+\kappa}, \zeta_n(t_{\nu-k+\kappa})) - f(t_{\nu-k+\kappa}, \overline{\zeta}(t_{\nu-k+\kappa}))] \\ &+ \sum_{\kappa=0}^{k} \beta_\kappa [f(t_{\nu-k+\kappa}, \overline{z}_n(t_{\nu-k+\kappa})) - f(t, \overline{z}_n(t))] \end{aligned}$$

in $(t_{\nu-1}, t_\nu)$ due to (4.6.7) and (4.1.8).

The first term on the right hand side of (4.6.7) may be estimated by

$$L \sum_{\kappa=0}^{k} |\beta_\kappa| \, \|\zeta_n(t_{\nu-k+\kappa}) - \bar{\zeta}_n(t_{\nu-k+\kappa})\| \le M_1 h$$

due to (4.6.6). For the second member, we use (2.3.2) to obtain the estimate

$$L \sum_{\kappa=0}^{k} |\beta_\kappa| \left\| \begin{pmatrix} t_{\nu-k+\kappa} - t \\ \bar{z}(t_{\nu-k+\kappa}) - \bar{z}_n(t) \end{pmatrix} \right\| \le M_2 h$$

since f is bounded uniformly in $B_R(z)$ and the variation of $\bar{z}$ is given by (4.6.7). Thus δ_n permits an estimate

$$\|\delta_n(t)\| \le (M_1 + M_2) h \quad \text{for } t > t_{k-1}.$$

Furthermore, there is a constant M_0 such that $\|\bar{z}_n(t_{k-2}) - z_n(t_{k-2})\| \le M_0 h$ due to (4.6.7), (4.6.2), and the consistency of $\mathfrak{M}_{\mathbb{T}}$. Thus (4.6.1) holds if h is so small that $\bar{\rho} := \rho - \bar{c}\bar{M} h > 0$ and

$$M_0 h < \bar{\delta}_0(\bar{\rho}), \qquad (M_1 + M_2) h < \bar{\delta}_1(\bar{\rho}), \tag{4.6.9}$$

with $\bar{\delta}_0$ and $\bar{\delta}_1$ from (2.3.7) and $\bar{c}\bar{M}$ from (4.6.6).

c) In using the Lipschitz continuity and boundedness of f we have relied on the fact that $\zeta_n(t_\nu)$ and $\bar{\zeta}_n(t_\nu)$ remain in $B_R(z)$ for all ν. This fact may be established inductively as follows:

Assume that h is also small enough that (4.6.3) implies $\bar{\zeta}(t_\nu) \in B_R(z)$ if $\zeta_n(t_{\nu-k+\kappa}) \in B_\rho(z)$, $\kappa = 0(1)k-1$, $\zeta_n(t_\nu) \in B_R(z)$, and $\bar{\zeta}_n(t_{\nu-1}) \in B_{\bar{\rho}}(z)$. Then the validity of (4.6.1) for $\nu < \nu_0$ implies $\zeta_n(t_{\nu_0}) \in B_R(z)$ (see Lemma 4.1.2) and thus, with $\bar{\zeta}_n(t_{\nu_0-1}) \in B_{\bar{\rho}}(z)$, $\bar{\zeta}_n(t_{\nu_0}) \in B_R(z)$. Hence, our argument in b) is justified at $t = t_{\nu_0}$ and we have $\bar{\zeta}_n(t_{\nu_0}) = \bar{z}_n(t_{\nu_0}) \in B_{\bar{\rho}}(z)$ and $\zeta_n(t_{\nu_0}) \in B_\rho(z)$ (cf. (4.6.6)). □

Corollary 4.6.2. *Under the situation of Theorem* 4.6.1, *the difference equations of* $\mathfrak{M}_{\mathbb{T}}(\{\text{IVP }1\}_{\mathbb{T}})$ *are totally stable for sufficiently large n.*

Proof. See proof of Corollary 3.5.2. □

Remark. Corollary 4.6.2 implies that even a computational solution of $\mathfrak{M}_{\mathbb{T}}(\{\text{IVP }1\}_{\mathbb{T}})$ remains arbitrarily close to the true solution z of the $\{\text{IVP }1\}_{\mathbb{T}}$ uniformly in $[0, \infty)$ if h is sufficiently small and if the local computing error is kept sufficiently small uniformly; see the formulation of Corollary 3.5.2. Remark 1 after Corollary 3.5.2 applies again.

For stable uniform k-step methods which are based on linear k-step schemes which are *not strongly D-stable*, Theorem 4.6.1 and Corollary 4.6.2 need not hold. A well-known example is furnished by the application of the 2-step midpoint or the Simpson method (Example 3, Section 4.1.1) to $y'+y=0$, $y_0=1$; see Henrici [1], Section 5.3-1. Here the solution of the discretization deviates arbitrarily far from e^{-1} as t becomes large no matter how small (but fixed) h is taken.

The result analogous to Theorem 3.5.3, that the discretization of an exponentially stable $\{IVP1\}_{\mathbb{T}}$ is exponentially stable for sufficiently small h, should also be true under the assumptions of Theorems 4.6.1 (strong D-stability) and 3.5.3 (Lipschitz continuous f_y). It seems, however, that a proof based upon Theorem 2.3.3 (see proof of Theorem 3.5.3) is not possible since the function $k_n(t, x)$ of (3.5.5) cannot be properly defined in the multistep situation[8].

In any case, one will rarely have $a=1$ in the exponential stability estimate (2.3.31). This is another reason why the analysis of the behavior of multistep discretization on infinite intervals is considerably more involved; cf. Corollary 3.5.4 and the remarks preceding it.

4.6.2 Stability Regions of Linear Multistep Methods

In Section 2.3.6 we defined stability regions to describe the stability behavior of discretizations of $\{IVP1\}_{\mathbb{T}} \in J_c$ (linear equations with constant coefficients). Since the discretization of such an $\{IVP1\}_{\mathbb{T}}$ by a uniform linear k-step method with equidistant grids reduces to a linear difference equation of order k with constant coefficients in the scalar case, the stability characteristics of the discretization depend on the location of the zeros of the characteristic polynomial of that difference equation, cf. the end of Section 2.3.6.

Def. 4.6.1. The polynomial

(4.6.10) $$\varphi(x, H) := \rho(x) - H\,\sigma(x)$$

is called *characteristic polynomial* of the uniform linear multistep-method based upon $\langle \rho, \sigma \rangle$.

Theorem 4.6.3. *For a given uniform linear k-step method $\mathfrak{M}_{\mathbb{T}}$ denote the zeros of its characteristic polynomial by $\bar{\xi}_\kappa(H)$, $\kappa=1(1)k$. The region $\mathfrak{H}_0$ of absolute stability for $\mathfrak{M}_{\mathbb{T}}$ is given by*

(4.6.11) $$\mathfrak{H}_0 := \{H \in \mathbb{C} : |\bar{\xi}_\kappa(H)| < 1,\ \kappa = 1(1)k\} \subset \mathbb{C}.$$

Similarly, the regions $\mathfrak{H}_\mu$ of μ-exponential stability for $\mathfrak{M}_{\mathbb{T}}$ are

(4.6.12) $$\mathfrak{H}_\mu := \{H \in \mathbb{C} : |\bar{\xi}_\kappa(H)| < e^{-\mu}\ \ \kappa = 1(1)k\}, \qquad \mu > 0.$$

(The regions $\mathfrak{H}_\mu$ may be empty for $\mu \geq \bar{\mu}$, with some $\bar{\mu} \geq 0$.)

---

[8] Added in proof: G. Schinner, a student of the author, has meanwhile succeeded in proving the result about the exponential stability for linear multistep methods.

Proof. For $y' - gy = 0$, the linear k-step discretization takes the form

$$[\rho(T_h) - hg\sigma(T_h)]\eta_{\nu-k} = 0, \qquad \nu \geq k. \tag{4.6.13}$$

Let

$$\bar{g} := S^{-1} g S = \left(\begin{array}{ccc|c} \gamma_1 & 1 & & \\ & \ddots & 1 & 0 \\ 0 & & \gamma_1 & \\ \hline & & & \ddots \\ & 0 & & & \gamma_r \end{array}\right)$$

be the Jordan normal form of g, then (4.6.13) transforms into

$$[\rho(T_h) - h\bar{g}\sigma(T_h)]\bar{\eta}_{\nu-k} = 0, \quad \text{with } \bar{\eta}_\nu := S^{-1}\eta_\nu. \tag{4.6.14}$$

If g has r different eigenvalues γ_i, $i = 1(1)r$, (4.6.14) contains at least r *scalar* difference equations of order k with constant coefficients the characteristic polynomials of which are

$$\rho(x) - h\gamma_i\sigma(x) = 0, \qquad i = 1(1)r.$$

The corresponding components of $\bar{\eta}$ decrease exponentially if and only if $h\gamma_i \in \mathfrak{H}_0$ as defined in (4.6.11). This in turn leads to an exponential decay for any further components of $\bar{\eta}$ associated with γ_i (if γ_i is a multiple eigenvalue of g with nonlinear elementary divisors).

Similarly, if all eigenvalues of g satisfy $h\gamma_i \in \mathfrak{H}_\mu$, then the eigenvector components of $\bar{\eta}_\nu$ show an exponential decay like $\exp(-\mu'\nu)$, with some $\mu' > \mu$. Hence, it is possible to find another constant μ'', $\mu' > \mu'' > \mu$, such that the principal vector components decrease like $\exp(-\mu''\nu)$ or faster. This proves the μ/h-stability of (4.6.14) and hence of (4.6.13). □

While every RK-method possesses a non-empty region of absolute stability (see Theorem 3.5.10) this is not the case for linear multistep methods.

Theorem 4.6.4. *A linear multistep method possesses a non-empty region $\mathfrak{H}_0$ of absolute stability (with* 0 *on its boundary) if its associated polynomial ρ has no essential zeros except $x_1 = 1$ or if the growth parameters λ_i of the essential zeros x_i satisfy* $\operatorname{Re}\lambda_i > 0$.

Proof. If the zeros x_i of ρ satisfy $|x_i| < 1$ for $i > 1$, then the corresponding zeros $\xi_i(H)$ of the characteristic polynomial φ satisfy $|\xi_i(H)| < 1$ for $|H| < c_0$, with some $c_0 > 0$. Hence, the existence of $\mathfrak{H}_0$ depends only on $\xi_1(H)$ and we may use the argument at the end of the proof of Theorem 3.5.10.

For an essential zero x_i we have

(4.6.15) $$\xi_i(H)=x_i[1+\lambda_i H+O(H^2)]$$

as follows from a substitution of (4.6.15) into $\varphi(x, H)$ and Def. 4.4.2 of λ_i. For $\operatorname{Re}\lambda_i>0$ the domain $\mathfrak{H}_0^i:=\{H\in\mathbb{C}:|\xi_i(H)|<1\}$ contains an interval $(-H_i, 0)$ on the negative real axis, $H_i>0$. Such an interval is trivially contained in $\mathfrak{H}_0^i$ if $|x_i|<1$. Thus, $\mathfrak{H}_0=\bigcap_i \mathfrak{H}_0^i$ contains the interval $\left(-\min_i H_i, 0\right)$ and cannot be empty. □

If the associated polynomial ρ of the linear k-step method $\mathfrak{M}_{\mathbb{T}}$ possesses at least one essential zero x_i with a real growth parameter $\lambda_i<0$, then the corresponding domain $\mathfrak{H}_0^i$ lies essentially in $\mathbb{C}_+$ and the imaginary axis is tangent to its boundary in 0. Hence for each $\gamma\in\mathbb{C}_-$ there must exist an $h_0>0$ such that $h\gamma\notin\mathfrak{H}_0^i$ and thus, $h\gamma\notin\mathfrak{H}_0$ for $0<h<h_0$. Hence, for each given exponentially stable $\{\text{IVP}\,1\}_{\mathbb{T}}\in J_c$ we may find a step h_0 such that its discretization by $\mathfrak{M}_{\mathbb{T}}$ is not exponentially stable for $h<h_0$. This proves the assertion of Theorem 4.5.1.

Normally the region $\mathfrak{H}_0$ is empty in the above situation. In any case, $\mathfrak{H}_0$ cannot contain a section $(0, a e^{i\varphi})$, $a>0$, on a ray $H=t e^{i\varphi}$, with $\pi/2<\varphi<3\pi/2$. This makes $\mathfrak{H}_0$ useless even if it is non-empty.

All uniform k-step methods whose linear k-step scheme $\langle\rho, \sigma\rangle$ satisfies the assumptions of Theorem 4.5.2 belong to this class of methods. In particular, this class includes all stable methods with even k and order $k+2$. Explicit symmetric linear k-step methods are always weakly stable since, due to $\beta_k=0$ and (4.5.12), $\varphi(x,H)=\alpha_k(x^k+\cdots-1)$ which implies $\prod|\xi_i(H)|\equiv 1$. Implicit symmetric methods may be "stabilized" (cf. (4.5.4)).

Example. For the explicit 2-step midpoint method, the characteristic polynomial is $\varphi(x,H)=\frac{1}{2}(x^2-Hx-1)$. Since $|\xi_1(H)|\cdot|\xi_2(H)|\equiv 1$, we cannot have $|\xi_1(H)|<1$ and $|\xi_2(H)|<1$ simultaneously so that $\mathfrak{H}_0$ is empty and the method is weakly stable.

The existence of A-stable linear k-step methods has been investigated by Dahlquist:

Lemma 4.6.5 (*Dahlquist* [3]). *A linear k-step method is A-stable if and only if $\rho(x)/\sigma(x)$ is regular and has a nonnegative real part for $|x|>1$.*

This observation implies several results among which the following is most prominent:

Theorem 4.6.6 (*Dahlquist* [3]). *There exists no A-stable linear k-step method of an order $p>2$. Among the A-stable linear k-step methods of order 2, the implicit trapezoidal method has the smallest error constant.*

A much simpler, though basic, result is:

Theorem 4.6.7. *An explicit linear k-step method cannot be A-stable or A(0)-stable (cf. Def. 2.3.14).*

Proof. The coefficients of $\varphi(x, H)$ are linear in H. For an explicit method the coefficient of x^k is constant and at least one of the zeros $\xi_i(H)$ must grow without bound in modulus as $H \to -\infty$ by Vieta's Theorem. □

The stability properties of the implicit trapezoidal method[9] have been analyzed to some extent in Section 3.5. The fact that $|\xi_1(H)|$—or $|\gamma(H)|$ in the notation of Section 3.5—tends to 1 as $H \to -\infty$ implies that the exponential damping becomes poor for "very negative" eigenvalues. This is also displayed by the fact that each of the regions $\mathfrak{H}_\mu$, $\mu > 0$, is finite.

The implicit Euler method has been recognized as a discretization method which retains the exponential stability of the $\{\text{IVP } 1\}_{\text{II}}$ under almost all circumstances. Its 2-step counterpart is the method based upon the 2-step differentiation scheme (see Example 2 in Section 4.1.1):

$$(4.6.16) \qquad \frac{1}{h}\left[\frac{3}{2}\eta_\nu - 2\eta_{\nu-1} + \frac{1}{2}\eta_{\nu-2} - hf(t_\nu, \eta_\nu)\right] = 0 .$$

Here we have $\varphi(x, H) = (\frac{3}{2} - H)x^2 - 2x + \frac{1}{2}$ and

$$\xi_{1,2}(H) = (2 \pm \sqrt{1+2H})^{-1}.$$

A simple analysis establishes that $|\xi_i(H)| < 1$ for $H \in \mathbb{C}_-$, $i = 1, 2$. Furthermore, $\max_i |\xi_i(H)|$ decreases from 1 to 0 as H goes from 0 to $-\infty$ along the real axis and for each $\mu > 0$ there exists a $\mu' > 0$ such that $\max_i |\xi_i(H)| < e^{-\mu}$ if $\operatorname{Re} H < -\mu'$. Thus (4.6.16) is L-stable in the sense of Def. 2.3.16.

According to Theorem 4.6.6, k-step differentiation methods (which are of order k) cannot be A-stable for $k > 2$. But for $k = 3(1)6$ they are at least $A(\alpha)$-stable, see Def. 2.3.14. Nørsett [1] has determined the following values for α:

k	2	3	4	5	6
α	$\pi/2$	1.544	1.278	0.905	0.328

For $k \geq 7$, $\mathfrak{H}_0$ does not contain the entire negative real axis. Gear [1] has advocated the use of k-step differentiation methods for the numerical treatment of stiff problems.

Another positive result has been found by Widlund:

Theorem 4.6.8 (*Widlund* [1]). *For each given* $\alpha \in [0, \pi/2)$ *there exist* $A(\alpha)$*-stable linear k-step methods with* $k = p = 3$ *and* $k = p = 4$.

---

[9] Which is a linear 1-step method and a RK-method at the same time.

In concluding this section, we emphasize that we have dealt only with *linear* multistep methods. The stability properties of *multistage* multistep methods, such as predictor-corrector methods, differ considerably from those of one-stage methods. Some results are given in Section 5.5.

4.6.3 Strong Stability for Arbitrary n

In establishing the exponential stability of RK-discretizations for arbitrary steps h_ν we encountered considerable difficulties, see Section 3.5.2. Thus we may expect that it will be very difficult to find classes of exponentially stable $\{\mathrm{IVP}\,1\}_{\mathbb{T}}$ beyond J_c w.r.t. which there exist strongly exponentially stable linear k-step methods, $k>1$. In any case, we have to limit ourselves of A-stable—or at least $A(\alpha)$-stable—methods which leaves only a few methods anyway according to the results of Section 4.6.2.

The only non-trivial result that we have been able to obtain is the following:

Theorem 4.6.9. *Consider a scalar* $\{\mathrm{IVP}^{(2)}\,1\}_{\mathbb{T}}$ *and its discretization by the 2-step differentiation method* (4.6.16), *with equidistant grids. If for an arbitrary fixed* h,

$$g_\nu := f_y(t_\nu, \zeta_n(t_\nu)) \le \bar{g} < 0 \quad \text{for all } \nu, \tag{4.6.17}$$

then the discretization is exponentially stable for that h.

Proof. In place of (4.6.16) we consider the linearization

$$\frac{1}{h}\left[\left(\frac{3}{2} - h g_\nu\right)\varepsilon_\nu - 2\varepsilon_{\nu-1} + \frac{1}{2}\varepsilon_{\nu-2}\right] = 0; \tag{4.6.18}$$

the exponential stability of (4.6.18) is equivalent to that of (4.6.16) according to Corollary 2.3.9.

We construct a Liapunov function $v(\xi, t)$, $\xi \in \mathbb{R}^2$, of the form

$$v(\xi, t) = \xi^T V \xi, \quad \text{with } V = \begin{pmatrix} a & b \\ b & c \end{pmatrix} \text{ positive definite.} \tag{4.6.19}$$

We take $a=1$ and show that it is possible for any given value of $\bar{H} := h\bar{g} < 0$ to choose b and c such that $\Delta_h v(\varepsilon_\nu, t_\nu)$ is negative definite for all ν if $h g_\nu \le \bar{H}$. Let

$$\gamma_\nu := (1 - \tfrac{2}{3} h g_\nu)^{-1}, \qquad \bar{\gamma} := (1 - \tfrac{2}{3}\bar{H})^{-1}.$$

(4.6.18) implies (see Section 2.3.3 for the notation)

$$\bar{\varepsilon}_\nu(\varepsilon_{\nu-1}, t_{\nu-1}) = \frac{\gamma_\nu}{3}(4\varepsilon_{\nu-1} - \varepsilon_{\nu-2})$$

so that

$$\begin{aligned}
h\Delta_h v(\boldsymbol{\varepsilon}_{\nu-1}, t_{\nu-1}) &= \boldsymbol{\varepsilon}_\nu^T V \boldsymbol{\varepsilon}_\nu - \boldsymbol{\varepsilon}_{\nu-1}^T V \boldsymbol{\varepsilon}_{\nu-1} \\
&= \frac{\gamma_\nu^2}{9}(4\varepsilon_{\nu-1} - \varepsilon_{\nu-2})^2 + 2b\frac{\gamma_\nu}{3}(4\varepsilon_{\nu-1} - \varepsilon_{\nu-2})\varepsilon_{\nu-1} + c\,\varepsilon_{\nu-1}^2 \\
&\quad - \varepsilon_{\nu-1}^2 - 2b\,\varepsilon_{\nu-1}\varepsilon_{\nu-2} - c\,\varepsilon_{\nu-2}^2 \\
&= \left[\tfrac{16}{9}\gamma_\nu^2 + \tfrac{8}{3}\gamma_\nu b + c - 1\right]\varepsilon_{\nu-1}^2 \\
&\quad - \left[\tfrac{8}{9}\gamma_\nu^2 - \tfrac{2}{3}\gamma_\nu b - 2b\right]\varepsilon_{\nu-1}\varepsilon_{\nu-2} + \left[\tfrac{1}{9}\gamma_\nu^2 - c\right]\varepsilon_{\nu-2}^2 \\
&=: A_\nu\varepsilon_{\nu-1}^2 + 2B_\nu\varepsilon_{\nu-1}\varepsilon_{\nu-2} + C_\nu\varepsilon_{\nu-2}^2.
\end{aligned}$$

We choose b such that B_ν vanishes for $\gamma_\nu = \bar{\gamma}$; the necessary conditions $A_\nu < 0$, $C_\nu < 0$ lead to

$$\frac{\gamma_\nu^2}{2} < c < 1 + \frac{64\gamma_\nu^3}{9(6+2\gamma_\nu)} - \frac{16}{9}\gamma_\nu^2. \tag{4.6.20}$$

Again we choose c such that (4.6.20) is satisfied for $\gamma_\nu = \bar{\gamma}$; it then holds for any $\gamma_\nu \in (0, \bar{\gamma})$ as the lower bound in (4.6.20) is increasing and the upper bound decreasing in γ_ν. It now remains to show that

$$B_\nu^2 - A_\nu C_\nu < 0 \quad \text{for } 0 < \gamma_\nu \le \bar{\gamma}$$

which may be established by a tedious calculation.

Thus there exists a constant $a_3 > 0$ such that

$$\Delta_h v(\boldsymbol{\varepsilon}_{\nu-1}, t_{\nu-1}) \le -a_3(\varepsilon_{\nu-1}^2 + \varepsilon_{\nu-2}^2) \quad \text{for all } \nu$$

and the exponential stability of (4.6.18) follows from Theorem 2.3.6. □

The proof of Theorem 4.6.9 shows that the theory of Section 2.3.3 is indeed applicable to k-step methods for $k > 1$. Note that the classical results on the asymptotic stability of difference equations of an order greater than 1 require that the coefficients tend to limiting values as $t \to \infty$. No such assumptions are used here.

Chapter 5

Multistage Multistep Methods

Having studied the peculiarities of multistage and multistep methods separately in Chapters 3 and 4 by analyzing their simplest representative classes we will now consider discretization methods for IVP 1 which combine the features of multistage and multisteps methods, cf. Section 2.1.3. We will, however, still restrict ourselves to forward step methods which compute only quantities which are supposed to be approximations to values of z and z'; other f.s.m. will be discussed in Chapter 6. The emphasis of the analysis will be on some important special classes of multistage multistep methods.

5.1 General Analysis

5.1.1 A General Class of Multistage Multistep Procedures

The forward step methods applicable to IVP 1 which we will consider in Chapter 5 will be characterized by forward step procedures of the form

$$\frac{1}{h}\sum_{\kappa=0}^{k}\left[\mathbf{A}_\kappa \boldsymbol{\eta}_{\nu-k+\kappa} - h\,\mathbf{B}_\kappa \mathbf{f}(\boldsymbol{\eta}_{\nu-k+\kappa})\right] = 0, \qquad \nu \geq k. \tag{5.1.1}$$

Here we have employed the notational conventions of Section 2.1.5, see (2.1.30) and (2.1.31); the norm $\mathfrak{N}_m$ in (2.1.32) will be the maximum norm. The $\mathbf{A}_\kappa$ and $\mathbf{B}_\kappa$ are $ms \times ms$-matrices derived from $m \times m$-matrices A_κ and B_κ, $\kappa = 0(1)k$, by a formalism analogous to (3.1.2):

$$\mathbf{A}_\kappa := \begin{pmatrix} a_{11}^\kappa I & a_{12}^\kappa I & \dots\dots & a_{1m}^\kappa I \\ \vdots & & & \vdots \\ a_{m1}^\kappa I & a_{m2}^\kappa I & \dots\dots & a_{mm}^\kappa I \end{pmatrix},$$

$$\mathbf{B}_\kappa := \begin{pmatrix} b_{11}^\kappa I & b_{12}^\kappa I & \dots\dots & b_{1m}^\kappa I \\ \vdots & & & \vdots \\ b_{m1}^\kappa I & b_{m2}^\kappa I & \dots\dots & b_{mm}^\kappa I \end{pmatrix}, \tag{5.1.2}$$

where I is the $s\times s$-identity matrix and $a^{\kappa}_{\mu\lambda}$, $b^{\kappa}_{\mu\lambda}$, $\mu,\lambda=1(1)m$, are the coefficients of the matrices A_κ and B_κ, $\kappa=0(1)k$.

The essential information on (5.1.1) is contained in these $2(k+1)$ $m\times m$-matrices A_κ and B_κ. Introducing a suitable equivalence concept, which would have to be an extension of the equivalence concepts for RK-procedures (Def. 3.2.2) and linear k-step procedures (Def. 4.1.2), we could thus define a *multistage k-step scheme* as an equivalence class of matrices of the form

$$\mathbf{M} := \begin{pmatrix} A_0 & A_1 & \cdots\cdots & A_k \\ B_0 & B_1 & \cdots\cdots & B_k \end{pmatrix} \tag{5.1.3}$$

However, in all the multistage multisteps methods which have been used or suggested so far, the matrices A_κ and B_κ have very special structures which make it easy to define in a natural fashion appropriate representatives from an equivalence class of matrices (5.1.3) of that structure. Thus, there is little value in pursuing formally this aspect of forward step procedures (5.1.1). Nevertheless we will use the term *m-stage k-step scheme* informally to denote the package of information which permits the construction of the *m-stage k-step procedure* (5.1.1) for each given IVP 1.

Also we will normally assume that (5.1.1) has been written in a form which does not involve unnecessary complications such as redundant stages or steps, cf. Sections 3.2.3 and 4.1.1. Since we have to have $\det(A_k)\neq 0$ in any case if (5.1.1) is to be solved for $\boldsymbol{\eta}_\nu$ we will require—in accordance with that policy—that A_k is a nonsingular lower triangular matrix:

Def. 5.1.1. The forward step procedure (5.1.1), with

$$a^{k}_{\mu\lambda}=0 \quad \text{for } \mu<\lambda, \qquad a^{k}_{\mu\mu}\neq 0, \qquad \mu=1(1)m, \tag{5.1.4}$$

$$\left(\sum_{\kappa=0}^{k} A_\kappa\right)e=0 \quad \text{where } e:=(1,\dots,1)^T\in\mathbb{R}^m, \tag{5.1.5}$$

is called an *m-stage k-step procedure.*

If B_k satisfies

$$b^{k}_{\mu\lambda}=0 \quad \text{for } \mu\leq\lambda$$

the procedure is called *explicit*, else *implicit*.

Remarks. 1. The condition (5.1.5) has been introduced for the same reason as the first condition (4.1.3) in Def. 4.1.1, see the remark after Def. 4.1.1.

2. If (5.1.1) is implicit we assume that there exists no equivalent explicit form of (5.5.1).

3. Any nontrivial m-stage k-step procedure is "*nonlinear*" in the sense that it employs resubstitutions into f.

If (5.1.1) is explicit it may immediately be solved for $\boldsymbol{\eta}_\nu$ by successively solving the m stages of (5.1.1) for η_ν^μ, $\mu=1(1)m$. In the implicit case, the existence of a unique value $\boldsymbol{\eta}_\nu$ which satisfies (5.1.1) for given $\boldsymbol{\eta}_{\nu-k+\kappa}$, $\kappa=0(1)k-1$, is guaranteed if h is sufficiently small and the $\eta_{\nu-k+\kappa}^\mu$, $\kappa=0(1)k-1$, $\mu=1(1)m$, are sufficiently close to a value $z(t)$ of the true solution. The result follows from the uniform Lipschitz condition (2.1.6) on f in complete analogy to Lemma 3.1.1 and Lemma 4.1.2; we omit the precise formulation since the global existence of a unique solution for sufficiently small h of the discretization generated by a consistent and stable multistage multistep method also follows from our general Theorem 1.2.3.

To indicate the scope of the discussion in this chapter we now give a number of typical examples of m-stage k-step procedures which have been suggested in the literature. The first two examples are to show that RK-procedures and linear k-step procedures form natural simple subclasses of the general procedure (5.1.1).

Examples. 1. *RK-procedures:* An m-stage (see Def. 3.1.1) RK-procedure with the $(m+1)\times m$ generating matrix[1] B is an $(m+1)$-stage 1-step procedure (5.1.1) with

$$A_0=\begin{pmatrix}0 & \dots & 0 & -1\\ \vdots & & \vdots & \vdots\\ 0 & \dots & 0 & -1\end{pmatrix}, \qquad A_1=I,$$

$$B_0=0, \qquad B_1=\begin{pmatrix} & & 0\\ & B & \vdots\\ & & 0\end{pmatrix}.$$

The definitions 3.1.2 and 5.1.1 of "explicit" and "implicit" coincide. If the first row of B vanishes (as it always does for explicit RK-procedures) a simpler representation of the same RK-procedure is given by the m-stage 1-step procedure with A_0 and A_1 as above ($m\times m$ only) but with

$$B_0=\begin{pmatrix}0 & \dots & 0 & b_{21}\\ \vdots & & \vdots & \vdots\\ 0 & \dots & 0 & b_{m+1,1}\end{pmatrix}, \qquad B_1=\begin{pmatrix}b_{22} & \dots & b_{2m} & 0\\ \vdots & & \vdots & \vdots\\ b_{m+1,2} & \dots & b_{m+1,m} & 0\end{pmatrix}$$

where $b_{\mu\lambda}$, $\mu=2(1)m+1$, $\lambda=1(1)m$, are the elements of the generating matrix B. Thus the historical way of counting the number of stages of RK-procedures is actually appropriate for all explicit and some implicit RK-procedures.

2. *Linear k-step procedures:* Trivially, these are 1-stage k-step procedures (5.1.1) with the 1×1-"matrices"

$$A_\kappa=\alpha_\kappa, \qquad B_\kappa=\beta_\kappa, \qquad \kappa=0(1)k.$$

---

[1] Obviously, it would have been more coherent—but contrary to the usage in the literature—to denote the generating matrices by B in Chapter 3.

3. *Predictor-corrector procedures:* The procedure (4.1.12), with $\eta_\nu^{(0)}$ computed from a k-step predictor with coefficients $\alpha_\kappa^{(0)}$, $\beta_\kappa^{(0)}$, is an $(m+1)$-stage k-step procedure (5.1.1) with

$$A_\kappa = \begin{pmatrix} 0 & \cdots & 0 & \alpha_\kappa^{(0)} \\ \vdots & & \vdots & \alpha_\kappa \\ \vdots & & \vdots & \vdots \\ 0 & \cdots & 0 & \alpha_\kappa \end{pmatrix}, \qquad \qquad A_k = \begin{pmatrix} \alpha_k^{(0)} & & & 0 \\ & \alpha_k & & \\ 0 & & \ddots & \\ & & & \alpha_k \end{pmatrix},$$

$$\kappa = 0(1)k-1,$$

$$B_\kappa = \begin{pmatrix} 0 & \cdots & 0 & \beta_\kappa^{(0)} \\ \vdots & & \vdots & \beta_\kappa \\ \vdots & & \vdots & \vdots \\ 0 & \cdots & 0 & \beta_\kappa \end{pmatrix}, \qquad \qquad B_k = \begin{pmatrix} 0 & & & 0 \\ \beta_k & \ddots & & \\ & \ddots & \ddots & \\ 0 & & \beta_k & 0 \end{pmatrix}.$$

Obviously, this is an *explicit* procedure according to Def. 5.1.1 although it employs a corrector; cf. Remark 2 after Lemma 4.1.2.

4. *Cyclic k-step procedures:* k successive steps of a k-cyclic linear k-step method (cf. Section 4.3.1) may be represented as one k-stage 1-step procedure (5.1.1), with

$$A_0 = \begin{pmatrix} \alpha_0^1 & \cdots\cdots & \alpha_{k-1}^1 \\ & \alpha_0^2 \; \cdots & \alpha_{k-2}^2 \\ 0 & \ddots & \vdots \\ & & \alpha_0^k \end{pmatrix}, \quad A_1 = \begin{pmatrix} \alpha_k^1 & & & 0 \\ \alpha_{k-1}^2 & \alpha_k^2 & & \\ \vdots & & \ddots & \\ \alpha_1^k & \cdots\cdots & & \alpha_k^k \end{pmatrix},$$

$$B_0 = \begin{pmatrix} \beta_0^1 & \cdots\cdots & \beta_{k-1}^1 \\ & \ddots & \vdots \\ 0 & \ddots & \vdots \\ & & \beta_0^k \end{pmatrix}, \quad B_1 = \begin{pmatrix} \beta_k^1 & & 0 \\ \vdots & \ddots & \\ \vdots & & \ddots \\ \beta_1^k & \cdots\cdots & \beta_k^k \end{pmatrix}.$$

5. The following explicit 3-stage 1-step procedure from Butcher [1] is of order 4 but needs only 3 f-evaluations per step:

$$A_0 = \begin{pmatrix} 0 & 0 & -1 \\ 0 & 0 & -1 \\ 0 & 0 & -1 \end{pmatrix}, \quad A_1 = I,$$

$$B_0 = \begin{pmatrix} -\frac{1}{4} & 0 & \frac{3}{4} \\ 1 & 0 & -2 \\ 0 & 0 & \frac{1}{6} \end{pmatrix}, \quad B_1 = \begin{pmatrix} 0 & 0 & 0 \\ 2 & 0 & 0 \\ \frac{2}{3} & \frac{1}{6} & 0 \end{pmatrix}.$$

5.1.2 Simple m-stage k-step Methods

Def. 5.1.2. A forward step method $\mathfrak{M} = \{E_n, E_n^0, \Delta_n, \Delta_n^0, \varphi_n\}_{n\in\mathbb{N}'}$ applicable to IVP1 is called a *simple m-stage k-step method* if

it is an m-stage k-step method in the sense of Defs. 2.1.8—2.1.10;

it has an equidistant grid sequence $\{\mathbb{G}_n\}_{n\in\mathbb{N}'}$, and the vectors $\mathbf{t}_\nu = (t_\nu^\mu) \in \mathbb{R}^m$ (see (2.1.33)) are given by

$$(5.1.6) \qquad t_\nu^\mu := t_{\nu-1} + \tau_\mu h = (\nu - 1 + \tau_\mu)h, \quad \mu = 1(1)m, \quad \nu = 0(1)n,$$

where $\tau = (\tau_1, \ldots, \tau_m)^T \in \mathbb{R}^m$ is a fixed vector associated with $\mathfrak{M}$;

(5.1.7) $$(\varphi_n(F)\eta)(t_\nu) = \frac{1}{h}\sum_{\kappa=0}^{k}[\mathbf{A}_\kappa \boldsymbol{\eta}(t_{\nu-k+\kappa}) - h\mathbf{B}_\kappa \mathbf{f}(\boldsymbol{\eta}(t_{\nu-k+\kappa}))], \quad \nu = k(1)n,$$

with fixed matrices A_κ and B_κ, $\kappa = 0(1)k$.

Remarks. 1. Simple m-stage k-step methods are "uniform" in the sense that they use one and the same forward step procedure (5.1.1), cf. Def. 3.3.1 and Def. 4.2.1. However, a considerable degree of non-uniformity may be incorporated into (5.1.1) as we have seen (Example 4 in Section 5.1.1). The restriction to equidistant grids is a simplifying assumption for the general discussion; for many special classes of m-stage k-step methods satisfactory step change procedures can be constructed.

2. Note that we associate the values of all stages of $(\Delta_n y)(t_\nu)$ with the gridpoint t_ν (see (2.1.15)) although they are values of y at $t_{\nu-1}+\tau_\mu h$. This is merely a notational matter.

For the derivation of the consistency condition it is convenient to introduce the vector (see (5.1.5) for e)

(5.1.8) $$c := \left(\sum_{\kappa=0}^{k} B_\kappa\right) e =: (c_1, \ldots, c_m)^T \in \mathbb{R}^m$$

and to define[2]

(5.1.9) $$\Delta_n^0 \begin{pmatrix} d_0 \\ d(t) \end{pmatrix}(t_\nu) = c \times d(t_{\nu-k}) \quad \text{for } \nu \geq k;$$

the $\times$-product of elements from $\mathbb{R}^m$ and $\mathbb{R}^s$ has been defined in (3.1.15).

Theorem 5.1.1. *A simple m-stage k-step method $\mathfrak{M}$ is consistent with an* IVP 1 *at any $y \in E$ (in the domain of f) if and only if the starting procedure of $\mathfrak{M}$ is consistent with the* IVP 1 *and*

(5.1.10) $$\sum_{\kappa=0}^{k} A_\kappa((\kappa-1)e+\tau) = \sum_{\kappa=0}^{k} B_\kappa e = c.$$

Proof. a) The consistency of the starting procedure (cf. Def. 2.1.9 and Def. 2.2.1) implies

(5.1.11) $$\lim_{n\to\infty} [F_n \Delta_n y - \Delta_n^0 F y](t_\nu) = 0 \quad \text{for } \nu = 0(1)k-1$$

with any reasonable choice of $\Delta_n^0 \begin{pmatrix} d_0 \\ d(t) \end{pmatrix}(t_\kappa)$, $\kappa = 0(1)k-1$. For $\nu \geq k$,

[2] This definition of Δ_n^0 is not an extension of the more refined definition (4.2.1) for linear k-step methods. It does not generate a suitable concept of *order* of consistency; instead we shall use the approach of Def. 5.1.3.

we have from (5.1.7), (5.1.9) and (2.1.15)/(5.1.6)

$$[F_n \Delta_n y - \Delta_n^0 F y](t_\nu)$$
$$= \frac{1}{h} \sum_{\kappa=0}^{k} [\mathbf{A}_\kappa \mathbf{y}(\mathbf{t}_{\nu-k+\kappa}) - h \mathbf{B}_\kappa \mathbf{f}(\mathbf{y}(\mathbf{t}_{\nu-k+\kappa}))] - c \times [y'(t_{\nu-k}) - f(y(t_{\nu-k}))]$$
$$= \frac{1}{h} \sum_{\kappa=0}^{k} \mathbf{A}_\kappa [e \times y(t_{\nu-k}) + h((\kappa-1)e+\tau) \times y'(t_{\nu-k}) + o(h)] - c \times y'(t_{\nu-k})$$
$$- \sum_{\kappa=0}^{k} \mathbf{B}_\kappa [e \times f(y(t_{\nu-k})) + O(h)] + c \times f(y(t_{\nu-k}))$$
$$= \left[\sum_{\kappa=0}^{k} A_\kappa((\kappa-1)e+\tau) - c \right] \times y'(t_{\nu-k}) + o(1),$$

where we have expanded all values of y and y' about $t_{\nu-k}$ and used the Lipschitz continuity of f and y and the continuity of y' as well as (5.1.5) and (5.1.8). Thus, by (5.1.10)

$$[F_n \Delta_n y - \Delta_n^0 F y](t_\nu) = o(1) \quad \text{for } \nu = k(1)n$$

which implies, together with (5.1.11), $\lim_{n\to\infty} \|F_n \Delta_n y - \Delta_n^0 F y\|_{E_n^0} = 0$.

b) If (5.1.10) does not hold we have, e. g., for $y' - 1 = 0$ and $y(t) = t$

$$\|F_n \Delta_n y - \Delta_n^0 F y\|_{E_n^0} \geq \frac{n-k}{n} \left\| \left[\sum_{\kappa=0}^{k} A_\kappa((\kappa-1)e+\tau) - c \right] \times \begin{pmatrix} 1 \\ \vdots \\ 1 \end{pmatrix} \right\|_{\mathbb{R}^{ms}} > 0 .$$

The necessity of the consistency of the starting procedure is trivial for the norm (2.2.3). □

Remark. We could have required $c = e$ in (5.1.8) and (5.1.10) to normalize each stage of the m-stage k-step procedure (5.1.5). While we will actually use this normalization in general, the normalization $a_{\mu\mu}^k = 1$ is widespread in the literature and used exclusively with RK-methods (see Example 1 in Section 5.1.1).

Examples (cf. the examples at the end of Section 5.1.1).

1. *RK-methods:* Here we have from Section 3.1.4

$$\tau = \Phi((1,1), B) = Be = c = \left(\sum_{\lambda=1}^{m} b_{\mu\lambda} \right)$$

and (5.1.10) becomes

$$(1 - \alpha(\psi))e + c = c$$

with $\alpha(\psi) = \Phi^{m+1}((1,1), B) = \hat{\Phi}_1^{(1)}(B) = \sum_\lambda b_{m+1,\lambda}$, the step factor of the underlying RK-scheme (Def. 3.2.3). Thus we have $\alpha(\psi) = 1$ for consistency as shown in Theorem 3.3.5.

2. *Uniform linear k-step methods:* Here we have $m=1$ and $\tau=1$ and (5.1.10) becomes simply

$$\sum_{\kappa=1}^{k} \kappa\alpha_\kappa = \sum_{\kappa=0}^{k} \beta_\kappa$$

or $\rho'(1)=\sigma(1)$ ($=1$ with our normalization), cf. (4.1.27) and Theorem 4.2.1.

Normally the various stages l_ν^μ, $\mu=1(1)m$, of the values of the *local discretization error*

$$\mathbf{l}_\nu = l_n(t_\nu) = \frac{1}{h} \sum_{\kappa=0}^{k} [\mathbf{A}_\kappa \mathbf{z}(\mathbf{t}_{\nu-k+\kappa}) - h\,\mathbf{B}_\kappa \mathbf{z}'(\mathbf{t}_{\nu-k+\kappa})]$$

of a simple m-stage k-step method will tend to zero like different powers of h.

Def. 5.1.3. The *μ-th stage* of a simple m-stage k-step method is said to be *consistent of order* q_μ, $\mu=1(1)m$, if

$$\|l_\nu^\mu\| = O(h^{q_\mu}) \quad \text{for } \nu=k(1)n \tag{5.1.12}$$

for any sufficiently smooth IVP1 and if the q_μ are the largest integers such that (5.1.12) holds.

Def. 5.1.4. For a simple m-stage k-step method let $q := \max_{\mu=1(1)m} q_\mu$, with q_μ from (5.1.12), and for some $y \in D^{(q+1)}$ form the expansion

$$\frac{1}{h}\sum_{\kappa=0}^{k}\left[A_\kappa\begin{pmatrix}\vdots\\ y(t+(\kappa-1+\tau_\mu)h)\\ \vdots\end{pmatrix} - h\,B_\kappa\begin{pmatrix}\vdots\\ y'(t+(\kappa-1+\tau_\mu)h)\\ \vdots\end{pmatrix}\right] = \begin{pmatrix}\vdots\\ h^{q_\mu} c_\mu C_\mu y^{(q_\mu+1)}(t)\\ \vdots\end{pmatrix} + \begin{pmatrix}\vdots\\ O(h^{q_\mu+1})\\ \vdots\end{pmatrix} \tag{5.1.13}$$

where the c_μ have been defined in (5.1.8). The constants C_μ, $\mu=1(1)m$, are called *error constants* of $\mathfrak{M}$.

Many m-stage k-step methods are constructed in such a fashion that the orders of approximation achieved by some or all of the stages of the solution ζ_n of the discretizations $\mathfrak{M}$(IVP1) are higher than some or all of the q_μ; this is well-known with RK-methods. The order of consistency as defined in Def. 1.1.6 is therefore of little value with m-stage methods except that it gives a lower bound on the order of convergence which is achieved by all of the stages.

5.1.3 Stability and Convergence of Simple m-stage k-step Methods

It is to be expected from the general results in Section 2.2.5 and from the analysis in Section 4.2.3 that the stability of a simple m-stage k-step method will depend only on the matrices A_κ, $\kappa=0(1)k$. More specifically it should depend on whether each solution of the system of m linear difference equations ($\det A_k \neq 0$)

$$\sum_{\kappa=0}^{k} A_\kappa \varepsilon_{\nu-k+\kappa} = 0, \qquad \nu \leq k, \tag{5.1.14}$$

with given initial values $\varepsilon_\kappa = \delta_\kappa$, $\kappa = 0(1)k-1$, remains bounded uniformly in ν in terms of the initial values:

$$\|\varepsilon_\nu\| \leq \Gamma_0 \max_{\kappa=0(1)k-1} \|\delta_\kappa\|, \qquad \nu \geq k. \tag{5.1.15}$$

Lemma 5.1.2. *A uniform bound* (5.1.15) *for the solutions of* (5.1.14) *exists if and only if the polynomial*

$$\rho(x) := \det\left(\sum_{\kappa=0}^{k} A_\kappa x^\kappa\right) \tag{5.1.16}$$

satisfies the following criterion:

(i) *There are no zeros of ρ outside the closed unit disk.*

(ii) *For each r_j-fold zero x_j of ρ on the unit circle there exist r_j linearly independent solutions $\{\xi_\nu^i\}_{\nu\geq 0}$ of* (5.1.14) *which satisfy*

$$\xi_\nu^i = x_j^\nu \xi_0^i, \qquad \xi_\nu^i \in \mathbb{R}^m. \tag{5.1.17}$$

Proof. Since $\det A_k \neq 0$ was assumed, we can write (5.1.14) in the form

$$\begin{pmatrix} \varepsilon_{\nu-k+1} \\ \vdots \\ \vdots \\ \varepsilon_{\nu-1} \\ \varepsilon_\nu \end{pmatrix} = \begin{pmatrix} 0 & I & & & 0 \\ & 0 & \ddots & & \\ & & \ddots & \ddots & \\ & & & & I \\ -A_k^{-1}A_0 & \cdots & \cdots & & -A_k^{-1}A_{k-1} \end{pmatrix} \begin{pmatrix} \varepsilon_{\nu-k} \\ \vdots \\ \vdots \\ \vdots \\ \varepsilon_{\nu-1} \end{pmatrix} =: Q \begin{pmatrix} \varepsilon_{\nu-k} \\ \vdots \\ \vdots \\ \vdots \\ \varepsilon_{\nu-1} \end{pmatrix}.$$

Thus, an estimate (5.1.15) holds if and only if the powers of Q are uniformly bounded and this is the case when there are no eigenvalues of Q outside the unit disk and the eigenvalues on the unit circle are simple or have linear elementary divisors if they are multiple. In the latter case, the number of linearly independent eigenvectors of Q for that eigenvalue x_i equals the multiplicity of x_i; but due to the structure of Q all these eigenvectors have to be of the form $(\xi_0^i, x_i\xi_0^i, \ldots, x_i^{k-1}\xi_0^i)^T$.

It follows from a well-known determinantal identity (or by an argument analogous to that used in the proof of Lemma 4.3.2) that

$$\begin{aligned} \rho(x) &= \det\left(\sum_{\kappa=0}^{k} A_\kappa x^\kappa\right) = \det(A_k)\det\left(\sum_{\kappa=0}^{k} A_k^{-1}A_\kappa x^\kappa\right) \\ &= \det(A_k)\det(xI - Q), \end{aligned} \tag{5.1.18}$$

i. e., the zeros of ρ are the eigenvalues of Q. Thus, the necessary and sufficient conditions (i) and (ii) on the zeros of ρ follow immediately from the above-stated conditions for the power-boundedness of Q. ▯

Def. 5.1.5. The polynomial (5.1.16) of degree mk is called the *associated polynomial* of the simple m-stage k-step method with coefficient matrices A_κ, $\kappa=0(1)k$.

Remark. Since powers of x which may be factored out from ρ do not affect the stability situation (see Lemma 5.1.2) one often drops factors x^l, $l>0$, in ρ and still calls the remaining polynomial the associated polynomial of the m-stage k-step method.

Lemma 5.1.3. *A simple m-stage k-step method* $\mathfrak{M}$ *is stable for the trivial IVP1* $\mathfrak{P}_0$ *(see* (4.2.24)) *if and only if its associated polynomial satisfies the criterion of Lemma* 5.1.2.

Proof. The stability of $\mathfrak{M}(\mathfrak{P}_0)$ is equivalent to the fact that the system of m inhomogeneous linear difference equations

$$\sum_{\kappa=0}^{k} A_\kappa \varepsilon_{\nu-k+\kappa} = h\delta_\nu, \qquad \nu \ge k, \tag{5.1.19}$$

with initial values $\varepsilon_\kappa=\delta_\kappa$, $\kappa=0(1)k-1$, admits a bound

$$\|\varepsilon_\nu\| \le \Gamma\left[\max_{\kappa=0(1)k-1} \|\delta_\kappa\| + h\sum_{\nu'=k}^{\nu} \|\delta_{\nu'}\|\right], \qquad \nu\ge k. \tag{5.1.20}$$

We proceed as in Section 4.2.2 and represent the solution of (5.1.19) by superposition of solutions of the homogeneous equation: If all δ_ν vanish except δ_{ν_0} we have to solve (5.1.14) with initial conditions

$$\varepsilon_{\nu_0-\kappa} = \begin{cases} 0, & \kappa=1(1)k-1, \\ h A_k^{-1}\delta_{\nu_0}, & \kappa=0, \end{cases}$$

so that $\|\varepsilon_\nu\| \le \Gamma_0 h \|A_k^{-1}\| \, \|\delta_{\nu_0}\|$ for $\nu \ge \nu_0$ by (5.1.15). Superposition yields (5.1.20) with $\Gamma=\Gamma_0 \max(\|A_k^{-1}\|, 1)$.

The necessity of our criterion follows from the necessity for the estimate (5.1.15) in Lemma 5.1.2. ▯

Theorem 5.1.4. *A simple m-stage k-step method is stable for an arbitrary IVP1 if and only if its associated polynomial satisfies the criterion of Lemma* 5.1.2.

Proof. The proof of Theorem 4.2.7 may be repeated almost literally: The mappings G_n are now given by

$$(G_n\eta)(t_\nu) = \begin{cases} s_{n\nu}^{(0)}(z_0) - s_{n\nu}(z_0), & \nu=0(1)k-1, \\ -\sum_{\kappa=0}^{k} \mathbf{B}_\kappa \mathbf{f}(\boldsymbol{\eta}_{\nu-k+\kappa}), & \nu=k(1)n; \end{cases} \tag{5.1.21}$$

they satisfy, for all $\eta^{(1)}, \eta^{(2)} \in B_R(z)$

$$\begin{aligned}(5.1.22)\qquad \|G_n\eta^{(1)} - G_n\eta^{(2)}\|_{E_n^0}^{[\nu]} &\leq h \sum_{\mu=k}^{\nu} \left\| \sum_{\kappa=0}^{k} \mathbf{B}_\kappa(\mathbf{f}(\boldsymbol{\eta}^{(1)}_{\mu-k+\kappa}) - \mathbf{f}(\boldsymbol{\eta}^{(2)}_{\mu-k+\kappa})) \right\| \\ &\leq Lh\left(\sum_{\kappa=0}^{k} \|B_\kappa\|\right) \sum_{\mu=0}^{\nu} \|\boldsymbol{\eta}^{(1)}_\mu - \boldsymbol{\eta}^{(2)}_\mu\|\end{aligned}$$

since $\|\mathbf{f}(\boldsymbol{\eta}^{(1)}_\mu) - \mathbf{f}(\boldsymbol{\eta}^{(2)}_\mu)\| \leq L\|\boldsymbol{\eta}^{(1)}_\mu - \boldsymbol{\eta}^{(2)}_\mu\|$ by (2.1.31), (2.1.32) and (2.1.6). Thus Corollary 2.2.7 implies the assertion as before. The necessity follows from Lemma 5.1.3. □

Examples.

1. *RK-methods:*

$$\rho(x) = \det\begin{pmatrix} x & 0 & & -1 \\ & \ddots & & \vdots \\ & & x & -1 \\ 0 & & & x-1 \end{pmatrix} = (x-1)x^m$$

which reconfirms the automatic stability of RK-methods.

2. *Uniform linear k-step methods:* ρ is the associated polynomial of the underlying linear k-step scheme; the conditions of Lemma 5.1.2 reduce to the root criterion for $m=1$.

It is an immediate consequence of Theorem 1.2.4 that a simple m-stage k-step method converges for an arbitrary IVP 1 if it satisfies the conditions of Theorems 5.1.1 and 5.1.4. Note that convergence means

$$(5.1.23)\qquad \lim_{n\to\infty} \max_{\nu=0(1)n} \|(\zeta_n)^\mu_\nu - z(t^\mu_\nu)\|_{\mathbb{R}^s} = 0, \qquad \mu = 1(1)m,$$

since we used the maximum norm in (2.1.32) for our analysis; i.e., each stage of the values of the solution of the discretization converges to the value of the true solution at the appropriate t^μ_ν.

Def. 5.1.6. For a given IVP 1 the *μ-th stage* of a simple m-stage k-step method is said to be *convergent of order* p_μ, $\mu = 1(1)m$, if

$$(5.1.24)\qquad \max_{\nu=0(1)n} \|(\zeta_n)^\mu_\nu - z(t^\mu_\nu)\|_{\mathbb{R}^s} = O(h^{p_\mu}).$$

It is clear from Theorem 1.2.4 that

$$p_\mu \geq \min_{\lambda=1(1)m} q_\lambda =: \bar{q}$$

where the q_λ are the stagewise orders of consistency, see Def. 5.1.3. However, with many m-stage methods some or all of the p_μ are strictly larger than $\bar{q}$ and it may even happen that some of the p_μ satisfy $p_\mu > \max q_\lambda$. For an important class of simple m-stage k-step methods we will obtain detailed results on the relation between the p_μ and the q_μ in Theorem 5.2.3. The analysis presented there may be extended to some other situations.

5.2 Predictor-corrector Methods

5.2.1 Characterization, Subclasses

Def. 5.2.1. An m-stage k-step procedure (5.1.1) is called an m-stage k-step *predictor-corrector procedure (PC-procedure)* if it is explicit and if

$$A_\kappa = \begin{pmatrix} 0 \dots 0 & a_{1m}^\kappa \\ \vdots \quad \vdots & \vdots \\ 0 \dots 0 & a_{mm}^\kappa \end{pmatrix}, \quad \kappa = 0(1)k-1, \qquad A_k = \begin{pmatrix} a_{11}^k & & 0 \\ & \ddots & \\ 0 & & a_{mm}^k \end{pmatrix}. \tag{5.2.1}$$

It is called a *straight* m-stage k-step PC-procedure if in addition

$$B_\kappa = \begin{pmatrix} 0 \dots 0 & b_{1m}^\kappa \\ \vdots \quad \vdots & \vdots \\ 0 \dots 0 & b_{mm}^\kappa \end{pmatrix}, \qquad \kappa = 0(1)k-1. \tag{5.2.2}$$

Def. 5.2.2. A simple m-stage k-step method which employs a (straight) m-stage k-step PC-procedure as its forward step procedure is called a *(straight)* m-stage k-step *predictor-corrector method (PC-method)* if $\tau_m = 1$ in (5.1.6).

Straight k-step PC-methods have the simplest structure of all m-stage k-step methods since only the values η_ν^m of the last stage are used in the further computation, see (5.2.1) and (5.2.2); they may thus also be considered as 1-stage k-step methods (but not as *simple* 1-stage k-step methods!). For $k=1$ we have precisely the explicit RK-methods (uniform and with equidistant grids), see Example 1 in Section 5.1.1; thus, straight PC-methods represent a generalization of the structure of explicit RK-methods to k-step methods.

If the PC-procedure is not straight, some values of $f(\eta_\nu^\mu)$, $\mu < m$, but not values of the η_ν^μ themselves are used in the further computation. In any case, the values ζ_ν^m of the m-stage of the solution ζ of a predictor-corrector discretization are normally regarded as *the* approximation to the values of the true solution.

Def. 5.2.3. An m-stage PC-method $\mathfrak{M}$ is said to be *of order* p if, for any IVP$^{(p)}$1, the m-th stage of $\mathfrak{M}$ is convergent of order p (cf. Def. 5.1.6).

Two widely used PC-procedures arise from solving a k-step corrector procedure in the iterative fashion (4.1.12):

Def. 5.2.4. For a given k-step predictor scheme ψ^0 and a k-step corrector scheme ψ the $(m+1)$-stage k-step PC-procedure with

$$A_\kappa = \begin{pmatrix} 0 \dots\dots 0 & \alpha_\kappa^0 \\ \vdots \qquad \vdots & \alpha_\kappa \\ \vdots \qquad \vdots & \vdots \\ 0 \dots \quad 0 & \alpha_\kappa \end{pmatrix}, \quad \kappa = 0(1)k-1, \qquad A_k = \begin{pmatrix} \alpha_k^0 & & & 0 \\ & \alpha_\kappa & & \\ & & \ddots & \\ 0 & & & \alpha_k \end{pmatrix}, \tag{5.2.3}$$

is called the $\left.\begin{matrix}\text{P(EC)}^m\text{E}\\ \text{P(EC)}^m\end{matrix}\right\}$-*procedure*[3] based on ψ^0 and ψ if

$$(5.2.4)\qquad B_\kappa=\begin{cases}\begin{pmatrix}0\dots & 0 & \beta_\kappa^0\\ \vdots & \vdots & \beta_\kappa\\ \vdots & \vdots & \vdots\\ 0\dots & 0 & \beta_\kappa\end{pmatrix}\\[2ex] \begin{pmatrix}0\dots 0 & \beta_\kappa^0 & 0\\ \vdots\quad\vdots & \beta_\kappa & \vdots\\ \vdots\quad\vdots & \vdots & \vdots\\ 0\dots 0 & \beta_\kappa & 0\end{pmatrix}\end{cases},\quad \kappa=0(1)k-1,\quad B_k=\begin{pmatrix}0 & & & 0\\ \beta_k & \ddots & & \\ & \ddots & \ddots & \\ 0 & & \beta_k & 0\end{pmatrix},$$

where

$$(5.2.5)\qquad \begin{pmatrix}\alpha_0^0 \dots\dots\dots & \alpha_k^0\\ \beta_0^0\dots\ \beta_{k-1}^0 & 0\end{pmatrix}\in\psi^0,\qquad \begin{pmatrix}\alpha_0\dots\dots\alpha_k\\ \beta_0\dots\dots\beta_k\end{pmatrix}\in\psi.$$

The same designation is used for k-step methods employing P(EC)^{m}E- or P(EC)^m-procedures and the vector (see (5.1.6))

$$(5.2.6)\qquad \tau=e\in\mathbb{R}^{m+1}.$$

Remarks. 1. To conform with common practice we will count the $m+1$ stages η_ν^μ of a P(EC)^{m}E- or P(EC)^m-procedure from 0 to m rather than from 1 to $m+1$. Thus, η_ν^0 will be the "predicted value" and η_ν^μ, $\mu=1(1)m$, the "corrected values".

2. To include the possibility of ψ^0 and ψ having different values of k we relax the last condition (4.1.3) on linear k-step schemes by permitting $\alpha_0^2=\beta_0^2=0$ in either the predictor or the corrector. Thus, one of the two generating matrices (5.2.5) may not be the standard representative M_0 of its linear k-step scheme, and the associated polynomials $\rho(x)=\sum_{\kappa=0}^{k}\alpha_\kappa x^\kappa$, $\sigma(x)=\sum_{\kappa=0}^{k}\beta_\kappa x^\kappa$, with coefficients from (5.2.5), may have the common divisor x^l; see the discussion in Section 4.1.1. In any case, the associated polynomials of ψ^0 and ψ are assumed to be of the *same* degree k in a k-step P(EC)^{m}E- or P(EC)^m-procedure.

Note that a P(EC)^{m}E-method is a straight PC-method while a P(EC)^m-method is not; but the storage requirements are the same for both kinds of PC-methods since the B_κ, $\kappa=0(1)k-1$, contain only one nonvanishing column in each case, see (5.2.4). On the other hand, one step of a P(EC)^{m}E-method requires $m+1$ evaluations of f against m

[3] The letters *P*, *E*, and *C* refer to *P*rediction, *E*valuation (of f), and *C*orrection, $P(EC)^m$ is short for $PECEC\dots EC$.

evaluations for a P(EC)^m-method in which the values $f(\eta_\nu^m)$ are never used and thus need not be evaluated. (P(EC)^m-methods are sometimes called "simplified" PC-methods for that reason.) In practical applications, m is typically 1 or 2.

A slight generalization of Def. 5.2.4 which may result in an increase in the size of the stability regions (see Section 5.5.2) is obtained by forming a final value η_ν^{m+1} as a linear combination of the η_ν^μ, $\mu=0(1)m$ (see Remark 1 above). Such predictor-corrector methods would be k-step extensions of such explicit RK-methods whose generating matrix has non-zero entries off the subdiagonal in the last row only, like the classical RK 4-method.

The most restrictive aspect of Def. 5.2.4 is really (5.2.6) which arises naturally from the fact that all the predicted and corrected values η_ν^μ approximate the true solution at the gridpoint t_ν due to the structure of (5.2.4). An important class of PC-methods which do not satisfy this restriction ("hybrid methods", "methods with off-step points") will be considered in more detail in Section 5.3.

5.2.2 Stability and Order of Predictor-corrector Methods

Theorem 5.2.1. *An m-stage k-step PC-method is stable for an arbitrary IVP1 if and only if the polynomial*

$$\rho_m(x) := \sum_{\kappa=0}^{k} a_{mm}^\kappa x^\kappa \tag{5.2.7}$$

satisfies the root criterion.

Proof. The associated polynomial of an m-stage k-step PC-method is (see (5.1.16) and (5.2.1))

$$\rho(x) = \det\begin{pmatrix} a_{11}^k x^k & & 0 & \sum_{\kappa=0}^{k-1} a_{1m}^\kappa x^\kappa \\ & \ddots & & \vdots \\ & & \alpha_{m-1,m-1}^k x^k & \sum_{\kappa=0}^{k-1} a_{m-1,m}^\kappa x^\kappa \\ 0 & & & \sum_{\kappa=0}^{k} a_{m,m}^\kappa x^\kappa \end{pmatrix}$$

$$= \left(\prod_{\mu-1}^{m-1} a_{\mu\mu}^k\right) x^{(m-1)k} \rho_m(x)$$

so that the assertion follows from Theorem 5.1.4. It is easily checked that the special situation in (ii) of Lemma 5.1.2 (which permits multiple zeros of ρ on the unit circle) cannot arise for PC-methods. $\square$

Theorem 5.2.1 shows immediately that the predictor scheme used in a $P(EC)^m E$- or $P(EC)^m$-method *need not be D-stable*. Also, in a more general PC-method the multistep procedures used in *all stages but the last one* need not be D-stable. On the other hand, the fact that the k-step procedure used in the last stage has to be D-stable brings back the restrictions of Theorem 4.2.11 in the case of $P(EC)^m E$- or $P(EC)^m$-methods which are subject to (5.2.6) so that the last stage of such methods cannot be consistent of an order greater than $2[k/2]+2$.

From the special structure (5.2.1) of the A_κ in a PC-method it follows that a perturbation which occurs in the μ-th stage, $\mu<m$, can affect the m-th stage only via the $b^\kappa_{m\mu}$ and is thus "damped" by powers of h. For a formal treatment of this situation we introduce the following notation: Let

$$|b| := (|b_\mu|) \quad \text{for a vector } b \in \mathbb{R}^m \text{ with components } b_\mu,$$
$$|B| := (|b_{\mu\lambda}|) \quad \text{for a matrix } B \text{ with elements } b_{\mu\lambda}.$$

For a given m-stage k-step predictor-corrector procedure let $\bar{B}$ and $\bar{B}'$ be the upper $(m-1)\times(m-1)$ minors of $\sum_{\kappa=0}^{k} |B_\kappa|$ and $\sum_{\kappa=0}^{k-1} |B_\kappa|$ resp., $\bar{b}$ and $\bar{b}'$ the vectors

$$\sum_{\kappa=0}^{k \text{ resp. } k-1} (|b^\kappa_{m1}|, |b^\kappa_{m2}|, \ldots, |b^\kappa_{m,m-1}|)^T \in \mathbb{R}^{m-1}.$$

Furthermore, define

$$\text{(5.2.8)} \qquad \begin{aligned} \bar{d}^T &:= h\bar{b}^T(I-h\bar{B})^{-1} =: (\bar{d}_1, \ldots, \bar{d}_{m-1}) \in \mathbb{R}^{m-1}, \\ \bar{d}^T_s &:= h\bar{d}^T\bar{B}' + h\bar{b}'^T =: (d'_1, \ldots, d'_{m-1}) \in \mathbb{R}^{m-1}. \end{aligned}$$

For sufficiently small $h>0$ the $\bar{d}_\mu$ and d'_μ are defined as rational functions of h which are regular at 0; hence there are maximal nonnegative integers γ_μ and γ'_μ such that

$$\text{(5.2.9)} \qquad \bar{d}_\mu = O(h^{\gamma_\mu}), \quad d'_\mu = O(h^{\gamma'_\mu}), \quad \mu = 1(1)m-1.$$

(If some $d'_\mu \equiv 0$ we set $\gamma'_\mu = \infty$ and define $h^\infty := 0$.) Finally, we set $\gamma_m := \gamma'_m := 0$.

Theorem 5.2.2. *Consider an m-stage k-step predictor-corrector method* $\mathfrak{M}$ *and the discretization* $\mathfrak{D} = \mathfrak{M}(\text{IVP}\,1) = \{E_n, E^0_n, F_n\}_{n\in\mathbb{N}'}$, *for some IVP1. If the associated polynomial* ρ_m *of* (5.2.7) *satisfies the root criterion, then there exists a constant S independent of n such that for any elements* $\eta_1, \eta_2 \in B_R(z)$ *(see* (2.1.6))

(5.2.10)
$$\max_{\nu'=0(1)\nu} \|(\eta_1)^m_{\nu'} - (\eta_2)^m_{\nu'}\| \le S\Bigg[\sum_{\mu=1}^{m} h^{\gamma'_\mu} \max_{\kappa=0(1)k-1} \|(\eta_1)^\mu_\kappa - (\eta_2)^\mu_\kappa\|$$
$$+ h \sum_{\nu'=k}^{\nu} \sum_{\mu=1}^{m} h^{\gamma_\mu} \|(F_n\eta_1)^\mu_{\nu'} - (F_n\eta_2)^\mu_{\nu'}\|\Bigg], \qquad \nu = k(1)n.$$

Proof. We proceed as in Section 2.2.5 and in various previous stability proofs; F_n is the mapping in $\mathfrak{M}(\text{IVP}\,1)$, G_n is defined by

$$(G_n\eta)(t_\nu) := \begin{cases} 0, & \nu = 0(1)k-1, \\ \sum_{\kappa=0}^{k} \mathbf{B}_\kappa \mathbf{f}(\boldsymbol{\eta}_{\nu-k+\kappa}), & \nu = k(1)n. \end{cases}$$

For A_κ from (5.2.1), the m-th "stage" of the difference equation (5.1.19) is independent of the others; thus, for $\eta_1, \eta_2 \in B_R(z)$ we have from

$$\sum_{\kappa=0}^{k} \mathbf{A}_\kappa((\boldsymbol{\eta}_1)_{\nu-k+\kappa} - (\boldsymbol{\eta}_2)_{\nu-k+\kappa}) = h((F_n + G_n)(\eta_1 - \eta_2))_\nu, \qquad \nu \ge k,$$

by (5.1.20)

$$\|(\eta_1)^m_\nu - (\eta_2)^m_\nu\| \le \Gamma\Bigg[\max_{\kappa=0(1)k-1} \|(\eta_1)^m_\kappa - (\eta_2)^m_\kappa\|$$
$$+ h \sum_{\nu'=k}^{\nu} \|((F_n+G_n)\eta_1)^m_{\nu'} - ((F_n+G_n)\eta_2)^m_{\nu'}\|\Bigg]$$
$$\le \Gamma\Bigg[\max_\kappa \|(\eta_1)^m_\kappa - (\eta_2)^m_\kappa\| + h \sum_{\nu'=k}^{\nu} \|(F_n\eta_1)^m_{\nu'} - (F_n\eta_2)^m_{\nu'}\|\Bigg]$$
$$+ h\Gamma \sum_{\nu'=k}^{\nu} \Bigg\|\Bigg(\sum_{\kappa=0}^{k} \mathbf{B}_\kappa[\mathbf{f}((\boldsymbol{\eta}_1)_{\nu'-k+\kappa}) - \mathbf{f}((\boldsymbol{\eta}_2)_{\nu'-k+\kappa})]\Bigg)^m\Bigg\|$$
$$\le \Gamma[\ldots] + h\Gamma L \sum_{\mu=1}^{m-1} \sum_{\kappa=0}^{k} |b^\kappa_{m\mu}| \sum_{\nu'=k}^{\nu} \|(\eta_1)^\mu_{\nu'-k+\kappa} - (\eta_2)^\mu_{\nu'-k+\kappa}\|$$
$$+ h\Gamma L \sum_{\kappa=0}^{k-1} |b^\kappa_{mm}| \sum_{\nu'=k}^{\nu} \|(\eta_1)^m_{\nu'-k+\kappa} - (\eta_2)^m_{\nu'-k+\kappa}\|$$

(5.2.11)
$$\le \Gamma\Bigg[\Bigg(1 + hLk \sum_{\kappa=0}^{k-1} |b^\kappa_{mm}|\Bigg) \max_\kappa \|(\eta_1)^m_\kappa - (\eta_2)^m_\kappa\|$$
$$+ hL\Bigg(\sum_{\kappa=0}^{k-1} |b^\kappa_{mm}|\Bigg) \sum_{\nu'=k}^{\nu-1} \|(\eta_1)^m_{\nu'} - (\eta_2)^m_{\nu'}\|$$
$$+ hL \sum_{\mu=1}^{m-1} \Bigg(k\Bigg(\sum_{\kappa=0}^{k-1} |b^\kappa_{m\mu}|\Bigg) \max_\kappa \|(\eta_1)^\mu_\kappa - (\eta_2)^\mu_\kappa\|$$

$$+\left(\sum_{\kappa=0}^{k}|b_{m\mu}^{\kappa}|\right)\sum_{\nu'=k}^{\nu}\|(\eta_1)_{\nu'}^{\mu}-(\eta_2)_{\nu'}^{\mu}\|\Bigg)$$
$$+h\sum_{\nu'=k}^{\nu}\|(F_n\eta_1)_{\nu'}^{m}-(F_n\eta_2)_{\nu'}^{m}\|\Bigg].$$

In order to estimate $\sum_{\nu'=k}^{\nu}\|(\eta_1)_{\nu'}^{\mu}-(\eta_2)_{\nu'}^{\mu}\|$, we consider the μ-th stage of $F_n\eta_1-F_n\eta_2$ at $t_{\nu'}$:

$$
\begin{aligned}
|a_{\mu\mu}^{k}|\,\|(\eta_1)_{\nu'}^{\mu}-(\eta_2)_{\nu'}^{\mu}\| \le \sum_{\kappa=0}^{k-1}\big[&|a_{\mu m}^{\kappa}|\,\|(\eta_1)_{\nu'-k+\kappa}^{m}-(\eta_2)_{\nu'-k+\kappa}^{m}\| \\
&+h\,|b_{\mu m}^{\kappa}|\,\|f((\eta_1)_{\nu'-k+\kappa}^{m})-f((\eta_2)_{\nu'-k+\kappa}^{m})\|\big] \\
&+h\sum_{\lambda=1}^{m-1}\sum_{\kappa=0}^{k}|b_{\mu\lambda}^{\kappa}|\,\|f((\eta_1)_{\nu'-k+\kappa}^{\lambda})f((\eta_2)_{\nu'-k+\kappa}^{\lambda})\| \\
&+h\|(F_n\eta_1)_{\nu'}^{\mu}-(F_n\eta_2)_{\nu'}^{\mu}\|;
\end{aligned}
\tag{5.2.12}
$$

after using the Lipschitz condition on f we sum this inequality from $\nu'=k$ to ν:

$$
\begin{aligned}
&|a_{\mu\mu}^{k}|\sum_{\nu'=k}^{\nu}\|(\eta_1)_{\nu'}^{\mu}-(\eta_2)_{\nu'}^{\mu}\| \\
&\quad\le\left(\sum_{\kappa=0}^{k-1}\big[|a_{\mu m}^{\kappa}|+hL|b_{\mu m}^{\kappa}|\big]\right)\sum_{\nu'=k}^{\nu}\max_{\rho=\nu'-k(1)\nu'-1}\|(\eta_1)_{\rho}^{m}-(\eta_2)_{\rho}^{m}\| \\
&\quad+hL\sum_{\lambda=1}^{m-1}\left(\sum_{\kappa=0}^{k}|b_{\mu\lambda}^{\kappa}|\right)\sum_{\nu'=k}^{\nu}\|(\eta_1)_{\nu'}^{\lambda}-(\eta_2)_{\nu'}^{\lambda}\| \\
&\quad+hL\sum_{\lambda=1}^{m-1}k\left(\sum_{\kappa=0}^{k-1}|b_{\mu\lambda}^{\kappa}|\right)\max_{\kappa=0(1)k-1}\|(\eta_1)_{\kappa}^{\lambda}-(\eta_2)_{\kappa}^{\lambda}\| \\
&\quad+h\sum_{\nu'=k}^{\nu}\|(F_n\eta_1)_{\nu'}^{\mu}-(F_n\eta_2)_{\nu'}^{\mu}\|, \qquad \mu=1(1)m-1.
\end{aligned}
\tag{5.2.13}
$$

Now let $h_0>0$ be sufficiently small so that the $(m-1)\times(m-1)$-matrix

$$h\hat{B}:=hL\left(\frac{\sum_{\kappa=0}^{k}|b_{\mu\lambda}^{\kappa}|}{|a_{\mu\mu}^{k}|}\right), \qquad \mu,\lambda=1(1)m-1,$$

has a spectral radius smaller than 1. Then the system (5.2.13) of linear inequalities for $\sum_{\nu'}\|(\eta_1)_{\nu'}^{\mu}-(\eta_2)_{\nu'}^{\mu}\|$ implies for $h\le h_0$, with a suitable constant D,

$$\sum_{\nu'=k}^{\nu} \|(\eta_1)_{\nu'}^{\mu} - (\eta_2)_{\nu'}^{\mu}\| \le D \sum_{\nu'=k}^{\nu-1} \max_{\rho=0(1)\nu'} \|(\eta_1)_{\rho}^{m} - (\eta_2)_{\rho}^{m}\|$$

$$(5.2.14) \qquad + hL \sum_{\lambda=1}^{m-1} \hat{b}_{\mu\lambda}^{(s)} \max_{\kappa=0(1)k-1} \|(\eta_1)_{\kappa}^{\lambda} - (\eta_2)_{\kappa}^{\lambda}\|$$

$$+ h \sum_{\lambda=1}^{m-1} \hat{b}_{\mu\lambda} \sum_{\nu'=k}^{\nu} \|(F_n\eta_1)_{\nu'}^{\lambda} - (F_n\eta_2)_{\nu'}^{\lambda}\|,$$

where $\hat{b}_{\mu\lambda}$, $\mu, \lambda = 1(1)m-1$, are the elements of $(I - h\hat{B})^{-1}\,\mathrm{diag}(1/a_{\mu\mu}^{k})$ and $\hat{b}_{\mu\lambda}^{(s)}$ the elements of the matrix $(I - h\hat{B})^{-1}\left(\sum_{\kappa=0}^{k-1} |b_{\mu\lambda}^{\kappa}|/|a_{\mu\mu}^{k}|\right)$.

Upon substitution of (5.2.14) into (5.2.11) we replace $\hat{B}$ by $\bar{B}$, etc., and the $\bar{d}_\mu$ and d'_μ by h^{γ_μ} and $h^{\gamma'_\mu}$ since we are interested only in the powers of h involved. In this fashion we obtain for $h \le h_0$, with some constants S_0, S_1, S_2,

$$\|(\eta_1)_{\nu}^{m} - (\eta_2)_{\nu}^{m}\| \le S_0 \sum_{\mu=1}^{m} h^{\gamma'_\mu} \max_{\kappa=0(1)k-1} \|(\eta_1)_{\kappa}^{\mu} - (\eta_2)_{\kappa}^{\mu}\|$$

$$+ hS_1 \sum_{\nu'=k}^{\nu} \sum_{\mu=1}^{m} h^{\gamma_\mu} \|(F_n\eta_1)_{\nu'}^{\mu} - (F_n\eta_2)_{\nu'}^{\mu}\|$$

$$+ hS_2 \sum_{\nu'=k}^{\nu-1} \max_{\rho=k(1)\nu'} \|(\eta_1)_{\rho}^{m} - (\eta_2)_{\rho}^{m}\|.$$

We can now apply Corollary 2.1.4 and (2.1.28) (with trivial modifications) to obtain the assertion (5.2.10). □

Remark. For a straight PC-method $\bar{B}'$ as well as $\bar{b}'$ vanish (see (5.2.2)) so that the d'_μ vanish and the estimate (5.2.10) does not depend on the initial values of the first $m-1$ stages as is to be expected since these initial values do not enter into the later computation.

We are now in a position to check how the order of consistency of the various stages of a PC-method affects the order of the method.

Theorem 5.2.3. *If the stages of a stable m-stage k-step PC-method* $\mathfrak{M}$ *are consistent of order* q_μ, $\mu = 1(1)m$, *for a sufficiently smooth IVP1, and if the starting procedures for the various stages are of order* q'_μ, $\mu = 1(1)m$, *then* $\mathfrak{M}$ *is at least of order*

$$(5.2.15) \qquad p = \min_{\mu=1(1)m} \min(q_\mu + \gamma_\mu, q'_\mu + \gamma'_\mu), \quad \textit{with } \gamma_\mu, \gamma'_\mu \textit{ from } (5.2.9).$$

Furthermore, the orders p_μ *of convergence of the first* $m-1$ *stages* (see Def. 5.1.6) *of* $\mathfrak{M}$ *are given by*

$$(5.2.16)\qquad (I-h\bar{B})^{-1}\left[\begin{pmatrix} h^{\bar{q}_1} \\ \vdots \\ h^{\bar{q}_{m-1}} \end{pmatrix} + hB'\begin{pmatrix} h^{q'_1} \\ \vdots \\ h^{q'_{m-1}} \end{pmatrix}\right] = \begin{pmatrix} O(h^{p_1}) \\ \vdots \\ O(h^{p_{m-1}}) \end{pmatrix}$$

where

$$(5.2.17)\qquad \bar{q}_\mu := \min(p, q_\mu+1), \quad \mu = 1(1)m-1.$$

Proof. To derive (5.2.15) we set $\eta_1 = \zeta_n$, $\eta_2 = \Delta_n z$ in (5.2.10); then $F_n\eta_1 = 0$ and, by Def. 5.1.3 and our assumptions,

$$\begin{aligned}(\eta_1)_\kappa^\mu - (\eta_2)_\kappa^\mu &= O(h^{q'_\mu}), \quad \kappa = 0(1)k-1,\\ (F_n\eta_2)_\nu^\mu &= O(h^{q_\mu}), \quad \nu = k(1)n,\end{aligned} \qquad \mu = 1(1)m,$$

which implies $\|\zeta_n - \Delta_n z\|_{E_n} = O(h^p)$, with p given by (5.2.15).

With the same η_1 and η_2 we have from (5.2.12) for $\mu = 1(1)m-1$

$$|a_{\mu\mu}^k| \, \|(\varepsilon_n)_{\nu'}^\mu\| \le hL \sum_{\lambda=1}^{m-1} \sum_{\kappa=0}^{k} |b_{\mu\lambda}^\kappa| \, \|(\varepsilon_n)_{\nu'-k+\kappa}^\lambda\| + O(h^p) + O(h^{q_\mu+1})$$

by (5.2.15). Taking $\max\limits_{\nu'}$ on both sides we obtain

$$\begin{aligned}|a_{\mu\mu}^k| \max_{\nu'=k(1)n} \|(\varepsilon_n)_{\nu'}^\mu\| \le{}& hL \sum_{\lambda=1}^{m-1} \left(\sum_{\kappa=0}^{k} |b_{\mu\lambda}^\kappa|\right) \max_{\nu'=k(1)n} \|(\varepsilon_n)_{\nu'}^\lambda\| \\ &+ hL \sum_{\lambda=1}^{m-1} \left(\sum_{\kappa=0}^{k-1} |b_{\mu\lambda}^\kappa|\right) O(h^{q'_\lambda}) + O(h^{\bar{q}_\mu})\end{aligned}$$

which yields the assertion (5.2.16) by the same reasoning as was employed in going from (5.2.13) to (5.2.14) and in the subsequent remarks. □

Remark. As $\gamma_m = \gamma'_m = 0$ while $\gamma_\mu \ge 1$, $\gamma'_\mu \ge 1$ for $\mu < m$, the order of a PC-method is normally determined by the order of consistency of its last stage. The orders q'_μ of starting procedures which are not needed for the method $\mathfrak{M}$ are irrelevant due to $\gamma'_\mu = \infty$ in (5.2.15) and due to multiplication by $\bar{B}'$ in (5.2.16). It also follows from (5.2.16) and (5.2.17) that

$$p_\mu \le \bar{q}_\mu = \min(p, q_\mu+1) \quad \text{for } \mu = 1(1)m-1.$$

Corollary 5.2.4. *In a* P(EC)^{m}E*-method* $\mathfrak{M}$ *let the predictor and corrector schemes be of orders* p_0 *and* p_c *resp., and let the starting procedure (for the values of the last stage) be of order* p_s. *Then* $\mathfrak{M}$ *is of order*

$$(5.2.18)\qquad p = \min(p_s, p_c, p_0+m).$$

and the μ-th *stage*[4], $\mu=0(1)m-1$, *is convergent of order*

$$p_\mu = \min(p, p_0+\mu+1). \tag{5.2.19}$$

The same result holds for a P(EC)^m*-method if the starting procedure for the* $(m-1)$*-st stage values is of order* p_s-1.

Proof. From (5.2.4) we find for a P(EC)^{m}E-method that $\vec{B}$ and $\vec{b}$ have nonvanishing elements only in the subdiagonal and in the last component resp., while $\vec{B}'$ and $\vec{b}'$ vanish completely. The evaluation of (5.2.8) gives $\vec{d}_\mu = O(h^{m-\mu})$, $\mu=0(1)m$, and (5.2.18) follows from (5.2.15). (5.2.19) follows from (5.2.16)—(5.2.18) and the observation that the elements in the first column of $(I-h\vec{B})^{-1}$ are of order $O(h^\mu)$, $\mu=0(1)m-1$.

For a P(EC)^m-method there are nonvanishing elements also in the last columns of $\vec{B}$ and $\vec{B}'$. But a short calculation shows that the first component of $\vec{d}$ as defined by (5.2.8) remains $O(h^m)$ while $\vec{d}_s$ now has a nonvanishing last component of order $O(h)$ so that $\gamma'_{m-1}=1$. Also the determination of the p_μ remains unaltered. □

Remarks. 1. Note that, even for $m=1$, the order of a P(EC)^m-method is not lower than that of the P(EC)^{m}E-method with the same predictor and corrector.

2. The fact that the order of the predictor may be less than that of the corrector is not very helpful since the predictor need not be D-stable so that high-order predictor schemes could easily be designed (see Theorem 4.1.8).

5.2.3 Analysis of the Discretization Error

From Corollary 4.4.3 we know that the solution of a uniform linear k-step discretization of an IVP$^{(p+1)}$1 admits an asymptotic expansion $\zeta_n = \Delta_n(z+h^p e_p)+O(h^{p+1})$ if the underlying linear k-step scheme is of exact order p and the starting procedure is of order $p+1$; e_p is the solution of the differential equation (4.4.17). We will now attempt to find under which conditions the m-th stage of the solution ζ_n of a PC-discretization of an IVP$^{(p+1)}$1 satisfies

$$\zeta_n^m(t_\nu) = z(t_\nu)+h^p e_p(t_\nu)+O(h^{p+1}), \qquad t_\nu \in \mathbb{G}_n, \tag{5.2.21}$$

where

$$e_p(0)=0, \qquad e'_p - f'(z(t))\,e_p = \varphi(z(t)), \tag{5.2.22}$$

with some function $\varphi\colon \mathbb{R}^s \to \mathbb{R}^s$. We will then call φ the principal error function of the PC-procedure in analogy with (4.4.17) and Def. 4.4.1.

In analogy with the notation introduced before Theorem 5.2.2 we now let (cf. (5.2.8)/(5.2.9)) for a given PC-method

[4] See Remark 1 after Def. 5.2.4.

B be the $(m-1)\times(m-1)$-matrix with elements $\sum_{\kappa=0}^{k}\frac{b_{\mu\lambda}^{\kappa}}{a_{\mu\mu}^{k}}$, $\mu,\lambda=1(1)m-1$,

(5.2.23)

$b\in\mathbb{R}^{m-1}$ be the vector with components $\sum_{\kappa=0}^{k}\frac{b_{m\mu}^{\kappa}}{a_{mm}^{k}}$, $\mu=1(1)m-1$.

We then form $d^T:=hb^T(I-hB)^{-1}$, expand its components in powers of h and denote by

(5.2.24) d_μ the factor of h^{p-q_μ} in the expansion of the μ-th component of d.

Note that $\bar{d}_\mu=O(h^{\gamma_\mu})$ by (5.2.9) and $p-q_\mu\le\gamma_\mu$ by (5.2.15); hence d_μ either vanishes or is the factor of the lowest power h^{γ_μ} in the μ-th component of d. Finally we set

$$d_m:=\begin{cases}1 & \text{if } p=q_m\\ 0 & \text{if } p<q_m\end{cases}.$$

Theorem 5.2.5. *Consider a stable m-stage k-step PC-method* $\mathfrak{M}$ *of order* $p\ge1$ *and assume that* (*see* (5.2.15))

(5.2.25)
$$q'_\mu+\gamma'_\mu>p,\qquad \mu=1(1)m$$

and
$$2p_\mu>p,\qquad \mu=1(1)m-1.$$

Then the solution ζ_n *of* $\mathfrak{M}(\mathrm{IVP}^{(p+1)}1)$ *satisfies* (5.2.21)/(5.2.22) *with*

$$\varphi(z(t))=-\sum_{\mu=1}^{m-1}d_\mu\frac{a_{mm}^{k}c_\mu}{a_{\mu\mu}^{k}c_m}C_\mu\underbrace{f'(z(t))\dots f'(z(t))}_{\gamma_\mu\text{-times}}z^{(q_\mu+1)}(t)-d_m C_m z^{(q_m+1)}(t);$$

(5.2.26)

the c_μ *and* C_μ *are from* (5.1.13).

Proof. We introduce the polynomials

$$\hat{\rho}_\mu(x):=-\frac{1}{a_{\mu\mu}^{k}}\sum_{\kappa=0}^{k-1}a_{\mu m}^{\kappa}x^\kappa,\qquad \sigma_{\mu\lambda}(x):=\frac{1}{a_{\mu\mu}^{k}}\sum_{\kappa=0}^{k}b_{\mu\lambda}^{\kappa}x^\kappa.$$

From (5.1.5) we have $\hat{\rho}_\mu(1)=1$, $\mu=1(1)m$, and from the assumed consistency and (5.1.10) we have

(5.2.27)
$$k-\hat{\rho}'_m(1)=\sum_{\lambda=1}^{m}\sigma_{m\lambda}(1)=\frac{c_m}{a_{mm}^{k}}.$$

Let $\varepsilon:=\zeta_n-\Delta_n z$; from (5.2.1)/(5.2.2) and (5.1.13) we obtain, for $\mu=1(1)m$,

(5.2.28)
$$\begin{aligned}\varepsilon_\nu^\mu&=\hat{\rho}_\mu(T_h)\,\varepsilon_{\nu-k}^m+h\,\sigma_{\mu m}(T_h)\,f'(z_{\nu-k})\,\varepsilon_{\nu-k}^m\\&\quad+h\sum_{\lambda=1}^{m-1}\sigma_{\mu\lambda}(T_h)\,f'(z_{\nu-k}^\lambda)\,\varepsilon_{\nu-k}^\lambda+h^{q_\mu+1}\left(\frac{c_\mu}{a_{\mu\mu}^{k}}\right)C_\mu z^{(q_\mu+1)}+O(h^{p+2}),\end{aligned}$$

where $z_\nu^\mu = z(t_\nu^\mu)$. We have omitted the further terms in the expansion of $f(\zeta_\nu^\lambda) - f(z_\nu^\lambda)$ on the basis of our assumption $2p_\mu > p$. Similarly, the remainder terms from (5.1.13) have been omitted on the basis of Theorem 5.2.2: A perturbation of the μ-th stage by $O(h^{q_\mu+1})$ affects the ε_ν^m only by $O(h^{q_\mu+\gamma_\mu+1}) \le O(h^{p+1})$, see (5.2.15).

Now we take (5.2.28) for $\mu = 1(1)m-1$ and substitute it into (5.2.28) for $\mu = m$:

$$\begin{aligned}\varepsilon_\nu^m = {} & \hat\rho_m(T_h)\,\varepsilon_{\nu-k}^m + h\,\sigma_{mm}(T_h)\,f'(z_{\nu-k})\,\varepsilon_{\nu-k}^m \\ & + h\sum_{\lambda=1}^{m-1} \sigma_{m\lambda}(T_h)\,f'(z_{\nu-k}^\lambda)\,\hat\rho_\lambda(T_h)\,\varepsilon_{\nu-2k}^m \\ & + h^2 \sum_{\lambda=1}^{m-1} \sigma_{m\lambda}(T_h)\,f'(z_{\nu-k}^\lambda) \sum_{\rho=1}^{m-1} \sigma_{\lambda\rho}(T_h)\,f'(z_{\nu-2k}^\rho)\,\varepsilon_{\nu-2k}^\rho \\ & - h^{q_m+1}\,\frac{c_m}{d_{mm}^k}\,C_m z_\nu^{(q_m+1)} \\ & - \sum_{\lambda=1}^{m-1} \sigma_{m\lambda}(T_h)\,f'(z_{\nu-k}^\lambda)\,h^{q_\lambda+2}\,\frac{c_\lambda}{d_{\lambda\lambda}^k}\,C_\lambda z_{\nu-k}^{(q_\lambda+1)} + O(h^{p+2});\end{aligned} \tag{5.2.29}$$

we have omitted the terms involving $h^2\varepsilon_{\nu'}^m$ since they are $O(h^{p+2})$. If $p_\mu < p$ for any $\mu < m$, we repeat this procedure; in each repetition the $\varepsilon_{\nu'}^\rho$ in the expression for ε_ν^m gain another factor h so that they finally disappear into the $O(h^{p+2})$-term. If we should reach the initial values due to the backward shift involved in the substitution we use our assumption (5.2.25) and (5.2.10) to see that the initial errors cannot contribute to the $O(h^p)$-part of the ε_ν^m.

For a moment, we neglect the $f'(z_{\nu-jk}^\lambda)$ and the $z_{\nu-jk}^{(q_\lambda+1)}$ in the result of this finite recursive process. Then the vector with components $h^{q_\mu+1} c_\mu C_\mu$ appears multiplied by $h b^T(I + hB + h^2B^2 + \cdots) = h b^T(I - hB)^{-1} = d^T$, cf. (5.2.23). The $O(h^{p+1})$-parts of this product arise from those terms in the expansion of the components of d^T which are $O(h^{p-q_\mu})$ as asserted in (5.2.24). When we now insert the quantities stemming from z, we can consider all of them at a suitable $t_{\nu-k}$ since this shift introduces only a factor $1 + O(h)$ under our differentiability assumptions. Finally we observe that, for each μ, either $d_\mu = 0$ or $p - q_\mu = \gamma_\mu$ so that precisely γ_μ substitutions were involved in the generation of the $O(h^{p+1})$-term containing $C_\mu z^{(q_\mu+1)}$; this implies γ_μ multiplications by $f'(z)$. Thus, our final difference equation for ε_ν^m has the form

$$\begin{aligned}(5.2.30)\qquad &\frac{1}{h}\Bigg[\varepsilon_\nu^m-\hat\rho_m(T_h)\,\varepsilon_{\nu-k}^m-h\,\sigma_{mm}(T_h)\,f'(z_{\nu-k})\,\varepsilon_{\nu-k}^m\\ &\qquad\qquad -h\sum_{\lambda=1}^{m-1}\sigma_{m\lambda}(T_h)\,\hat\rho_\lambda(T_h)\,f'(z_{\nu-2k})\,\varepsilon_{\nu-2k}^m\Bigg]\\ &\quad=h\,\frac{c_m}{a_{mm}^k}\,\varphi(z_\nu)+O(h^{p+1}),\end{aligned}$$

with φ from (5.2.26); the initial values satisfy

$$(5.2.30\,\text{a})\qquad \varepsilon_\kappa^m=O(h^{p+1}),\qquad \kappa=0(1)k-1,$$

due to (5.2.25).

After the usual replacement of ε_ν^m by $h^p\bar\varepsilon_\nu^m$, division of (5.2.30) by h^p, and omission of the remainder terms we have a consistent and stable linear $2k$-step discretization of (5.2.22): At first we observe that the right-hand side fits our discretization due to (5.2.27) and $\hat\rho_\lambda(1)=1$. Furthermore, the associated polynomials $\rho(x)=x^{2k}-\hat\rho_m(x)\,x^k$ and $\sigma(x)=\sigma_{mm}(x)\,x^k+\sum_{\lambda=1}^{m-1}\sigma_{m\lambda}(x)\,\hat\rho_\lambda(x)$ satisfy $\rho(1)=0$ and

$$\rho'(1)=2k-k\,\hat\rho_m(1)-\hat\rho_m'(1)=k-\hat\rho_m'(1)=\sum_{\lambda=1}^{m}\sigma_{m\lambda}(1)=\sigma(1),$$

see (5.2.27); this implies consistency. The fact that ρ satisfies the root criterion follows from the stability of $\mathfrak{M}$ and Theorem 5.2.1. Thus, the $\bar\varepsilon_\nu^m$ approximate the $e(t_\nu)$ of (5.2.22) with an error of $O(h)$ and due to stability the omission of the remainder term in (5.2.30) has only an effect of the same order which establishes (5.2.21). □

Remark. The assumption (5.2.25) on the order of the starting procedure of $\mathfrak{M}$ is necessary to avoid the complications arising from the k-step structure of $\mathfrak{M}$ (see Section 4.4); the assumption $2p_\mu>p$ serves merely to prevent the appearance of quadratic terms in the ε_ν^μ within the $O(h^p)$-term. An inclusion of such terms presents no fundamental difficulties.

Def. 5.2.5. In the situation of Theorem 5.2.5, the mapping defined by (5.2.26) is called the *principal error function* of $\mathfrak{M}$.

Corollary 5.2.6. *In the situation of Theorem 5.2.5 if*

$$(5.2.31)\qquad q_\mu+\gamma_\mu>p\quad \textit{for}\ \ \mu=1(1)m-1$$

the principal error function of $\mathfrak{M}$ reduces to

$$(5.2.32)\qquad \varphi(z(t))=-C_m\,z^{(p+1)}(t)$$

Proof. By (5.2.24), $d_\mu=0$ for $\mu=1(1)m-1$, and $p=q_m$ by (5.2.15). □

Corollary 5.2.7. *For a* P(EC)^{m}E- *or* P(EC)^m-*method, if* (5.2.25) *is satisfied and* (*cf. Corollary* 5.2.4)

$$p_0 + m > p_c = p$$

then (5.2.32) *holds with* $C_m = C_c$, *the error constant of the corrector scheme.*

If $p_0 + m = p_c = p$ *we have for an arbitrary normalization of predictor and corrector*

$$\varphi(z(t)) = -C_c z^{(p+1)}(t) - \left(\frac{\beta_k}{\alpha_k}\right)^{m-1} \frac{\beta^*}{\alpha_k^0} \frac{c_0}{c_m} C_0 \underbrace{f'(z(t)) \dots f'(z(t))}_{m\text{-times}} z^{(p_0+1)}(t), \tag{5.2.33}$$

where

$$\beta^* := \begin{cases} \sum\limits_{\kappa} \beta_\kappa & \\ \beta_k & \end{cases} \textit{for a} \left.\begin{matrix} \mathrm{P(EC)}^m\text{-} \\ \mathrm{P(EC)}^m\mathrm{E}\text{-} \end{matrix}\right\} \textit{method}. \tag{5.2.34}$$

Proof. The result for $p_0 + m > p_c$ follows immediately from Corollary 5.2.6 since $\gamma_0 = m$, $\gamma_\mu = m - \mu$.

In the case $p_0 + m = p$ we find from (5.2.4) and (5.2.23) that for both types of PC-methods the element in the lower left corner of $(I - hB)^{-1}$ is $(h\beta_k/\alpha_k)^{m-1}$. Since $b^T = (0, \dots, 0, \beta^*/\alpha_k)$ and $d_\mu = 0$ for $\mu = 1(1)m-1$, (5.2.33) is an immediate consequence of (5.2.26). □

The derivation of a rigorous assertion about asymptotic expansions to an order higher than p for m-stage k-step PC-methods will not be attempted. From the analysis in this and the preceding section we may expect that the situation is similar to that in Theorems 4.4.5 and 4.4.7, with $\rho_m(x) = \sum\limits_{\kappa}^{k} a_{mm}^{\kappa} x^\kappa$ taking the place of the associated polynomial ρ of ψ, since the behavior of the solutions of PC-discretizations depends principally on their m-th stage. Thus, under the conditions and with the precautions explained in Section 4.4, Richardson extrapolation should be possible for PC-methods.

On the other hand, it seems highly improbable that there are non-trivial PC-discretizations which have a solution with a "symmetric" asymptotic expansion of the type (4.5.23).

5.2.4 Estimation of the Local Discretization Error

Again—as in Sections 3.4.4 and 4.5.5—we try to obtain an estimate of the *generated contribution in the current step* to the m-th stage of the global discretization error, i. e., of $h^{p+1}\varphi(z(t_\nu))$ rather than $h(l_n)_\nu^m$. We will assume as in Section 4.5.5 that we are dealing with a discretization of an $\mathrm{IVP}^{(p+2)}1$ and that the m-th stage of ζ_n satisfies (4.5.33); the x_i are now the essential zeros of ρ_m, cf. the discussion at the end of the previous section.

If we have the situation of Corollary 5.2.6, we may apply Theorem 4.5.11 with trivial modifications according to (5.2.32). However, if the PC-method is not straight and the $f(\zeta_\nu^m)$ have not been computed at all—as for example in a $\mathrm{P(EC)}^m$-method—one will wish to use the f-values at some previous stage $\zeta_\nu^{\mu_0}$, $\mu_0 < m$. If $\tau_{\mu_0} = 1$ and the $\zeta_\nu^{\mu_0}$ differ from the ζ_ν^m only by $O(h^{p+1})$ then $\zeta_\nu^{\mu_0} = z(t_\nu) + h^p e_p(t_\nu) + O(h^{p+1})$ and Theorem 4.5.11 still holds. The order of $\zeta_\nu^{\mu_0} - \zeta_\nu^m = \varepsilon_\nu^{\mu_0} - \varepsilon_\nu^m$ may be obtained from (5.2.29) and (5.2.16).

If the principal error function of a PC-method includes other terms besides $-C_m z^{(p+1)}(t)$, i.e., if some of the d_μ, $\mu = 1(1)m$, do not vanish in (5.2.26), one has to use $\bar{p} > p$ in Theorem 4.5.11 (see also Remark 2 after this theorem). The observation concerning the use of $f(\zeta_\nu^{\mu_0})$ in place of $f(\zeta_\nu^m)$ applies as above.

We formulate the result for $\mathrm{P(EC)}^m\mathrm{E}$- and $\mathrm{P(EC)}^m$-methods only:

Theorem 5.2.8. *If* (4.5.33) *holds for the final stage* ζ_n^m (*cf. Remark* 1 *after Def.* 5.2.4) *of the solution* ζ_n *of the discretization of an* $\mathrm{IVP}^{(p+2)}1$ *by a* $\mathrm{P(EC)}^m\mathrm{E}$- *or* $\mathrm{P(EC)}^m$-*method and if* $p_0 + m > p_c = p$ *then*

$$h^{p+1}\varphi(z(t_\nu)) = \frac{C_c}{C_c - \bar{C}}\left[\bar{\rho}(T_h)(\zeta_n)_{\nu-k}^m - h\bar{\sigma}(T_h)\begin{cases} f((\zeta_n)_{\nu-k}^m) \\ f((\zeta_n)_{\nu-k}^{m-1}) \end{cases}\right] + O(h^{p+2}) \tag{5.2.35}$$

for any linear k-step scheme $\langle\bar{\rho}, \bar{\sigma}\rangle$ *compatible with the corrector scheme* ψ *of the PC-method.*

If $p_0 + m = p_c$, (5.2.35) *remains valid provided that* $\bar{p} > p$ (*so that* $\bar{C} = 0$, *see* (4.5.34)) *and the values of* $f((\zeta_n)_\nu^m)$ *are used.*

Proof. With $p_0 + m > p_c$ we have $\varphi = -C_c z^{(p+1)}$ and $\varepsilon_\nu^{m-1} - \varepsilon_\nu^m = O(h^{p+1})$ from (5.2.29) and (5.2.19); the proof of Theorem 4.5.11 may be repeated literally. When $p_0 + m = p_c$, $\bar{p} > p_c$, and the values of $f((\zeta_n)_\nu^m)$ are used only trivial modifications are necessary. □

The classical approach in the case of $\mathrm{P(EC)}^m\mathrm{E}$- or $\mathrm{P(EC)}^m$-methods is to use a predictor of the same order as the corrector and to use the predictor scheme $\langle\rho_0, \sigma_0\rangle$ for $\langle\bar{\rho}, \bar{\sigma}\rangle$ in Theorem 5.2.8 (cf. Remark 1 after Corollary 4.5.12):

Corollary 5.2.9. *In the situation of Theorem* 5.2.8, *if* $p_0 = p_c$ *and the predictor scheme* ψ^0 *is compatible with the corrector scheme* ψ *then*

$$
\begin{aligned}
h^{p+1}\,\varphi(z(t_\nu)) &= \frac{C_c}{C_c - C_0}\,\alpha_k^0((\zeta_n)_\nu^m - (\zeta_n)_\nu^0) + O(h^{p+2}) \\
&= \frac{C_c}{C_c - C_0}\,\alpha_k^0((\zeta_n)_\nu^\mu - (\zeta_n)_\nu^0) + O(h^{p+2}), \qquad \mu \geq 1.
\end{aligned} \tag{5.2.36}
$$

Proof. For $\mu = m$, (5.2.36) follows immediately from the proofs of Corollary 4.5.12 and Theorem 5.2.8. For $1 \leq \mu < m$, we note that $(\zeta_n)_\nu^\mu - (\zeta_n)_\nu^m = O(h^{p+2})$ under our assumptions, see (5.2.29) and (5.2.19). □

Remark. (5.2.36) holds alike for P(EC)^{m}E- and P(EC)^m-methods. In the literature, more restrictive assumptions on the relation between ψ^0 and ψ are often made, see, e.g., Henrici [1], Theorem 5.13. In our formulation, the result applies also to unstable high-order predictors.

Using a PEC- or PECE-method with $p_0 = p$, one has two different approximations ζ_ν^0 and ζ_ν^1 to $z(t_\nu)$ which are both of the same order p. Thus, one might think of forming a combination of ζ_ν^0 and ζ_ν^1 such that the principal parts of the local discretization errors of the two stages cancel. Such a 3-stage method consisting of a predictor stage, a corrector stage and a linear combination stage has been suggested by Hamming [1]; see also the end of Section 5.2.1.

Let the predictor $\langle \rho_0, \sigma_0 \rangle$ and the corrector $\langle \rho, \sigma \rangle$ both be of order p and have error constants C_0 and C_c resp., $C_0 \neq C_c$. According to the approach outlined above we would compute ζ_ν^0 and ζ_ν^1 as in a PEC-procedure and then form

$$
\zeta_\nu^2 := \frac{1}{C_c - C_0}\,[C_c \zeta_\nu^0 - C_0 \zeta_\nu^1] = \zeta_\nu^1 - \frac{C_c}{C_c - C_0}(\zeta_\nu^1 - \zeta_\nu^0);
$$

ζ_ν^2 would then be used as the final value at t_ν, ζ_ν^2 and $f(\zeta_\nu^2)$ would be used in the following steps.

Writing down the coefficient matrices of such a PC-procedure, we see immediately that the above procedure is equivalent to the PECE-procedure based upon $\langle \rho_0, \sigma_0 \rangle$ and a scheme $\langle \bar{\rho}, \bar{\sigma} \rangle$ with

$$
\bar{\rho} := \frac{1}{C_c - C_0}\,[C_c \rho_0 - C_0 \rho], \qquad \bar{\sigma} := \frac{1}{C_c - C_0}\,[C_c \sigma_0 - C_0 \sigma]
$$

which is of order $p+1$ so that the corresponding PECE-method is also of order $p+1$ by Corollary 5.2.4. However, the effective scheme $\langle \bar{\rho}, \bar{\sigma} \rangle$ will be D-stable only if $p+1 \leq 2[k/2]+2$ according to Theorem 4.2.11, where k refers to the "longer" of the two schemes $\langle \rho_0, \sigma_0 \rangle$ and $\langle \rho, \sigma \rangle$, see Remark 2 after Def. 5.2.4. Thus, we are faced with the same restrictions as before.

Hamming's approach seemed advantageous since he used a $k+1$-step predictor and a k-step corrector both of order $k+1$ to obtain an overall order $k+2$ for what appeared to be a k-step PC-method. But the method is really a $k+1$-step method for which an order $k+2$ is standard.

5.2.5 Estimation of the Global Discretization Error

It is a remarkable property of PC-methods that the differential equation (5.2.22) for the principal part of the global discretization error can be *exact* if the predictor and corrector schemes are chosen properly; compare Lemma 3.4.7 and Theorem 3.4.8 for RK-methods. In fact, the application of the estimation technique based upon the exactness of (5.2.22) is even simpler for PC-methods than for RK-methods.

Lemma 5.2.10. *Consider a* PECE- *or* PEC-*procedure of order* p, *with a predictor scheme of order* $p-1$. *If and only if the error constants* C_0 *and* C_c *of the predictor and corrector schemes satisfy*[5]

(5.2.38) $$\beta^* C_0 + \alpha_k^0 C_c = 0, \quad \text{with } \beta^* \text{ from (5.2.34)},$$

the principal error function φ *of the PC-procedure satisfies* (3.4.21) *with*

(5.2.39) $$\Phi(z(t)) = -C_c z^{(p)}(t).$$

Proof. For Φ from (5.2.39) we have

$$\begin{aligned}\frac{d}{dt}\Phi(z(t)) - f'(z(t))\,\Phi(z(t)) &= -C_c z^{(p+1)}(t) + C_c f'(z(t))\, z^{(p)}(t)\\ &= -C_c z^{(p+1)}(t) - \frac{\beta^*}{\alpha_k^0} C_0\, f'(z(t)) z^{(p)}(t) = \varphi(z(t))\end{aligned}$$

by Corollary 5.2.7. □

Theorem 5.2.11. *Consider a stable k-step* PECE- *or* PEC-*method* $\mathfrak{M}$ *the PC-procedure of which satisfies the assumptions of Lemma* 5.2.10. *If the starting procedure of* $\mathfrak{M}$ *satisfies*

(5.2.40) $$s_{n\kappa}^1(z_0) = z(\kappa h) - C_c z^{(p)}(\kappa h) h^p + O(h^{p+1}), \qquad \kappa = 0(1)k-1,$$

and if $s_{n\kappa}^0(z_0) - z(\kappa h)$ *is of order* p *in the case of a* PEC-*method, then the solution* ζ_n *of the discretization of an* $IVP^{(p+1)}$ *1 by* $\mathfrak{M}$ *satisfies*

(5.2.41) $$(\zeta_n)_\nu^1 - z(t_\nu) = \beta^* [(\zeta_n)_\nu^1 - (\zeta_n)_\nu^0] + O(h^{p+1}),$$

with β^* *from* (5.2.34).

---

[5] Here and in the remainder of this section we assume again that we use our standard normalization (4.1.8) which implies $c_\mu = 1$ in (5.1.13).

Proof. According to Lemma 3.4.7 and Lemma 5.2.10 the function e_p in (5.2.21) must be a solution of the differential equation (5.2.22) with initial condition (see (3.4.22) and (5.2.39))

$$e_p(0) = -C_c z^{(p)}(0) \tag{5.2.42}$$

if it is to satisfy

$$e_p(t) = -C_c z^{(p)}(t) \quad \text{for all } t \geq 0. \tag{5.2.43}$$

The fact that (except for the factor h^p) the difference equation (5.2.30) satisfied by $\varepsilon_\nu^1 = (\zeta_n)_\nu^1 - z(t_\nu)$ is a consistent and stable discretization of the differential equation (5.2.22) was obtained in the proof of Theorem 5.2.5. (Note once more that we have stages numbered 0 and 1 with a PEC(E)-method in place of 1 and 2.) If we replace the assumption (5.2.25) on the starting procedure of $\mathfrak{M}$ by (5.2.40)—and if $q_0' = p$ so that $q_0' + \gamma_0' = p+1 > p$ in the case of a PEC-method—we find that the new initial condition

$$e_\kappa^1 = h^p[-C_c z^{(p)}(\kappa h) + O(h)], \qquad \kappa = 0(1)k-1,$$

for (5.2.30) is (without the factor h^p) a consistent discretization of the initial condition (5.2.42), cf. (2.2.7). Thus, (5.2.21) holds with (5.2.43).

We now form, using (5.2.21),

$$\begin{aligned} &\rho_0(T_h)(\zeta_n)_{\nu-k}^1 - h\sigma_0(T_h) f((\zeta_n)_{\nu-k}^1) \\ &= \rho_0(T_h) z(t_{\nu-k}) - h\sigma_0(T_h) z'(t_{\nu-k}) + h^p \rho_0(T_h) e_p(t_{\nu-k}) + O(h^{p+1}) \\ &= h^p C_0 z^{(p)}(t_\nu) + O(h^{p+1}) \end{aligned} \tag{5.2.44}$$

since $\rho_0(1)=0$ and e_p is differentiable; on the other hand the first stage of the PECE-procedure may written as

$$\alpha_k^0((\zeta_n)_\nu^0 - (\zeta_n)_\nu^1) + \rho_0(T_h)(\zeta_n)_{\nu-k}^1 - h\sigma_0(T_h) f((\zeta_n)_{\nu-k}^1) = 0. \tag{5.2.45}$$

(5.2.44) and (5.2.45) imply

$$\begin{aligned} \beta^*((\zeta_n)_\nu^1 - (\zeta_n)_\nu^0) &= h^p \frac{\beta^*}{\alpha_k^0} C_0 z^{(p)}(t_\nu) + O(h^{p+1}) \\ &= -h^p C_c z^{(p)}(t_\nu) + O(h^{p+1}) \quad \text{by (5.2.38)} \\ &= (\zeta_n)_\nu^1 - z(t_\nu) + O(h^{p+1}) \quad \text{by (5.2.21) and (5.2.43).} \end{aligned}$$

In the case of a PEC-method we have to take the f-values in (5.2.45) at $(\zeta_n)_\nu^0$ in place of $(\zeta_n)_\nu^1$ but this causes only a change of $O(h^{p+1})$ according to (5.2.19). □

Theorem 5.2.11 shows that PC-methods with exact error equations permit the estimation of the global discretization error as a by-product *without any additional effort*.

The case of the PEC-method where $\beta^*=1$ is particularly remarkable. Here (5.2.41) implies

(5.2.46) $$(\zeta_n)_\nu^0 = z(t_\nu) + O(h^{p+1}),$$

i.e., the error of the *predicted* value is $O(h^{p+1})$ while that of the corrected value is only $O(h^p)$. In other words, the method produces at the same time an $O(h^p)$- and an $O(h^{p+1})$-approximation to $z(t_\nu)$ at each gridpoint. (Of course, this presents the old dilemma: One may either use the difference of the two values as an estimate on the error, or use the more accurate value without knowledge of the error.)

In the case of the PECE-method we may analogously produce an $O(h^{p+1})$-approximation to $z(t_\nu)$ by forming

(5.2.47) $$\bar{\zeta}_\nu := (\zeta_n)_\nu^1 - \beta_k((\zeta_n)_\nu^1 - (\zeta_n)_\nu^0).$$

Thus, the PC-methods of Theorem 5.2.11 are the precise analogues of Butcher's RK-methods of effective order $p+1$, cf. Sections 3.3.5 and 3.4.3. While it is not so easy to construct RK-methods with an exact differential equation (3.4.22) due to the complicated structure of the principal error functions of RK-methods (see (3.4.1)), the analogous construction is quite simple for PECE- and PEC-methods:

Theorem 5.2.12. *For any given corrector scheme ψ of order $p\geq 2$ there exist predictor schemes ψ_0 of order $p-1$ such that* (5.2.38) *is satisfied.*

Proof. Take, with arbitrary constants a_μ, $\mu=2(1)p$,

(5.2.48) $$\rho_0(x) = (x-1) + \sum_{\mu=2}^{p} a_\mu (x-1)^\mu =: \sum_{\kappa=0}^{p} \alpha_\kappa^0 x^\kappa$$

and define σ_0 as the unique polynomial of degree $p-1$ satisfying

(5.2.49) $$\sigma_0(x) = \sum_{\kappa=0}^{p-1} \beta_\kappa^0 x^\kappa = \frac{\rho_0(x)}{\log x} + \frac{a_p C_c}{\beta^*}(x-1)^{p-1} + O((x-1)^p)$$

(5.2.48) and (5.2.49) imply, as $\alpha_k^0 = \alpha_p^0 = a_p$,

$$\frac{\rho_0(x)}{\log x} - \sigma_0(x) = -\frac{\alpha_k^0}{\beta^*} C_c (x-1)^{p-1} + O((x-1)^p);$$

thus, $\langle\rho_0, \sigma_0\rangle$ is of order $p-1$ and has the error constant $C_0 = -(\alpha_k^0/\beta^*)\, C_c$. □

Remark. For $p\geq 4$, it is possible to take ρ_0 of degree $p-1$ and choose the α_μ, $\mu=2(1)p-1$, such that the predictor scheme $\langle\rho_0, \sigma_0\rangle$ defined in an analogous fashion satisfies (5.2.38) but has $k=p-1$ which is the normal k to expect for a predictor of order $p-1$.

Example. Take, as a 1-step corrector of order $p=2$, the "trapezoidal rule"

$$\frac{1}{h}\left[\eta_\nu-\eta_{\nu-1}-\frac{h}{2}(f(\eta_\nu)+f(\eta_{\nu-1}))\right]=0,$$

the error constant of which is $C_c=-\frac{1}{12}$. If we choose $\rho_0(x)=x^2-x=(x-1)+(x-1)^2$ we find from (5.2.49)

$$\sigma_0(x)=1+\frac{3}{2}(x-1)-\frac{1}{12\beta^*}(x-1)=\begin{cases}\frac{4}{3}x-\frac{1}{3} & \text{for } \beta^*=\frac{1}{2},\\ \frac{17}{12}x-\frac{5}{12} & \text{for } \beta^*=1.\end{cases}$$

Thus the PEC(E)-method which is based upon the trapezoidal rule as corrector and uses

$$\begin{cases}\eta_\nu^0:=\eta_{\nu-1}+\frac{h}{3}[4f(\eta_{\nu-1})-f(\eta_{\nu-2})] & \text{(PECE)}\\ \eta_\nu^0:=\eta_{\nu-1}+\frac{h}{12}[17f(\eta_{\nu-1})-5f(\eta_{\nu-2})] & \text{(PEC)}\end{cases}$$

for the prediction, permits the estimation

$$(\zeta_n)_\nu^1-z(t_\nu)=\left.\begin{matrix}\frac{1}{2}\\1\end{matrix}\right\}((\zeta_n)_\nu^1-(\zeta_n)_\nu^0)+O(h^3)$$

if it is started such that

$$(\zeta_n)_\kappa^1=z(\kappa h)+\frac{h^2}{12}z''(0)+O(h^3),\qquad \kappa=0,1. \tag{5.2.50}$$

Actually, since $(\zeta_n)_0^1$ is used only in the f-values of the predictor, it is sufficient to have (5.2.50) for $\kappa=1$ which may be achieved, e. g., by the 2-stage procedure

$$\zeta_1^0:=z_0+hf(z_0),$$

$$\zeta_1^1:=z_0+\frac{h}{12}[7f(\zeta_1^0)+5f(z_0)],$$

as is easily checked.

The above example shows that the unorthodox starting procedure (5.2.40) may be realized by standard PC-procedures.

5.3 Predictor-corrector Methods with Off-step Points

5.3.1 Characterization

The only advantages over implicit linear multistep methods offered by the *classical* PC-methods (i. e. $P(EC)^m E$- and $P(EC)^m$-methods) are their explicitness and the possibilities of the estimation techniques of Sections 5.2.4 and 5.2.5 for the local and the global discretization errors. The attainable order for a stable method, on the other hand, is still fully subject to the restrictions of Theorem 4.2.11 due to the fact that the vector τ of (5.1.6) has all components equal to 1, see (5.2.6).

It has been realized independently by several authors (Gragg and Stetter [1], Gear [2], Butcher [9]) that the abandoning of (5.2.6) removes these restrictions on the order of a stable method. The resultant methods can fully utilize the number of parameters present in the last stage to attain high orders without becoming unstable (at least for k not too large, see Section 5.3.3).

Def. 5.3.1. An m-stage k-step PC-method with $t_\nu^\mu \neq t_\nu$, i.e. $\tau_\mu \neq 1$, for l different values of μ, $1 \leq l \leq m-1$, is called a PC-method *with l off-step points*. (Note that τ_m has to equal 1 according to Def. 5.2.2.)

Typically, in an m-stage method with $l = m-2$ off-step points we have

$$\begin{aligned} &0 < \tau_\mu < 1 \quad \text{for } \mu = 1(1)m-2, \\ &\tau_{m-1} = \tau_m = 1, \end{aligned} \tag{5.3.1}$$

and the matrix B_k in (5.1.1) is a full lower subtriangular matrix (the diagonal elements have to vanish by Def. 5.2.1). However, it is also feasible to have $\tau_\mu = i(\mu)$, with *integers* $i(\mu) > 1$ for one or several μ as suggested by Urabe [1] and Timlake [1]. Here the "off-step" points are actually gridpoints but they lie ahead of the gridpoint at which the solution is currently computed.

Normally, an m-stage k-step PC-method with off-step points will be a straight PC-method, i.e., only the f-values of the last stage will be used in the *following* steps. Such methods are the proper multistep generalization of explicit m-stage RK-methods which may be classified as "straight m-stage 1-step PC-methods with off-step points" in our terminology, cf. Example 1 in Section 5.1.1. For this reason, PC-methods with off-step points are also called *hybrid methods* as they combine the features of RK-methods and linear multistep methods.

If $\tau_{m-1} = 1$ (cf. (5.3.1)) and $p_{m-1} = p$ (cf. Theorem 5.2.3), one could also use the values of $f((\zeta_n)_\nu^{m-1})$ in place of $f((\zeta_n)_\nu^m)$ in the subsequent steps and thus save one evaluation of f without decreasing the order p of the method as may be checked by Theorem 5.2.2. Since we have quite carefully displayed the relation between $\mathrm{P(EC)}^m\mathrm{E}$- and $\mathrm{P(EC)}^m$-methods throughout Section 5.2, we will restrict ourselves now to straight PC-methods with off-step points.

Of course, one could also consider using some $f((\zeta_n)_\nu^\mu)$ for which $\tau_\mu \neq 1$ in the subsequent steps, i.e., one could admit nonvanishing elements in the B_κ, $\kappa = 0(1)k-1$, at arbitrary positions. The simplest extension of classical PC-methods in this direction would have 2 stages, the first one being "off-step", and would use *only* f-values at the off-step points:

$$B_\kappa = \begin{pmatrix} b_{11}^\kappa & 0 \\ b_{21}^\kappa & 0 \end{pmatrix}, \quad \kappa = 0(1)k-1, \qquad B_k = \begin{pmatrix} 0 & 0 \\ b_{21}^k & 0 \end{pmatrix},$$

with $\tau_1 \neq 1$. We will shortly consider such methods in Section 5.3.3.

Example. The following typical example of a 4-stage 2-step PC-method with 2 off-step points has been taken from Butcher [10], the method is of order 6, $\tau_1=\frac{1}{3}$, $\tau_2=\frac{2}{3}$, $\tau_3=\tau_4=1$. (The coefficients are not normalized for $\sum_\kappa B_\kappa e=e$.)

$$\eta_\nu^1 := \frac{16\eta_{\nu-1}+11\eta_{\nu-2}}{27}+h\frac{16f_{\nu-1}+4f_{\nu-2}}{27},$$

$$\eta_\nu^2 := \frac{47\eta_{\nu-1}-20\eta_{\nu-2}}{27}+h\frac{27f_\nu^1-22f_{\nu-1}-7f_{\nu-2}}{27},$$

$$\eta_\nu^3 := \frac{-13\eta_{\nu-1}+23\eta_{\nu-2}}{10}+h\frac{108f_\nu^2-189f_\nu^1+284f_{\nu-1}+64f_{\nu-2}}{80},$$

$$\eta_\nu := \eta_\nu^4 := \frac{48\eta_{\nu-1}+\eta_{\nu-2}}{49}+h\frac{160f_\nu^3+648f_\nu^2+405f_\nu^1+280f_{\nu-1}+7f_{\nu-2}}{1470}.$$

(For simplicity we have set $\eta_\nu := \eta_\nu^4$, $f_\nu^\mu := f(\eta_\nu^\mu)$, $f_\nu := f(\eta_\nu^4)$.)

5.3.2 Determination of the Coefficients and Attainable Order

The multistep procedures to be used in the second and further stages of a PC-method with off-step points differ from those discussed in Sections 4.1.2 and 4.1.3 in the following respect: While we admitted the occurrence of a value of y at each of the $t+r_\kappa h$ in the general difference-differential operators L_h of (4.1.16), we now need operators L_h which, at certain points, contain only a derivative value but no function value.

We will first analyze the situation with *one off-step point*; the arguments $t+\bar{r}_\kappa h$, $\kappa=0(1)k$, are supposed to represent the gridpoints, i.e., $\bar{r}_\kappa=\kappa$, $\kappa=0(1)k$, while $\bar{r}_{k+1}$ will refer to the off-step point. Aside from

$$\bar{r}_{k+1} \neq \bar{r}_\kappa, \qquad \kappa=0(1)k, \tag{5.3.2}$$

we impose no restriction on $\bar{r}_{k+1}$ at present; in particular, we do not consider the D-stability of the generated multistep schemes until Section 5.3.3. As the notational inconvenience is negligible, we retain the notation $\bar{r}_\kappa$ in place of κ since the results hold for general step ratio vectors which may be important in the construction of step change procedures with off-step points.

Theorem 5.3.1. *For given* $\bar{r}_\kappa$, $\kappa=0(1)k+1$, *it is possible to choose coefficients* α_κ, $\kappa=0(1)k$, *and* β_κ, $\kappa=0(1)k+1$, *such that*

$$\frac{1}{h}\sum_{\kappa=0}^{k}\left[\alpha_\kappa y(t+\bar{r}_\kappa h)-h\beta_\kappa y'(t+\bar{r}_\kappa h)\right]-\beta_{k+1}y'(t+\bar{r}_{k+1}h)=O(h^{2k+1})$$

(5.3.3)

for $y \in D^{(2k+1)}$. *Furthermore, if* $\bar{r}_{k+1}$ *is chosen such that*

$$(5.3.4) \qquad \sum_{\kappa=0}^{k} (\bar{r}_\kappa - \bar{r}_{k+1})^{-1} = 0$$

then (5.3.3) *is* $O(h^{2k+2})$ *for* $y \in D^{(2k+2)}$.

Proof. We use the approach of Theorem 4.1.8 and form

$$(5.3.5) \qquad \psi(x) = K h^{2k+1} \left(\frac{c_{k+1}}{x - \bar{r}_{k+1} h} - \frac{h}{2(x - \bar{r}_{k+1} h)^2} \right) \prod_{\kappa=0}^{k} (x - \bar{r}_\kappa h)^{-2}.$$

We have to choose c_{k+1} such that the residual of ψ at $x = \bar{r}_{k+1} h$ vanishes in order to keep $y(t + r_{k+1} h)$ from appearing in the differential-difference operator $L_h y(t) = \frac{1}{2\pi i} \oint \psi(x-t) y(x) dx$ (cf. Lemma 4.1.7). To compute $\operatorname{Res} \psi(x)$ we observe that

$$\prod_{\kappa=0}^{k} (x - \bar{r}_\kappa h)^{-2} = h^{-(2k+2)} \prod_{\kappa=0}^{k} (\bar{r}_{k+1} - \bar{r}_\kappa)^{-2}$$
$$\times \left[1 - \frac{2}{h} (x - \bar{r}_{k+1} h) \sum_{\kappa=0}^{k} (\bar{r}_{k+1} - \bar{r}_\kappa)^{-1} + O((x - \bar{r}_{k+1} h)^2) \right]$$

so that, for (5.3.5),

$$(5.3.6) \qquad \operatorname*{Res}_{x = \bar{r}_{k+1} h} \psi(x) = \frac{K}{h} \left[c_{k+1} - \sum_{\kappa=0}^{k} (\bar{r}_\kappa - \bar{r}_{k+1})^{-1} \right] \prod_{\kappa=0}^{k} (\bar{r}_\kappa - \bar{r}_{k+1})^{-2}$$

which vanishes if we choose

$$(5.3.7) \qquad c_{k+1} = \sum_{\kappa=0}^{k} (\bar{r}_\kappa - \bar{r}_{k+1})^{-1}.$$

On the other hand, we see from (5.3.5) that, as $|x| \to \infty$,

$$\psi(x) = \begin{cases} O(|x|^{-2(k+1)-1}) & \text{if } c_{k+1} \neq 0 \\ O(|x|^{-2(k+1)-2}) & \text{if } c_{k+1} = 0 \end{cases}$$

which proves the assertion via Lemma 4.1.7. □

Remarks. 1. It is trivial that (5.3.4) cannot hold if $\bar{r}_{k+1} > \bar{r}_\kappa$ for all $\kappa = 0(1)k$.

2. By the approach of Theorem 4.1.9, we could, of course, have constructed operators L_h of the necessary type for prescribed α_κ, choosing them equal to 0 at the appropriate arguments. But the order obtained in this fashion would have been below $2k+1$ in general.

Corollary 5.3.2. *In the situation of Theorem* 5.3.1, *if* β_k *is to vanish we can choose the other coefficients such that* (5.3.3) *is of order* $O(h^{2k})$, *and of order* $O(h^{2k+1})$ *if*

$$\sum_{\kappa=0}^{k-1} (\bar{r}_\kappa - \bar{r}_{k+1})^{-1} + \tfrac{1}{2}(\bar{r}_k - \bar{r}_{k+1})^{-1} = 0\,. \tag{5.3.8}$$

Proof. Replace the factor $(x-\bar{r}_k h)^{-2}$ by $(1/h)(x-\bar{r}_k h)^{-1}$ in (5.3.5) and modify the remainder of the proof accordingly. □

The coefficients α_κ, β_κ for the procedures which attain the maximal orders guaranteed by Theorem 5.3.1 and Corollary 5.3.2 are found by a partial fraction decomposition of (5.3.5) (with (5.3.7)) as in Section 4.1.3:

$$\begin{aligned} \frac{1}{h}\alpha_\kappa &= \operatorname*{Res}_{x=\bar{r}_\kappa h} \psi(x), \\ \beta_\kappa &= \lim_{x\to\bar{r}_\kappa h} (x-\bar{r}_\kappa h)^2 \psi(x), \end{aligned} \qquad \kappa = 0(1)k; \tag{5.3.9}$$

the constant K has to be chosen so as to achieve the desired normalization.

If we use the coefficients constructed according to Theorem 5.3.1 for the third stage of a 3-stage k-step PC-method $\mathfrak{M}$ with $\tau_1 = \bar{r}_{k+1} + 1 - k$, $\tau_2 = \tau_3 = 1$, we want to have the first 2 stages sufficiently accurate for the order of $\mathfrak{M}$ to become $2k+1$. As both components of $\vec{d}$ in (5.2.8) become $O(h)$ for such a method we have $\gamma_1 = \gamma_2 = 1$ in (5.2.15), i.e., we should have an order $2k$ of consistency for the first and second stage. But we can only achieve an order $2k-1$ for the first predictor according to Theorem 4.1.8 while we could obtain order $2k$ for the second predictor (which uses the f-value at η_ν^1) by Corollary 5.3.2.

The following suggestion due to Butcher [9] remedies this situation: Take both the first and the second stage procedures of order $2k-1$; the method $\mathfrak{M}$ is then of order $2k$ and its principal error function can be determined by Theorem 5.2.5. From (5.2.24) we easily obtain $d_1 = b_{31}^k = \beta_{k+1}$, $d_2 = b_{32}^k = \beta_k$, $d_3 = 0$, where the β_k refer to the notation in Theorem 5.3.1. Thus (see (5.2.26))

$$\varphi(z(t)) = -\frac{1}{c_3}(\beta_{k+1} c_1 C_1 + \beta_k c_2 C_2) f'(z(t)) z^{(2k)}(t) \tag{5.3.10}$$

if we assume that the coefficients of all stages have been normalized by $d_{\mu\mu}^k = 1$, $\mu = 1(1)3$.

As there is a one-parameter family of procedures of order $2k-1$ for the second stage we can now choose our particular second stage procedure such that $c_2 C_2 = -(\beta_{k+1}/\beta_k) c_1 C_1$ which annihilates $\varphi(z(t))$ and makes $\mathfrak{M}$ of order $2k+1$. In certain exceptional cases (depending on the choice of $\bar{r}_{k+1}$) where the annihilation of (5.3.10) cannot be achieved in this fashion we need only exchange the roles of the first and the second stage: Predict at t_ν in the first stage and at $t_{\nu-k+\bar{r}_{k+1}}$ in the second using the value of $f(\eta_\nu^1)$. Thus, there do exist 3-stage k-step methods (with one off-step point) of order $2k+1$ if the corresponding third stage schemes are D-stable. This is the case under certain conditions which will be discussed in Section 5.3.3.

If we choose $\bar{r}_{k+1}$ such that the third stage is of order $2k+2$ (cf. Theorem 5.3.1) sufficiently accurate procedures for the first two stages cannot be obtained without reaching back beyond $t_{\nu-k}$. On the other hand, many possibilities exist for a 3-stage k-step method of order $2k$ or lower.

The situation with *two off-step points* is characterized by

Theorem 5.3.3. *For given $\bar{r}_\kappa$, $\kappa=0(1)k+2$, it is possible to choose coefficients α_κ, $\kappa=0(1)k$, and β_κ, $\kappa=0(1)k+2$, such that*

$$\begin{aligned}\frac{1}{h}\sum_{\kappa=0}^{k}\left[\alpha_\kappa y(t+\bar{r}_\kappa h)-h\beta_\kappa y'(t+\bar{r}_\kappa h)\right]-\beta_{k+1}y'(t+\bar{r}_{k+1}h)\\ -\beta_{k+2}y'(t+\bar{r}_{k+2}h)=O(h^{2k+2})\end{aligned} \tag{5.3.11}$$

for $y\in D^{(2k+2)}$ if

$$\sum_{\kappa=0}^{k}(\bar{r}_\kappa-\bar{r}_{k+\lambda})^{-1}\neq 0,\qquad \lambda=1,2. \tag{5.3.12}$$

Proof. Following Butcher [10] we write ψ in the form

$$\begin{aligned}\psi(x)=K h^{2k+2}\prod_{\kappa=0}^{k}(x-\bar{r}_\kappa h)^{-2}\\ \times\Bigg[(x-\bar{r}_{k+1}h)^{-1}-\frac{h}{2c_{k+1}}(x-\bar{r}_{k+1}h)^{-2}-(x-\bar{r}_{k+2}h)^{-1}\\ +\frac{h}{2c_{k+2}}(x-\bar{r}_{k+2}h)^{-2}\Bigg]\end{aligned} \tag{5.3.13}$$

with $c_{k+\lambda}:=\sum_{\kappa=0}^{k}(\bar{r}_\kappa-\bar{r}_{k+\lambda})^{-1}\neq 0$ according to (5.3.12).

As in the proof of Theorem 5.3.1 we find

$$\operatorname*{Res}_{x=\bar{r}_{k+\lambda}h}\psi(x)=0,\qquad \lambda=1,2,$$

and due to the choice of the signs in the bracket of (5.3.13) we have $\psi(x)=O(|x|^{-2(k+1)-2})$ as $|x|\to\infty$. □

Remark. Of course, the result corresponding to Corollary 5.3.2 holds analogously. Also it is possible to make the bracket in (5.3.13) of order $O(|x|^{-3})$ by a special choice of $\bar{r}_{k+1}$ and $\bar{r}_{k+2}$ so that the resulting difference operator (5.3.11) is of order $O(h^{2k+3})$. We omit the details.

In the construction of a 4-stage k-step PC-method with 2 off-step points and $\tau_\lambda = \bar{r}_{k+\lambda}+1-k$, $\lambda=1,2$, $\tau_3=\tau_4=1$, with a final stage procedure of order $2k+2$ constructed according to Theorem 5.3.3, the previous difficulties are encountered in an even worse form; the first stage is now 2 orders too low if it uses the same k. But one can actually choose the predictors of the first three stages of order $2k-1$ each in such a fashion that their $O(h^{2k})$- and $O(h^{2k+1})$-contributions to the error of the final stage cancel and the method becomes of order $2k+2$ after all. For details see Butcher [10].

A family of 5-stage 2-step PC-methods with 3 off-step points which achieves $p=7=2k+3$ has also been constructed by Butcher [10]. While the general approach remains the same, the technical details become cumbersome for larger k and l. Naturally there is an immense variety of m-stage k-step PC-methods with l off-step points if one refrains from aiming at the maximal order which generally seems to be $2k+l$ for $m \geq l+2$.

5.3.3 Stability of High Order PC-methods with Off-step Points

So far we have not worried about the stability of the m-stage k-step PC-methods which we have constructed in Section 5.3.2 although their high orders are naturally worthless without stability. According to Theorem 5.2.1 this stability is equivalent to the D-stability of the procedure of the last stage; hence we need only check whether the associated polynomials ρ of the difference operators constructed in Theorem 5.3.1 and 5.3.3 satisfy the root criterion.

We therefore consider the zeros x_i of the associated polynomials for difference operators of order $O(h^{2k+l})$ as a function of k and the number l of off-step points and of the position of the off-step points characterized by $\bar{r}_{k+\lambda}$, $\lambda=1(1)l$, and define

$$R(k,l;\bar{r}_{k+1},\ldots,\bar{r}_{k+l}) := \max_{i=2(1)k} |x_i|;$$

we assume as before that $\bar{r}_\kappa = \kappa$ for $\kappa=1(1)k$.

In the degenerate case $l=0$ (linear k-step schemes) we have $R(1,0)=0$, $R(2,0)=1$ (Simpson-scheme), but $R(k,0)>1$ for $k>2$ according to Theorem 4.2.11.

For $l=1$, Butcher [9] found by numerical computation that there exist non-empty intervals $I_{k,1}$ such that $R(k,1;\bar{r}_{k+1})<1$ for $\bar{r}_{k+1} \in I_{k,1}$

for $k=1(1)7$; for $k>7$ these intervals are empty. Thus there exist 3-stage k-step PC-methods with one off-step point which are *convergent* of order $2k+1$ for $k=1(1)7$ but not for a larger k. Table 5.1 is intended to give an impression of the situation:

Table 5.1

k	stable for $<\bar{r}_{k+1}-k<$		$\bar{r}^*_{k+1}-k$
1	$-\infty$	$+\infty$	$-.50000$
2	-1	$+\infty$	$-.42265$
3	$-.634$	$.137$	$-.38197$
4	$-.472$	$-.001$	$-.35557$
5	$-.395$	$-.245$	$-.33655$
6	$-.354$	$-.303$	$-.32196$
7	$-.330$	$-.312$	$-.31028$

The intervals $I_{k,1}$, which define the range from which $\bar{r}_{k+1}$ may be chosen for a stable method, shrink with increasing k; except for small k, the off-step point has to lie between the "old" gridpoint $t_{\nu-1}$ and the "new" gridpoint t_ν which is somehow natural. $\bar{r}_{k+1}=k+1$ leads to a stable method with maximal order $2k+1$ only for $k=1$ and 2; for higher k the order has to be reduced if stability is to be retained. It is also interesting that for $k=1(1)6$ at least one of the zeros $\bar{r}^*_{k+1}$ of (5.3.4) lies in $I_{k,1}$ (see Table 5.1); thus for $k\leq 6$ there exist k-step schemes with 1 off-step point which are of order $2k+2$ and D-stable.

For $l=2$, Butcher [10] states that the regions in the $\bar{r}_{k+1},\bar{r}_{k+2}$-plane in which $R(k,2;\bar{r}_{k+1},\bar{r}_{k+2})<1$ are non-empty for $k=1(1)15$ but not for $k>15$. Again these regions confine the off-step points largely to the interval $(t_{\nu-1},t_\nu)$ if stable methods of maximal order are desired.

For $l=3$, the case $k=1$ is interesting for a different reason: It is clear that for an arbitrary choice of $\bar{r}_2,\bar{r}_3,\bar{r}_4$ (distinct and different from $\bar{r}_0=0$ and $\bar{r}_1=1$) coefficients β_κ, $\kappa=0(1)4$, can be found such that

$$\frac{y(t+h)-y(t)}{h}+\sum_{\kappa=0}^{4}\beta_\kappa y'(t+\bar{r}_\kappa h)=O(h^5)=O(h^{2k+3})$$

(see Theorem 4.1.9); also stability is no problem. But the necessary accuracy in the prediction obviously cannot be achieved in 4 prediction stages by the arrangement described in Section 5.3.2 for $l=1$ and 2, for otherwise we would have a 5-stage RK-method of order 5. Even more remarkable is the above-mentioned construction of 5-stage 2-step PC-methods with 3 off-step points of order 7; these methods also turn out to be stable, see Butcher [10].

A systematic study of the *stability regions* of high order PC-methods with off-step points has not yet been made. Note that it is particularly important for high order methods to be well-behaved also for values of the step h which are not excessively small, since otherwise their superiority may be wasted.

5.4 Cyclic Forward Step Methods

5.4.1 Characterization

Almost universally in the discussion of f.s.m., the tacit assumption is made that—except at step changes—the same forward step procedure is used throughout. Therefore, a f.s.m. is normally characterized by *its* forward step procedure. We have called attention to this assumption by calling such f.s.m. *uniform*; most of our effort has been spent on their analysis.

The most natural extension of this concept consists of the *cyclic* use of a fixed number of different forward step procedures within a f.s.m.; this extension is suggested by analogous extensions in the iterative solution of linear algebraic equations. That the use of such cyclic f.s.m. may lead to new achievements was seen in Section 4.3.3 for cyclic linear k-step methods: With $k-1$, resp. k different linear k-step procedures we could reach orders of convergence $2k-1$ and $2k$ resp. We also found that cyclic RK-methods may have larger regions of absolute stability, see the example at the end of Section 3.5.3. In what follows we will use the theory of m-stage k-step methods to gain further insight into cyclic f.s.m.

A general m-cyclic k-step method is a k-step method (Def. 2.1.8) which uses m different mappings Φ_n^μ, $\mu=1(1)m$, in a cyclic fashion:

$$\Phi_{n\nu}(\eta_\nu,\ldots,\eta_{\nu-k}) = \Phi_n^\mu(\eta_\nu,\ldots,\eta_{\nu-k}), \qquad \mu=\nu-k+1 \bmod m, \tag{5.4.1}$$

cf. (2.1.13). For simplicity we will always assume equidistant grids in our treatment of cyclic multistep methods and use the parameter $h=1/n$ in most places.

An m-cyclic $\bar{m}$-stage k-step method $\mathfrak{M}$—where the Φ_ν^μ in (5.4.1) are $\bar{m}$-stage k-step procedures in the sense of Def. 5.1.1—may always be regarded as a *uniform* $m\bar{m}$-stage multistep method $\bar{\mathfrak{M}}$. We need only introduce the quantities

$$\bar{\boldsymbol{\eta}}_{\bar{\nu}} = \begin{pmatrix} \boldsymbol{\eta}_{\bar{\nu}m} \\ \vdots \\ \boldsymbol{\eta}_{(\bar{\nu}+1)m-1} \end{pmatrix} \in \mathbb{R}^{m\bar{m}s}, \tag{5.4.2}$$

where $\boldsymbol{\eta}_\nu\in\mathbb{R}^{\bar{m}s}$ are the values of the elements $\eta\in E_n$ from $\mathfrak{M}$, and then rewrite the procedures (5.4.1) in terms of the $\bar{\boldsymbol{\eta}}_{\bar{\nu}}$. It is clear that $\mathfrak{M}$ becomes uniform: The computation of $\bar{\boldsymbol{\eta}}_{\bar{\nu}}$ from $\bar{\boldsymbol{\eta}}_{\bar{\nu}-1},\ldots,\bar{\boldsymbol{\eta}}_{\bar{\nu}-\bar{k}}$ follows the same pattern as that of $\bar{\boldsymbol{\eta}}_{\bar{v}-1}$ from $\bar{\boldsymbol{\eta}}_{\bar{v}-2},\ldots,\bar{\boldsymbol{\eta}}_{\bar{v}-\bar{k}-1}$ since we go through a full cycle of $\mathfrak{M}$ to advance from $\bar{\boldsymbol{\eta}}_{\bar{v}-1}$ to $\bar{\boldsymbol{\eta}}_{\bar{v}}$. In this new formulation the number $\bar{k}$ of previous values $\bar{\boldsymbol{\eta}}_{\bar{v}'}$ used in the computation of the next value $\bar{\boldsymbol{\eta}}_{\bar{v}}$ will normally be smaller than k due to the greater amount of information in one value (5.4.2). The transcription of the starting procedures is evident.

Example. See Example 4 in Section 5.1.1.

In the following we will restrict the $\bar{m}$-stage k-step procedures in an m-cyclic f.s.m. to procedures (5.1.1) which may be regarded as one-stage procedures.

Def. 5.4.1. An $\bar{m}$-stage k-step procedure (5.1.1) is called *effectively* 1-*stage* if it can be written in the form

$$\frac{1}{h}\sum_{\kappa=0}^{k}\alpha_\kappa\eta^{\bar{m}}_{v-k+\kappa}-\Psi(h;\eta^{\bar{m}}_{v-1},\ldots,\eta^{\bar{m}}_{v-k})=0. \tag{5.4.3}$$

The *increment function* $\Psi\colon \mathbb{R}\times(\mathbb{R}^s)^k\to\mathbb{R}^s$ (which naturally depends on the mapping f of the IVP1 under consideration) arises from formally solving the $\bar{m}$ stages of (5.1.1) for $\bar{\eta}^m_v$. In order that this be possible, the A_κ and B_κ, $\kappa=0(1)k-1$, must have nonvanishing elements in their last column only. The process is an immediate generalization of the construction of the increment function of a RK-procedure (3.1.3), see Section 3.1.2; it is clear that the analogue of Lemma 3.1.2 holds, i.e., that Ψ is Lipschitz continuous w.r.t. its k last arguments and picks up any further smoothness properties of f.

It is obvious that RK-procedures and straight k-step PC-procedures (Def. 5.2.1) are effectively 1-stage procedures. Such procedures *need not be explicit,* however, as might be suggested by the choice of the arguments in Ψ; it is not necessary that Ψ can be expressed explicitly in terms of repeated applications of f, see the discussion after Def. 3.1.4.

Let m different procedures (5.4.3), where we will omit the superscript $\bar{m}$ from now on, be characterized by the coefficients α^μ_κ, $\kappa=0(1)k$, and the increment functions Ψ^μ, $\mu=1(1)m$; k may be supposed to be the same for each of the m procedures (5.4.3) if we allow that the left hand side of (5.4.3) may not depend on all of the η_{v-k}. As in Section 4.3.1 we form the $m\times(m+k)$-matrix

$$A=\begin{pmatrix}\alpha_0^1 & \alpha_1^1 & \cdots & \alpha_k^1 & & & 0\\ & \alpha_0^2 & \cdots & \cdots & \alpha_k^2 & & \\ & & \ddots & & & \ddots & \\ & 0 & & \alpha_0^m & \cdots & \cdots & \alpha_k^m\end{pmatrix}, \tag{5.4.4}$$

and the $m \times m$-matrices $A_0, \ldots, A_{j_0}$,

(5.4.5) $$j_0 := \left[\frac{k-1}{m}\right] + 1,$$

by partioning A into $m \times m$-submatrices, starting from the right edge and adding zero columns on the left if necessary (see the illustration above Lemma 4.3.2). We then introduce

(5.4.6) $$\boldsymbol{\eta}_{\bar{\nu}} := \begin{pmatrix} \eta_{k+(\bar{\nu}-j_0)m} \\ \vdots \\ \eta_{k+(\bar{\nu}-j_0+1)m-1} \end{pmatrix} =: \begin{pmatrix} \eta_{\bar{\nu}}^1 \\ \vdots \\ \eta_{\bar{\nu}}^m \end{pmatrix} \in \mathbb{R}^{ms}, \qquad \eta_{\bar{\nu}}^{\mu} \in \mathbb{R}^s,$$

and write one full cycle of m procedures (5.4.3) in the form

(5.4.7) $$\frac{1}{h} \sum_{j=0}^{j_0} \mathbf{A}_j \boldsymbol{\eta}_{\bar{\nu}-j_0+j} - \boldsymbol{\Psi}(h; \boldsymbol{\eta}_{\bar{\nu}}, \ldots, \boldsymbol{\eta}_{\bar{\nu}-j_0}) = 0, \qquad \bar{\nu} \geq j_0.$$

where the $\mathbf{A}_j$ are formed from the A_j defined above as in (5.1.2) and

(5.4.8) $$\boldsymbol{\Psi}(h; \boldsymbol{\eta}_{\bar{\nu}}, \ldots, \boldsymbol{\eta}_{\bar{\nu}-j_0}) = \begin{pmatrix} \Psi^1(h; \eta_{\nu-1}^m, \ldots) \\ \vdots \\ \Psi^m(h; \eta_{\nu}^{m-1}, \ldots) \end{pmatrix}.$$

Def. 5.4.2. A k-step method which cyclically employs m different effectively 1-stage k-step procedures

(5.4.9) $$\frac{1}{h} \sum_{\kappa=0}^{k} \alpha_\kappa^\mu \eta_{\nu-k+\kappa} - \Psi^\mu(h; \eta_{\nu-1}, \ldots, \eta_{\nu-k}) = 0, \quad \mu \equiv \nu - k + 1 \bmod m,$$

is called a *straight m-cyclic k-step method.*

Def. 5.4.3. For a given straight m-cyclic k-step method $\mathfrak{M}$, the uniform m-stage j_0-step method $\bar{\mathfrak{M}}$ with forward step procedures (5.4.7), where the A_j, j_0, and $\boldsymbol{\eta}_{\bar{\nu}}$ are defined by (5.4.4)—(5.4.6) and $\boldsymbol{\Psi}$ by (5.4.8), with

(5.4.10) $$\mathbf{t}_{\bar{\nu}} := \begin{pmatrix} (k+(\bar{\nu}-j_0)m)\,h \\ \vdots \\ (k+(\bar{\nu}-j_0+1)m-1)\,h \end{pmatrix} =: \begin{pmatrix} t_{\bar{\nu}}^1 \\ \vdots \\ t_{\bar{\nu}}^m \end{pmatrix},$$

and with the starting procedure

(5.4.11) $$\boldsymbol{\eta}_j - \begin{pmatrix} \zeta_{k+(j-j_0)m} \\ \vdots \\ \zeta_{k+(j-j_0+1)m-1} \end{pmatrix} = 0, \qquad j = 0(1)j_0 - 1,$$

where the ζ_κ are the values defined by the starting procedure of $\mathfrak{M}$, is called the *equivalent m-stage method* of $\mathfrak{M}$; $\boldsymbol{\Psi}$ is called its *increment function.*

Remark. The subscripts in (5.4.6) have been chosen such that $\boldsymbol{\eta}_{j_0}^1 = \eta_k$ so that (5.4.7) for $\bar{\nu} = j_0$ represents the first cycle of the original m-cyclic

k-step method. If k is not a multiple of m, the first few components of $\boldsymbol{\eta}_0$ do not correspond to η_κ's of the original method; but these components of $\boldsymbol{\eta}_0$ are not used in (5.4.7) due to the 0-columns on the left edge of A_0 so that they may be left undefined in (5.4.11).

It is immediately clear that, for an arbitrary IVP 1, the values ζ_ν of a solution of $\mathfrak{M}$(IVP 1) and $\boldsymbol{\zeta}_{\bar{\nu}}$ of a solution of $\bar{\mathfrak{M}}$(IVP 1) *obtained for the same value of h* satisfy the relation (5.4.6). Thus, the analysis of $\mathfrak{M}$ may be completely replaced by that of the equivalent uniform m-stage method $\bar{\mathfrak{M}}$.

Example. k-cyclic linear k-step methods (cf. Example 4 in Section 5.1.1):
Here $j_0=1$ and $\boldsymbol{\Psi}(h;\boldsymbol{\eta}_\nu,\boldsymbol{\eta}_{\nu-1})=\mathbf{B}_0\boldsymbol{\eta}_{\nu-1}+\mathbf{B}_1\boldsymbol{\eta}_\nu$; the matrices A_0, A_1, B_0, B_1 have been displayed in Section 5.1.1.

We conclude this section by extending two concepts from RK-procedures to essentially 1-stage k-step procedures (5.4.3):

Def. 5.4.4. An effectively 1-stage k-step procedure is said to be *of order p* if, for each IVP$^{(p)}$1,

$$\frac{1}{h}\sum_{\kappa=0}^{k}\alpha_\kappa\bar{z}(-(k-\kappa)h,\xi)-\Psi(h;\bar{z}(-h,\xi),\ldots,\bar{z}(-kh,\xi))=O(h^p) \tag{5.4.12}$$

where $\bar{z}$ is the local solution of the IVP$^{(p)}$1 (see Def. 3.1.6) and $\xi\in\mathbb{R}^s$ is such that $\bar{z}(h,\xi)$ is defined for sufficiently small $|h|$.

Def. 5.4.5. For an effectively 1-stage k-step procedure of exact order p, the mapping $\varphi:\mathbb{R}^s\to\mathbb{R}^s$ defined for a given IVP$^{(p+1)}$1 by

$$\varphi(\xi):=-\lim_{h\to 0}\frac{1}{h^p}\left[\frac{1}{h}\sum_{\kappa=0}^{k}\alpha_\kappa\bar{z}(-(k-\kappa)h,\xi)-\Psi(h;\bar{z}(-h,\xi),\ldots)\right] \tag{5.4.13}$$

is called the *principal error function* of the procedure.

Remark. It is obvious that Def. 5.4.5 leads to the previous definitions of principal error functions in the case of RK-procedures (Def. 3.4.1) and linear k-step procedures (Def. 4.4.1). For straight m-stage k-step PC-procedures (Def. 5.2.5) this is not immediately evident but it may be seen by an analysis of the proof of Theorem 5.2.5 for this particularly simple special case. Thus, Def. 5.4.5 generalizes the previous definitions of principal error functions (except in the case of Def. 5.2.5 which also covers non-straight PC-procedures which are not covered by Def. 5.4.5).

5.4.2 Stability and Error Propagation

The Lipschitz continuity of the increment functions Ψ^μ in (5.4.9) (see remarks after Def. 5.4.1) carries over to the increment function $\boldsymbol{\Psi}$ of the equivalent m-stage method $\bar{\mathfrak{M}}$ if we use a norm (2.1.32). Thus the stability discussion in Section 5.1.3 for simple m-stage methods may be extended to cover m-stage methods with forward step procedures (5.4.7): Lemma 5.1.2 and Lemma 5.1.3 do not depend on the increment function at all and the proof of Theorem 5.1.4 uses only the Lipschitz condition (5.1.22) on the mappings G_n which holds in an analogous form for the procedures (5.4.7).

Def. 5.4.6. The polynomial of degree $j_0 m$

$$\rho(x) := \det\left(\sum_{j=0}^{j_0} A_j x^j\right), \quad \text{with } A_j \text{ from (5.4.4)}, \tag{5.4.14}$$

is called the *associated polynomial* of the straight m-cyclic k-step method $\mathfrak{M}$ as well as of its equivalent m-stage method. The remark after Def. 5.1.5 applies again.

Note that Def. 5.4.6 agrees with Def. 4.3.2 of the associated polynomial of an m-cyclic linear k-step method, with the reservation about dropping common powers of x in ρ. According to the proof of Lemma 4.3.2, (5.4.14) always contains a factor $x^{j_0 m-k}$ so that the reduced form of ρ is of degree k.

Corollary 5.4.1. *A straight m-cyclic k-step method is stable for an arbitrary IVP1 if and only if its associated polynomial satisfies the criterion of Lemma* 5.1.2.

Proof. See the discussion at the beginning of this section. □

Let us now look more closely at the way in which the inhomogeneities in

$$\frac{1}{h}\sum_{j=0}^{j_0} A_j \varepsilon_{\bar{\nu}-j_0+j} = \delta_{\bar{\nu}}, \qquad \bar{\nu} \geq j_0, \tag{5.4.15}$$

are propagated for A_j from (5.4.4), $\varepsilon_{\bar{\nu}}$, $\delta_{\bar{\nu}} \in \mathbb{R}^m$. Using the linear superposition approach we find that each particular $\delta_{\bar{\nu}}$ is transformed into an initial condition

$$\varepsilon_{\bar{\nu}-j_0+j} = \begin{cases} 0, & j = 0(1)j_0 - 1, \\ h A_{j_0}^{-1} \delta_{\bar{\nu}}, & j = j_0, \end{cases} \tag{5.4.16}$$

which is then decomposed into its components w.r.t. a fundamental system $\{\xi_{\bar{\nu}}^i\}$ of (5.4.15). The $\xi_{\bar{\nu}}^i$ associated with zeros x_i of ρ inside the unit disk will decrease exponentially with increasing $\bar{\nu}$ while the ones associated with essential zeros will remain constant in norm. Thus if the decomposition of (5.4.16) does not generate components w.r.t. these "essential fundamental solutions" of (5.4.15) the effect of the perturbation $\delta_{\bar{\nu}}$ will decay exponentially. It is this mechanism which permits m-cyclic f.s.m. to be convergent of order $p+1$ although their individual forward step procedures are only of order p.

Assume that the polynomial ρ of (5.4.14) for the system (5.4.15) of difference equations satisfies the criterion of Lemma 5.1.2; we form a fundamental system of solutions $\{\zeta_{\bar{\nu}}^i\}_{\bar{\nu}\geq 0}$, $i=1(1)mj_0$ for the homogeneous equation (5.4.15) in the following fashion:

For the q essential zeros x_i oj ρ (counting r-fold zeros r times) we take the solutions $\{\zeta_{\bar{\nu}}^i\}$, $i=1(1)q$, which satisfy (5.1.17); see the proof of Lemma 5.1.2. The remaining mj_0-q solutions $\{\zeta_{\bar{\nu}}^i\}$ in the fundamental system are then related to non-essential zeros of ρ only; thus the $\|\zeta_{\bar{\nu}}^i\|$, $i>q$, grow at most like $\max_{i>q} \bar{\nu}^{r_i-1}|x_i|^{\bar{\nu}}$ where r_i is the multiplicity of x_i. If we choose $\hat{x}$ such that

(5.4.17) $$\max_{i=q+1(1)mj_0} |x_i| < \hat{x} < 1$$

then there exists a constant K such that

(5.4.18) $$\|\zeta_{\bar{\nu}}^i\| \leq K\hat{x}^{\bar{\nu}} \quad \text{for } i=q+1(1)mj_0, \quad \bar{\nu}\geq 0.$$

Lemma 5.4.2. *Consider the system of difference equations*

(5.4.19) $$\sum_{j=0}^{j_0} (A_j - hG_{\bar{\nu}j})\bar{\varepsilon}_{\bar{\nu}-j_0+j} = 0, \quad \bar{\nu}\geq j_0,$$

with a polynomial (5.4.14) *satisfying the criterion of Lemma* 5.1.2 *and* $m\times m$*-matrices* $G_{\bar{\nu}j}$ *which are uniformly bounded:*

(5.4.20) $$\|G_{\bar{\nu}j}\| \leq G, \quad j=0(1)j_0, \quad \bar{\nu}\geq j_0.$$

With the fundamental system for (5.4.15) *described above, assume that the initial values* $\bar{\varepsilon}_{\bar{\nu}}\in\mathbb{R}^m$, $\bar{\nu}=0(1)j_0-1$, *of* (5.4.19) *satisfy*

(5.4.21) $$\begin{pmatrix} \bar{\varepsilon}_0 \\ \vdots \\ \bar{\varepsilon}_{j_0-1} \end{pmatrix} = \sum_{i=q+1}^{mj_0} d_i \begin{pmatrix} \zeta_0^i \\ \vdots \\ \zeta_{j_0-1}^i \end{pmatrix}.$$

Then there exist constants $\bar{\Gamma}$, $\bar{K}$ *independent of* $\bar{\nu}$ *and* h, *and an* $h_0>0$ *such that the solution of* (5.4.19)—(5.4.21) *satisfies, for* $0<h\leq h_0$,

(5.4.22) $$\|\bar{\varepsilon}_{\bar{\nu}}\| \leq [\bar{K}\hat{x}^{\bar{\nu}} + \bar{\Gamma}h] \max_{j=0(1)j_0-1} \|\bar{\varepsilon}_j\|.$$

Proof. The difference between the solution $\varepsilon_{\bar{\nu}}$ of

(5.4.23) $$\sum_{j=0}^{j_0} A_j\varepsilon_{\bar{\nu}-j_0+j} = 0$$

and the solution $\bar{\varepsilon}_{\bar{\nu}}$ of (5.4.19), both with initial values (5.4.21), satisfies

(5.4.24) $$\sum_{j=0}^{j_0} (A_j - hG_{\bar{\nu}j})(\bar{\varepsilon}-\varepsilon)_{\bar{\nu}-j_0+j} = h\sum_{j=0}^{j_0} G_{\bar{\nu}j}\varepsilon_{\bar{\nu}-j_0+j}.$$

If the left-hand side of (5.4.24) did not contain the $G_{\bar{\nu}j}$-terms we would have

$$\|\bar{\varepsilon}_{\bar{\nu}}-\varepsilon_{\bar{\nu}}\| \le h\Gamma \sum_{\bar{\nu}'=j_0}^{\bar{\nu}} \left\| \sum_{j=0}^{j_0} G_{\bar{\nu}'j}\varepsilon_{\bar{\nu}'-j_0+j} \right\| \tag{5.4.25}$$

by Lemma 5.1.3, since the initial values of (5.4.24) vanish. By an application of Lemma 2.1.3 the details of which we omit it follows that an estimate (5.4.25), with a different constant Γ, also holds for the solution of (5.4.24) for $0<h\le h_0$, with a suitable constant h_0 independent of $\bar{\nu}$. ((5.4.24) is a "neighboring difference equation" to that without the $G_{\bar{\nu}j}$ in the sense of Section 2.2.5.)

Now let $\max_{i=q+1(1)mj_0} |d_i| \le d \max_{j=0(1)j_0-1} \|\varepsilon_j\|$ for (5.4.21) so that the solution $\varepsilon_{\bar{\nu}}$ of (5.4.23)/(5.4.21) satisfies

$$\|\varepsilon_{\bar{\nu}}\| \le K\,d(mj_0-q)\hat{x}^{\bar{\nu}} \max_{j=0(1)j_0-1} \|\varepsilon_j\| =: \bar{K}\hat{x}^{\bar{\nu}} \max_j \|\varepsilon_j\| \tag{5.4.26}$$

by (5.4.18). The combination of the estimates (5.4.25) and (5.4.26) then leads, with (5.4.20), to

$$\|\bar{\varepsilon}_{\bar{\nu}}\| \le \|\varepsilon_{\bar{\nu}}\| + \|\bar{\varepsilon}_{\bar{\nu}}-\varepsilon_{\bar{\nu}}\| \le \bar{K}\left[\hat{x}^{\bar{\nu}} + h\Gamma(j_0+1)G\sum_{\nu'=0}^{\bar{\nu}} \hat{x}^{\nu'}\right] \max_{j=0(1)j_0-1} \|\varepsilon_j\|$$

which is equivalent to (5.4.22) since $\sum_{\nu'} \hat{x}^{\nu'}$ is bounded uniformly due to $\hat{x}<1$. □

For the formulation of the main result we suppose that the m procedures (5.4.9) are all at least of order p and denote their principal error functions by

$$\varphi^{\mu}(\xi) = \begin{pmatrix} \varphi_1^{\mu}(\xi) \\ \vdots \\ \varphi_s^{\mu}(\xi) \end{pmatrix} \in \mathbb{R}^s, \qquad \mu = 1(1)m,$$

but set $\varphi^{\mu}(\xi)\equiv 0$ if the μ-th procedure is actually of an order greater than p. We then define

$$\boldsymbol{\varphi}(\xi) := \begin{pmatrix} \varphi^1(\xi) \\ \vdots \\ \varphi^m(\xi) \end{pmatrix} \in \mathbb{R}^{ms}; \qquad \boldsymbol{\varphi}_{\sigma}(\xi) := \begin{pmatrix} \varphi_{\sigma}^1(\xi) \\ \vdots \\ \varphi_{\sigma}^m(\xi) \end{pmatrix} \in \mathbb{R}^m, \quad \sigma = 1(1)s. \tag{5.4.27}$$

Theorem 5.4.3. *Consider a stable straight m-cyclic k-step method* $\mathfrak{M}$ *with forward step procedures* (5.4.9) *which are of order* p. *If the starting procedure of* $\mathfrak{M}$ *is of order* $p+1$ *and if, with the notation* (5.4.27),

$$\begin{pmatrix} 0 \\ \vdots \\ 0 \\ A_{j_0}^{-1}\boldsymbol{\varphi}_{\sigma}(z(t)) \end{pmatrix} = \sum_{i=q+1}^{mj_0} d_{i\sigma}(t) \begin{pmatrix} \xi_0^i \\ \vdots \\ \xi_{j_0-1}^i \end{pmatrix}, \qquad \sigma = 1(1)s, \tag{5.4.28}$$

where the $\{\zeta_{\bar{\nu}}^i\}$, $i=1(1)mj_0$, *form the fundamental system described above, then* $\mathfrak{M}$ *is convergent of order* $p+1$ *for each* $IVP^{(p+1)}1$.

Proof. The global discretization error $\varepsilon := \zeta_n - \Delta_n z$ of the equivalent m-stage method $\bar{\mathfrak{M}}$ of $\mathfrak{M}$ satisfies ($\boldsymbol{\varepsilon}_\nu \in \mathbb{R}^{ms}$)

$$(5.4.29)\quad \begin{aligned} &\sum_{j=0}^{j_0} \mathbf{A}_j \boldsymbol{\varepsilon}_{\bar{\nu}-j_0+j} - h[\boldsymbol{\psi}(h; \boldsymbol{\zeta}_{\bar{\nu}}, \ldots, \boldsymbol{\zeta}_{\bar{\nu}-j_0}) - \boldsymbol{\psi}(h; \mathbf{z}(\mathbf{t}_{\bar{\nu}}), \ldots, \mathbf{z}(\mathbf{t}_{\bar{\nu}-j_0}))] \\ &\qquad = h^{p+1} \begin{pmatrix} \varphi^1(z(t_{\bar{\nu}}^1)) \\ \vdots \\ \varphi^m(z(t_{\bar{\nu}}^m)) \end{pmatrix} + O(h^{p+2}) \end{aligned}$$

according to Def. 5.4.5, with $\mathbf{t}_{\bar{\nu}}$ from (5.4.10). For an $\mathrm{IVP}^{(p+1)}1$, $p \ge 1$, the Ψ are Lipschitz-continuously differentiable and $\|\boldsymbol{\varepsilon}_\nu\| = O(h^p)$ by the stability and consistency of order p of $\bar{\mathfrak{M}}$; furthermore, the φ^μ are Lipschitz-continuous. We may therefore replace (5.4.29) by

$$(5.4.30)\qquad \sum_{j=0}^{j_0} (\mathbf{A}_j - h\mathbf{G}_{\bar{\nu}j}) \boldsymbol{\varepsilon}_{\bar{\nu}-j_0+j} = h^{p+1} \boldsymbol{\varphi}(\mathbf{z}(t_{\bar{\nu}})) + O(h^{p+2})$$

where the $\mathbf{G}_{\bar{\nu}j}$ satisfy (5.4.20) with some constant G independent of h.

If the $\mathbf{G}_{\bar{\nu}j}$ did not interrelate the different σ-components of the $\varepsilon_{\bar{\nu}}^\mu \in \mathbb{R}^s$ (e. g. for an IVP 1 with diagonal f) we would immediately have the situation of Lemma 5.4.2 for each σ-component of (5.4.30). But it is clear that the result of Lemma 5.4.2 extends to our situation since the estimate (5.4.26) may be obtained for each σ-component separately and the indicated application of Lemma 2.1.3 can just as well be made to the system (5.4.30) of ms difference equations.

We now use the approach explained for (5.4.15)/(5.4.16) to obtain, with "initial conditions" at $\bar{\nu}' - j_0 + 1, \ldots, \bar{\nu}'$, $\bar{\nu}' \ge j_0$, an estimate for the effect of $\boldsymbol{\varphi}$ on the solution of (5.4.30); the starting errors and the remainder term in (5.4.30) contribute only an $O(h^{p+1})$-term to $\boldsymbol{\varepsilon}_{\bar{\nu}}$ due to the assumed stability and the order $p+1$ of the starting procedure. With a bound $\bar{A}$ for $A_{j_0}^{-1}$ in (5.4.28) we thus obtain (see (5.4.22))

$$\|\varepsilon_{\bar{\nu}}\| \le \sum_{\bar{\nu}'=j_0}^{\bar{\nu}} [K\hat{\varkappa}^{\bar{\nu}'} + \Gamma h] \bar{A} h^{p+1} \max_{t \in [0,1]} \|\boldsymbol{\varphi}(\mathbf{z}(t))\| + O(h^{p+1})$$

which implies ($\hat{\varkappa} < 1$)

$$\|\varepsilon\|_{\bar{E}_n} = \max_{\bar{\nu}} \|\varepsilon_{\bar{\nu}}\| = O(h^{p+1});$$

this is the assertion due to the equivalence of $\mathfrak{M}$ and $\bar{\mathfrak{M}}$. □

Remark. Theorem 5.4.3 shows that the global discretization error of $\mathfrak{M}$ is $O(h^{p+1})$ *at each stage* of the cyclic process.

Corollary 5.4.4. *If a straight m-cyclic k-step method* $\mathfrak{M}$ *with k-step procedures* (5.4.9) *of order p converges with order* $p+1$ *due to the satisfaction of* (5.4.28), *then the* $m-1$ *further m-cyclic k-step methods which use the same forward step procedures* (5.4.9) *in the same cyclic order but start the cycle with a different procedure, are also convergent of order* $p+1$.

Proof. One may either show that (5.4.28) remains valid under cyclic exchanges of the procedures (5.4.9), or one may define the equivalent m-stage methods for the rearranged m-cyclic methods in such a way that they employ the previous m-stage procedure (5.4.7). This may be achieved by including some of the procedures (5.4.9) of the first cycle of the cyclic method in the starting procedure for the m-stage method; the starting procedure remains of order $p+1$ if it was of that order originally. □

Example. As at the end of Sections 4.3.1 and 4.3.2.

$$A=\left(\begin{array}{cc|cc} -\frac{1}{2} & 0 & \frac{1}{2} & 0 \\ 0 & 0 & -1 & 1 \end{array}\right), \qquad k=m=2, \qquad j_0=1,$$

$p=3$ but $\varphi_\sigma^1(\xi)\equiv 0$ as the first 2-step procedure of the cycle is the Simpson procedure. We have $mj_0=2$, $\rho(x)=\frac{1}{2}(x-1)x$, hence $q=1$, $x_1=1$, $x_2=0$ and $\xi_0^1=\begin{pmatrix}1\\1\end{pmatrix}$, $\xi_0^2=\begin{pmatrix}0\\1\end{pmatrix}$.

$$A_1^{-1}\begin{pmatrix}0\\ \varphi_\sigma^2\end{pmatrix}=\begin{pmatrix}0\\ \varphi_\sigma^2\end{pmatrix}=\varphi_\sigma^2\begin{pmatrix}0\\1\end{pmatrix}$$

so that (5.4.28) is satisfied; this proves the convergence of order 4 for this method.

5.4.3 Primitive m-cyclic k-step Methods

Def. 5.4.7. A straight m-cyclic k-step method is called *primitive* if all the forward step procedures (5.4.9) have the same coefficients α_κ, $\kappa=0(1)k$:

$$\frac{1}{h}\sum_{\kappa=0}^{k}\alpha_\kappa\eta_{\nu-k+\kappa}-\Psi^\mu(h;\eta_{\nu-1},\ldots,\eta_{\nu-k})=0, \qquad \mu=\nu-k+1 \bmod m. \tag{5.4.31}$$

Remark. Without loss of generality we can assume the normalization $\sum_{\kappa=1}^{k}\kappa\alpha_\kappa=1$ for (5.4.31) since it is easy to see that $\sum_\kappa\kappa\alpha_\kappa$ has to be different from zero for a consistent discretization of an IVP 1.

Example. Straight m-cyclic 1-step methods (= m-cyclic RK-methods) are necessarily primitive as $\alpha_1=1$, $\alpha_0=-1$ is the only choice.

When we form the equivalent m-stage method $\overline{\mathfrak{M}}$ of a primitive m-cyclic k-step method we observe that the fundamental solutions[6] $\{\xi_{\bar\nu}^i\}$, $\xi_{\bar\nu}^i\in\mathbb{R}^m$, of

[6] We have chosen the notation $\xi_{\bar\nu}^i$ to distinguish between the fundamental solutions of (5.4.32) and (5.4.33).

(5.4.32) $$\sum_{j=0}^{j_0} A_j \boldsymbol{\varepsilon}_{\bar{v}-j_0+j} = 0, \qquad \boldsymbol{\varepsilon}_{\bar{v}} \in \mathbb{R}^m,$$

are identical with the fundamental solutions $\{\xi_v^i\}$, $\xi_v^i \in \mathbb{R}$, of

(5.4.33) $$\sum_{\kappa=0}^{k} \alpha_\kappa \varepsilon_{v-k+\kappa} = 0$$

in the following sense:

If k is a multiple of m so that $mj_0 = k$ (see (5.4.5)) we have simply

(5.4.34) $$\{\boldsymbol{\xi}_{\bar{v}}^i\} = \left\{ \begin{pmatrix} \xi_0^i \\ \vdots \\ \xi_{m-1}^i \end{pmatrix}, \begin{pmatrix} \xi_m^i \\ \vdots \\ \xi_{2m-1}^i \end{pmatrix}, \ldots\ldots \right\}, \qquad i = 1(1)mj_0$$

as is clear from

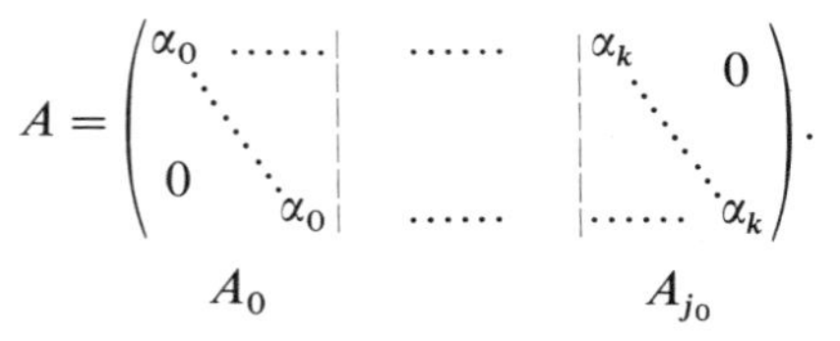

If $mj_0 = k + l$, $l > 0$, A_0 has l 0-columns on its left edge. Here we use the fundamental system

(5.4.34 a) $$\{\boldsymbol{\xi}_{\bar{v}}^i\} := \left\{ \begin{pmatrix} 0 \\ \vdots \\ 0 \\ \xi_0^i \\ \vdots \\ \xi_{m-l-1} \end{pmatrix}, \begin{pmatrix} \xi_{m-l}^i \\ \vdots \\ \vdots \\ \xi_{2m-l-1}^i \end{pmatrix}, \ldots\ldots \right\}, \qquad i = 1(1)k,$$

(5.4.34 b) $$\{\boldsymbol{\xi}_{\bar{v}}^{k+\lambda}\} := \left\{ \begin{pmatrix} 0 \\ \vdots \\ 1 \\ 0 \\ \vdots \\ 0 \end{pmatrix} \leftarrow \lambda, \begin{pmatrix} 0 \\ \vdots \\ \vdots \\ \vdots \\ 0 \end{pmatrix}, \ldots\ldots \right\}, \qquad \lambda = 1(1)l.$$

Of course, these $\boldsymbol{\xi}_{\bar{v}}^i$, $i > k$, are never actually used in the construction of a solution of (5.4.15).

Lemma 5.4.5. *A primitive m-cyclic k-step method is stable for an arbitrary IVP1 if and only if the associated polynomial of* (5.4.33) *satisfies the root criterion.*

Proof. Follows from the above construction of the fundamental system and from the proof of Theorem 4.3.3. □

The application of the considerations of Section 5.4.2 to primitive m-cyclic k-step methods rests upon

Lemma 5.4.6. *Let the A_j in (5.4.32) be from a primitive m-cyclic k-step method and define a fundamental system of (5.4.32) by (5.4.34). Then*

$$(5.4.35) \qquad e^T \begin{pmatrix} A_{j_0} & & 0 \\ \vdots & \ddots & \\ A_1 & \cdots & A_{j_0} \end{pmatrix} \begin{pmatrix} \xi_0^i \\ \vdots \\ \xi_{j_0-1}^i \end{pmatrix} = \begin{cases} 1 & \text{for } i=1, \\ 0 & \text{for } i=2(1)mj_0, \end{cases}$$

where $\{\xi_{\bar{\nu}}^1\}$ is the fundamental solution associated with the zero $x_1=1$ of $\rho(x)=\sum_\kappa \alpha_\kappa x^\kappa$.

Proof. If k is a multiple of m, (5.4.35) is simply (cf. (5.4.4) and (5.4.34))

$$(5.4.36) \qquad \sum_{\kappa=0}^{k-1} \left(\sum_{\lambda=\kappa+1}^{k} \alpha_\lambda \right) \xi_\kappa^i = \begin{cases} 1, & i=1, \\ 0, & i=2(1)k. \end{cases}$$

From $\rho(1)=\sum_\kappa \alpha_\kappa = 0$ it follows that

$$\sum_{\kappa=0}^{k} \alpha_\kappa x^\kappa = (x-1) \sum_{\kappa=0}^{k-1} \left(\sum_{\lambda=\kappa+1}^{k} \alpha_\lambda \right) x^\kappa;$$

hence all zeros $x_i, i>1$, of ρ are zeros of $\sum_\kappa \left(\sum_{\lambda=\kappa+1}^{k} \alpha_\lambda \right) x^\kappa$ and the associated fundamental solutions satisfy (5.4.36). For $\{\xi_\nu^1\}=\{1,1,\ldots\}$, the left-hand side of (5.4.36) becomes $\sum_{\lambda=1}^{k} \lambda \alpha_\lambda$ which is 1 by our normalization (see remark after Def. 5.4.7).

If $mj_0=k+l$ it is easily seen that the first l columns of the matrix in (5.4.35) sum to 0 whereas the following ones lead again to (5.4.36) by the construction (5.4.34) of the $\xi_{\bar{\nu}}^i$. □

Theorem 5.4.7. *Consider a stable primitive m-cyclic k-step method $\mathfrak{M}$, with forward step procedures (5.4.31) which are at least of order p and with a starting procedure of order $p+1$. If the polynomial $\rho(x)=\sum_\kappa \alpha_\kappa x^\kappa$ satisfies the strong root criterion, then $\mathfrak{M}$ is convergent of order $p+1$ for an arbitrary IVP 1 if and only if*

$$(5.4.37) \qquad \sum_{\mu=1}^{m} \varphi^\mu(z(t)) \equiv 0$$

where the principal error functions φ^μ are defined as in (5.4.27) (i.e., $\varphi^\mu \equiv 0$ if the μ-th procedure (5.4.31) is of an order greater than p).

Proof. We consider the equivalent m-stage method $\bar{\mathfrak{M}}$ of $\mathfrak{M}$; according to Theorem 5.4.3 we have to show that the coefficients $d_{1\sigma}$ in the decomposition

$$\begin{pmatrix} 0 \\ \vdots \\ 0 \\ A_{j_0}^{-1}\varphi_\sigma \end{pmatrix} = \sum_{i=1}^{mj_0} d_{i\sigma} \begin{pmatrix} \xi_0^i \\ \vdots \\ \xi_{j_0-1}^i \end{pmatrix} \tag{5.4.38}$$

vanish. A pre-multiplication of (5.4.38) by $e^T \begin{pmatrix} A_{j_0} & & 0 \\ \vdots & \ddots & \\ A_1 & \dots & A_{j_0} \end{pmatrix}$ leads to

$$d_{1\sigma} = \sum_{\mu=1}^{m} \varphi_\sigma^\mu$$

by Lemma 5.4.6. The necessity of (5.4.37) follows from counterexamples chosen such that the $\varphi^\mu(z(t))$ are constants. □

Remark. It is clear from Theorem 5.4.3 that a primitive m-cyclic k-step method with procedures (5.4.31) of order p may also be convergent of order $p+1$ if ρ possesses extraneous essential zeros; but then the φ^μ have to satisfy further conditions besides (5.4.37).

Example. m-cyclic RK-methods.

As $\rho = x-1$, the situation of Theorem 5.4.7 always arises. Example 1 in Section 2.2.4 is now trivial since (2.2.16) is a 2-cyclic RK-method with $\varphi^1(z(t)) = -\frac{1}{2}z''(t)$, $\varphi^2(z(t)) = \frac{1}{2}z''(t)$ so that $\varphi^1 + \varphi^2$ vanishes.

We will later indicate a different approach to cyclic RK-methods.

The fact that this example has been previously used in connection with *Spijker's norm* (2.2.14) is not a coincidence: Theorem 5.4.7 may also be obtained by using the stability concept based on Spijker's norm; see Section 2.2.4.

Alternative proof of Theorem 5.4.7. The proof of Theorem 4.2.7 can immediately be extended to primitive m-cyclic k-step methods since we need only the Lipschitz condition (4.2.28) on the mappings G_n which are now defined as

$$(G_n\eta)(t_\nu) = -\Phi^\mu(\eta_{\nu-1}, \dots, \eta_{\nu-k}), \qquad \mu = \nu - k + 1 \bmod m, \qquad \nu \geq k.$$

Hence we have the result that a primitive m-cyclic k-step method $\mathfrak{M}$ is stable w.r.t. Spijker's norm if and only if ρ is strongly D-stable.

Furthermore, $\mathfrak{M}$ is consistent of order $p+1$ with an IVP$^{(p+1)}$ 1 w.r.t. Spijker's norm if it satisfies (5.4.37): The local discretization error l_n of $\mathfrak{M}$ satisfies (cf. (5.4.12)—(5.4.13))

$$l_n(t_\nu) = \begin{cases} O(h^{p+1}), & \nu = 0(1)k-1, \\ -h^p\varphi^\mu(z(t_\nu)) + O(h^{p+1}), & \mu = \nu - k + 1 \bmod m, \qquad \nu \geq k. \end{cases}$$

Thus we have from (2.2.14)

$$\|l_n\|^*_{E_n^0} = \max_{\kappa=0(1)k-1} \|l_n(t_\kappa)\| + h \max_{\nu=k(1)n} \left\| \sum_{\nu'=k}^{\nu} l_n(t_{\nu'}) \right\|$$

$$\leq h^{p+1} \max_{\nu=k(1)n} \left\| \sum_{\bar{\nu}=j_0}^{\hat{\nu}} \sum_{\mu=1}^{m} \varphi^\mu(z(t^\mu_{\bar{\nu}})) + \sum_{\mu=1}^{\hat{\mu}} \varphi^\mu(z(t_{\nu-\hat{\mu}+\mu})) \right\| + O(h^{p+1}),$$

(5.4.39)

with $\nu-k+1 =: \hat{\nu}m+\hat{\mu},\ 0 \leq \hat{\mu} < m$.

The first μ-sums in (5.4.39) are $O(h)$ due to (5.4.37) and the Lipschitz-continuity of the φ^μ so that their total contribution is $O(1)$ since there are $O(n)$ of them; the second μ-sum is $O(1)$. Thus, $\|l_n\|^*_{E_n^0}$ is $O(h^{p+1})$ and the assertion of Theorem 4.3.7 follows from Theorem 1.2.4. □

Remarks. 1. Spijker was obviously aware of this application since he gives an example of a primitive 7-cyclic 3-step method in [1] which, by (5.4.37), is of one order higher than its procedures (5.4.31).

2. The further conditions on $\boldsymbol{\varphi}$ in the case of extraneous essential zeros of ρ cannot easily be obtained from Spijker's approach.

A fully developed theory of *m-cyclic RK-methods* is actually contained in our theory of RK-schemes and their composition, cf. Section 3.2.1: The m-cyclic RK-method $\mathfrak{M}$ with the forward-step procedures

$$\frac{\eta_\nu - \eta_{\nu-1}}{h} - \Psi_\mu(h, \eta_{\nu-1}) = 0, \qquad \mu = \nu \bmod m, \qquad \nu \geq 1,$$

with $\Psi_\mu = \Psi_\mu[f]$, is equivalent to the uniform RK-method $\bar{\mathfrak{M}}$ based upon the RK-scheme $\bar{\psi} = \psi_1 \psi_2 \ldots \psi_m$. (One step h of this method advances the numerical solution by mh as for the method $\bar{\mathfrak{M}}$ defined in Def. 5.4.3, see (5.4.10).)

The values of the $\hat{\Phi}^{(r)}_\lambda(\bar{\psi})$ can be derived from those of the $\hat{\Phi}^{(r)}_\lambda(\psi_\mu)$ by the composition law (3.2.5) for the generating matrices and (3.1.17)/(3.1.20). A general theory of these "composition formulas" for the $\hat{\Phi}^{(r)}_\lambda$ is contained in Butcher [3] and explicit formulas are listed for $\lambda = 1(1)n_r$, $r = 1(1)5$. We restrict our account of this theory to the following result which ties in with Theorem 5.4.7:

Lemma 5.4.8. *For any* $r \geq 1$ *and* $\lambda \in \{1, \ldots, n_r\}$

$\hat{\Phi}^{(r)}_\lambda(\psi_1 \psi_2) = \hat{\Phi}^{(r)}_\lambda(\psi_1) + \hat{\Phi}^{(r)}_\lambda(\psi_2) +$ *a polynomial in lower order Φ-values of ψ_1 and ψ_2.*

Proof. We use the inductive reasoning inherent in (3.1.17): For $r=\lambda=1$ we have from (3.2.5), with the vectors b^T introduced in (3.3.12a)

$$A_{12}=\begin{pmatrix}\bar{A}_1 & 0\\ e_2 b_1^T & \bar{A}_2\\ b_1^T & b_2^T\end{pmatrix} \quad\text{and}\quad \Phi((1,1),A_{12})=\begin{pmatrix}\bar{A}_1 e_1\\ \bar{A}_2 e_2+(b_1^T e_1)e_2\\ b_1^T e_1+b_2^T e_2\end{pmatrix}$$

so that $\hat{\Phi}_1^{(1)}(\psi_1\psi_2)=\hat{\Phi}_1^{(1)}(\psi_1)+\hat{\Phi}_1^{(1)}(\psi_2)$ which also follows from Theorem 3.2.1, 2 since $\Phi_1^{(1)}(\psi)=\alpha(\psi)$.

Assume that, for all $r\le r_0$, $\Phi((r,\lambda),A_{12})$ is of the form

$$\text{(5.4.40)}\qquad \begin{pmatrix}\overline{\Phi((r,\lambda),A_1)}\\ \overline{\Phi((r,\lambda),A_2)}+\cdots\\ \hat{\Phi}_\lambda^{(r)}(\psi_1)+\hat{\Phi}_\lambda^{(r)}(\psi_2)+\cdots\end{pmatrix}$$

where the bar indicates (as with A_μ) the omission of the last component (resp. row) and ... indicates an expression in $\Phi((r',\lambda),A_1)$ and $\Phi((r',\lambda),A_2)$ with $r'<r$ (including 0-order expressions like $(b_1^T e_1)e_2$). Then it follows immediately from (3.1.17) that the $\Phi((r_0+1,\lambda),A_{12})$ are also of the form (5.4.40). ▯

Lemma 5.4.8 trivially implies that

$$\text{(5.4.41)}\qquad \hat{\Phi}_\lambda^{(r)}(\psi_1,\ldots,\psi_m)=\sum_{\mu=1}^{m}\hat{\Phi}_\lambda^{(r)}(\psi_\mu)+\text{polynomial in lower order }\hat{\Phi}\text{-values of the }\psi_\mu.$$

Now assume that the ψ_μ are of order $p+1$, $\mu=1(1)m$, so that by (3.3.4)

$$\text{(5.4.42)}\qquad \hat{\Phi}_\lambda^{(r)}(\psi_\mu)=\frac{1}{\gamma_\lambda^{(r)}},\qquad \lambda=1(1)n_r,\qquad r=1(1)p+1,$$

and, by Theorem 3.3.7 and the additivity of the step factor α,

$$\text{(5.4.43)}\qquad \hat{\Phi}_\lambda^{(p+1)}(\psi_1,\ldots,\psi_m)=\frac{m^{p+1}}{\gamma_\lambda^{(p+1)}},\qquad \lambda=1(1)n_{p+1}.$$

According to (5.4.41) the validity of (5.4.43)—which implies the convergence of order $p+1$ of the m-cyclic RK-method $\mathfrak{M}$ based upon $\psi_1,\ldots,\psi_m$—remains unaltered if the ψ_μ *do not satisfy* (5.4.42) for $r=p+1$ but their sum equals the previous value

$$\text{(5.4.44)}\qquad \sum_{\mu=1}^{m}\hat{\Phi}_\lambda^{(p+1)}(\psi_\mu)=\frac{m^{p+1}}{\gamma_\lambda^{(p+1)}},\qquad \lambda=1(1)n_{p+1},$$

Thus if the ψ_μ are of order p and satisfy (5.4.44) our m-cyclic method $\mathfrak{M}$ is convergent of order $p+1$, Hence, (5.4.44) is equivalent to (5.4.37).

It is clear that the theory of RK-schemes may be used to get even farther reaching results than Theorem 5.4.7; compare Section 3.3.5.

5.4.4 General Straight m-cyclic k-step Methods

The possibilities of primitive m-cyclic k-step methods are rather restricted since their effectively 1-stage k-step procedures (5.4.31) have to be D-stable according to Lemma 5.4.5. On the other hand, we have seen in Section 4.3.3 that the associated polynomial (5.4.14) of a general straight m-cyclic k-step method may satisfy the root criterion, although none of the associated polynomials ρ_μ of the procedures (5.4.9) do so!

Lemma 5.4.9. *Given two k-step procedures* (5.4.9) *with associated polynomials* $\rho_\mu(x)=\sum \alpha_\kappa^\mu x^\kappa$, $\mu=1,2$. *The coefficients* α_κ *of the associated polynomial* (5.4.14) *of the generated 2-cyclic k-step method are given by*

$$\alpha_\kappa = \sum_{i+j=2\kappa} (-1)^i \alpha_i^2 \alpha_j^1, \qquad \kappa=0(1)k. \tag{5.4.45}$$

Proof. We have $j_0=[(k+1)/2]$. For even k,

$$A=\begin{pmatrix} \alpha_0^1 & \alpha_1^1 & \Big| & \cdots & \Big| & \alpha_k^1 & 0 \\ 0 & \alpha_0^2 & \Big| & \cdots & \Big| & \alpha_{k-1}^2 & \alpha_k^2 \end{pmatrix}$$

and

$$\rho(x)=\det \sum_{j=0}^{\frac{k}{2}} A_j x^j = \det \begin{pmatrix} \sum_{j=0}^{\frac{k}{2}} \alpha_{2j}^1 x^j & \sum_{j=1}^{\frac{k}{2}-1} \alpha_{2j+1}^1 x^j \\ \sum_{j=0}^{\frac{k}{2}} \alpha_{2j-1}^2 x^j & \sum_{j=0}^{\frac{k}{2}} \alpha_{2j}^2 x^j \end{pmatrix}$$

from which (5.4.45) is easily obtained. For odd k, the validity of (5.4.45) is established in an analogous manner. □

(5.4.45) permits the recursive computation of the coefficients of ρ for a general straight m-cyclic k-step method.

Example. $k=2$, $\rho_\mu(x)=(x-1)(x-x_\mu)$, $\mu=1(1)m$.

From $\alpha_2^\mu=1$, $\alpha_1^\mu=-(1+x_\mu)$, $\alpha_0^\mu=x_\mu$ we obtain
$\alpha_2=1$, $\alpha_1=-(1+\prod x_\mu)$, $\alpha_0=\prod x_\mu$ for any $m\geq 2$

so that $\rho(x)=(x-1)\left(x-\prod_{\mu=1}^{m} x_\mu\right)$.

(5.4.45) also shows that, for given polynomials ρ_1 and ρ of degree k, it is normally possible to determine a polynomial ρ_2 of degree k such that the relation of Lemma 5.4.9 holds. Mischak [1] has shown that (5.4.45) interpreted as a linear system for the α_κ^2 has a non-singular matrix if ρ_1 is not "degenerate". Here a polynomial is called degenerate if it has a vanishing zero or a pair of zeros with a vanishing sum. If ρ_1 is degenerate, ρ also has to be degenerate in order that ρ_2 exists.

This analysis of (5.4.45) shows that we may normally choose $m-1$ polynomials ρ_μ arbitrarily (with $\rho_\mu(1)=0$) and then determine ρ_m such that the associated polynomial ρ of the m-cyclic method satisfies the criterion of Lemma 5.1.2. One may then choose the increment functions Ψ_μ so as to achieve the highest possible order p with all of the procedures (5.4.9). The m-cyclic method constructed in this fashion is, with a sufficiently accurate starting procedure, convergent of order p. However, for the construction of $(k-1)$-cyclic linear k-step schemes of order $2k-1$ (see the first part of Section 4.3.3) one cannot choose the ρ_μ independently because one has to achieve an order $2k-1$ for $\langle\rho_m, \sigma_m\rangle$ which is not possible for prescribed ρ_m according to Theorem 4.1.9. The simultaneous construction of the $\langle\rho_\mu, \sigma_\mu\rangle$ has been analyzed by Mischak [1].

Of course, we need not restrict our attention to linear k-step procedures. Our theory shows that the replacement of the linear procedures of Section 4.3 by the corresponding P(EC)^{m}E-procedures does not change the situation as long as the principal error functions of the PC-procedures are the same as those of the correctors, cf. Corollary 5.2.7. But we might also use PC-procedures with off-step points if we want further to increase the accessible order.

If one selects the increment functions Ψ_μ such that the principal error functions φ_μ of the individual procedures (5.4.9) of order p satisfy (5.4.28), then the generated m-cyclic method is even convergent of order $p+1$ by Theorem 5.4.3. This corresponds to the constructions in the second part of Section 4.3.3 where k-cyclic linear k-step methods of order $2k$ were obtained. Whereas in Section 4.3.3 we could not offer a formal proof for the fact that only *one* further condition on the linear k-step schemes of the k-cyclic method was sufficient for the increase in order, this fact is now a consequence of Theorem 5.4.3. Obviously, condition (4.3.24) of Donelson and Hansen [1] is equivalent to (5.4.28), with $q=1$, in the case of k linear k-step schemes of order $2k-1$.

We do not want to indicate further the relations between the theory in Section 4.3 and the theory based on the m-stage approach since the technical details are very cumbersome.

Example. $m=k=2$, cf. the example at the end of Section 4.3.2. If we parametrize the linear 2-step schemes ψ_μ of order 3 by the second zero x_μ of ρ_μ we obtain as a 2-cyclic linear 2-step method based upon ψ_1 and ψ_2

$$A=\left(\begin{array}{cc|cc} \dfrac{x_1}{1-x_1} & -\dfrac{1+x_1}{1-x_1} & \dfrac{1}{1-x_1} & 0 \\ 0 & \dfrac{x_2}{1-x_2} & -\dfrac{1+x_2}{1-x_2} & \dfrac{1}{1-x_2} \end{array}\right),$$

$$\varphi(\mathbf{z}(t)) = \tfrac{1}{24}\begin{pmatrix} \frac{1+x_1}{1-x_1} \\ \frac{1+x_2}{1-x_2} \end{pmatrix} z^{\mathrm{IV}}(t), \qquad A_1^{-1}\varphi = \tfrac{1}{24}\begin{pmatrix} 1+x_1 \\ (1+x_2)(2+x_1) \end{pmatrix} z^{\mathrm{IV}}.$$

The fundamental sequences $\{\zeta_{\bar{\nu}}^{i}\}$ for the zeros $\bar{x}_1=1$ and $\bar{x}_2=x_1x_2$ of the associated polynomial ρ of A are

$$\left\{\begin{pmatrix}1\\1\end{pmatrix}, \begin{pmatrix}1\\1\end{pmatrix}, \dots\right\} \quad \text{and} \quad \left\{\begin{pmatrix}1+x_1\\ x_1(1+x_2)\end{pmatrix}, \; x_1x_2\begin{pmatrix}1+x_1\\ x_1(1+x_2)\end{pmatrix}, \dots\right\}$$

so that (5.4.28) becomes

$$\tfrac{1}{24}(1+x_1)z^{\mathrm{IV}} = d_1 + (1+x_1)d_2,$$
$$\tfrac{1}{24}(1+x_2)(2+x_1)z^{\mathrm{IV}} = d_1 + x_1(1+x_2)d_2,$$

and the condition for $d_1=0$ is easily found to be

$$(1+x_1)(1+x_2)=0.$$

Thus we have to take one of the x_μ as -1 (Simpson scheme); then the other x_μ may be arbitrarily taken from $(-1, +1)$ and the generated 2-cyclic 2-step method will be stable and convergent of order 4 (with a starting procedure of order 4) for an $\mathrm{IVP}^{(4)}\,1$.

5.5 Strong Stability

Due to the complicated structure of general multistage multistep methods we will restrict our analysis of their strong stability properties to the presentation of their characteristic polynomial and the discussion of special cases.

5.5.1 Characteristic Polynomial, Stability Regions

Theorem 5.5.1. *The characteristic polynomial of a simple m-stage k-step method with the forward step procedure* (5.1.1) *is given by*

$$\varphi(x,H) := \det\left[\sum_{\kappa=0}^{k} (A_\kappa - H\,B_\kappa)x^\kappa\right]. \tag{5.5.1}$$

Proof. We have to show that the simple m-stage k-step method $\mathfrak{M}$ (see Def. 5.1.2) is exponentially stable for an $\{\mathrm{IVP}\,1\}_{\mathrm{T}} \in J_c$ with a matrix g such that the eigenvalues of hg are inside

$$\mathfrak{H}_0 := \{H\in\mathbb{C}\colon \text{ all zeros } x_i(H) \text{ of } \varphi(x,H) \text{ satisfy } |x_i(H)|<1\}. \tag{5.5.2}$$

Let us first consider the case of a scalar equation $y'-gy=0$, $g\in\mathbb{C}$. With $H:=hg\in\mathbb{C}$ as usual we have the system of m linear difference equations of order k

$$\sum_{\kappa=0}^{k} [A_\kappa - H\,B_\kappa]\boldsymbol{\eta}_{\nu-k+\kappa} = 0. \tag{5.5.3}$$

We may assume that $A_k - H\,B_k$ is nonsingular for $H\in\mathfrak{H}_0$. (For H-values in the vicinity of an H which makes $A_k-H\,B_k$ singular, (5.5.3) would have very rapidly growing solution.) Therefore we may proceed as in the proof of Lemma 5.1.2 and replace (5.5.3) by

$$\begin{pmatrix} \boldsymbol{\eta}_{\nu-k+1} \\ \vdots \\ \vdots \\ \boldsymbol{\eta}_{\nu} \end{pmatrix} = \begin{pmatrix} 0 & I & & 0 \\ & \ddots & \ddots & \\ 0 & & 0 & I \\ -C_k^{-1} C_0 & \cdots & \cdots & -C_k^{-1} C_{k-1} \end{pmatrix} \begin{pmatrix} \boldsymbol{\eta}_{\nu-k} \\ \vdots \\ \vdots \\ \boldsymbol{\eta}_{\nu-1} \end{pmatrix} = Q(H) \begin{pmatrix} \boldsymbol{\eta}_{\nu-k} \\ \vdots \\ \vdots \\ \boldsymbol{\eta}_{\nu-1} \end{pmatrix}$$

(5.5.4)

where $C_\kappa(H) := A_\kappa - H B_\kappa$, $\kappa = 0(1)k$. Clearly, we have exponential stability if and only if $\|[Q(H)]^\nu\|$ decreases like c^ν, for some $c<1$; this occurs when all eigenvalues of $Q(H)$ have modulus smaller than 1 (irrespective of their multiplicity). But the eigenvalues of $Q(H)$ are the zeros of $\varphi(x,H)$, cf. (5.1.18).

If $s>1$, so that g is an $s \times s$-matrix, we proceed as in the proof of Theorem 4.6.3: We transform g into its Jordan normal form $\bar{g}$. For each of the r eigenvalues γ_i of g we obtain a "scalar" problem (5.5.3) (i.e., with a scalar H) for the respective components of the $\bar{\eta}_\nu^\mu = S^{-1} \eta_\nu^\mu$, see (4.6.14). This system of m linear difference equations possesses only exponentially decreasing solutions if $H_i = h\gamma_i \in \mathfrak{H}_0$ as shown above. In the case of a multiple eigenvalue γ_i without a full eigenvector space, those components of the $\bar{\eta}_\nu^\mu$ which correspond to principal vectors of g w.r.t. γ_i satisfy an inhomogeneous system (5.5.3), with $H = h\gamma_i$ and an exponentially decreasing inhomogeneity (for $h\gamma_i \in \mathfrak{H}_0$). Thus these components also decrease exponentially for $h\gamma_i \in \mathfrak{H}_0$. □

Remark. The characteristic polynomial $\varphi(x,H)$ of a simple m-stage k-step method $\mathfrak{M}$ is normally of degree mk in x and of degree m in H. In many special cases the degree of φ in x is reducible since powers of x may be factored out, which obviously does not influence $\mathfrak{H}_0$ (see (5.5.2)). The polynomial without this factor x^l is also called the characteristic polynomial of $\mathfrak{M}$. Similarly, multiplication of $\varphi(x,H)$ by a function of H does not change $\mathfrak{H}_0$ and is therefore permitted.

If only $m-l$ of the mk different $\eta_\kappa^\mu \in \mathbb{R}^s$, $\kappa = 0(1)k-1$, $\mu = 1(1)m$, enter into the computation of the η_k^μ, the characteristic polynomial must possess the common factor x^l and the reduced characteristic polynomial is of degree $m-l$ only. For examples see Section 5.5.2.

As in the proof of Theorem 4.6.3 we may also show that the regions $\mathfrak{H}_\mu$ of μ-exponential stability are given by

$$\mathfrak{H}_\mu := \{H \in \mathbb{C}\colon \text{ all zeros of } \varphi \text{ satisfy } |x_i(H)| < e^{-\mu}\}.$$

Naturally, these regions may be empty.

It is clear that the stability regions provide only a qualitative guide in the case of general $\{\text{IVP 1}\}_{\mathbb{T}} \notin J_c$; see the discussion in Sections 3.5.4 and 3.5.5.

Example. Example 5 from Section 5.1.1. For Butcher's explicit 3-stage 1-step method of order 4 we obtain

$$\varphi(x,H) = \begin{vmatrix} x+\frac{H}{4} & 0 & -\left(1+\frac{3}{4}H\right) \\ -H(2x+1) & x & -1+2H \\ -\frac{2}{3}Hx & -\frac{H}{6}x & x-1-\frac{H}{6} \end{vmatrix}$$

$$= x^3 - x^2\left(1+\frac{3}{4}H+\frac{1}{2}H^2+\frac{1}{4}H^3\right) - x\frac{H}{4}\left(1+H+\frac{H^2}{6}\right).$$

The characteristic polynomial is effectively only of degree 2 in x as the old second stage is not used in the computation of the values at the next gridpoint. The stability region $\mathfrak{H}_0$ ends somewhere between -2 and -2.5 along the negative real H-axis which is typical for explicit one-step methods.

5.5.2 Stability Regions of PC-methods

The degree in x of the characteristic polynomial φ of a PC-method is usually much smaller than mk since the initial iteration stages are not used in further computation. In particular, φ must be reducible to degree k for a *straight* m-stage k-step PC-method, see Def. 5.2.1 and 5.2.2.

For P(EC)^{m}E- and P(EC)^m-methods (see Def. 5.2.4) the characteristic polynomial may be expressed in terms of the associated polynomials of the predictor and the corrector:

Theorem 5.5.2. *Consider a k-step* P(EC)^{m}E- *or* P(EC)^m-*method based on the predictor scheme* $\langle\rho^0,\sigma^0\rangle$ *and the corrector scheme* $\langle\rho,\sigma\rangle$. *The (reduced) characteristic polynomial of the* P(EC)^{m}E-*method is given by*

$$\varphi(x,H) = \left[\frac{1}{\alpha_k}\sum_{\mu=0}^{m-1}\left(\frac{H\beta_k}{\alpha_k}\right)^{\mu}\right](\rho(x)-H\sigma(x)) + \frac{1}{\alpha_k^0}\left(\frac{H\beta_k}{\alpha_k}\right)^{m}(\rho^0(x)-H\sigma^0(x)),$$

(5.5.5)

that of the P(EC)^m-*method by*

$$\varphi(x,H) = \left[\frac{1}{\alpha_k}\sum_{\mu=0}^{m-1}\left(\frac{H\beta_k}{\alpha_k}\right)^{\mu}\right](\rho(x)-H\sigma(x))x^k + \frac{H}{\alpha_k^0\alpha_k}\left(\frac{H\beta_k}{\alpha_k}\right)^{m-1}(\rho^0(x)\sigma(x)-\rho(x)\sigma^0(x)). \tag{5.5.6}$$

Proof. a) From Theorem 5.5.1 and Def. 5.2.4 we have for the characteristic polynomial of the $\mathrm{P(EC)}^m\mathrm{E}$-method the expression

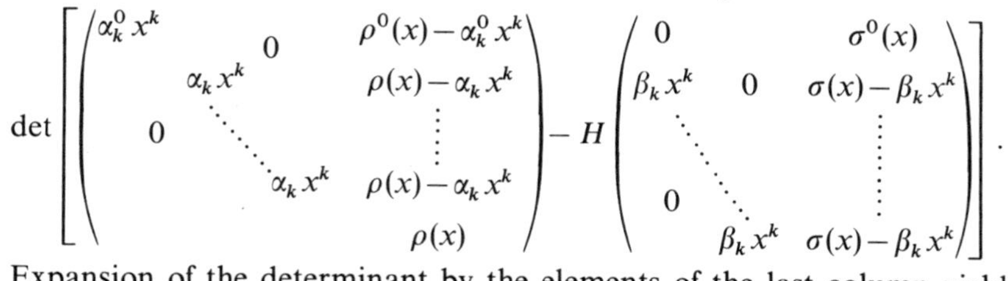

Expansion of the determinant by the elements of the last column yields

$$\begin{aligned}
&(\rho^0(x)-H\,\sigma^0(x)-\alpha_k^0\,x^k)\,(H\,\beta_k\,x^k)^m\\
&\quad+(\rho(x)-H\,\sigma(x)-(\alpha_k-H\,\beta_k)\,x^k)\,(H\,\beta_k\,x^k)^{m-1}(\alpha_k^0\,x^k)+\cdots\\
&\quad+(\rho(x)-H\,\rho(x)-(\alpha_k-H\,\beta_k)\,x^k)\,(H\,\beta_k\,x^k)\,(\alpha_k^0\,x^k)\,(\alpha_k\,x^k)^{m-2}\\
&\quad+(\rho(x)-H\,\sigma(x)+H\,\beta_k\,x^k)\,(\alpha_k^0\,x^k)\,(\alpha_k\,x^k)^{m-1}\\
&\quad=x^{mk}\,\alpha_k^0\,(\alpha_k)^m\,\varphi(x,H),\quad \text{with } \varphi \text{ as in (5.5.5).}
\end{aligned}$$

b) The derivation of (5.5.6) from the general expression (5.5.1) is more cumbersome since the last two columns of $\sum_\kappa (A_\kappa - H\,B_\kappa)$ are now non-zero. The details are purely technical, and thus we restrict ourselves to verifying (5.5.6) for $m=2$:

$$\begin{aligned}
&\det\left[\begin{pmatrix}\alpha_k^0\,x^k & 0 & \rho^0(x)-\alpha_k^0\,x^k\\ 0 & \alpha_k\,x^k & \rho(x)-\alpha_k\,x^k\\ 0 & 0 & \rho(x)\end{pmatrix}-H\begin{pmatrix}0 & \sigma^0(x) & 0\\ \beta_k\,x^k & \sigma(x)-\beta_k\,x^k & 0\\ 0 & \sigma(x) & 0\end{pmatrix}\right]\\
&=x^k\begin{vmatrix}\alpha_k^0 & -H\,\sigma^0(x) & \rho^0(x)-\alpha_k^0\,x^k\\ -H\,\beta_k & \alpha_k\,x^k-H\,\sigma(x)+H\,\beta_k\,x^k & \rho(x)-\alpha_k\,x^k\\ 0 & -H\,\sigma(x) & \rho(x)\end{vmatrix}\\
&=\alpha_k^2\,\alpha_k^0\,x^k\left\{\left(\frac{\rho^0(x)}{\alpha_k^0}-x^k\right)\frac{H\,\beta_k}{\alpha_k}\,\frac{H\,\sigma(x)}{\alpha_k}+\left(\frac{\rho(x)}{\alpha_k}-x^k\right)\frac{H\,\sigma(x)}{\alpha_k}\right.\\
&\qquad\left.+\frac{\rho(x)}{\alpha_k}\left(x^k-\frac{H\,\sigma(x)}{\alpha_k}+\frac{H\,\beta_k}{\alpha_k}\,x^k\right)-\frac{\sigma(x)}{\alpha_k}\,\frac{H\,\beta_k}{\alpha_k}\,\frac{H\,\rho^0(x)}{\alpha_k^0}\right\}\\
&=\alpha_k^2\,\alpha_k^0\,x^k\left\{\frac{1}{\alpha_k}\left(1+\frac{H\,\beta_k}{\alpha_k}\right)(\rho(x)-H\,\sigma(x))\,x^k\right.\\
&\qquad\left.+\frac{H}{\alpha_k^0\,\alpha_k}\,\frac{H\,\beta_k}{\alpha_k}\,(\rho^0(x)\,\sigma(x)-\rho(x)\,\sigma^0(x))\right\}. \qquad \square
\end{aligned}$$

Remark. Note that both $\langle\rho^0,\sigma^0\rangle$ and $\langle\rho,\sigma\rangle$ have been assumed to be k-step schemes. This may always be formally achieved, see Remark 2 after Def. 5.2.4.

Corollary 5.5.3. *Denote the zeros of the characteristic polynomial of a* $\left.\begin{matrix}\mathrm{P(EC)^m E}\\ \mathrm{P(EC)^m}\end{matrix}\right\}$*-method by* $x_i^{(m)}(H)$, $i=1(1)\begin{cases}k\\2k\end{cases}$, *and the zeros of the characteristic polynomial* $\rho(x)-H\sigma(x)$ *of the corrector method by* $x_i(H)$, $i=1(1)k$. *If and only if*

$$\left|\frac{H\beta_k}{\alpha_k}\right|<1 \tag{5.5.7}$$

we have (with a suitable numbering of the zeros)

$$\lim_{m\to\infty} x_i^{(m)}(H)=x_i(H), \qquad i=1(1)k,$$

$$\lim_{m\to\infty} x_i^{(m)}(H)=0, \qquad i=k+1(1)2k.$$

Proof. The sufficiency of (5.5.7) follows from the continuous dependence of the zeros on the coefficients of the polynomial. The necessity follows from the fact that the coefficients of (5.5.5) and (5.5.6) do not tend towards a limit as $m\to\infty$ if (5.5.7) is violated. □

According to Corollary 5.5.3, one must not expect that the stability regions of a PC-method will resemble those of the corrector more closely when the corrector has been iterated more often. This would be true if and only if the region $\mathfrak{H}_0$ of the corrector were to lie totally inside the disk in the H-plane determined by (5.5.7). $|\alpha_k/\beta_k|$ is typically close to 2 whereas corrector methods, being implicit, have large stability regions; thus, in all practical cases, our assumption is not satisfied.

Extensive computations of stability regions have shown that the stability regions of $\mathrm{P(EC)^m E}$-methods as well as of $\mathrm{P(EC)^m}$-methods normally remain completely within the disk $|H|<|\alpha_k/\beta_k|$. For given predictor and corrector schemes their detailed shape depends strongly on m and on whether a final evaluation E is made; the approximate size of the regions, however, remains largely independent of these parameters, see, e. g., Stetter [6].

In the same paper[7], it is shown that the stability regions of PC-methods are appreciably enlarged if the final value η_ν^m is formed as a suitable linear combination of the η_ν^μ, $\mu=0(1)m-1$. There is practically no extra effort and normally no change in accuracy. No systematic study of these slight generalizations of $\mathrm{P(EC)^m E}$-methods has been made so far.

Lambert [2] has observed that the characteristic polynomial of a k-step $\mathrm{P(EC)^m}$-method is identical to that of a $\mathrm{P(EC)^{m-1}E}$-method with

[7] The second paragraph of "Remark" after Theorem 2.2 is *incorrect* in this paper, cf. Theorem 4.6.6.

the same corrector and a certain $2k$-step predictor. To see this we write (5.5.6) in the form

$$\left[\frac{1}{\alpha_k}\sum_{\mu=0}^{m-2}\left(\frac{H\beta_k}{\alpha_k}\right)^{\mu}\right](\rho(x)-H\sigma(x))x^k$$

(5.5.8)

$$+\frac{1}{\alpha_k^0}\left(\frac{H\beta_k}{\alpha_k}\right)^{m-1}\frac{1}{\alpha_k}\left[\alpha_k^0\rho(x)x^k-H\big(\sigma(x)(\alpha_k^0x^k-\rho^0(x))+\rho(x)\sigma^0(x)\big)\right],$$

which is (5.5.5) with

$$\bar{\rho}(x)=\rho(x)x^k,\qquad \bar{\rho}^0(x)=\frac{\alpha_k^0}{\alpha_k}\rho(x)x^k,$$

$$\bar{\sigma}(x)=\sigma(x)x^k,\qquad \bar{\sigma}^0(x)=\frac{1}{\alpha_k}\left[\sigma(x)(\alpha_k^0x^k-\rho^0(x))+\rho(x)\sigma^0(x)\right].$$

Obviously, $\bar{\rho}^0$ and $\bar{\sigma}^0$ have degrees $2k$ and $2k-1$ resp. For $m=1$, (5.5.8) reduces to the characteristic polynomial of the predictor based on the linear $2k$-step scheme $\langle\bar{\rho}^0,\bar{\sigma}^0\rangle$. Lambert concludes that, for a given corrector, the maximal stability region for P(EC)^m-methods cannot be larger than for P(EC)^{m}E-methods if arbitrary predictors as well as an arbitrary number m of iterations are considered.

5.5.3 Stability Regions of Cyclic Methods

In the discussion of cyclic f.s.m. in Sections 4.3 and 5.4 the main emphasis was on the achievable order of convergence. Even so, it had become evident that the cyclic use of different forward step procedures tends to "increase" stability: It has been shown that there exist stable cyclic multistep methods all "component methods" of which are outright unstable, see Section 4.3.3.

Furthermore, in the example at the end of Section 4.3.1 we saw that the cyclic use of a Simpson procedure and a 2-step Adams corrector leads to a method which is no longer weakly stable but still of order 4. We are now in a position to determine the region of absolute stability for this 2-cyclic 2-step method. Writing it as a 2-stage 1-step method (see Example 4 in Section 5.1.1) we have

$$A_0=\begin{pmatrix}-1&0\\0&0\end{pmatrix},\quad A_1=\begin{pmatrix}1&0\\-1&1\end{pmatrix},\quad B_0=\begin{pmatrix}\frac{1}{3}&\frac{4}{3}\\0&-\frac{1}{12}\end{pmatrix},\quad B_1=\begin{pmatrix}\frac{1}{3}&0\\\frac{2}{3}&\frac{5}{12}\end{pmatrix},$$

and from (5.5.1)

$$\varphi(x,H)=\begin{vmatrix}x-1-\dfrac{H}{3}(x+1) & -\dfrac{4}{3}H\\ -x\left(1+\dfrac{2}{3}H\right) & x\left(1-\dfrac{5}{12}H\right)+\dfrac{H}{12}\end{vmatrix}$$

$$=x^2(1-\tfrac{3}{4}H+\tfrac{5}{36}H^2)-x(1+\tfrac{7}{6}H+\tfrac{7}{9}H^2)-\tfrac{1}{12}(H+\tfrac{1}{3}H^2).$$

It turns out that $\mathfrak{H}_0$ extends to -3 along the negative real axis which is exactly halfway between 0 for the Simpson method and -6 for the 2-step implicit Adams method.

Much better results are to be expected when the choice of the components of the cyclic f.s.m. is made with the aim of achieving a large region of absolute stability rather than a high order of convergence. An analysis of this situation is in progress; we will indicate its scope with a discussion of the simplest case.

We wish to combine cyclically m explicit RK-procedures of order 1 such that the resulting method is only convergent of order 1 but has an enlarged region of absolute stability. The simplest explicit RK-procedure of order 1 is the Euler-procedure, its only parameter being its stepsize. Thus, our problem is reduced to finding an m-cycle of stepsizes which will make the Euler method "more stable".

If we consider one cycle as one step of an m-stage 1-step method and denote the steps of the individual Euler procedures by $a_\mu h$, $\mu = 1(1)m$, $\sum_\mu a_\mu = 1$, we obtain the characteristic polynomial

$$\varphi(x, H) = x - (1 + a_1 H)(1 + a_2 H) \cdots (1 + a_m H) =: x - \gamma(H).$$

When we consider only the extension of $\mathfrak{H}_0$ along the negative real axis we simply have to choose the growth function

$$\gamma(H) \approx T_m\left(\frac{H}{m^2} + 1\right)$$

where $T_m(x) = \cos(m \arccos x)$ is the m-th Chebyshev polynomial. If we take $T_m(H/m^2 + 1)$ itself then the region $\mathfrak{H}_0$ breaks up into m disjoint components whose boundaries are connected at the real values of H at which the Chebyshev polynomial has its etrema of modulus 1; the "endpoint" of the boundary of $\mathfrak{H}_0$ is $-2m^2$ in this case. The condition $\sum a_\mu = 1$ is automatically satisfied since $\gamma'(0) = 1$, which is the condition for the cyclic method to be of order 1.

More reasonable regions $\mathfrak{H}_0$ are obtained when we employ the polynomial γ_e of degree m which has $m-1$ extrema of modulus $e < 1$ with alternating signs and satisfies

$$\gamma_e(0) = \gamma_e'(0) = 1\,. \tag{5.5.9}$$

Naturally γ_e is also a transformed Chebyshev polynomial, viz.

$$\gamma_e(H) = e\, T_m\left(\frac{H}{a} + b\right) \tag{5.5.10}$$

where a and b are easily obtained from the conditions (5.5.9). For $e \approx 0.8$ we obtain stability regions for our m-cyclic Euler method which are nearly convex and still extend close to $-2m^2$ along the negative real axis.

In comparing these stability regions with the region $|H+1|<1$ of the uniform Euler method we have to reduce H by a factor m since we have performed m Euler steps in one cycle. This means that, by the simple trick of using different steps in a cyclic fashion, the size of $\mathfrak{H}_0$ has increased by a factor of almost m.

It is obvious that there is some loss in accuracy to be expected since the generated error in one step is $O(h^2)$ and a cyclic variation in stepsize will tend to increase the sum of the local errors. The above discussion is intended mainly as an indication of the potential of cyclic f.s.m. and not as a suggestion for practical use.

Of course, a similar construction for reducing the error propagation is well-known in the context of iteration methods for the solution of linear algebraic equations. It is common practice to use cycles of different overrelaxation factors in an SOR-method (see, e.g., Varga [1]). The use of this trick for the numerical solution of initial value problems (for parabolic partial differential equations) was proposed and analyzed by G. Wachspress [1]; the construction of the γ_e is also due to him.

Chapter 6

Other Discretization Methods for IVP 1

6.1 Discretization Methods with Derivatives of f

In Chapters 3—5, we considered forward step procedures for IVP 1 which use evaluations of f only. In recent years it has been found that the automatic computation of values of higher derivatives of the local solution of an IVP 1 is feasible for large classes of IVP 1. In Section 6.1, we will survey some of the principal approaches which yield f. s. m. using such higher derivatives.

6.1.1 Recursive Computation of Higher Derivatives of the Local Solution

For given $f: \mathbb{R}^s \to \mathbb{R}^s$ the local solution $\bar{z}(h, \xi)$ of the associated IVP 1 has been defined in Def. 3.1.6.

Def. 6.1.1. For a given IVP$^{(p)}$ 1 and ξ from the domain of f we define $\bar{z}^{(r)}: \mathbb{R}^s \to \mathbb{R}^s$ for $r = 0(1)p+1$ by

$$\bar{z}^{(r)}(\xi) := \frac{1}{r!} \frac{\partial^r}{\partial h^r} \bar{z}(0, \xi). \tag{6.1.1}$$

From $\bar{z}(0, \xi) = \xi$, $(\partial/\partial h)\bar{z}(h, \xi) = f(\bar{z}(h, \xi))$ (see (3.1.22)) we obtain on the one hand

$$\frac{\partial}{\partial \xi} \bar{z}(0, \xi) f(\xi) = f(\xi),$$

$$\frac{\partial}{\partial h}\left[\frac{\partial}{\partial \xi} \bar{z}(h, \xi) f(\xi)\right] = f'(\bar{z}(h, \xi))\left[\frac{\partial}{\partial \xi} \bar{z}(h, \xi) f(\xi)\right],$$

on the other hand

$$\frac{\partial}{\partial h} \bar{z}(0, \xi) = f(\xi),$$

$$\frac{\partial}{\partial h}\left[\frac{\partial}{\partial h} \bar{z}(h, \xi)\right] = f'(\bar{z}(h, \xi))\left[\frac{\partial}{\partial h} \bar{z}(h, \xi)\right];$$

this implies $(\partial/\partial h)\bar{z}(h,\xi)=(\partial/\partial\xi)\bar{z}(h,\xi)\,f(\xi)$. Similarly one obtains

$$\frac{\partial^r}{\partial h^r}\bar{z}(h,\xi)=\frac{\partial}{\partial\xi}\left[\frac{\partial^{r-1}}{\partial h^{r-1}}\bar{z}(h,\xi)\right]f(\xi), \qquad r=1(1)p+1\,,$$

which implies

(6.1.2) $$r\bar{z}^{(r)}(\xi)=\bar{z}^{(r-1)\prime}(\xi)\,f(\xi), \qquad r=1(1)p+1\,.$$

From (3.1.23), (3.1.26), (3.1.27), and Theorem 3.1.4 we have

(6.1.3) $$\bar{z}^{(r)}(\xi)=\frac{1}{(r-1)!}\sum_{\lambda=1}^{n_r}\frac{\beta_\lambda^{(r)}}{\gamma_\lambda^{(r)}}F_\lambda^{(r)}(\xi), \qquad r\geq 1\,,$$

where the $F_\lambda^{(r)}$ are the elementary differentials of f, see Section 3.1.3. For wide classes of functions f it is possible to compute values of $\bar{z}^{(r)}$ without the explicit differentiation processes represented by the $F_\lambda^{(r)}$ in (6.1.3); this has been pointed out by R. Moore [1], Fehlberg [4], Leavitt [1], and various other authors.

The basic approach is always the same: It is assumed that the evaluation of each component $f_\sigma:\mathbb{R}^s\to\mathbb{R}$ of f may be decomposed into a sequence of arithmetic operations and evaluations of elementary functions involving the arguments and constants. For example,

$$f_2(\xi_1,\ldots,\xi_4)=\frac{[(\alpha_1\xi_1+\alpha_2\xi_2)\exp\xi_3-\xi_4^3]^{\frac{1}{2}}}{(1+\cos\xi_2)}$$

may be decomposed into

$$\begin{array}{llll} q_1:=\alpha_1\xi_1; & q_2:=\alpha_2\xi_2; & q_3:=q_1+q_2; & \\ q_4:=\exp(\xi_3); & q_5:=q_3q_4; & q_6:=(\xi_4)^3; & q_7:=q_5-q_6; \\ q_8:=\sqrt{q_7}; & q_9:=\cos(\xi_2); & q_{10}:=1+q_9; & f_2:=\dfrac{q_8}{q_{10}}. \end{array}$$

Each of the intermediate results q_i in the decomposition of the f_σ, $\sigma=1(1)s$, is itself a function of the arguments of f and we may define, for given fixed $\xi\in\mathbb{R}^s$,

(6.1.4) $$q_{im}:=\frac{1}{m!}\frac{\partial^m}{\partial h^m}q_i(\bar{z}_1(h,\xi),\ldots,\bar{z}_s(h,\xi))\bigg|_{h=0}, \qquad m=0,1,\ldots$$

The following recursion formulas hold for $m=1,2,\dots$ (proof by induction):

$$
\begin{array}{ll}
q_i := q_j \pm q_k & \text{implies } q_{im} := q_{jm} \pm q_{km} \\
q_i := q_j q_k & q_{im} := \sum_{\mu=0}^{m} q_{j\mu} q_{k,m-\mu} \\
q_i := q_j / q_k & q_{im} := \left(q_{jm} - \sum_{\mu=0}^{m-1} q_{i\mu} q_{k,m-\mu} \right) \Big/ q_k \\
q_i := \text{const} & q_{im} := 0 \\
q_i := \exp q_k & q_{im} := \frac{1}{m} \sum_{\mu=0}^{m-1} (m-\mu) q_{i\mu} q_{k,m-\mu} \\
\left.\begin{array}{l} q_i := \sin q_k \\ q_j := \cos q_k \end{array}\right\} & \begin{array}{l} q_{im} := \frac{1}{m} \sum_{\mu=0}^{m-1} (m-\mu) q_{j\mu} q_{k,m-\mu} \\ q_{jm} := -\frac{1}{m} \sum_{\mu=0}^{m-1} (m-\mu) q_{i\mu} q_{k,m-\mu} \end{array} \\
q_i := (q_k)^c & q_{im} := \frac{1}{m q_k} \sum_{\mu=0}^{m-1} (c(m-\mu)-\mu) q_{i\mu} q_{k,m-\mu}, \\
 & \text{etc.}
\end{array}
$$

The relation

$$
(6.1.5) \quad q_i := \xi_\sigma = \bar{z}_\sigma^{(0)}(\xi_1, \dots, \xi_s) \quad \text{implies} \quad q_{im} := \frac{1}{m} f_{\sigma,m-1} = \bar{z}_\sigma^{(m)}(\xi_1, \dots, \xi_s)
$$

permits the recursive concatenation of the formulas which result from the decomposition of f, so that the components of the $\bar{z}^{(r)}(\xi)$ may be computed.

Example. $f\begin{pmatrix}\xi_1\\ \xi_2\end{pmatrix} = \begin{pmatrix}\xi_1^2+\xi_2^2 \\ 1+\dfrac{\xi_1}{\xi_2}\end{pmatrix}$, compute $\bar{z}^{(3)}\begin{pmatrix}\xi_1\\ \xi_2\end{pmatrix}$

$$
q_1 := \xi_1 \xi_1, \quad q_2 := \xi_2 \xi_2, \quad f_1 := q_1 + q_2,
$$

$$
q_3 := \frac{\xi_1}{\xi_2}, \quad f_2 := 1 + q_3.
$$

Starting from $\xi_{10} := \xi_1$, $\xi_{20} := \xi_2$ and $\xi_{11} := f_{10} = \xi_1^2 + \xi_2^2$, $\xi_{21} := f_{20} = 1 + \xi_1/\xi_2$, one has

$$
q_{11} := 2\xi_{10}\xi_{11}, \quad q_{21} := 2\xi_{20}\xi_{21}, \quad f_{11} := q_{11} + q_{21};
$$

$$
q_{31} := \frac{\xi_{11} - q_{30}\xi_{21}}{\xi_{20}}, \quad f_{21} := q_{31};
$$

$$
\xi_{12} := \tfrac{1}{2} f_{11}, \quad \xi_{22} := \tfrac{1}{2} f_{21};
$$

$$
q_{12} := 2\xi_{10}\xi_{12} + \xi_{11}\xi_{11}, \quad q_{22} := 2\xi_{20}\xi_{22} + \xi_{21}\xi_{21};
$$

$$f_{12} := q_{12} + q_{22};$$

$$q_{32} := \frac{\xi_{12} - q_{30}\xi_{22} - q_{31}\xi_{21}}{\xi_{20}}, \qquad f_{22} := q_{32};$$

$$\xi_{13} := \tfrac{1}{3} f_{12} := \bar{z}_1^{(3)}\begin{pmatrix}\xi_1\\ \xi_2\end{pmatrix}, \qquad \xi_{23} := \tfrac{1}{3} f_{22} := \bar{z}_2^{(3)}\begin{pmatrix}\xi_1\\ \xi_2\end{pmatrix}.$$

When the decomposition of f into admissible operations has been specified, the compilation of the program for the computation of the $\bar{z}^{(r)}(\xi)$ may be executed automatically by the computer. Obviously, all the intermediate q_{im} have to be saved. For a set of routines which implement the recursion formulas and the recursive procedure see Wanner [1].

Since the number of operations required by most of the recursion formulas increases proportionally as the order m of the derivative to be computed, the total number of arithmetic operations for the computation of $\bar{z}^{(\rho)}(\xi)$, $\rho = 1(1)r$, increases like r^2. This is a very modest growth compared with the growth of the number n_r of different elementary differentials of order r and with the growth in the complexity of these differentials, cf. Section 3.1.3.

6.1.2 Power Series Methods

After we have found it feasible to compute the values $\bar{z}^{(p)}(\xi)$ of the derivatives of the local solution of an IVP 1 it becomes a natural idea to generate a discretization of an IVP 1 by using truncated *local Taylor expansions*.

Def. 6.1.2. A *power series method of order p* for an IVP 1 is given by E_n, E_n^0, Δ_n, Δ_n^0 as in Def. 3.3.1 and

$$(6.1.6) \quad [\varphi_n(F)\eta](t_\nu) = \begin{cases} \eta_0 - z_0, & \nu = 0, \\ \dfrac{1}{h_\nu}\left[\eta_\nu - \sum\limits_{\mu=0}^{p} h_\nu^\mu \bar{z}^{(\mu)}(\eta_{\nu-1})\right], & \nu = 1(1)n, \end{cases} \qquad t_\nu \in \mathbb{G}_n.$$

Example. The forward step procedure

$$\frac{1}{h_\nu}[\eta_\nu - \eta_{\nu-1} - h_\nu f(\eta_{\nu-1}) - h_\nu^2 \bar{z}^{(2)}(\eta_{\nu-1})] = 0, \qquad \nu \geq 1,$$

is from a power series method of order 2. A power series method of order 1 is an Euler method.

Theorem 6.1.1. *A power-series method of order p is convergent of order p for an IVP$^{(p)}$ 1.*

Proof. It is a trivial consequence of (cf. (6.1.1))

$$(6.1.7) \qquad \frac{1}{\mu!} z^{(\mu)}(t_{\nu-1}) = \bar{z}^{(\mu)}(z(t_{\nu-1}))$$

that the local discretization error of (6.1.6) is $O(n^{-p})$.

The Lipschitz condition on the non-principal part of (6.1.6) follows from (6.1.3) and the definition of an IVP$^{(p)}$ 1. Thus the assumptions of Corollary 2.2.7 are satisfied and a power series method of order p is stable for an IVP$^{(p)}$ 1. (The details are analogous to those in the proof of Theorem 3.3.1.) The assertion is now a consequence of Theorem 1.2.4. □

As the computation of $\bar{z}^{(\mu)}$ is roughly equivalent to μ evaluations of f for many choices of f—see the end of Section 6.1.1—a power series method of order p will normally require a higher computational effort per step than a RK-method of order p for $p>1$. On the other hand, power series methods are very straightforward and transparent.

An *estimation of the local discretization error* may naturally be obtained by consideration of the next term $h^p \bar{z}^{(p+1)}(\eta_{\nu-1})$ in the Taylor expansion (6.1.6). Thus, a stepsize control mechanism may be used in the traditional manner.

Through the use of *interval analysis techniques* one may even be able to compute rigorous lower and upper bounds for the components of the true solution z. For an account of these possibilities the reader may consult Moore [1], Chapter 13.

The analysis of *strong stability* properties of power series methods proceeds along the same lines as for RK-methods. The characteristic polynomial of a power series method of order p is simply

$$\varphi(x,H) = x - \sum_{\mu=0}^{p} \frac{H^\mu}{\mu!}. \tag{6.1.8}$$

Being fully explicit, power series methods are not suitable for the numerical integration of stiff IVP 1.

6.1.3 The Perturbation Theory of Groebner-Knapp-Wanner

In his investigations on Lie series Groebner has derived a representation for the solution of an IVP 1 in terms of the solution of a neighboring IVP 1. Knapp and Wanner have modified Groebner's theory with the aim of applying it to the numerical solution of IVP 1. See Groebner-Knapp [1] and Knapp-Wanner [1], [2].

Theorem 6.1.2. *Consider an IVP$^{(r+1)}$ 1, with $f: \mathbb{R}^s \to \mathbb{R}^s$ and initial value z_0, and with solution z and local solution $\bar{z}$. Let $\hat{z}$ be the solution of the IVP 1 (of the same dimension s)*

$$\hat{z}(0) = z_0; \quad \hat{z}'(t) - \hat{f}(\hat{z}(t)) = 0, \ t>0. \tag{6.1.9}$$

Then, for any $t>0$,

$$z(t)-\hat{z}(t)=\sum_{\rho=1}^{r}\int_0^t (t-\tau)^{\rho-1}\left[\rho\overline{z}^{(\rho)}(\hat{z}(\tau))-\overline{z}^{(\rho-1)\prime}(\hat{z}(\tau))\,\hat{f}(\hat{z}(\tau))\right]d\tau$$
(6.1.10)
$$+\int_0^t (t-\tau)^r\left[(r+1)\overline{z}^{(r+1)}(z(\tau))-\overline{z}^{(r)\prime}(\hat{z}(\tau))\,\hat{f}(\hat{z}(\tau))\right]d\tau$$

where $\overline{z}^{(\rho)}: \mathbb{R}^s\to\mathbb{R}^s$ *has been defined in Def.* 6.1.1.

Proof. (6.1.10) holds for $r=0$:

$$z(t)-\hat{z}(t)=\int_0^t[z'(\tau)-\hat{z}'(\tau)]\,d\tau=\int_0^t\left[\overline{z}^{(1)}(z(\tau))-\hat{f}(\hat{z}(\tau))\right]d\tau$$

due to (6.1.2); this equals (6.1.10) since $\overline{z}^{(0)\prime}(\xi)$ is the identity for arbitrary ξ.

Assume (6.1.10) to be correct for $r-1$, then trivially

$$z(t)-\hat{z}(t)=\sum_{\rho=1}^{r}\int_0^t (t-\tau)^{\rho-1}\left[\rho\overline{z}^{(\rho)}(\hat{z}(\tau))-\overline{z}^{(\rho-1)\prime}(\hat{z}(\tau))\,\hat{f}(\hat{z}(\tau))\right]d\tau$$
(6.1.11)
$$+\int_0^t (t-\tau)^{r-1}\,r\left[\overline{z}^{(r)}(z(\tau))-\overline{z}^{(r)}(\hat{z}(\tau))\right]d\tau.$$

Partial integration of the remainder term yields

$$-(t-\tau)^r\left[\overline{z}^{(r)}(z(\tau))-\overline{z}^{(r)}(\hat{z}(\tau))\right]\Big|_0^t$$
$$+\int_0^t (t-\tau)^r\left[\overline{z}^{(r)\prime}(z(\tau))\,z'(\tau)-\overline{z}^{(r)\prime}(\hat{z}(\tau))\,\hat{z}'(\tau)\right]d\tau.$$

The first term vanishes at both ends since $z(0)=\hat{z}(0)=z_0$; in the bracket of the integral the first term is $(r+1)\overline{z}^{(r+1)}(z(\tau))$ by (6.1.2) and (6.1.1) while $\hat{z}'(\tau)=\hat{f}(\hat{z}(\tau))$ by (6.1.9). □

Remark. The perturbation formula with the remainder in the form of (6.1.11) is naturally valid for an $\mathrm{IVP}^{(r)}1$.

Example. For $\hat{z}(\tau)\equiv z_0$, $\hat{f}(\xi)\equiv 0$, we obtain

$$\overline{z}^{(\rho)}(\hat{z}(\tau))=\overline{z}^{(\rho)}(z_0)=\frac{1}{\rho!}z^{(\rho)}(0)\qquad\text{(cf. (6.1.7))}$$

and (6.1.10) becomes the Taylor formula

$$z(t)-z_0=\sum_{\rho=1}^{r}\frac{t^\rho}{\rho!}z^{(\rho)}(0)+\int_0^t\frac{(t-\tau)^r}{r!}z^{(r+1)}(\tau)\,d\tau.$$

This shows that Theorem 6.1.2 is a nontrivial generalization of the Taylor formula.

Knapp and Wanner have observed (see Knapp-Wanner [1], Wanner [1]) that values of the quantities $\bar{z}^{(\rho-1)\prime}(\xi)\hat{f}(\xi)$ may be recursively computed just as easily as those of the quantities $\bar{z}^{(\rho)}(\xi)=(1/\rho)\bar{z}^{(\rho-1)\prime}(\xi)\,f(\xi)$. Obviously, one merely has to replace the relation $\xi_{\sigma 1}=f_{\sigma 0}$ of (6.1.5) by

$$\xi_{\sigma 1}=\hat{z}_{\sigma 1}=\hat{f}_{\sigma 0} \tag{6.1.12}$$

in the *last* differentiation level of the recursion.

However, since most of the recursion formulas in Section 6.1.1 have been derived in an indirect manner, this leads to nontrivial changes in the recursion formulas. As an example we derive the new formula for the case of the quotient:

Let, for given fixed $\xi\in\mathbb{R}^s$,

$$\hat{q}_{im}:=q'_{im}(\xi)\,\hat{f}(\xi)=\sum_{\sigma=1}^{s}\frac{\partial q_{im}}{\partial\xi_\sigma}\hat{f}_\sigma(\xi) \qquad \text{(cf. (6.1.4))}. \tag{6.1.13}$$

For $q_i:=q_j/q_k$ we obtain via $q_j=q_i q_k$

$$q_{jm}=\sum_{\mu=0}^{m}q_{i\mu}q_{k,m-\mu}, \qquad \hat{q}_{jm}=\sum_{\mu=0}^{m}(\hat{q}_{i\mu}q_{k,m-\mu}+q_{i\mu}\hat{q}_{k,m-\mu}).$$

Thus

$$\hat{q}_{im}:=\frac{\left(\hat{q}_{jm}-\sum_{\mu=0}^{m-1}\hat{q}_{i\mu}q_{k,m-\mu}-\sum_{\mu=0}^{m}q_{i\mu}\hat{q}_{k,m-\mu}\right)}{q_k}.$$

In a similar fashion one obtains the complementary set of formulas, see e. g. Wanner [1]:

$q_i:=q_j\pm q_k$ implies $\hat{q}_{im}:=\hat{q}_{jm}\pm\hat{q}_{km}$

$q_i:=q_j q_k$ $\qquad \hat{q}_{im}:=\sum_{\mu=0}^{m}(\hat{q}_{j\mu}q_{k,m-\mu}+q_{j\mu}\hat{q}_{k,m-\mu})$

$q_i:=\frac{q_j}{q_k}$ $\qquad \hat{q}_{im}:=\frac{\left(\hat{q}_{jm}-\sum_{\mu=0}^{m-1}\hat{q}_{i\mu}q_{k,m-\mu}-\sum_{\mu=0}^{m}q_{im}\hat{q}_{k,m-\mu}\right)}{q_k}$

$q_i:=\text{const}$ $\qquad \hat{q}_{im}:=0$

$q_i:=\exp q_k$ $\qquad \hat{q}_{im}:=\frac{1}{m}\sum_{\mu=0}^{m-1}(m-\mu)(\hat{q}_{i\mu}q_{k,m-\mu}+q_{i\mu}\hat{q}_{k,m-\mu}), \quad m>0,$

$\qquad \hat{q}_{i0}:=\hat{q}_{k0}q_i,$

$$\left.\begin{array}{l} q_i := \sin q_k \\ \\ q_j := \cos q_k \end{array}\right\} \qquad \begin{array}{l} \hat{q}_{im} := \dfrac{1}{m} \sum\limits_{\mu=0}^{m-1} (m-\mu)(\hat{q}_{j\mu} q_{k,m-\mu} + q_{j\mu} \hat{q}_{k,m-\mu}), \\ \hat{q}_{jm} := -\dfrac{1}{m} \sum\limits_{\mu=0}^{m-1} (m-\mu)(\hat{q}_{i\mu} q_{k,m-\mu} + q_{i\mu} \hat{q}_{k,m-\mu}), \\ \hat{q}_{i0} := \hat{q}_{k0} q_j, \qquad \hat{q}_{j0} := -\hat{q}_{k0} q_i \end{array}$$

$$q_i := (q_k)^c \qquad \begin{array}{l} \hat{q}_{im} := \dfrac{1}{m q_k} \sum\limits_{\mu=0}^{m-1} (c(m-\mu)-\mu)(\hat{q}_{i\mu} q_{k,m-\mu} + q_{i\mu} \hat{q}_{k,m-\mu}) \\ \qquad\quad - \dfrac{\hat{q}_{k0}}{q_k} \sum\limits_{\mu=0}^{m-1} q_{i\mu}, \\ \hat{q}_{i0} := c \dfrac{\hat{q}_k}{q_k} q_i . \end{array}$$

If the q_{im} and $\hat{q}_{i,m-1}$ for the decomposition of f are computed concurrently for $m=1,2,\ldots,r$, with (6.1.5) supplemented by

$$q_i := \xi_\sigma \quad \text{implies } \hat{q}_{i0} := \hat{f}_\sigma(\xi), \tag{6.1.14}$$

one obtains the components of $\overline{z}^{(\rho-1)\prime}(\xi)\hat{f}(\xi)$ as $\hat{\xi}_{\sigma,\rho-1}$.

6.1.4 Groebner-Knapp-Wanner Methods

The perturbation formula (6.1.10) of Groebner-Knapp-Wanner (with the remainder term omitted) may be turned into a forward step procedure by the specification of a particular approximate local solution $\hat{z}$ and by the use of a quadrature formula for the evaluation of the integrals.

The most immediate choice for $\hat{z}$ is a truncated Taylor series:

$$\hat{z}_w(t;\xi) := \xi + \sum_{\omega=1}^{w} t^\omega \overline{z}^{(\omega)}(\xi), \qquad w \geq 0. \tag{6.1.15}$$

Since

$$\hat{z}'_w(t;\xi) = \sum_{\omega=0}^{w-1} (\omega+1) t^\omega \overline{z}^{(\omega+1)}(\xi) =: \hat{f}_w(t;\xi) \quad \text{for fixed } \xi \in \mathbb{R}^s, \tag{6.1.16}$$

the function $\hat{f}$ of (6.1.9) depends explicitly on t—only on t, in fact. Therefore we would have to consider the independent variable as a separate component (cf. (2.1.2)) in the use of (6.1.9)/(6.1.10):

$$\begin{pmatrix} t \\ \hat{z}_w \end{pmatrix}'(\tau) = \begin{pmatrix} 1 \\ \hat{f}_w(\tau) \end{pmatrix} =: \hat{f}\begin{pmatrix} \tau \\ \hat{z}_w(\tau) \end{pmatrix}$$

where $\hat{f}$ is the function of (6.1.9). However, since the corresponding extension of f would also possess a first component 1, this component cancels in the formation of

$$\rho\overline{z}^{(\rho)}(\hat{z}_w(\tau)) - \overline{z}^{(\rho-1)\prime}(\hat{z}_w(\tau))\,\hat{f}\begin{pmatrix}\tau\\ \hat{z}_w(\tau)\end{pmatrix} = \overline{z}^{(\rho-1)\prime}(\hat{z}_w(\tau))\left[f\begin{pmatrix}\tau\\ \hat{z}_w(\tau)\end{pmatrix} - \hat{f}\begin{pmatrix}\tau\\ \hat{z}_w(\tau)\end{pmatrix}\right]$$

(see (6.1.2)) and we can use $\hat{f}_w$ from (6.1.16) in place of $\hat{f}\begin{pmatrix}\tau\\ \hat{z}_w(\tau)\end{pmatrix}$ in (6.1.10).

The choice of the quadrature method for the integrals in (6.1.10) should take into account the fact that the integrand vanishes like a power of τ near the lower end:

Lemma 6.1.3. *Let* $\hat{z}$ *of* (6.1.9) *satisfy*

$$\hat{z}(t) - z(t) = O(t^{w+1}). \tag{6.1.17}$$

Then, for an $IVP^{(p)}$ *1,*

$$\begin{aligned} &\rho\overline{z}^{(\rho)}(\hat{z}(t)) - \overline{z}^{(\rho-1)\prime}(\hat{z}(t))\,\hat{f}(\hat{z}(t)) \\ &= \overline{z}^{(\rho-1)\prime}(\hat{z}(t))\,[f(\hat{z}(t)) - \hat{f}(\hat{z}(t))] = O(t^w), \qquad \rho = 1(1)p+1. \end{aligned} \tag{6.1.18}$$

Proof. (6.1.17) and the continuous differentiability of $\hat{z}$ and z imply, for some $\bar{t}>0$,

$$\hat{z}(t) - z(t) = t^{w+1} q(t) \quad \text{for } t \in [0, \bar{t}]$$

where q is a differentiable function in $[0, \bar{t}]$. Hence

$$\hat{f}(\hat{z}(t)) - f(z(t)) = \hat{z}'(t) - z'(t) = (w+1)\,t^w q(t) + t^{w+1} q'(t) = O(t^w)$$

and

$$\begin{aligned} \|f(\hat{z}(t)) - \hat{f}(\hat{z}(t))\| &= \|f(\hat{z}(t)) - f(z(t))\| + \|\hat{f}(\hat{z}(t)) - f(z(t))\| \\ &= O(t^{w+1}) + O(t^w) = O(t^w) \end{aligned}$$

through the Lipschitz continuity of f and (6.1.17). Now (6.1.18) follows from the boundedness of $\overline{z}^{(\rho-1)\prime}$. □

The result of Lemma 6.1.3 suggests the use of a Gaussian quadrature rule whose k nodes are the zeros of the k-th degree polynomial from the system of polynomials orthogonal on the interval of integration $[0, t]$ with weight function τ^w.

For the interval $[0,1]$ and given values of w and k let these quadrature formulas be given by

$$\int_0^1 x(\tau)\,d\tau \approx \sum_{\kappa=1}^{k} c_\kappa^{w,k}\, x(a_\kappa^{w,k}). \tag{6.1.19}$$

Then, for $x(\tau)=O(\tau^w)$,

$$\int_0^t (t-\tau)^{\rho-1} x(\tau)d\tau = t^\rho \sum_{\kappa=1}^{k} (1-a_\kappa^{w,k})^{\rho-1} c_\kappa^{w,k} x(a_\kappa^{w,k} t)+O(t^{w+2k+1})$$

(6.1.20)

by the well-known theory of Gaussian integration; see, e. g., Krylov [1].

With parameters w, r, and k from $\mathbb{N}$ we may thus turn (6.1.10)/(6.1.18) into the following forward step procedure

$$\frac{1}{h_\nu}\Big\{\eta_\nu-\hat{z}_w(h_\nu;\eta_{\nu-1})$$

$$(6.1.21)\qquad -\sum_{\rho=1}^{r} h_\nu^\rho \sum_{\kappa=1}^{k} (1-a_\kappa^{w,k})^{\rho-1} c_\kappa^{w,k} \overline{z}^{(\rho-1)'}(\hat{z}_w(a_\kappa^{w,k} h_\nu;\eta_{\nu-1}))$$

$$[f(\hat{z}_w(a_\kappa^{w,k} h_\nu;\eta_{\nu-1}))-\hat{f}_w(a_\kappa^{w,k} h_\nu;\eta_{\nu-1})]\Big\}=0,$$

where $\hat{z}_w$ and $\hat{f}_w$ have been defined in (6.1.15)—(6.1.16) and the $a_\kappa^{w,k}$ and $c_\kappa^{w,k}$ in (6.1.19).

Def. 6.1.3. A f.s.m. applicable to IVP 1, with E_n, E_n^0, Δ_n, Δ_n^0 as in Def. 3.3.1 and with a forward step procedure (6.1.21) is called a *G(roebner)-K(napp)-W(anner)-method* or *Lie series method.*

Theorem 6.1.4. *A GKW-method with parameters w, r, k is convergent of order*

$$(6.1.22)\qquad p=w+\min(r,2k)$$

for an IVP$^{(p)}$ 1.

Proof. From (6.1.11) we obtain for the values of the local discretization error of a GKW-method with forward step procedure (6.1.21)

$$\frac{1}{h_\nu}\left[\int_0^{h_\nu} (h_\nu-\tau)^{r-1} r[\overline{z}^{(r)}(z(\tau))-\overline{z}^{(r)}(\hat{z}_w(\tau;z(t_{\nu-1})))]d\tau + \text{quadrature error}\right]$$

$$=O(h_\nu^{r-1})O(h_\nu^{w+1})+O(h_\nu^{w+2k})$$

due to $\hat{z}_w(\tau;z(t_{\nu-1}))-z(\tau)=O(\tau^{w+1})$ and the Lipschitz continuity of $\overline{z}^{(r)}$ on the one hand and (6.1.20) on the other hand. Thus the GKW-method is consistent of order p as given by (6.1.22).

Stability follows once more from the Lipschitz continuity of the increment function in (6.1.21) w.r.t. $\eta_{\nu-1}$ so that the assertion follows, as usual, from Theorem 1.2.4. □

The result on the order of a GKW-method raises the question of whether this method gives any advantage over a simple power series method of order $w+r$. According to our observation at the end of Section 6.1.1 the computational effort for the evaluation of the coefficients $\bar{z}^{(\omega)}(\xi)$ of $\hat{z}_w(t;\xi)$ is proportional to w^2. Similarly, we find from the considerations in Section 6.1.3 that the computational effort for the evaluation of the terms in the sum in (6.1.21) is proportional to $2r^2$ for each node. Thus the total effort for the computation of η_ν from (6.1.21) is roughly proportional to w^2+2kr^2. Due to (6.1.22) we should choose $k \geq r/2$ so that we have an effort like w^2+r^3 for a convergence order $p=w+r$.

This indicates that (for larger w) the highest order for a given effort is not achieved with $r=0$ —i. e., the power series method—but roughly with $3r^2=2w$. Thus, the following choices of w, r, k should be reasonable:

w	r	k
1— 4	1	1
4—10	2	1 or 2
10—20	3	2

Furthermore, Wanner [1] reports that for constant $p=w+r$ the actual errors tend to decrease with increasing r and correspondingly decreasing w.

An *estimation of the locally generated discretization error* may be based upon the remainder term in (6.1.11) if $2k>r$ so that the quadrature error is of higher order than this remainder term, cf. proof of Theorem 6.1.4. We put

$$\begin{aligned}(6.1.23)\qquad &\bar{z}^{(r)}(\bar{z}(\tau,\eta_{\nu-1}))-\bar{z}^{(r)}(\hat{z}_w(\tau;\eta_{\nu-1}))\\ &=\left(\frac{\tau}{h_\nu}\right)^{w+1}\left[\bar{z}^{(r)}(\bar{z}(h_\nu,\eta_{\nu-1}))-\bar{z}^{(r)}(\hat{z}_w(h_\nu;\eta_{\nu-1}))\right]+O(\tau^{w+2})\end{aligned}$$

which is justified by (6.1.15). The integration of the remainder integral in (6.1.11) with (6.1.23) leads to

$$\left(\int_0^{h_\nu}(h_\nu-\tau)^{r-1}\left(\frac{\tau}{h_\nu}\right)^{w+1}d\tau\right)r\left[\bar{z}^{(r)}(\bar{z}(h_\nu,\eta_{\nu-1}))-\bar{z}^{(r)}(\hat{z}_w(h_\nu;\eta_{\nu-1}))\right]+O(h_\nu^{w+r+2})$$

$$(6.1.24)\qquad =\frac{r!(w+1)!}{(r+w+1)!}\left[\bar{z}^{(r)}(\eta_\nu)-\bar{z}^{(r)}(\hat{z}_w(h_\nu;\eta_{\nu-1}))\right]h_\nu^r+O(h_\nu^{w+r+2})$$

where we have replaced $\bar{z}(h_\nu,\eta_{\nu-1})$ by $\eta_\nu=\bar{z}(h_\nu,\eta_{\nu-1})+O(h_\nu^{w+r+1})$; the resultant error of order $O(h_\nu^{w+2r+1})$ is absorbed by the term $O(h_\nu^{w+r+2})$ for $r\geq 1$. The principal part of (6.1.24) may be evaluated by the same technique as is used for the evaluation of (6.1.21). The use of (6.1.24) is described by Wanner [1] and is reported to give very good estimates for reasonably small h_ν.

A complete FORTRAN program for the GKW-method may be found in Knapp-Wanner [2] or Wanner [1].

6.1.5 Runge-Kutta-Fehlberg Methods

Knowledge of an approximate local solution $\hat{z}(t,\eta_{\nu-1})$ may also be utilized in the computation of η_ν by means of a combination with the Runge-Kutta approach: The differential equation for $\tilde{z}(t,\eta_{\nu-1}) := \bar{z}(t,\eta_{\nu-1}) - \hat{z}(t,\eta_{\nu-1})$ is "solved" by a special RK-procedure which takes into account the vanishing of the lower derivatives of $\tilde{z}$ at $t=0$. This idea has been suggested and elaborated by Fehlberg in several papers (e.g. [5], [6]).

Lemma 6.1.5. *Assume that the solution* $\tilde{z}$ *of* $\tilde{z}(0)=0$, $\tilde{z}'(t)=\tilde{f}(t,\tilde{z}(t))$, *with* $\tilde{f}\colon \mathbb{R}\times\mathbb{R}^s\to\mathbb{R}^s$, *satisfies*

$$\tilde{z}'(0)=\tilde{z}''(0)=\cdots=\tilde{z}^{(w)}(0)=0, \qquad w\geq 1. \tag{6.1.25}$$

Define $f\colon \mathbb{R}^{s+1}\to\mathbb{R}^{s+1}$ *by*

$$f\begin{pmatrix} t\\ \xi\end{pmatrix}=\begin{pmatrix}1\\ \tilde{f}(t,\xi)\end{pmatrix}, \qquad \xi\in\mathbb{R}^s. \tag{6.1.26}$$

Then all elementary differentials of f *of orders* 2 *to* w *vanish at* $\begin{pmatrix}0\\0\end{pmatrix}$. *For the orders* $r=w+1(1)2w+2$ *the only elementary differentials not vanishing at* $\begin{pmatrix}0\\0\end{pmatrix}$ *are of the form*

$$\{\ldots\{\{\{f^{w+\omega_1}\}f^{\omega_2}\}f^{\omega_3}\}\ldots f^{\omega_J}\} \tag{6.1.27}$$

with $\omega_j\geq 0$, $j=1(1)J$, $\sum_{j=1}^{J}\omega_j=r-w-J$, *and there are exactly* 2^{r-w-1} *different such elementary differentials.*

Proof. a) We prove the first assertion by induction: $\{f\}\begin{pmatrix}0\\0\end{pmatrix}=\begin{pmatrix}0\\ z'(0)\end{pmatrix} =\begin{pmatrix}0\\0\end{pmatrix}$. If all elementary differentials of order $2(1)r-1$, $r\leq w$, have been shown to vanish at $\begin{pmatrix}0\\0\end{pmatrix}$, all elementary differentials $\{F_1\ldots F_j\}$

of order r with at least one F_j of an order greater than one have to vanish at $\begin{pmatrix}0\\0\end{pmatrix}$ as well; see Def. 3.1.5. This leaves $\{f^{r-1}\}$, which also has to vanish at $\begin{pmatrix}0\\0\end{pmatrix}$ since

$$\tilde{z}^{(r)}(0)=r\sum_{\lambda=1}^{n_r}\frac{\beta_\lambda^{(r)}}{\gamma_\lambda^{(r)}}F_\lambda^{(r)}\begin{pmatrix}0\\0\end{pmatrix}=0 \quad \text{by (6.1.25)},$$

with all $\beta_\lambda^{(r)}\neq 0$; cf. Theorems 3.1.4 and 3.1.6.

b) An elementary differential $\{F_1\ldots F_j\}$ of an order $r>w$ may not vanish at $\begin{pmatrix}0\\0\end{pmatrix}$ only if the F_j are of order 1 or of an order greater w. For $r\leq 2w+2$, only one F_j can be of an order greater w which leads to (6.1.27). Here it is easily seen that for each of the possibilities $J=1(1)r-w$ there are $\binom{r-w-1}{J-1}$ different choices for the ω_j, $j=1(1)J$, and that the sum of these binomial coefficients is 2^{r-w-1}. □

We now consider a given IVP 1 with $f:\mathbb{R}^s\to\mathbb{R}^s$ and assume that we know an approximate local solution $\hat{z}(t,\xi)$ which satisfies (differentiation is w.r.t. the first argument t)

$$\hat{z}^{(\omega)}(0,\xi)=\bar{z}^{(\omega)}(0,\xi) \quad \text{for } \omega=0(1)w. \tag{6.1.28}$$

Then the function $\tilde{z}:\mathbb{R}\times\mathbb{R}^s\to\mathbb{R}^s$ defined by

$$\tilde{z}(t,\xi):=\bar{z}(t,\xi)-\hat{z}(t,\xi) \tag{6.1.29}$$

satisfies the assumption (6.1.25) of Lemma 6.1.5 for fixed $\xi\in\mathbb{R}^s$. Furthermore,

$$\begin{aligned}\tilde{z}'(t,\xi)&=f(\bar{z}(t,\xi))-\hat{z}'(t,\xi)\\&=f(\tilde{z}(t,\xi)+\hat{z}(t,\xi))-\hat{z}'(t,\xi)=:\tilde{f}(t,\tilde{z}(t,\xi)).\end{aligned} \tag{6.1.30}$$

If we are able to evaluate $\hat{z}(t,\eta_{\nu-1})$ at given values of t we may apply a RK-procedure to compute approximately $\tilde{z}(h_\nu,\eta_{\nu-1})$ from (6.1.30) and $\tilde{z}(0,\eta_{\nu-1})=0$ and set

$$\eta_\nu:=\tilde{z}(h_\nu,\eta_{\nu-1})+\hat{z}(h_\nu,\eta_{\nu-1}). \tag{6.1.31}$$

The main distinction between such a RK-procedure and the ones considered in Section 3.1 stems from the fact that t now plays a distinct role. The general recipe (2.1.2) for making a differential equation formally autonomous would destroy the important characteristic (6.1.25) of (6.1.30) since the first component of (6.1.26) does not vanish at $t=0$. Hence, we have to admit the possibility that the points of evaluation for t in the various stages of our RK-procedure are not trivially correlated with the coefficients $a_{\mu\lambda}$ of the procedure.

Thus, an m-stage[1] RKF-procedure with a specified approximate local solution $\hat{z}$ and with parameters

$$\alpha = \begin{pmatrix} \alpha_1 \\ \vdots \\ \alpha_m \\ 1 \end{pmatrix} \quad \text{and} \quad A = \begin{pmatrix} a_{11} & \cdots\cdots & a_{1m} \\ \vdots & & \vdots \\ a_{m+1,1} & \cdots & a_{m+1,m} \end{pmatrix}$$

may be constructed in the following fashion: Let

$$\hat{\mathbf{z}}_\alpha(h,\xi) := \begin{pmatrix} \vdots \\ \hat{z}(\alpha_\mu h,\xi) \\ \vdots \end{pmatrix} \in \mathbb{R}^{(m+1)s} \text{ and } \hat{\mathbf{z}}'_\alpha(h,\xi) := \begin{pmatrix} \vdots \\ \hat{z}'(\alpha_\mu h,\xi) \\ \vdots \end{pmatrix}, \tag{6.1.32}$$

and compute $\eta_\nu = \eta_\nu^{m+1}$ from

$$\boldsymbol{\eta}_\nu := \hat{\mathbf{z}}_\alpha(h_\nu, \eta_{\nu-1}) + h_\nu \mathbf{A}[\mathbf{f}(\boldsymbol{\eta}_\nu) - \hat{\mathbf{z}}'_\alpha(h_\nu, \eta_{\nu-1})], \tag{6.1.33}$$

with the notation of Section 3.1.1.

For implicit A (cf. Def. 3.1.2) it may be shown as in Lemma 3.1.1 that $\boldsymbol{\eta}_\nu$ is well defined for sufficiently small h_ν. Thus (6.1.33) defines an increment function $\Psi: \mathbb{R} \times \mathbb{R}^s \to \mathbb{R}^s$ via

$$\Psi(h_\nu, \eta_{\nu-1}) := \begin{cases} \dfrac{1}{h_\nu}(\eta_\nu^{m+1} - \eta_{\nu-1}), & h_\nu \neq 0, \\ f(\eta_{\nu-1}), & h_\nu = 0, \end{cases} \quad \text{see (3.1.9).} \tag{6.1.34}$$

Remark. In the definition of Ψ for $h_\nu = 0$ we have assumed that the approximate solution $\hat{z}$ satisfies (6.1.28) for a $w \geq 1$. Then $(\eta_\nu^{m+1} - \eta_{\nu-1})/h_\nu \to \hat{z}'(0, \eta_{\nu-1}) = \bar{z}'(0, \eta_{\nu-1}) = f(\eta_{\nu-1})$ for $h_\nu \to 0$.

Def. 6.1.4. A f.s.m. applicable to IVP 1, with E_n, E_n^0, Δ_n, Δ_n^0 as in Def. 3.3.1 and with a forward step procedure (3.1.11) where Ψ has been defined via (6.1.33)/(6.1.34) is called a *R(unge)-K(utta)-F(ehlberg)-method.*

Example. Take $\hat{z}_w$ from (6.1.15), $w \geq 1$, for the approximate local solution. Then the 3-stage RKF-procedure with

$$\alpha = \begin{pmatrix} 1 \\ \dfrac{w+1}{w+3} \\ 1 \\ 1 \end{pmatrix}, \quad A = \begin{pmatrix} 0 & 0 & 0 \\ \dfrac{(w+1)^w}{(w+3)^{w+1}} & 0 & 0 \\ -\dfrac{1}{w+1} & 2\dfrac{(w+3)^w}{(w+1)^{w+1}} & 0 \\ 0 & \dfrac{1}{2(w+2)}\left(\dfrac{w+3}{w+1}\right)^{w+1} & \dfrac{1}{2(w+2)} \end{pmatrix}$$

[1] Cf. footnote in Def. 3.1.1.

generates a RK-method of order $w+3$, see Fehlberg [5]. (The choice $w=0$ leads back to an ordinary RK-procedure.) α was chosen in the above fashion to reduce the number of different evaluations of $\hat{z}$ and $\hat{z}'$.

In the analysis of the consistency of RKF-methods we may consider the IVP 1 of Lemma 6.1.5 and proceed as in Sections 3.1.4, 3.1.5, and 3.3.2. However, an analysis of (6.1.33) shows that we have to replace the definition of $\Phi((1,1), A)$ in (3.1.17) by

$$\Phi((1,1),(A,\alpha)) := \alpha$$

while the further recursive definition of $\Phi(d,(A,\alpha))$ remains unaltered; e.g., $\Phi(2,1),(A,\alpha)) = A\alpha$ (in place of $A(Ae)$), etc.

In checking the conditions for order p:

$$\text{(6.1.35)} \qquad \hat{\Phi}_\lambda^{(r)}((A,\alpha)) = \frac{1}{\gamma_\lambda^{(r)}}, \qquad \begin{matrix} \lambda = 1(1)n_r, \\ r = 1(1)p, \end{matrix}$$

(cf. Def. 3.3.2 and Theorem 3.3.5) we may—according to Lemma 6.1.5—neglect $r=1(1)w$ and all differentials which are not of the form (6.1.27), since the associated terms vanish in the expansions of the local solution of the RKF-procedure and of the exact local solution. Due to the structure of (6.1.27) the remaining conditions (6.1.35) are rather simple and may easily be written down. With (see Section 3.3.3)

$$b = (a_{m+1,1}, \ldots, a_{m+1,m})^T, \qquad \bar{A} = \begin{pmatrix} a_{11} & \ldots & a_{1m} \\ \vdots & & \vdots \\ a_{m1} & \ldots & a_{mm} \end{pmatrix}, \qquad c = \begin{pmatrix} \alpha_1 \\ \vdots \\ \alpha_m \end{pmatrix},$$

and the notation (3.1.16) we obtain for the differential (6.1.27)

$$\text{(6.1.36)} \qquad b^T(\ldots \bar{A}(\bar{A}(\bar{A}\, c^{w+\omega_1} \cdot c^{\omega_2}) \cdot c^{\omega_3}) \ldots \cdot c^{\omega_J}) = \prod_{j=1}^{J} \left(m + \sum_{i=1}^{j} \omega_i + j \right)^{-1}.$$

The right-hand side of (6.1.36) follows from $\gamma_1^{(1)} = 1$ and the recursion (3.1.28).

Example. For $p = w+2$ we have the 3 conditions

$$b^T c^w = \frac{1}{w+1}, \qquad b^T c^{w+1} = \frac{1}{w+2}, \qquad b^T \bar{A}\, c^w = \frac{1}{(w+1)(w+2)}.$$

For *explicit* RKF-procedures, the following numbers m of stages are needed for an order p ($w \geq 1$):

p	$w+1$	$w+2$	$w+3$	$w+4$
m	1	2	3	5

Obviously, the computational effort necessary for a desired increase in order is smaller with RKF-methods than with GKW-methods for $p>w+1$.

For *implicit* RKF-procedures, Wanner has proved the extension of Theorem 3.3.11:

Theorem 6.1.6 (*Wanner* [1]). *For each w and each k there exist RKF-methods which are consistent of order $w+2k$ (with a sufficiently smooth IVP 1).*

The proof of Theorem 6.1.6 is an extension of Butcher's proof for Theorem 3.3.11; the α_μ, $\mu=1(1)m$, for the optimal order methods have to be the zeros of the m-th degree polynomial from the system of orthogonal polynomials with weight function t^w in $[0,1]$.

The verification of the stability and hence of the convergence of order p for IVP$^{(p)}$ 1 of RKF-methods proceeds by standard arguments and is left to the reader.

6.1.6 Multistep Methods with Higher Derivatives

When procedures for the automatic computation of $\bar{z}^{(\mu)}(\xi)$ from ξ are available (cf. Section 6.1.1), it becomes feasible to include higher derivatives into the multistep approach of Chapter 4. Thus we may consider multistep procedures of the form

$$\begin{aligned}(6.1.37)\qquad \frac{1}{bh}\sum_{\kappa=0}^{k}\big[&\alpha_\kappa\eta_{\nu-k+\kappa}-h\beta_\kappa^{(1)}\bar{z}^{(1)}(\eta_{\nu-k+\kappa})\\ &-h^2\beta_\kappa^{(2)}\bar{z}^{(2)}(\eta_{\nu-k+\kappa})-\cdots-h^w\beta_\kappa^{(w)}\bar{z}^{(w)}(\eta_{\nu-k+\kappa})\big]=0,\qquad \nu\geq k.\end{aligned}$$

(6.1.37) is *explicit* in η_ν only if all the $\beta_k^{(\omega)}$, $\omega=1(1)w$, vanish. Otherwise (6.1.37) may be solved for η_ν in the usual predictor-corrector fashion with the help of an explicit procedure (6.1.37), see Section 5.2.1. For $w=1$ we have the situation of Def. 4.1.1 since $\bar{z}^{(1)}(\xi)=f(\xi)$.

Multistep procedures with higher derivatives have been suggested by various authors; no systematic general analysis seems to exist. Actually, the developments of Chapter 4 may be, to a large extent, generalized to multistep procedures with higher derivatives. We will only indicate some of the basic considerations and assume a constant step h throughout.

Let us associate with (6.1.37) the polynomials

$$(6.1.38)\qquad \rho(x):=\sum_{\kappa=0}^{k}\alpha_\kappa x^\kappa,\qquad \sigma_\omega(x):=\sum_{\kappa=0}^{k}\beta_\kappa^{(\omega)}x^\kappa,\qquad \omega=1(1)w,$$

and the difference operator (cf. Def. 4.1.6)

(6.1.39) $$L_h y(t) := \frac{1}{bh}\left[\rho(T_h)y(t) - \sum_{\omega=1}^{w} \frac{h^\omega}{\omega!}\sigma_\omega(T_h)y^{(\omega)}(t)\right].$$

Then we have for (6.1.37)

$$L_h y(t) = O(h^p) \quad \text{for } y \in D^{(p)}$$

as the condition for consistency of order p, cf. Def. 4.1.7. Equivalently we may write (cf. Theorem 4.1.3):

(6.1.40) $$\frac{1}{bh}\left[\rho(e^h) - \sum_{\omega=1}^{w} \frac{h^\omega}{\omega!}\sigma_\omega(e^h)\right] = O(h^p) \quad \text{as } h \to 0.$$

Examples. 1. The implicit 1-step procedure

$$\frac{\eta_\nu - \eta_{\nu-1}}{h} - \frac{1}{2}[\overline{z}^{(1)}(\eta_\nu) + \overline{z}^{(1)}(\eta_{\nu-1})] + \frac{h}{6}[\overline{z}^{(2)}(\eta_\nu) - \overline{z}^{(2)}(\eta_{\nu-1})] = 0$$

is of order 4. (Formulas of this type are called Obreshkov formulas by some authors.)
2. The explicit 2-step procedure

$$\frac{\eta_\nu - \eta_{\nu-1}}{h} - \frac{1}{2}[-\overline{z}^{(1)}(\eta_{\nu-1}) + 3\overline{z}^{(1)}(\eta_{\nu-2})] + \frac{h}{6}[17\overline{z}^{(2)}(\eta_{\nu-1}) + 7\overline{z}^{(2)}(\eta_{\nu-2})] = 0$$

is of order 4.

From a k-step starting procedure and a k-step forward step procedure (6.1.37) we may define a *linear k-step method with higher derivatives* in analogy to Def. 4.2.1. Theorems 4.2.1 and 4.2.2 extend trivially. Theorem 4.2.7 also applies immediately to such methods, which are thus stable if (and only if) the polynomial ρ of (6.1.38) satisfies the root criterion (cf. Def. 4.2.2). Supposedly this will imply restrictions on the maximal order of a D-stable k-step procedure (6.1.37) similar to those derived in Theorem 4.2.11. No general results are known at this time.

On the other hand, when the coefficients α_κ are prescribed it seems that the remaining parameters $\beta_\kappa^{(\omega)}$ may be fully utilized to increase the order. A partial result along these lines is the following extension of Theorem 4.1.9:

Theorem 6.1.7. *For given* α_κ, $\kappa = 0(1)k$, $\sum \alpha_\kappa = 0$, *there exist unique coefficients* $\beta_\kappa^{(1)}$ *and* $\beta_\kappa^{(2)}$, $\kappa = 0(1)k'$, *such that the procedure* (6.1.37) *is of order* $2(k'+1)$.

Proof. According to (6.1.40), with $e^h = x$, $h = \log x$, we have to choose the polynomials σ_1 and σ_2 of degree k' such that

(6.1.41) $$\frac{\rho(x)}{\log x} - \sigma_1(x) - \frac{1}{2}\sigma_2(x)\log x = O((x-1)^{2(k'+1)}).$$

Let

$$\frac{\rho(x)}{\log x} =: \sum_{\mu=0}^{\infty} c_\mu (x-1)^\mu \quad \text{and} \quad \sigma_\omega(x) =: \sum_{\kappa=0}^{k'} b_\kappa^{(\omega)} (x-1)^\kappa.$$

We may choose the $b_\kappa^{(2)}$, $\kappa=0(1)k'$, to be the solution of the linear system

$$\text{(6.1.42)} \qquad \frac{1}{2} \sum_{\kappa=0}^{k'} b_\kappa^{(2)} \frac{(-1)^{\mu-\kappa+1}}{\mu-\kappa} = c_\mu, \qquad \mu = k'+1(1)2k'+1,$$

since the matrix occuring in (6.1.42) is regular for arbitrary k'. If we now choose

$$b_\kappa^{(1)} = c_\kappa - \frac{1}{2} \sum_{\lambda=0}^{\kappa-1} b_\lambda^{(2)} \frac{(-1)^{\kappa-\lambda+1}}{\kappa-\lambda}, \qquad \kappa = 0(1)k',$$

we have determined σ_1 and σ_2 in accordance with (6.1.41). □

Example. Example 2 above, $k=2$, $k'=1$. Here $\rho(x)=x^2-x$, $\rho(x)/\log x = 1+\frac{3}{2}(x-1) + \frac{5}{12}(x-1)^2 - \frac{1}{24}(x-1)^3 + O((x-1)^4)$.
The $b_\kappa^{(2)}$ are determined from

$$-\tfrac{1}{2} b_0^{(2)} + b_1^{(2)} = \tfrac{5}{6}, \qquad \tfrac{1}{3} b_0^{(2)} - \tfrac{1}{2} b_1^{(2)} = -\tfrac{1}{12}$$

as $b_0^{(2)}=4$, $b_1^{(2)}=\frac{17}{6}$, and the $b_\kappa^{(1)}$ are immediately found to be $b_0^{(1)}=1$, $b_1^{(1)}=-\frac{1}{2}$, which produces the coefficients $\beta_\kappa^{(1)}$, $\beta_\kappa^{(2)}$ indicated previously.

Except for the fact that they are multistep methods and thus subject to the difficulties of starting and step change, k-step methods with higher derivatives seem quite attractive for moderate $w>1$. No systematic experiences with such methods have yet been reported.

6.2 General Multi-value Methods

The main distinction between classical one-step and multi-step methods concerns the amount of information on the previous computation which is saved and enters into the next step of the computation. It has been observed that this feature is really independent of the one- or multi-step character: Several independent values which are saved and used subsequently may as well refer to just the last gridpoint as to several preceding gridpoints. This makes a distinction between "one-value" and "multi-value" methods more reasonable than the normal "one-step" and "multi-step" distinction.

6.2.1 Nordsieck's Approach

In his well-known papers, Nordsieck [1], [2] suggested the following approach to the numerical solution of an IVP 1:

In the standard k-step methods, the past history of the numerical integration enters into the computation of η_ν through the saved values $\eta_{\nu-\kappa}$, $f_{\nu-\kappa}$, $\kappa=1(1)k$. A similar amount of information could be provided by approximate values of the true solution z and its higher derivatives at the most recent gridpoint $t_{\nu-1}$ (the first derivative is known immediately from the differential equation). This would make the method effectively a one-step method, with the associated ease of step change.

Contrary to the considerations in Section 6.1, the approximate values of the derivatives of z at t_ν are not to be computed from η_ν via an evaluation of $\bar{z}^{(\rho)}(\eta_\nu)$ but jointly with η_ν in a combined forward step procedure: Denote by

$$a_{\rho\nu} \quad \text{an approximation to } \frac{h_\nu^\rho}{\rho!} z^{(\rho)}(t_\nu), \qquad \rho=1,2,\dots$$

and let

$$(6.2.1) \qquad \boldsymbol{\eta}_\nu := \begin{pmatrix} \eta_\nu \\ a_{1\nu} \\ \vdots \\ a_{r\nu} \end{pmatrix} \in \mathbb{R}^{(r+1)s}.$$

Assume that $\boldsymbol{\eta}_{\nu-1}$ is known; then a trivial truncated Taylor-expansion leads to a "predicted" value

$$(6.2.2) \qquad \boldsymbol{\eta}_\nu^{(0)} := \mathbf{P}\boldsymbol{\eta}_{\nu-1}.$$

Here, the $(r+1)\times(r+1)$-matrix $P=(p_{\mu\lambda})$ is defined by

$$(6.2.3) \qquad p_{\mu\lambda} := \begin{cases} \binom{\lambda}{\mu}, & \lambda \ge \mu, \\ 0, & \lambda < \mu, \end{cases} \qquad \mu,\lambda = 0(1)r;$$

$\mathbf{P}$ denotes (as previously, cf. (3.1.2)) the $(r+1)s\times(r+1)s$-matrix with $s\times s$ block elements $p_{\mu\lambda}I$.

Example. For $r=4$ we have

$$\begin{pmatrix} \eta_\nu \\ a_{1\nu} \\ a_{2\nu} \\ a_{3\nu} \\ a_{4\nu} \end{pmatrix}^{(0)} := \begin{pmatrix} I & I & I & I & I \\ & I & 2I & 3I & 4I \\ & & I & 3I & 6I \\ & 0 & & I & 4I \\ & & & & I \end{pmatrix} \begin{pmatrix} \eta_{\nu-1} \\ a_{1,\nu-1} \\ a_{2,\nu-1} \\ a_{3,\nu-1} \\ a_{4,\nu-1} \end{pmatrix}.$$

Since this extrapolation has not used the IVP1 at all we will generally not have $a_{1\nu}^{(0)} = h_\nu f(\eta_\nu^{(0)})$; hence we correct $\boldsymbol{\eta}_\nu^{(0)}$ by adding a suitable multiple of $h_\nu f(\eta_\nu) - a_{1\nu}^{(0)}$ and require (with the $\times$-notation of (3.1.15))

$$\boldsymbol{\eta}_\nu = \boldsymbol{\eta}_\nu^{(0)} + l \times (h_\nu f(\eta_\nu) - a_{1\nu}^{(0)}) \tag{6.2.4}$$

where $l \in \mathbb{R}^{r+1}$ has a second component 1 to make

$$a_{1\nu} = h_\nu f(\eta_\nu). \tag{6.2.5}$$

Let $\mathbf{p}_1^T$ be the second block row of $\mathbf{P}$; then (6.2.2) and (6.2.4) may be combined into the implicit forward step procedure

$$\frac{1}{h_\nu} [\boldsymbol{\eta}_\nu - \mathbf{P} \boldsymbol{\eta}_{\nu-1} - l \times (h_\nu f(\eta_\nu) - \mathbf{p}_1^T \boldsymbol{\eta}_{\nu-1})] = 0 \tag{6.2.6}$$

which we will call a *Nordsieck procedure*.

In order to define a discretization method applicable to an $\mathrm{IVP}^{(r)}1$ on the basis of (6.2.6) we could choose for E_n and E_n^0 the $(r+1)$-fold Cartesian products of the spaces E_n, resp. E_n^0 of Def. 3.3.1, with an extension (2.1.32) of the norms. With equidistant grids and $h = 1/n$ we could take for the Δ_n

$$(\Delta_n y)(t_\nu) = \begin{pmatrix} y(t_\nu) \\ h y'(t_\nu) \\ \vdots \\ \frac{h^r}{r!} y^{(r)}(t_\nu) \end{pmatrix}. \tag{6.2.7}$$

Correspondingly, the natural *starting procedure* for (6.2.6) is

$$\boldsymbol{\eta}_0 := \begin{pmatrix} z_0 \\ \overline{z}^{(1)}(z_0) \\ \vdots \\ \overline{z}^{(r)}(z_0) \end{pmatrix} \tag{6.2.8}$$

or a sufficiently accurate approximation of (6.2.8). Other starting procedures which do not use the derivatives of the true solution z at 0 (except $z'(0) = f(z_0)$) have been described by Nordsieck [1] (see also Osborne [1] and Gear [3]).

With a suitable choice of Δ_n^0, the f.s.m. defined in this fashion would be trivially consistent of order r due to the structure of P irrespectively of the choice of l. There is precisely one choice of l which would make it consistent of order $r+1$ (with an $\mathrm{IVP}^{(r+1)}1$) but it turns out that this l makes the method unstable for $r > 2$. However, the following result saves the situation.

Theorem 6.2.1. *For $r \geq 2$, if $l_1 = 1$ and the components $l_2, \ldots, l_r$ of l are chosen arbitrarily ($l_r \neq 0$) there is a unique choice of l_0 and a unique definition of the mappings Δ_n such that the corresponding f.s.m. is consistent of order $r+1$ with an $IVP^{(r+1)}$ 1.*

Proof. We take, for $y \in D^{(r+1)}$, with $y'(t) = f(y(t))$,

$$(\Delta_n y)(t_\nu) := \begin{pmatrix} y(t_\nu) \\ h y'(t_\nu) \\ \vdots \\ \vdots \\ \frac{h^r}{r!} y^{(r)}(t_\nu) \end{pmatrix} + \begin{pmatrix} 0 \\ 0 \\ \gamma_2 \\ \vdots \\ \gamma_r \end{pmatrix} \times \frac{h^{r+1}}{(r+1)!} y^{(r+1)}(t_\nu) \tag{6.2.9}$$

and choose the γ_ρ, $\rho = 2(1)r$, and l_0 such that (cf. (6.2.6))

$$\frac{1}{h}\left\{(\Delta_n y)_\rho(t_\nu) - \sum_{\sigma=\rho}^{r} \binom{\sigma}{\rho} (\Delta_n y)_\sigma(t_{\nu-1}) - l_\rho \left[h y'(t_\nu) - \sum_{\sigma=1}^{r} \sigma (\Delta_n y)_\sigma(t_{\nu-1})\right]\right\} = O(h^{r+1}), \quad \rho = 0(1)r. \tag{6.2.10}$$

(6.2.10) holds for $\rho = 2(1)r$ if the γ_ρ satisfy

$$\sum_{\sigma=\rho+1}^{r} \binom{\sigma}{\rho} \gamma_\sigma - l_\rho \sum_{\sigma=2}^{r} \sigma \gamma_\sigma = \binom{r+1}{\rho} - l_\rho (r+1), \quad \rho = 2(1)r. \tag{6.2.11}$$

This is a triangular linear system for the quantities $\gamma_3, \ldots, \gamma_r$, and $\sum \sigma \gamma_\sigma$, with diagonal coefficients $3, \ldots, r, l_r$; thus, it has a unique solution for $l_r \neq 0$. With this choice of the γ_ρ we obtain for the "correction quantity"

$$\begin{aligned} h y'(t_\nu) &- \sum_{\sigma=1}^{r} \sigma (\Delta_n y)_\sigma(t_{\nu-1}) \\ &= \left[(r+1) - \sum_{\sigma=2}^{r} \sigma \gamma_\sigma\right] \frac{h^{r+1}}{(r+1)!} y^{(r+1)}(t_{\nu-1}) + O(h^{r+2}) \\ &= \frac{r+1}{l_r} \frac{h^{r+1}}{(r+1)!} y^{(r+1)}(t_{\nu-1}) + O(h^{r+2}) \quad \text{by (6.2.11) with } \rho = r, \end{aligned}$$

and (6.2.10) holds for $\rho = 0$ if we choose l_0 such that

$$\sum_{\sigma=2}^{r} \gamma_\sigma + l_0 \frac{r+1}{l_r} = 1. \tag{6.2.12}$$

Due to $l_1 = 1$, (6.2.10) holds automatically for $\rho = 1$.

From a comparison of (6.2.8) and (6.2.9) we find that the starting procedure (6.2.8) generates an error of order $O(h^{r+1})$; but this is compatible with consistency of order $r+1$. □

Remark. The admission of $l_1 \neq 1$ would necessitate the introduction of a $\gamma_1 \neq 0$ which is determined by requiring the validity of (6.2.10) for $\rho=1$. No higher order of consistency can be achieved.

Example. $r=3$, $l_2=\frac{3}{4}$, $l_3=\frac{1}{6}$. The system (6.2.11) becomes

$$3\gamma_3 - \tfrac{3}{4}(2\gamma_2+3\gamma_3) = 3,$$
$$-\tfrac{1}{6}(2\gamma_2+3\gamma_3) = \tfrac{10}{3},$$

which yields $\gamma_2=\gamma_3=-4$. (6.2.12) produces $l_0=\frac{3}{8}$ as the correct choice for a procedure (6.2.6) which is consistent of order 4.

The proof of Theorem 6.2.1 shows that (6.2.9) is the appropriate definition of Δ_n in connection with a Nordsieck procedure (6.2.6) of order $r+1$. At first sight it seems that this destroys the possibility of changing the stepsize by simply multiplying the ρ-th component of $\boldsymbol{\eta}_\nu$ by $(h_{\nu+1}/h_\nu)^\rho$. With (6.2.9), this natural stepchange procedure will add a quantity of order $O(h^{r+1})$ to $\eta_\nu-(\Delta_n z)(t_\nu)$ since the extra terms in (6.2.9) are not taken into account. However, in a coherent grid sequence there are only a fixed finite number of stepchanges so that the above stepchange procedure does not decrease the order of consistency of the method.

On the basis of Corollary 2.2.7 it suffices to analyze the *stability* of Nordsieck methods for the trivial IVP1 $\mathfrak{P}_0$ of (4.2.24), cf. Section 4.2.3. Obviously the stability for $\mathfrak{P}_0$ of a discretization based upon (6.2.6) depends upon the location of the eigenvalues of the $(r+1)\times(r+1)$-matrix $P-lp_1^T$. Thus, a Nordsieck method is stable if and only if the characteristic polynomial of $P-lp_1^T$ satisfies the root criterion. The details are left to the reader.

Theorem 6.2.2. *For each choice of* $\lambda_1,\dots,\lambda_r$ *(real or in conjugate-complex pairs) there exist components* $l_1,\dots,l_r$ *for* l *such that* $P-lp_1^T$ *has the eigenvalues* $1, \lambda_1,\dots,\lambda_r$. *If one of the* λ_ρ *vanishes* l_1 *has to be* 1.

Proof. The form $\begin{pmatrix}1\\0\\\vdots\\0\end{pmatrix}$ of the first column of $P-lp_1^T$ implies that one eigenvalue is 1 and that the remaining eigenvalues are those of the lower right $r\times r$-minor which does not depend upon l_0.

Furthermore, the matrix $P-lp_1^T=(I-le_1^T)P$ is similar to $P(I-le_1^T)=P-(Pl)e_1^T$ in which the components of l occur linearly and in the second column only. Hence the vector of the coefficients of the polynomial $\prod_{\rho=1}^{r}(\lambda-\lambda_\rho)$ is a linear transform of the vector $(l_1,\ldots,l_r)^T$. The regularity of this transformation has been shown by Gear [4]; this implies that the l_ρ, $\rho=1(1)r$, are uniquely determined by the λ_ρ, $\rho=1(1)r$. From the structure of P (see (6.2.3)) one easily finds that

$$\prod_{\rho=1}^{r}\lambda_\rho=\det(P-lp_1^T)=1-l_1$$

which implies the final assertion. □

Theorems 6.2.1 and 6.2.2 show that we may, for each $r\geq 1$, construct Nordsieck methods which are consistent of order $r+1$ with an $\mathrm{IVP}^{(r+1)}1$ and for which $P-lp_1^T$ has prescribed eigenvalues $\lambda_2,\ldots,\lambda_r$ (while $\lambda_0=1$ and $\lambda_1=0$). If we choose the λ_ρ, $\rho\geq 2$, on the closed unit disk, with $\lambda_\rho\neq 1$ and no two λ_ρ coinciding on the unit circle, the corresponding Nordsieck method is stable and hence convergent of order $r+1$ for an $\mathrm{IVP}^{(r+1)}1$. The remark after Theorem 6.2.1 shows that the choice $\lambda_1=0$ and $l_1=1$ is no restriction of generality.

Example. For $r=3$ and $\lambda_1=\lambda_2=\lambda_3=0$ we obtain $l_1=1$ and $l_2=\frac{3}{4}$, $l_3=\frac{1}{6}$. Thus the method of the example below Theorem 6.2.1 is stable and hence convergent of order 4.

6.2.2 Nordsieck Predictor-corrector Methods

So far we have not taken into account the implicitness of (6.2.4) but assumed that (6.2.6) is somehow solved for η_ν. The iterative procedure

$$(6.2.13)\qquad \begin{cases}\boldsymbol{\eta}_\nu^{(0)}:=\mathbf{P}\boldsymbol{\eta}_{\nu-1}\\ \boldsymbol{\eta}_\nu^{(\mu)}:=\boldsymbol{\eta}_\nu^{(\mu-1)}+l\times(h\,f(\eta_\nu^{(\mu-1)})-a_{1\nu}^{(\mu-1)}),\qquad \mu=1,2,\ldots\end{cases}$$

represents a simple-minded approach towards the solution of (6.2.6).

Lemma 6.2.3. *For* $l_1=1$, *the sequence* $\boldsymbol{\eta}_\nu^{(\mu)}$ *defined by* (6.2.13) *converges towards the solution* $\boldsymbol{\eta}_\nu$ *of* (6.2.6) *for sufficiently small* h *and Lipschitz-continuous* f.

Proof. At first it is seen that a unique solution of (6.2.6) exists for sufficiently small h and Lipschitz-continuous f; cf. Lemmas 3.1.1 and 4.1.2.

Now $l_1=1$ implies

$$(6.2.14)\qquad a_{1\nu}^{(\mu)}=h\,f(\eta_\nu^{(\mu-1)}),\qquad \mu\geq 1,$$

so that for $\mu \geq 2$

$$
\begin{aligned}
&\boldsymbol{\eta}_\nu^{(\mu)} - \boldsymbol{\eta}_\nu^{(\mu-1)} = h\, l \times [f(\eta_\nu^{(\mu-1)}) - f(\eta_\nu^{(\mu-2)})] \\
&\|\boldsymbol{\eta}_\nu^{(\mu)} - \boldsymbol{\eta}_\nu^{(\mu-1)}\| \leq h\|l\|\, \|f(\eta_\nu^{(\mu-1)}) - f(\eta_\nu^{(\mu-2)})\| \\
&\qquad \leq h\, L\|l\|\, \|\boldsymbol{\eta}_\nu^{(\mu-1)} - \boldsymbol{\eta}_\nu^{(\mu-2)}\|
\end{aligned} \tag{6.2.15}
$$

with $\|\boldsymbol{\eta}_\nu\| := \max\left(\|\eta_\nu\|, \max_\rho \|a_{\rho\nu}\|\right)$; L is the Lipschitz constant of f. For $hL\|l\| < 1$ this implies the existence of $\overline{\boldsymbol{\eta}}_\nu := \lim_{\mu\to\infty} \boldsymbol{\eta}_\nu^{(\mu)}$.

Successive resubstitution into (6.2.13) yields, with (6.2.14),

$$
\boldsymbol{\eta}_\nu^{(\mu)} = \boldsymbol{\eta}_\nu^{(0)} + l \times (h\, f(\eta_\nu^{(\mu-1)}) - a_{1\nu}^{(0)}), \qquad \mu \geq 1, \tag{6.2.16}
$$

so that $\overline{\boldsymbol{\eta}}_\nu$ satisfies (6.2.6). □

On the basis of Lemma 6.2.3 it seems reasonable to define a PC-procedure for (6.2.6) as a truncated iteration (6.2.13):

$$
\left\{
\begin{aligned}
&\boldsymbol{\eta}_\nu^{(0)} := \mathbf{P}\boldsymbol{\eta}_{\nu-1} \\
&\boldsymbol{\eta}_\nu^{(\mu)} := \boldsymbol{\eta}_\nu^{(\mu-1)} + l \times (h f(\eta_\nu^{(\mu-1)}) - a_{1\nu}^{(\mu-1)}), \qquad \mu = 1(1)m, \\
&\boldsymbol{\eta}_\nu := \begin{cases} \boldsymbol{\eta}_\nu^{(m)}, & \text{for } \mathrm{P(EC)}^m, \\ \boldsymbol{\eta}_\nu^{(m)} + \begin{pmatrix} 0 \\ 1 \\ 0 \\ \vdots \\ 0 \end{pmatrix} \times (h f(\eta_\nu^{(m)}) - a_{1\nu}^{(m)}) & \text{for } \mathrm{P(EC)}^m\mathrm{E}. \end{cases}
\end{aligned}
\right. \tag{6.2.17}
$$

We have chosen the denotations $\mathrm{P(EC)}^m$ and $\mathrm{P(EC)}^m\mathrm{E}$ as indicated in (6.2.17) in analogy with Def. 5.2.4. The supplementary computation in the $\mathrm{P(EC)}^m\mathrm{E}$-method consists of the evaluation $f(\eta_\nu^{(m)})$ and the replacement of $a_{1\nu}^{(m)}$ by $hf(\eta_\nu^{(m)})$ so that (6.2.5) holds. For the η_ν produced by a $\mathrm{P(EC)}^m$-method (6.2.5) is generally not satisfied.

For any $m \geq 1$ the consistency and stability properties of such Nordsieck PC-methods are completely equivalent to those of the underlying implicit method:

Theorem 6.2.4. *Assume that the Nordsieck method based upon* (6.2.6) *is consistent of order* $r+1$, *with* Δ_n *defined by* (6.2.9), *and stable. Then, for any* $m \geq 1$, *the associated Nordsieck* $P(EC)^m$- *and* $P(EC)^m E$-*methods are also consistent of order* $r+1$ *and stable.*

Proof. The consistency assumption implies that the true solution z of an $\mathrm{IVP}^{(r+1)}1$ satisfies, with $(\Delta_n z)_\nu := (\Delta_n z)(t_\nu)$,

$$
\frac{1}{h}[(\Delta_n z)_\nu - \mathbf{P}(\Delta_n z)_{\nu-1} - l \times (h f(z_\nu) - \mathbf{p}_1^T(\Delta_n z)_{\nu-1})] = O(h^{r+1}). \tag{6.2.18}
$$

On the other hand the choice of P guarantees that

$$z_\nu - \mathbf{p}_0^T(\Delta_n z)_{\nu-1} = O(h^{r+1}) \tag{6.2.19}$$

so that by the Lipschitz continuity of f

$$hf(z_\nu) - hf(\mathbf{p}_0^T(\Delta_n z)_{\nu-1}) = O(h^{r+2}).$$

Therefore we may replace $f(z_\nu)$ by $f(\mathbf{p}_0^T(\Delta_n z)_{\nu-1})$ in (6.2.18) without changing the order of the left hand side; this implies the consistency result for the PEC-procedure.

From this result, (6.2.19), and (6.2.15) it follows via (6.2.16) that an increase in m does not change the situation. The supplementary evaluation in the P(EC)^{m}E-procedure simply annihilates the second component of the local discretization error.

With regard to stability we take again $f \equiv 0$ (cf. the proof of Theorem 6.2.2) and find that the matrix $(I - le_1^T)^m P$ has to be power-bounded in order that the P(EC)^m-method be stable. Since $I - le_1^T$ is idempotent for $l_1 = 1$,

$$(I - le_1^T)^m P = (I - le_1^T)P = P - lp_1^T$$

which is the stability matrix for the implicit method. In the case of the P(EC)^{m}E-method we have to premultiply $(I - le_1^T)^m P$ by $I - e_1 e_1^T$ which cannot destroy the stability. □

Theorem 6.2.4 shows that *for sufficiently small h* the Nordsieck PEC-method is practically as effective as the underlying implicit Nordsieck method. However, if f' has a large spectral radius at the true solution z—as in stiff systems—(6.2.13) will not converge for reasonable steps h while on the other hand it is necessary to preserve the implicitness of (6.2.6) sufficiently well. Therefore, the use of a Newton-method for the approximate solution of (6.2.6) is indicated.

The first step of the classical Newton-method for

$$\boldsymbol{\eta}_\nu - \boldsymbol{\eta}_\nu^{(0)} - l \times [hf(\eta_\nu) - a_{1\nu}^{(0)}] = 0 \tag{6.2.20}$$

with the initial approximation $\boldsymbol{\eta}_\nu^{(0)}$ becomes

$$\boldsymbol{\eta}_\nu^{(1)} - \boldsymbol{\eta}_\nu^{(0)} - l \times [hf(\eta_\nu^{(0)}) + hf'(\eta_\nu^{(0)})(\eta_\nu^{(1)} - \eta_\nu^{(0)}) - a_{1\nu}^{(0)}] = 0 \tag{6.2.21}$$

or

$$\eta_\nu^{(1)} - \eta_\nu^{(0)} = l_0[I - hl_0 f'(\eta_\nu^{(0)})]^{-1}(hf(\eta_\nu^{(0)}) - a_{1\nu}^{(0)}) \tag{6.2.22}$$

where I is the $s \times s$-unit matrix. Substitution of (6.2.22) into (6.2.21) yields

$$\boldsymbol{\eta}_\nu^{(1)} := \boldsymbol{\eta}_\nu^{(0)} + l \times [I - hl_0 f'(\eta_\nu^{(0)})]^{-1}(hf(\eta_\nu^{(0)}) - a_{1\nu}^{(0)}).$$

Inductively it may be shown that the further stages of a Newton-method for (6.2.20) take exactly the same form so that the Nordsieck $P(EN)^m$- resp. $P(EN)^m E$-procedure is given by

(6.2.23)

$$\begin{cases} \boldsymbol{\eta}_\nu^{(0)} := \mathbf{P}\boldsymbol{\eta}_{\nu-1}, \\ \boldsymbol{\eta}_\nu^{(\mu)} := \boldsymbol{\eta}^{(\mu-1)} + l \times [I - h l_0 f'(\eta_\nu^{(\mu-1)})]^{-1} (h f(\eta_\nu^{(\mu-1)}) - a_{1\nu}^{(\mu-1)}), \quad \mu = 1(1)m, \\ \boldsymbol{\eta}_\nu \;:= \begin{cases} \boldsymbol{\eta}_\nu^{(m)}, \\ \boldsymbol{\eta}_\nu^{(m)} + e_1 \times (h f(\eta_\nu^{(m)}) - a_{1\nu}^{(m)}). \end{cases} \end{cases}$$

Various modifications of this strict Newton procedure may be constructed: Approximations to f' may be used in place of f' and the re-evaluation of f' need not be carried out at each stage of the iteration and each step of the f.s.m. Gear [1] suggests updating $[I - h l_0 f']^{-1}$ only when the difference between $\boldsymbol{\eta}_\nu^{(m-1)}$ and $\boldsymbol{\eta}_\nu^{(m)}$ begins to exceed a prescribed quantity for given fixed m.

6.2.3 Equivalence of Generalized Nordsieck Methods

The Nordsieck PC-procedures of Section 6.2.2 may be generalized in the following fashion:

a) The meaning of the components $a_{\rho\nu}$, $\rho = 1(1)r$, of the "data vector" $\boldsymbol{\eta}_\nu$ may differ from the one assumed previously. The "predictor matrix" P which extrapolates the data without use of the differential equation has to be changed accordingly and the definition of the Δ_n has to be adapted.

b) The "residual function" $\mathbf{F}(\boldsymbol{\eta}_\nu) = [h f(\eta_\nu) - a_{1\nu}]$ of Section 6.2.2 may be replaced by some other function $\mathbf{F}: \mathbb{R}^{(r+1)s} \to \mathbb{R}^{qs}$ which satisfies

$$\mathbf{F}((\Delta_n z)(t_\nu)) = 0, \quad \nu = 1(1)n, \tag{6.2.24}$$

for the solution z of the IVP 1. Correspondingly, the vector $l: \mathbb{R}^1 \to \mathbb{R}^{r+1}$ has to be replaced by an appropriate "relaxation matrix" $L: \mathbb{R}^q \to \mathbb{R}^{r+1}$ which may depend on t_ν and on the stage μ of the iteration. Thus a *generalized Nordsieck procedure* is of the form

$$\begin{cases} \boldsymbol{\eta}_\nu^{(0)} := \mathbf{A}\boldsymbol{\eta}_{\nu-1}, \\ \boldsymbol{\eta}_\nu^{(\mu)} := \boldsymbol{\eta}_\nu^{(\mu-1)} + \mathbf{L}_{(\mu\nu)} \mathbf{F}(\boldsymbol{\eta}_\nu^{(\mu-1)}), \quad \mu = 1(1)m, \\ \boldsymbol{\eta}_\nu \;:= \boldsymbol{\eta}_\nu^{(m)}. \end{cases} \tag{6.2.25}$$

Since any non-singular linear transform of the data vector carries the same amount of information we may write (6.2.25) in a multitude of equivalent ways. Let Q be some regular $(r+1) \times (r+1)$-matrix and define

$$\bar{A} := Q A Q^{-1}, \quad \bar{L} := Q L, \quad \bar{\mathbf{F}}(\xi) := \mathbf{F}(\mathbf{Q}^{-1}\xi). \tag{6.2.26}$$

It is clear that the solution $\bar{\boldsymbol{\eta}}_\nu$ of the generalized Nordsieck method with the starting procedure $\bar{\boldsymbol{\eta}}_0 := Q\boldsymbol{\eta}_0$ and the forward step procedure

$$(6.2.27) \qquad \begin{cases} \bar{\boldsymbol{\eta}}_\nu^{(0)} := \bar{\mathbf{A}}\bar{\boldsymbol{\eta}}_{\nu-1} \\ \bar{\boldsymbol{\eta}}_\nu^{(\mu)} := \bar{\boldsymbol{\eta}}_\nu^{(\mu-1)} + \bar{\mathbf{L}}_{(\mu\nu)}\bar{\mathbf{F}}(\bar{\boldsymbol{\eta}}_\nu^{(\mu-1)}), \qquad \mu = 1(1)m, \\ \bar{\boldsymbol{\eta}}_\nu \ := \bar{\boldsymbol{\eta}}_\nu^{(m)}, \end{cases}$$

will satisfy throughout the computation (except for computing errors)

$$\bar{\boldsymbol{\eta}}_\nu = Q\boldsymbol{\eta}_\nu$$

where $\boldsymbol{\eta}_\nu$ is the solution of the method based upon (6.2.25) and started with $\boldsymbol{\eta}_0$. If we further define for (6.2.27)

$$(6.2.28) \qquad (\bar{\Delta}_n y)(t_\nu) := Q(\Delta_n y)(t_\nu)$$

the consistency and stability properties of the two methods will be the same.

Def. 6.2.1. Two generalized Nordsieck methods whose characteristic quantities are related by (6.2.26) and (6.2.28) with a regular matrix Q are called *equivalent*.

By means of such equivalence transformations it is possible to clarify the meaning and structure of generalized Nordsieck methods. The most interesting result in this direction concerns the Nordsieck methods of order $r+1$ based upon (6.2.17) with l such that all eigenvalues of $P - lp_1^T$ vanish except the one at 1 (cf. Theorems 6.2.1, 6.2.2, and 6.2.4, and the examples in Section 6.2.1). As Nordsieck realized in his original publication, these methods are equivalent to the classical r-step Adams PC-methods.

To understand this equivalence we formulate the Adams PC-methods (see Example 1 in Section 4.1.1 and Def. 5.2.4) as generalized Nordsieck methods. Take

$$(\Delta_n y)(t_\nu) := \begin{pmatrix} y(t_\nu) \\ hy'(t_\nu) \\ hy'(t_{\nu-1}) \\ \vdots \\ hy'(t_{\nu-r+1}) \end{pmatrix} \in \mathbb{R}^{(r+1)s}$$

so that the first component η_ν of $\boldsymbol{\eta}_\nu$ approximates $z(t_\nu)$ as usual while the further components $a_{\rho\nu}$, $\rho = 1(1)r$, approximate $hz'(t_{\nu-\rho+1}) = hf(z(t_{\nu-\rho+1}))$. The Adams predictor expresses $\eta_\nu^{(0)}$ in terms of $\boldsymbol{\eta}_{\nu-1}$:

$$\eta_\nu^{(0)} := 1 \cdot \eta_{\nu-1} + \beta_{r-1}^{(0)} a_{1,\nu-1} + \cdots + \beta_0^{(0)} a_{r,\nu-1};$$

the $\beta_\rho^{(0)}$ are from the Adams predictor scheme. Polynomial extrapolation for y' leads to

$$a_{1\nu}^{(0)} := \binom{r}{1} a_{1,\nu-1} - \binom{r}{2} a_{2,\nu-1} + \cdots + (-1)^{r-1} \binom{r}{r} a_{r,\nu-1}$$

while the further components of $\boldsymbol{\eta}_\nu^{(0)}$ are simply copied from $\boldsymbol{\eta}_{\nu-1}$ with a shift towards the bottom. This leads to the predictor matrix

$$A = \begin{pmatrix} 1 & \beta_{r-1}^{(0)} & \beta_{r-2}^{(0)} & \cdots & \beta_0^{(0)} & \\ 0 & r & -\binom{r}{2} & \cdots & (-1)^{r-1} & \\ \vdots & 1 & \ddots & & 0 & \\ \vdots & & & \ddots & & \\ 0 & & & & 1 & 0 \end{pmatrix}. \tag{6.2.29}$$

Since the coefficients of the Adams predictor and corrector schemes can be shown to satisfy

$$\beta_\rho^{(0)} + (-1)^{r-\rho} \binom{r}{\rho} \beta_r = \beta_\rho, \qquad \rho = 0(1)r-1,$$

it is possible to represent the classical correction step

$$\eta_\nu^{(1)} := \eta_{\nu-1} + \beta_r h f(\eta_\nu^{(0)}) + \beta_{r-1} h f_{\nu-1} + \cdots + \beta_0 h f_{\nu-r}$$

in the form

$$\eta_\nu^{(1)} := \eta_\nu^{(0)} + \beta_r [h f(\eta_\nu^{(0)}) - a_{1\nu}^{(0)}]. \tag{6.2.30}$$

If we supplement (6.2.30) by

$$a_{1\nu}^{(1)} := a_{1\nu}^{(0)} + [h f(\eta_\nu^{(0)}) - a_{1\nu}^{(0)}] \tag{6.2.31}$$

we have the form (6.2.25) with the residual function $\mathbf{F}(\boldsymbol{\eta}_\nu) = h f(\eta_\nu) - a_{1\nu}$ as previously and the relaxation vector $l = (\beta_r 1 0 \ldots 0)^T$. It is obvious from (6.2.30)—(6.2.31) that further iterations in (6.2.25) correspond to further iterations of the corrector in the Adams PC-method.

With the aid of appropriate numerical differentiation formulas it is simple to construct the matrix Q which takes

$$\begin{pmatrix} y(t_\nu) \\ h y'(t_\nu) \\ \vdots \\ h y'(t_{\nu-r+1}) \end{pmatrix} \quad \text{into} \quad \begin{pmatrix} y(t_\nu) \\ h y'(t_\nu) \\ \vdots \\ \frac{h^r}{r!} y^{(r)}(t_\nu) \end{pmatrix} + O(h^{r+1}).$$

It turns out that for this Q and A of from (6.2.29)

$$QAQ^{-1}=P$$

with elements (6.2.3). Furthermore, Q reproduces the first components of $\boldsymbol{\eta}_\nu$ so that $\bar{\mathbf{F}}=\mathbf{F}$ and the transformed vector $\bar{l}=Ql$ is such that $(I-le_1^T)P$ has the eigenvalues $1,0,\ldots,0$. This establishes the desired equivalence.

Example. For $r=3$, we have $\beta_2^{(0)}=\frac{23}{12}$, $\beta_1^{(0)}=-\frac{4}{3}$, $\beta_0^{(0)}=\frac{5}{12}$, so that

$$A=\begin{pmatrix}1 & \frac{23}{12} & -\frac{4}{3} & \frac{5}{12}\\ 0 & 3 & -3 & 1\\ 0 & 1 & 0 & 0\\ 0 & 0 & 1 & 0\end{pmatrix}.$$

$l=\begin{pmatrix}\frac{3}{8}\\ 1\\ 0\\ 0\end{pmatrix}$ produces the Adams corrector since

$$\beta_3=\tfrac{3}{8},\quad \beta_2=\tfrac{23}{12}-3\cdot\tfrac{3}{8}=\tfrac{19}{24},\quad \beta_1=-\tfrac{4}{3}+3\cdot\tfrac{3}{8}=-\tfrac{5}{24},\quad \beta_0=\tfrac{5}{12}-\tfrac{3}{8}=\tfrac{1}{24}.$$

Transformation by

$$Q=\begin{pmatrix}1 & 0 & 0 & 0\\ 0 & 1 & 0 & 0\\ 0 & \frac{3}{4} & -1 & \frac{1}{4}\\ 0 & \frac{1}{6} & -\frac{1}{3} & \frac{1}{6}\end{pmatrix}$$

yields

$$QAQ^{-1}=\begin{pmatrix}1 & 1 & 1 & 1\\ & 1 & 2 & 3\\ & & 1 & 3\\ & 0 & & 1\end{pmatrix}=P,\qquad Ql=\begin{pmatrix}\frac{3}{8}\\ 1\\ \frac{3}{4}\\ \frac{1}{6}\end{pmatrix},$$

$$Q\begin{pmatrix}y(t_\nu)\\ hy'(t_\nu)\\ hy'(t_{\nu-1})\\ hy'(t_{\nu-2})\end{pmatrix}=\begin{pmatrix}y(t_\nu)\\ hy'(t_\nu)\\ \frac{h^2}{2}y''(t_\nu)-\frac{h^4}{6}y^{\mathrm{IV}}(t_\nu)+O(h^5)\\ \frac{h^3}{6}y'''(t_\nu)-\frac{h^4}{6}y^{\mathrm{IV}}(t_\nu)+O(h^5)\end{pmatrix}.$$

The residual function F remains unaltered.

A comparison with the examples in Section 6.2.1 shows that we have not only obtained the correct "parameters" A, l and F but also the correct γ_ρ for (6.2.9).

By a suitable equivalence transformation, Gear has reformulated the $(2k-1)$-step Adams PC-methods as k-step generalized Nordsieck PC-methods of order $2k$. These methods which—like classical k-step methods—use approximations to k values of the solution and of its derivative for the data vector η_ν manage to attain their high order without loss of stability (cf. Theorem 4.2.11) by modifying also the stored η-values in the correction step; see Gear [3], [4].

Example. The classical 3-step Adams PC-procedure of the above example becomes, with

$$(\Delta_n y)(t_\nu) := \begin{pmatrix} y(t_\nu) \\ h y'(t_\nu) \\ y(t_{\nu-1}) \\ h y'(t_{\nu-1}) \end{pmatrix},$$

a generalized Nordsieck-procedure (6.2.25) with

$$A = \begin{pmatrix} -4 & 4 & 5 & 2 \\ 12 & -12 & 8 & 5 \\ 1 & 0 & 0 & 0 \\ 0 & 1 & 0 & 0 \end{pmatrix}, \quad l = \begin{pmatrix} \frac{3}{8} \\ 1 \\ -\frac{1}{24} \\ 0 \end{pmatrix},$$

and $\mathbf{F}(\boldsymbol{\eta}_\nu) = h f(\eta_\nu) - a_{1\nu}$ as usual. All extraneous eigenvalues are at 0 and the order is 4 although it is a "2-step" method.

Kohfeld and Thompson [1] have given a Nordsieck-type reformulation of multistep procedures with an off-step point (cf. Section 5.3.1).

6.2.4 Appraisal of Nordsieck Methods

The main algorithmic advantage of the (original) Nordsieck formulation (6.2.6) of multistep methods is the ease with which the step may be changed. If

$$\boldsymbol{\eta}_\nu = \begin{pmatrix} \eta_\nu \\ a_{1\nu} \\ \vdots \\ a_{r\nu} \end{pmatrix} \approx \begin{pmatrix} y(t_\nu) \\ h_\nu y'(t_\nu) \\ \vdots \\ \frac{h_\nu^r}{r!} y^{(r)}(t_\nu) \end{pmatrix} \tag{6.2.32}$$

the change to $h_{\nu+1} = c h_\nu$ requires simply the multiplication of $a_{\rho\nu}$ by c^ρ. As pointed out in Section 6.2.1, this introduces an extra error $O(h^{r+1})$ independently of the order of consistency of the method. This simplicity of the step change procedure more or less disappears when approximations to values at previous gridpoints appear in the data vector $\boldsymbol{\eta}_\nu$.

A *stepsize control* on the basis of the values of the local discretization error is possible for Nordsieck methods just as for classical methods; see Nordsieck [1] and Gear [1], [5].

A numerical advantage of the storage of the data in the form (6.2.32) rather than in the form of values at previous gridpoints is the following: If (partial) double or multiple precision is needed in standard multistep methods, all the $\eta_{\nu-\kappa}$ in the data vector have to be stored and manipulated in that precision. On the other hand it may be assumed that in (6.2.32) the $|a_{\rho\nu}|$ decrease in size with increasing ρ so that a lower precision is sufficient for higher ρ. (At the same time, the absence of such a decrease is a strong warning that the steps are too large for a proper representation of the solution z.)

Beyond algorithmic and computational aspects, however, a fundamentally new idea lies in Gear's interpretation of (generalized) Nordsieck PC-methods as *relaxation methods* as indicated in the beginning of Section 6.2.3. By a suitable choice of the residual function $\mathbf{F}$ and the relaxation matrix L we have been able to implement Newton's method in place of a contraction iteration without changing the formalism (see Section 6.2.2); with the same conceptual ease more practical modifications of Newton's methods which perhaps use only approximations to the Jacobian f' may be designed.

Far more important, the residual function $\mathbf{F}$ need not incorporate the IVP1 in its classical form $y' - f(y)$. Instead a form which is implicit in y' may just as well be used, as long as (6.2.24) holds; thus we may be able to treat implicit systems of ordinary differential equations of first order without solving for y' in each step. In this fashion we may even accomodate systems which contain one or more *algebraic equations* in place of differential equations. For more details of this important aspect of Nordsieck's formalism we refer the reader to Gear [4].

Finally, we wish to draw attention to an implementation of the classical multistep methods based upon differentiation schemes (see Example 2 in Section 4.1.1) in Nordsieck form. Using the $A(\alpha)$-stability of the differentiation methods (cf. Section 4.6.3), Gear [1] has devised an algorithm for the treatment of stiff systems of differential equations which quite automatically adapts itself to local requirements.

6.3 Extrapolation Methods

6.3.1 The Structure of an Extrapolation Method

With initial value problems it is a natural idea to use Richardson extrapolation not over the complete interval of integration—which may not be specified in advance anyway—but rather on successive subintervals $[\bar{t}_{\mu-1}, \bar{t}_\mu]$: To obtain the value $\chi(\bar{t}_\mu)$ of an approximate solution from the value $\chi(\bar{t}_{\mu-1})$ we treat the interval $[\bar{t}_{\mu-1}, \bar{t}_\mu]$ as our interval of integration and $\chi(\bar{t}_{\mu-1})$ as our initial value; with some discretization method $\mathfrak{M}$ we generate values $\zeta_{n_\rho}(\bar{t}_\mu)$ for various $n_\rho \in \mathbb{N}'$ and apply Richardson extrapolation to these $\zeta_{n_\rho}(\bar{t}_\mu)$ to produce the value of $\chi(\bar{t}_\mu)$, see Sections 1.4.1 to 1.4.3. The process is now repeated on $[\bar{t}_\mu, \bar{t}_{\mu+1}]$ with $\chi(\bar{t}_\mu)$ as the initial value.

Thus, in order to specify an extrapolation method EM we have to prescribe:

a) the *grid* $\bar{\mathbb{G}} := \{\bar{t}_\mu, \mu=1(1)m\}$. As in Section 2.1.2 we assume that $\bar{\mathbb{G}}$ is a grid on a fixed finite interval, say $[0,1]$; but it is clear that this is for formal reasons only.

b) the *forward step method* $\mathfrak{M}$. $\mathfrak{M}$ has to be applicable to IVP1 on each interval $[\bar{t}_{\mu-1}, \bar{t}_\mu]$, $\mu=1(1)m$; for the purposes of the computation of $\zeta_n(\bar{t}_\mu)$ the interval $[\bar{t}_{\mu-1}, \bar{t}_\mu]$ plays the role of our normalized interval $[0,1]$ in previous chapters. Naturally, the global discretization error of $\mathfrak{M}$ has to possess a $\hat{\Delta}$-reduced asymptotic expansion to some order J for a sufficiently smooth IVP1 where $\hat{\Delta}$ is the reduction to the endpoint of the interval of integration (i.e., 1 for the normalized interval), see Def. 1.4.1 and 1.4.2.

c) the *sequence* $\mathfrak{N} := \{n_\rho \in \mathbb{N}', \rho=0(1)r\}$, with $\mathbb{N}'$ from $\mathfrak{M}$. $\mathfrak{N}$ contains those values of the discretization parameter n for which the discretization $\mathfrak{M}$(IVP1) is solved on each interval $[\bar{t}_{\mu-1}, \bar{t}_\mu]$.

d) the *class* $\mathfrak{C}_0$ of interpolation functions which are used in the Richardson extrapolation procedure, see Section 1.4.1. Naturally the choice of $\mathfrak{C}_0$ has to be consistent with the structure of the asymptotic expansion for $\mathfrak{M}$.

For a given sufficiently smooth IVP1, EM generates a function $\chi: \bar{\mathbb{G}} \to \mathbb{R}^s$ which is an approximation to z on the points of $\bar{\mathbb{G}}$. It is defined by the process indicated at the beginning of this section:

Define

$$\chi(0) := z_0;$$

recursively for $\mu=1(1)m$:

solve the discretization generated by $\mathfrak{M}$ for the IVP1

$$(6.3.1) \qquad \begin{cases} y(\bar{t}_{\mu-1}) - \chi(\bar{t}_{\mu-1}) = 0, \\ \qquad y' - f(y) = 0, \quad t \in [\bar{t}_{\mu-1}, \bar{t}_\mu], \end{cases}$$

for each $n_\rho \in \mathfrak{N}$ to obtain $\zeta_{n_\rho}(\bar{t}_\mu)$, $\rho=0(1)r$; then

interpolate the $\zeta_{n_\rho}(\bar{t}_\mu)$ by means of the class $\mathfrak{C}_0$ and obtain $\chi(\bar{t}_\mu)$ as the value of the interpolation function at $n=\infty$.

Def. 6.3.1. A quadruple $\{\bar{\mathfrak{G}}, \mathfrak{M}, \mathfrak{N}, \mathfrak{C}_0\}$, specified as above, is called an *extrapolation method* (EM) for IVP1. The function $\chi: \bar{\mathfrak{G}} \to \mathbb{R}^s$ defined by the described application of EM to a given IVP1 is called the *solution of the EM* for that IVP1.

Example. $\bar{\mathfrak{G}} := \{0, \frac{1}{2}, 1\}$,
$\mathfrak{M} :=$ Euler method with equidistant grids,
$\mathfrak{N} := \{1, 2, 3\}$,
$\mathfrak{C}_0 :=$ polynomials in $\frac{1}{n}$ (of degree 2).

We apply this EM to the scalar IVP1

$$y(0) - 1 = 0; \quad y' - g y = 0, \quad g \in \mathbb{R}.$$

On the subinterval $[0, \frac{1}{2}]$ we obtain

$$\zeta_1\left(\frac{1}{2}\right) = 1 + \frac{g}{2}, \quad \zeta_2\left(\frac{1}{2}\right) = \left(1 + \frac{g}{4}\right)^2, \quad \zeta_3\left(\frac{1}{2}\right) = \left(1 + \frac{g}{6}\right)^3;$$

polynomial Richardson extrapolation (see Section 1.4.2) yields

$$\chi\left(\frac{1}{2}\right) = 0.5\,\zeta_1\left(\frac{1}{2}\right) - 4\,\zeta_2\left(\frac{1}{2}\right) + 4.5\,\zeta_3\left(\frac{1}{2}\right) = 1 + \frac{g}{2} + \frac{g^2}{8} + \frac{g^3}{48}.$$

Due to the simplicity of our IVP1 we obtain for the second subinterval $[\frac{1}{2}, 1]$ simply the above value multiplied by $\chi(\frac{1}{2})$, i.e.,

$$\chi(1) = \left[\chi\left(\frac{1}{2}\right)\right]^2 = \left[1 + \frac{g}{2} + \frac{g^2}{8} + \frac{g^3}{48}\right]^2 \approx e^g.$$

In practical applications of EM, the computation of the $\zeta_{n_\rho}(\bar{t}_\mu)$ and the extrapolation process are interleaved in the usual fashion: After the computation of each further $\zeta_{n_\rho}(\bar{t}_\mu)$ the complete associated diagonal in the extrapolation table is generated by the algorithms described in Sections 1.4.2 and 1.4.3. The agreement of the values at the far end of the current table is used to judge the accuracy which has been obtained so far. The computation of further ζ_{n_ρ} is discontinued when some criterion based upon this agreement is satisfied.

Thus the set $\mathfrak{N}$ need not be fully exhausted, and is rather considered as an (increasing) sequence which is used as far as needed. More details are given in Section 6.3.4.

6.3.2 Gragg's Method

The f.s.m. $\mathfrak{M}$ which is used in an extrapolation method has to be chosen rather carefully if the EM is to be efficient:

a) The $\hat{\Delta}$-reduced asymptotic expansion whose existence was stipulated should be in even powers of n only, in order that the extrapolation process be more effective.

b) $\mathfrak{M}$ should be explicit to permit fast computation, since the discretization on each subinterval has to be solved $r+1$ times for the computation of the values $\zeta_{n_\rho}(\bar{t}_\mu)$, $\rho=0(1)r$.

c) $\mathfrak{M}$ should have favorable strong stability properties so that there is no undue error propagation for an IVP1 with decaying solution components.

Requirements a) and b) are contradictory for RK-methods (Corollary 3.2.11 and Theorem 3.4.6); thus the use of a symmetric linear multistep method would seem indicated (Section 4.5.3). However, the associated polynomials ρ of symmetric linear k-step schemes possess only essential zeros (Theorem 4.5.4) and thus a $\hat{\Delta}$-reduced asymptotic expansion exists only under special circumstances (Section 4.4.4). Furthermore, this asymptotic expansion is in even powers only if a starting procedure satisfying (4.5.20) is used (Corollary 4.5.8). Finally, requirements b) and c) are contradictory for symmetric linear multistep methods as these methods are only weakly stable (discussion after Theorem 4.6.4).

Under these circumstances it is rather surprising that there exists a very simple f.s.m. which satisfies all three requirements and which is "almost" a linear multistep method; it simply incorporates a symmetric smoothing operator at the end of the interval of integration. This method $\mathfrak{M}_G$ has been suggested by Gragg [2] for the purpose of Richardson extrapolation; we have used it previously in various examples. For an IVP1 on $[0,1]$ the f.s.m. $\mathfrak{M}_G$ is characterized thus:

$\mathfrak{M}_G$ uses equidistant grids with $h=1/n$, $n\in\mathbb{N}':=\{2i, i\in\mathbb{N}\}$. The spaces E_n, E_n^0 and the mappings Δ_n, Δ_n^0 are as for a linear 2-step method (see Section 4.2.1). The discretization $\mathfrak{M}_G(\text{IVP}1)$ is defined by

$$(6.3.2)\quad (F_n\eta)_\nu := \begin{cases} \eta_0 - z_0, \\ \eta_1 - z_0 - h f(z_0), \\ \dfrac{\eta_\nu - \eta_{\nu-2}}{2h} - f(\eta_{\nu-1}), \qquad \nu = 2(1)n-1, \\ \dfrac{2}{3h}\left[\eta_n - \dfrac{1}{2}\eta_{n-1} - \dfrac{1}{2}\eta_{n-2}\right. \\ \qquad \left. - \dfrac{h}{2} f(\eta_{n-2} + 2h f(\eta_{n-1})) - h f(\eta_{n-1})\right]. \end{cases}$$

With the exception of the last step $\nu=n$, $\mathfrak{M}_G$ is identical with the linear 2-step method based upon the explicit midpoint scheme (see Example 3 at the end of Section 4.1.1), with an explicit starting procedure which satisfies the symmetry condition (4.5.20). The last step is non-linear in f but explicit, it incorporates the use of a smoothing operator as is immediately obvious when we write $(F_n\eta)_n=0$ in the form

$$(6.3.3)\qquad \begin{cases} \dfrac{\eta_n^0-\eta_{n-2}}{2h}-f(\eta_{n-1})=0,\\[2ex] \dfrac{\eta_{n+1}^0-\eta_{n-1}}{2h}-f(\eta_n^0)=0,\\[2ex] \eta_n:=\dfrac{1}{2}\left(\eta_n^0+\dfrac{\eta_{n-1}+\eta_{n+1}^0}{2}\right); \end{cases}$$

see the example at the end of Section 4.5.2.

Theorem 6.3.1. *The solution* ζ_n *of* $\mathfrak{M}_G(\mathrm{IVP}^{(2J+1)}1)$ *satisfies*

$$(6.3.4)\qquad \zeta_n(1)=z(1)+\sum_{j=1}^{J}h^{2j}\tilde{e}_{2j}+O(h^{2J+1})$$

where $\tilde{e}_2:=e_2(1)$ *and* $e_2:[0,1]\to\mathbb{R}^s$ *is the solution of*

$$(6.3.5)\qquad \begin{cases} e_2(0)=0,\\ e_2'-f'(z(t))e_2=\frac{1}{12}z'''(t)-\frac{1}{4}f'(z(t))z''(t), \qquad t\in[0,1]. \end{cases}$$

Proof. Consider $\zeta_n^0(t_\nu)$ defined by (6.3.2) for $\nu=0(1)n-1$ and by the first two lines of (6.3.3) for $\nu=n,\ n+1$ [2]. From Theorem 4.5.7 and the properties of $\mathfrak{M}_G$ we have

$$(6.3.6)\qquad \zeta_n^0(t_\nu)=z(t_\nu)+\sum_{j=1}^{J}h^{2j}[e_{2j,1}(t_\nu)+(-1)^\nu e_{2j,2}(t_\nu)]+O(h^{2J+1})$$

where the $e_{2j,i}$, $i=1,2$, satisfy differential equations of the type

$$(6.3.7)\qquad e_{2j,i}'+(-1)^i f'(z(t))e_{2j,i}=b_{2j,i}(t).$$

(Note that the associated polynomial $\rho=\frac{1}{2}(z^2-1)$ of the explicit midpoint scheme has the essential zeros $+1$ and -1, with growth parameters $+1$ and -1 resp.)

The fact that the symmetric smoothing operator of (6.3.3) retains the evenness of the asymptotic expansion (6.3.6) follows from Theorem 4.5.9 and Corollary 4.5.10. Furthermore, by its construction (see Theo-

---

[2] The use of t_{n+1}, which is actually outside $[0,1]$, could have been avoided but the analysis is more intuitive in this manner.

rem 4.5.3) the smoothing operator annihilates the contribution of e_{22} to $\tilde{e}_2$ and achieves

$$\tilde{e}_2 = e_{21}(1) + \tfrac{1}{4} z''(1).$$

From (4.4.43) and (4.4.44) we have

$$\begin{cases} e_{21}(0) = -\tfrac{1}{4} z''(0), \\ e'_{21} - f'(z(t)) e_{21} = -\tfrac{1}{6} z'''(t), \quad t \in [0,1]. \end{cases} \tag{6.3.8}$$

(6.3.8) implies that $e_2(t) = e_{21}(t) + \tfrac{1}{4} z''(t)$ satisfies (6.3.5). □

There exists an interesting alternative for the formulation of Gragg's algorithm (Stetter [7]): In place of our usual IVP1 we consider the system of $2s$ differential equations

$$\begin{cases} u(0) = v(0) = z_0, \\ u' = f(v), \quad v' = f(u), \end{cases} \tag{6.3.9}$$

the solution of which is naturally $u(t) = v(t) = z(t)$. We now apply a one-step method to (6.3.9) which treats the two "components" of (6.3.9) differently. Let φ_ν and ψ_ν be the approximations to $u(t_\nu)$ and $v(t_\nu)$ resp., with $t_\nu = \nu \bar{h} = 2\nu h$, defined by

$$\varphi_0 := \psi_0 := z_0,$$

$$\frac{\varphi_\nu - \varphi_{\nu-1}}{\bar{h}} - f\left(\frac{\psi_\nu + \psi_{\nu-1}}{2} - \frac{\bar{h}}{4}\big(f(\varphi_\nu) - f(\varphi_{\nu-1})\big)\right) = 0, \tag{6.3.10}$$

$$\frac{\psi_\nu - \psi_{\nu-1}}{\bar{h}} - \frac{1}{2}\big(f(\varphi_\nu) + f(\varphi_{\nu-1})\big) = 0. \tag{6.3.11}$$

The implicit appearance of (6.3.10)/(6.3.11) is deceptive; the explicit character of the forward step procedure becomes evident when it is written in the form

$$\begin{cases} \psi_{\nu-\frac{1}{2}} := \psi_{\nu-1} + \dfrac{\bar{h}}{2} f(\varphi_{\nu-1}), \\ \varphi_\nu := \varphi_{\nu-1} + \bar{h} f(\psi_{\nu-\frac{1}{2}}), \\ \psi_\nu := \psi_{\nu-\frac{1}{2}} + \dfrac{\bar{h}}{2} f(\varphi_\nu). \end{cases} \tag{6.3.12}$$

The equivalence of (6.3.10)/(6.3.11) and (6.3.12) is displayed by addition and subtraction resp., of the first and last lines of (6.3.12) which leads to (6.3.11) and

$$\psi_{\nu-\frac{1}{2}} = \frac{\psi_\nu + \psi_{\nu-1}}{2} - \frac{\bar{h}}{4}\big(f(\varphi_\nu) - f(\varphi_{\nu-1})\big). \tag{6.3.13}$$

Theorem 6.3.2. *The solution* η *of* (6.3.2) *for even n satisfies*

$$\eta_n = \tfrac{1}{2}[\varphi_{n/2} + \psi_{n/2}]. \tag{6.3.14}$$

Proof. Follows immediately from the observation that

(6.3.15)
$$\varphi_\nu = \eta_{2\nu}, \quad \nu = 0(1)\frac{n}{2} - 1; \qquad \varphi_{n/2} = \eta_n^0; \qquad \psi_{\nu-\frac{1}{2}} = \eta_{2\nu-1}, \quad \nu = 1(1)\frac{n}{2};$$

the smoothing procedure (6.3.3) may be written as

$$\eta_n := \tfrac{1}{2}(\eta_n^0 + \eta_{n-1} + hf(\eta_n^0)) = \tfrac{1}{2}(\varphi_{n/2} + \psi_{n/2}). \qquad \square$$

(6.3.11) and the second line of (6.3.12) may be interpreted as an application of a trapezoidal procedure to the second component of (6.3.9) and of a midpoint procedure to the first component. This formulation of Gragg's method shows that its starting and smoothing procedures are natural counterparts, the actual smoothing being reduced to an averaging of the two different approximations φ and ψ of the true solution.

By a trivial extension of Theorems 3.2.14 and 3.4.6 to RK-methods which treat different components of an IVP 1 differently we obtain the evenness of the asymptotic expansion for the global discretization error of (6.3.10)/(6.3.11). Similarly an extension of Theorem 3.4.5 yields the system

$$\begin{aligned} c_2'(t) - f'(z(t))\,d_2(t) &= p_\varphi(t), \\ d_2'(t) - f'(z(t))\,c_2(t) &= p_\psi(t), \end{aligned} \qquad c_2(0) = d_2(0) = 0, \tag{6.3.16}$$

for the $\bar{h}^2$-terms in

$$\begin{aligned} \varphi(t_\nu) &= z(t_\nu) + \bar{h}^2 c_2(t_\nu) + O(\bar{h}^4), \\ \psi(t_\nu) &= z(t_\nu) + \bar{h}^2 d_2(t_\nu) + O(\bar{h}^4); \end{aligned} \tag{6.3.17}$$

here p_φ and p_ψ are the principal error functions of (6.3.10) and (6.3.11) resp. By (6.3.14) and (6.3.17),

$$\eta_n = z(1) + \frac{\bar{h}^2}{2}(c_2(1) + d_2(1)) + O(\bar{h}^4);$$

from (6.3.16) we have

$$\tfrac{1}{2}(c_2 + d_2)(0) = 0,$$

$$\tfrac{1}{2}(c_2 + d_2)' - f'(z(t))\tfrac{1}{2}(c_2 + d_2) = \tfrac{1}{2}(p_\varphi(t) + p_\psi(t)),$$

and a short calculation yields

$$\tfrac{1}{2}(p_\varphi(t) + p_\psi(t)) = \tfrac{1}{48}z'''(t) - \tfrac{1}{16}f'(z(t))z''(t).$$

This establishes the validity of (6.3.5) in a more immediate fashion without the complicated theory of Section 4.4.4.

Corollary 6.3.3. *In the situation of Theorem* 6.3.1, $\tilde{e}_{2j}=e_{2j}(1)$ *for* $j=1(1)J$, *where the* $e_{2j}\colon[0,1]\to\mathbb{R}^s$ *satisfy differential equations of the form*

$$\text{(6.3.18)}\qquad \begin{cases} e_{2j}(0)=0\,, \\ e_{2j}'-f'(z(t))\,e_{2j}=b_{2j}(t)\,, \qquad t\in[0,1]\,. \end{cases}$$

Proof. By an immediate extension of the argument following Theorem 6.3.2. □

Corollary 6.3.3 shows that *no* term of the asymptotic expansion (6.3.4) satisfies a differential equation (6.3.7) with $i=2$; the parasitic terms enter merely into the inhomogeneities of (6.3.7) for $i=1$.

6.3.3 Strong Stability of $\mathfrak{M}_G$

In order to apply our concept of strong stability to Gragg's method $\mathfrak{M}_G$ we have to define its application on longer and longer intervals. For simplicity we will assume that the smoothing is performed after every interval of length 1, i.e., at each integer t; the arising $\mathbb{T}$-sequence of f.s.m. will be denoted by $\{\mathfrak{M}_G\}_{\mathbb{T}}$. Of course, $\mathfrak{M}_G$ is not intended to be used in this fashion but rather as the f.s.m. of an extrapolation method which may then be extended to longer and longer intervals. Nevertheless, the following discussion will give some insight into the mechanism of strong and weak stability.

First we consider a much simpler method than $\{\mathfrak{M}_G\}_{\mathbb{T}}$. Define a 2-step discretization $\mathfrak{D}_{\mathbb{T}}$ of an $\{\text{IVP }1\}_{\mathbb{T}}$ (cf. Section 2.3.2), with $h=1/n$ and $n\in\mathbb{N}$, by

$$\text{(6.3.19)}\qquad \begin{cases} \eta_0-z_0=0\,, \\[1ex] \dfrac{\eta_\nu-\eta_{\nu-2}}{2h}-f(t_{\nu-1},\eta_{\nu-1})=0\,, \quad \text{for } \nu\not\equiv 1 \bmod n\,, \\[2ex] \dfrac{\eta_\nu-\eta_{\nu-1}}{h}-f(t_{\nu-1},\eta_{\nu-1})=0\,, \quad \text{for } \nu\equiv 1 \bmod n\,. \end{cases}$$

On $[0,1]$, (6.3.19) simply describes an application of the explicit 2-step midpoint method with the starting procedure $\eta_1=z_0+hf(0,z_0)$ of $\mathfrak{M}_G$. In the extension to longer intervals, however, the midpoint algorithm is "restarted" after each interval of length 1. This method will be denoted by $\{\bar{\mathfrak{M}}_G\}_{\mathbb{T}}$.

Theorem 6.3.4. *Consider an exponentially stable* $\{IVP^{(2)}1\}_{\mathbb{T}}$ *which admits an estimate* (2.3.9) *such that*

$$\text{(6.3.20)}\qquad a e^{-\mu}<1\,.$$

Furthermore assume that $f_t + f_y f$ *is uniformly bounded in* $B_R(z)$ *by* $M_{f'}$.

Then for each $\rho > 0$ *there exists an* n_0 *such that the solution* η *of* (6.3.19) *satisfies*

$$\|\eta(t_\nu) - z(t_\nu)\| < \rho \quad \textit{for all } \nu$$

if $n \geq n_0$.

Proof. We consider the application of $\mathfrak{M}_G$ on successive intervals $[m-1, m]$, $m = 1, 2, \ldots,$ to the IVP 1's

$$(6.3.21) \qquad \begin{cases} y(m-1) - z(m-1) = \delta_{m-1}, \\ \qquad y' - f(t, y) = 0, \quad t \in [m-1, m]; \end{cases}$$

the solution of (6.3.21) is called $z_m: [m-1, m] \to \mathbb{R}^s$, and the solution of the discretization of (6.3.21) by $\mathfrak{M}_G$, with a fixed given $n \in \mathbb{N}$, is called $\bar{\eta}_m: \{\nu/n, \nu = (m-1)n(1)mn\} \to \mathbb{R}^s$.

If $\|\delta_{m-1}\| < R/a$, $z_m(t)$ remains in $B_R(z)$ according to (6.3.20). Under this assumption, the global discretization error of $\mathfrak{M}_G$ for (6.3.21) may be bounded by $(c/n) M_{f'}$ in $[m-1, m]$ uniformly in m; c is some constant. Hence $\delta_m := \bar{\eta}_m(m) - z(m)$ satisfies

$$(6.3.22) \quad \|\delta_m\| \leq \|\bar{\eta}_m(m) - z_m(m)\| + \|z_m(m) - z(m)\| \leq \frac{c}{n} M_{f'} + a \|\delta_{m-1}\| e^{-\mu},$$

according to (6.3.21) and (2.3.9). Due to $\delta_0 = 0$, (6.3.22) implies

$$\|\delta_m\| \leq \frac{c}{n} M_{f'} \sum_{\lambda=0}^{m-1} (a e^{-\mu})^\lambda < \frac{1}{n} \frac{c M_{f'}}{1 - a e^{-\mu}}$$

which may be made arbitrarily small uniformly in m by choosing n sufficiently large. ▯

Remark. It is well known that Theorem 6.3.4 does *not* hold for the extension of the usual explicit 2-step midpoint method to longer and longer intervals, see Section 4.6.1.

Theorem 6.3.5. $\{\mathfrak{M}_G\}_{\mathbb{T}}$ *is strongly exponentially stable for sufficiently large n w.r.t. linear* $\{IVP\,1\}_{\mathbb{T}}$ *which satisfy the assumptions of Theorem* 6.3.4.

Proof. Let our $\{\text{IVP}\,1\}_{\mathbb{T}}$ be given by $y' - g(t) y = 0$. We have to show that the solution of the 2-step difference equation

$$(6.3.23) \qquad \begin{cases} \begin{pmatrix} \varepsilon_{\nu_0} \\ \varepsilon_{\nu_0 - 1} \end{pmatrix} - \delta_{\nu_0} = 0, & & \\ \dfrac{\varepsilon_\nu - \varepsilon_{\nu-2}}{2h} - g(t_{\nu-1}) \varepsilon_{\nu-1} = 0, & \nu \not\equiv 1 \bmod n, & \\ & & \nu > \nu_0, \\ \dfrac{\varepsilon_\nu - \varepsilon_{\nu-1}}{h} - g(t_{\nu-1}) \varepsilon_{\nu-1} = 0, & \nu \equiv 1 \bmod n, & \end{cases}$$

satisfies an estimate (with some $\mu'>0$)

(6.3.24) $$\|\varepsilon_\nu\| \le \alpha \|\boldsymbol{\delta}_{\nu_0}\| \, e^{-\mu'(t_\nu - t_{\nu_0})} \quad \text{for all } \nu>\nu_0 ,$$

uniformly in ν_0 if $h=1/n$ is sufficiently small.

Let $[(\nu_0-1)/n]+1=: m_0$. Write the second line of (6.3.23) in the form

$$\begin{pmatrix} \varepsilon_\nu \\ \varepsilon_{\nu-1} \end{pmatrix} = \begin{pmatrix} 2hg(t_{\nu-1}) & I \\ I & 0 \end{pmatrix} \begin{pmatrix} \varepsilon_{\nu-1} \\ \varepsilon_{\nu-2} \end{pmatrix};$$

it is then clear that—with $\|\boldsymbol{\varepsilon}_\nu\| = \max(\|\varepsilon_\nu\|, \|\varepsilon_{\nu-1}\|)$—

$$\|\boldsymbol{\varepsilon}_\nu\| \le (1+2hL)\, \|\boldsymbol{\varepsilon}_{\nu-1}\| \quad \text{for } \nu=\nu_0+1(1)nm_0 ,$$

where $L=\sup_\nu \|g(t_\nu)\|$ which exists according to (2.3.2). Hence

(6.3.25) $$\|\varepsilon_{nm_0}\| \le \|\boldsymbol{\varepsilon}_{nm_0}\| < e^{2L} \|\boldsymbol{\delta}_{\nu_0}\| .$$

For $m=m_0+1, m_0+2, \ldots,$ we now regard (6.3.23) as a discretization of

$$e(m-1)-\varepsilon_{n(m-1)}=0,$$
$$e'-g(t)e=0, \qquad t\in[m-1,m],$$

the solution of which satisfies

(6.3.26) $$\|e(t)\| \le a\, \|\varepsilon_{n(m-1)}\| \exp(-\mu(t-m+1)), \qquad t\in[m-1,m].$$

For a homogeneous linear IVP 1 the global discretization error of a convergent "linear" method is proportional to the initial value and there exists an increasing continuous function $\rho:(0,h_0]\to \mathbb{R}_+$, with $\lim_{h\to 0} \rho(h)=0$, such that

(6.3.27) $$\|\varepsilon(t_\nu)-e(t_\nu)\| \le \rho(h)\, \|\varepsilon_{n(m-1)}\| \quad \text{for } t_\nu\in[m-1,m].$$

(6.3.26) and (6.3.27) imply

(6.3.28) $$\|\varepsilon_{nm}\| \le (ae^{-\mu}+\rho(h))\, \|\varepsilon_{n(m-1)}\| \quad \text{for all } m>m_0 .$$

Because of (6.3.20) we may choose h small enough so that $ae^{-\mu}+\rho(h)<1$, whence (6.3.25) and (6.3.28) imply the existence of an estimate (6.3.24). (The growth of ε_ν in the interval $[[t_\nu], t_\nu]$ may be treated like the growth in $[t_{\nu_0}, m_0]$; see the derivation of (6.3.25).) □

Theorems 6.3.4 and 6.3.5 show that the essential agent for the generation of strong stability with weakly stable multistep methods is the breaking of the propagation cycle of the components associated with the parasitic essential zeros. Naturally, the combination of the "restarting" with a smoothing procedure which eliminates a good deal of the accumulated contribution of the parasitic components further improves the situation.

Corollary 6.3.6. *Theorems* 6.3.4 *and* 6.3.5 *also hold for* $\{\mathfrak{M}_G\}_{\mathbb{T}}$.

Proof. Immediate transcription of the previous proofs. □

At first sight, one might think that the one-step formulation (6.3.10)/(6.3.11) of $\mathfrak{M}_G$ would immediately imply strong exponential stability for sufficiently large n according to Theorem 3.5.3. However, we now have a one-step algorithm not for the original $\{\text{IVP}\,1\}_{\mathbb{T}}$ but for the system (6.3.9) and this system is never exponentially stable: Take $f(y)=gy$, $g\in\mathbb{C}_-$; then (6.3.9) has the fundamental solutions $\begin{pmatrix}\exp(gt)\\ \exp(gt)\end{pmatrix}$ and $\begin{pmatrix}\exp(-gt)\\ -\exp(-gt)\end{pmatrix}$! Thus, our one-step formulation merely transfers the stability defect to the original problem.

The fact that the discretization of an exponentially stable $\{\text{IVP}\,1\}_{\mathbb{T}}$ by $\{\mathfrak{M}_G\}_{\mathbb{T}}$ can only be exponentially stable for sufficiently large n is clear from the explicitness of $\mathfrak{M}_G$. Some quantitative results have been reported by Stetter [8]. Let $\bar{g}_n$ be the smallest real number such that the discretization of $y'-gy=0$ by $\{\mathfrak{M}_G\}_{\mathbb{T}}$ is exponentially stable for n for real $g\in(\bar{g}_n, 0)$. Table 6.1 gives the approximate size of $\bar{g}_n$ for some n.

Table 6.1

n	2	4	6	8	12	16
$\bar{g}_n$	−3.09	−4.00	−4.62	−5.20	−6.00	−6.72

6.3.4 The Gragg-Bulirsch-Stoer Extrapolation Method

The only extrapolation method which has become widely used as of this date is the one suggested by Bulirsch and Stoer [3]. It is characterized by the following elements (see Def. 6.3.1):

a) The grid $\bar{\mathfrak{G}}$ is generated in the course of the computation in a fashion discussed below.

b) The f.s.m. $\mathfrak{M}$ is Gragg's method $\mathfrak{M}_G$.

c) The set $\mathfrak{N}$ is a section from the sequence

$$2, 4, 6, 8, 12, 16, 24, 32, 48, \ldots,$$

whose selection is also explained below[3].

d) The class $\mathfrak{C}_0$ of interpolation functions is the class of rational functions in n^{-2} described in Section 1.4.3; each component of the solution vector is interpolated separately, see the remark after Def. 1.4.3.

[3] There exist many variations; we will restrict ourselves to a discussion of the basic philosophy.

From experience it has been found that the application of Richardson extrapolation to values with an asymptotic expansion of the structure (6.3.4) tends to be most effective if r is about 6. (Here and in the following it is assumed that $J \geq 6$. If this assumption is not satisfied one has to keep $r \leq J$, of course.) Hence, one should attempt to choose $\bar{t}_\mu - \bar{t}_{\mu-1} =: \Delta \bar{t}_\mu$ such that the extrapolated value $\chi(\bar{t}_\mu)$ is just sufficiently accurate when it has been computed from $\zeta_{n_\rho}(\bar{t}_\mu)$, $n_\rho \in \mathfrak{N}$, $\rho = 0(1)6$.

As in Section 1.4.3 (see also (1.4.7)/(1.4.8)) we use the notation

$$T_{-1}^{\rho} := 0, \qquad T_0^{\rho} := \zeta_{n_\rho}(\bar{t}_\mu), \qquad \rho = 0(1)r, \tag{6.3.29}$$

and the recursion

$$T_k^i := T_{k-1}^{i+1} + \left[\left(\frac{n_{i+k}}{n_i} \right)^2 \left(1 - \frac{T_{k-1}^{i+1} - T_{k-1}^i}{T_{k-1}^{i+1} - T_{k-2}^{i+1}} \right) - 1 \right]^{-1} (T_{k-1}^{i+1} - T_{k-1}^i), \quad i+k \leq r; \tag{6.3.30}$$

for a fixed r this implies $\chi(\bar{t}_\mu) := T_r^0$. But—as indicated at the end of Section 6.3.1—r is not fixed in advance but is increased successively: After the computation of ζ_{n_ρ} for the next larger ρ the values of T_k^i, $i = \rho - k$, $k = 1(1)\rho$, are computed from (6.3.29)/(6.3.30) to yield

$$\chi_\rho(\bar{t}_\mu) := T_\rho^0 . \tag{6.3.31}$$

$\chi_\rho(\bar{t}_\mu)$ may now be compared to $\chi_{\rho-1}(\bar{t}_\mu)$ and the relative difference between the two values may be used as an indication of the relative accuracy of $\chi_\rho(\bar{t}_\mu)$; see Section 1.5.4.

Thus we may formalize our above idea about the choice of $\bar{t}_\mu$ as follows: We should have

$$\left\| \left(\frac{\chi_\rho(\bar{t}_\mu) - \chi_{\rho-1}(\bar{t}_\mu)}{\chi_\rho(\bar{t}_\mu)} \right) \right\| \begin{cases} > \bar{\varepsilon} & \text{for } \rho < 6, \\ < \bar{\varepsilon} & \text{for } \rho = 6, \end{cases} \tag{6.3.32}$$

where $\left(\frac{\cdots\cdots}{\cdots\cdots} \right)$ denotes the s-vector the components of which have been formed from the components of χ_ρ and $\chi_{\rho-1}$ in the indicated manner. $\bar{\varepsilon}$ is an accuracy parameter which must not be too small relative to the computing error; $\bar{\varepsilon}$ is assumed to be given and fixed.

If (6.3.32) is not satisfied for a trial choice of $\Delta \bar{t}_\mu$ one may disregard the computation of the $\chi_\rho(\bar{t}_\mu)$ and take a more appropriate value of $\Delta \bar{t}_\mu$, i. e., a larger one if $< \bar{\varepsilon}$ holds for some $\rho < 6$ and a smaller one if $< \bar{\varepsilon}$ does not hold for $\rho = 6$. Various suggestions have been made as to how much to increase or decrease $\Delta \bar{t}_\mu$ in these cases; see, e. g., Bulirsch and Stoer [3].

On the other hand, if the left hand side of (6.3.32) has become smaller than $\bar{\varepsilon}$ for $\rho=\rho_0<6$ one may simply terminate the computation at that ρ_0 and set $\chi(\bar{t}_\mu):=\chi_{\rho_0}(\bar{t}_\mu)$; the discrepancy between ρ_0 and 6 may then be utilized in the choice of the length $\Delta\bar{t}_{\mu+1}$ of the next subinterval. This will normally be the more economical strategy except when ρ_0 is very small, say $\rho_0<4$.

If the left hand side of (6.3.32) still exceeds $\bar{\varepsilon}$ for $\rho=6$ one may further increase ρ instead of repeating the computation with a smaller $\Delta\bar{t}_\mu$. Bulirsch and Stoer [3] suggest that k should not exceed 6 and that one should define

$$\chi_\rho(\bar{t}_\mu):=T_6^{\rho-6} \quad \text{for } \rho>6. \tag{6.3.33}$$

If a satisfaction of the accuracy requirement can be achieved in this fashion for a ρ not much greater than 6 (say 7 or 8) this should again be a more economical and efficient strategy than an immediate change of Δt_μ upon a failure of (6.3.32). However, if the solution of the IVP1 changes too abruptly in $[\bar{t}_{\mu-1},\bar{t}_\mu]$ so that the further increase of ρ does not help, one cannot but choose a new value for $\Delta\bar{t}_\mu$ and repeat the complete cycle.

Finally it is evident that a deviation from the criterion (6.3.32) is necessary if a solution component becomes very small in modulus as may happen when it changes sign.

Independent of the many possible variations and refinements it is an essential characteristic of the G(ragg)-B(ulirsch)-S(toer) method that $\Delta\bar{t}_\mu$ is determined in such a way that it depends on the *numerical structure of the current* (or previous) *extrapolation table.* As the "speed of convergence" of the extrapolation process indicates the local smoothness of the IVP1, this means that the lengths of the subintervals are adjusted to this local smoothness, in a natural and automatic manner. This results in a very satisfactory stepsize control mechanism.

The preference for rational over polynomial interpolation in the GBS-method is due to the results of extensive experiments where rational interpolation normally gave better results. No deeper analysis of the relative advantages seems to exist. The fact that a singularity may arise in rational extrapolation seems not to have caused any trouble so far.

Finally, let us discuss the strong stability properties of the GBS-method. From the discussion in Sections 6.3.2 and 6.3.3 it is clear that the weak stability effects of the midpoint method on a fixed interval $[\bar{t}_{\mu-1},\bar{t}_\mu]$ are the better eliminated the larger the value of n_ρ (see Table 6.1 at the end of Section 6.3.3). Although rational interpolation is a nonlinear process, it is intuitively clear that the values $\zeta_{n_\rho}(\bar{t}_\mu)$ with a larger n_ρ influence the value of $\chi(\bar{t}_\mu)$ stronger than those with a small

n_ρ. (With polynomial interpolation this follows immediately from the size of the $\gamma_{r\rho}^0$ in (1.4.10).) Thus the extrapolation process tends further to reinforce the stabilizing effect of the smoothing procedure of $\mathfrak{M}_G$.

If both effects are not sufficient for the elimination of the parasitic components in the application of the GBS-method to an exponentially stable IVP 1, this will become evident through the deteriorating convergence of the extrapolation. By the procedure for the choice of $\Delta\bar{t}_\mu$ this will lead to smaller sub-intervals $[\bar{t}_{\mu-1}, \bar{t}_\mu]$ in which the growth of the parasitic components is less pronounced. Thus the mechanism (6.3.32) for the choice of $\Delta\bar{t}_\mu$ also has a stabilizing effect.

A general quantitative analysis of the situation is virtually impossible due to the nonlinearity of the extrapolation procedure. For the case of polynomial extrapolation one may define stability regions for the GBS-method which are analogous to the regions of absolute stability for standard f.s.m., see Stetter [8]. The sizes of these regions—which are comparable to those of standard *explicit* f.s.m.—indicate that the GBS-method will not suffer from weak stability effects for normal IVP 1. However, for stiff systems it should prove just as ineffective as any explicit discretization method, cf. Section 2.3.7. These conclusions are well supported by practical experience.

Experiments by many different investigators have established the GBS-method as one of the best—if not *the* best—*general-purpose discretization method* for the numerical solution of IVP 1 (see Hull et al. [1]).

6.3.5 Extrapolation Methods for Stiff Systems

It may be expected that an extrapolation method will be able to handle stiff $\{\text{IVP}\,1\}_\mathbb{T}$ effectively only if it is based on an implicit f.s.m. $\mathfrak{M}$ which is A-stable, cf. Section 2.3.7. The natural choices for $\mathfrak{M}$ are the implicit one-step trapezoidal and midpoint methods which also satisfy the requirement a) of Section 6.3.2 for an asymptotic expansion in even powers of n; see the example at the end of Section 3.2.5 and Theorem 3.4.6.

In his fundamental paper on the strong exponential stability of f.s.m. Dahlquist [3] made the observation that an extrapolation method based on the trapezoidal rule may not itself by A-stable: If $\mathfrak{N}=\{1,2\}$ and even polynomial extrapolation is used then the associated growth function (cf. Section 3.2.4 and Theorem 3.5.10) of that extrapolation method is

$$\gamma(H)=\frac{4}{3}\left(\frac{1+\dfrac{H}{4}}{1-\dfrac{H}{4}}\right)^2-\frac{1}{3}\left(\frac{1+\dfrac{H}{2}}{1-\dfrac{H}{2}}\right).$$

Obviously, $\lim_{H\to\infty} \gamma(H) = \frac{5}{3} > 1$; along the real negative H-axis $\gamma(H)$ begins to exceed 1 near -26. This situation changes, however, when $\mathfrak{N}$ contains only even numbers; then each individual growth function for the ζ_{n_ρ} has $+1$ as its limit at ∞ and polynomial interpolation retains that limit.

To analyze the situation further we consider an extrapolation method with subintervals of length Δt, based upon the implicit trapezoidal or[4] the implicit midpoint method with equal steps of length $h_\rho := \Delta t / n_\rho$, with an arbitrary set $\mathfrak{N}$ of *even* numbers, and with even polynomial interpolation. We will call this method IEM (*I*mplicit EM).

The application of IEM to the scalar problem $y' - gy = 0$, $g \in \mathbb{C}_-$, with $z_0 = 1$, leads to

$$\zeta_{n_\rho}(\Delta t) = \left(\frac{1 + \frac{H_\rho}{2}}{1 - \frac{H_\rho}{2}} \right)^{n_\rho}, \quad \text{with } H_\rho := \frac{g \Delta t}{n_\rho} \in \mathbb{C}_-,$$

and

$$\chi(\Delta t) = \sum_{\rho=0}^{r} \gamma_{r\rho} \left(\frac{1 + \frac{H_\rho}{2}}{1 - \frac{H_\rho}{2}} \right)^{n_\rho}, \tag{6.3.34}$$

where the interpolation coefficients $\gamma_{r\rho}$ are defined by

$$\gamma_{r\rho} := \prod_{\rho'=0}^{r}{}' \left[1 - \left(\frac{n_{\rho'}}{n_\rho} \right)^2 \right]^{-1}; \tag{6.3.35}$$

see Section 1.4.2. For any $m \in \mathbb{N}$ we have $\chi(m \Delta t) = [\chi(\Delta t)]^m$ so that it suffices to consider the dependence of $\chi(\Delta t)$ upon $\mathfrak{N}$.

For the present purpose we will consider our extrapolation method as a one-step method with steps Δt and apply our Def. 2.3.12 of the regions $\mathfrak{H}_0$ of absolute stability. At first sight this approach seems to permit the occurence of perturbations only at the $\bar{t}_\mu$, which would be unrealistic. However, a perturbation which is introduced (see (2.3.30)) at some $t_{\nu 0} \in [\bar{t}_{\mu-1}, \bar{t}_\mu]$ during the computation of one of the ζ_{n_ρ} can only lead to a perturbation of $\chi(\bar{t}_\mu)$ of a modulus not larger than $a \|\delta_{\nu 0}\|$, with some uniform constant a. (This is true even when the IEM is applied to a general $\{\text{IVP } 1\}_{\mathbb{T}}$.) Thus the presence of exponential stability depends solely on the further growth behavior which is characterized by $\mathfrak{H}_0$.

---

[4] In an application to linear IVP1 with constant coefficients the two methods are equivalent.

For a given set $\mathfrak{N}$ of $r+1$ natural numbers n_ρ, $\rho=0(1)r$, we define $\gamma_{\mathfrak{N}}: \mathbb{C} \to \mathbb{C}$ by

$$\gamma_{\mathfrak{N}}(H) := \sum_{\rho=0}^{r} \gamma_{r\rho} \left(\frac{1+\dfrac{H}{2n_\rho}}{1-\dfrac{H}{2n_\rho}} \right)^{n_\rho}, \tag{6.3.36}$$

with $\gamma_{r\rho}$ from (6.3.35). Then the IEM with the set $\mathfrak{N}$ has the region of absolute stability

$$\mathfrak{H}_0 := \{H \in \mathbb{C}: |\gamma_{\mathfrak{N}}(H)| < 1\};$$

the proof of this fact follows precisely the proof of Theorem 3.5.10.

Lemma 6.3.7. *For any set $\mathfrak{N}$ which contains only even natural numbers, $|\gamma_{\mathfrak{N}}(H)|$ approaches 1 from below if H approaches ∞ through $\mathbb{C}_-$ in a direction not parallel to the imaginary axis.*

Proof. For even n_ρ, $\lim_{H\to\infty} ((1+H/2n_\rho)/(1-H/2n_\rho))^{n_\rho} = 1$ so that $\lim_{H\to\infty} \gamma_{\mathfrak{N}}(H) = 1$ due to $\sum_\rho \gamma_{r\rho} = 1$ which follows from the interpolation property for constant functions. To obtain the behavior of $\gamma_{\mathfrak{N}}$ in the vicinity of ∞ we define $\bar{\gamma}_{\mathfrak{N}}(x) := \gamma_{\mathfrak{N}}(1/x)$ for $x \in \mathbb{C}$ and form

$$\begin{aligned}
\bar{\gamma}'_{\mathfrak{N}}(0) &= \frac{d}{dx}\left[\sum_{\rho=0}^{r} \gamma_{r\rho}\left(\frac{2xn_\rho+1}{2xn_\rho-1}\right)^{n_\rho}\right]_{x=0} \\
&= \sum_{\rho=0}^{r} \gamma_{r\rho} n_\rho \left(\frac{2xn_\rho+1}{2xn_\rho-1}\right)^{n_\rho-1} \frac{-4n_\rho}{(2xn_\rho-1)^2}\Bigg|_{x=0} \\
&= -4\sum_{\rho=0}^{r} \gamma_{r\rho}(-1)^{n_\rho-1} n_\rho^2 = 4\sum_{\rho=0}^{r} \gamma_{r\rho} n_\rho^2 \quad \text{for even } n_\rho\,.
\end{aligned}$$

Assume that the n_ρ are ordered increasingly so that $n_\rho > n_{\rho-1}$, $\rho=1(1)r$; then it follows trivially from (6.3.35) that the $\gamma_{r\rho}$ alternate in sign and $\gamma_{rr} > 0$. The positivity of $\bar{\gamma}'_{\mathfrak{N}}(0)$ now follows from $|\gamma_{r\rho}| > |\gamma_{r,\rho-1}|$ which may also be established from (6.3.35).

Thus the vicinity of 0 in the x-plane is mapped similarly and without a rotation onto a vicinity of 1 in the $\bar{\gamma}_{\mathfrak{N}}$-plane so that $|\bar{\gamma}_{\mathfrak{N}}(x)| < 1$ along all x-paths approaching 0 in $\mathbb{C}_-$ which are not tangential to the imaginary axis at 0. Returning to the H-plane via $H = 1/x$ we have the assertion for $\gamma_{\mathfrak{N}}$. □

For many natural choices of $\mathfrak{N}$ it is also possible to show that $|\gamma_{\mathfrak{N}}(H)| < 1$ all along the negative real axis. (Presumably this is true for all even $\mathfrak{N}$.) Extensive numerical calculations have established that the typical shape for the region $\mathfrak{H}_0$ of an IEM is the following:

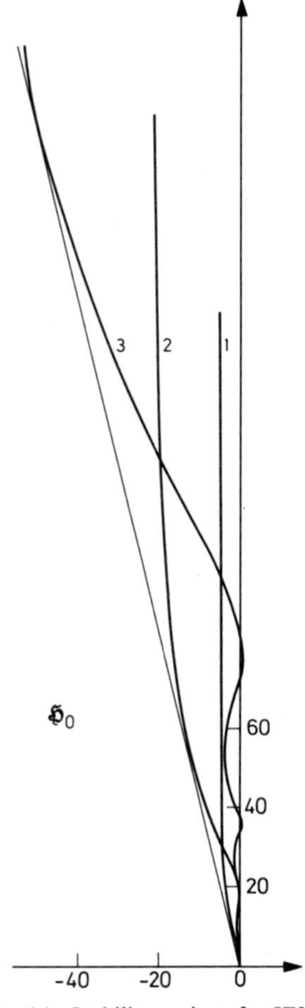

Fig. 6.1. Stability regions for IEM

In the terminology of Def. 2.3.14 this means that an IEM is $A(\alpha)$-stable which is very acceptable for the treatment of stiff equations; see the remark after Def. 2.3.14.

For the sets $\mathfrak{N}_r := \{2^{\rho+1}, \rho = 0(1)r\}$ the following approximate values were obtained for $\alpha_0 := \alpha/(\pi/2)$

r	0	1	2	3	4	5	6	7	8
α_0	1	0.881	0.861	0.856	0.855	0.855	0.855	0.855	0.855

While α_0 remains almost stationary when r is increased and obviously tends to a limit, the "bulge" of $\mathfrak{H}_0$ towards the ray through 0 moves towards infinity by a factor ≈ 4 per unit increase of r.

According to the discussion preceding Theorem 6.3.7, we have thus established that an IEM (with a reasonable set $\mathfrak{N}$) is strongly exponentially stable for arbitrary steps Δt w.r.t. those $\{\text{IVP } 1\}_{\mathbb{T}} \in J_c$ which have all eigenvalues of g in the sector $S(\alpha)$.

The practical use of IEM for stiff systems requires an effective resolution of two difficulties:

a) Due to the implicitness of our basic discretization methods the computation of the $\zeta_{n_\rho}(\bar{t}_\mu)$ becomes problematic. The introduction of a predictor-corrector approach would destroy the implicitness and hence the A-stability of the basic method, and it must therefore not be used. A Newton-type iteration is admissible but requires considerable effort. In particular, it requires sufficiently good *starting values* to keep the number of iterations small.

b) We want to use A- or $A(\alpha)$-stable methods with stiff systems so that we can keep the steps large in the presence of eigenvalues with large negative real parts. This requires that we recognize how large we may take each step without lowering the accuracy. The extrapolation table provides a good tool for judging the appropriateness of the current step $\Delta\bar{t}_\mu$ as has been explained in Section 6.3.4. However, in the application of IEM to stiff problems it may happen that—after the transient components have died away—$\Delta\bar{t}_\mu$ may be increased by a factor of the order of 100 under a criterion like (3.3.32) but that this fact is not indicated by the extrapolation table of the previous step. If this possibility of an enormous increase in stepsize is not recognized but the steps are built up say by factors of 2, then the total computational effort is considerably increased. Thus, more powerful stepsize control *and prediction* mechanisms are necessary in this situation.

It seems that a suitable version of an IEM which takes care of these difficulties may become a very strong competitor to any of the general discretization methods for stiff systems presently known.

Bibliography

Babuška, I., Práger, M., Vitásek, E.: (1) Numerical processes in differential equations. London: Interscience 1966.

Brunner, H.: (1) Stabilisierung optimaler Differenzenverfahren zur numerischen Integration gewöhnlicher Differentialgleichungen. Thesis, ETH Zürich 1969.

Bulirsch, R.: (1) Bemerkungen zur Romberg-Integration. Numer. Math. **6,** 6—16 (1964).

Bulirsch, R., Stoer, J.: (1) Über Fehlerabschätzung und Extrapolation mit rationalen Funktionen bei Verfahren vom Richardson-Typus. Numer. Math. **6**, 413—427 (1964).

— — (2) Asymptotic upper and lower bounds for results of extrapolation methods. Numer. Math. **8,** 93—104 (1966).

— — (3) Numerical treatment of ordinary differential equations by extrapolation methods. Numer. Math. **8,** 1—13 (1966).

Butcher, J. C.: (1) On the convergence of numerical solutions to ordinary differential equations. Math. Comp. **20,** 1—10 (1966).

— (2) Coefficients for the study of Runge-Kutta integration processes. J. Austral. Math. Soc. **3,** 185—201 (1963).

— (3) An algebraic theory of integration methods. Math. Comp. **26**, 79—106 (1972).

— (4) A convergence criterion for a class of integration methods. Math. Comp. **26,** 107—117 (1972).

— (5) On Runge-Kutta processes of high order. J. Austral. Math. Soc. **4,** 179—194 (1964).

— (6) Implicit Runge-Kutta processes. Math. Comp. **18,** 50—64 (1964).

— (7) Integration processes based on Radau quadrature formulas. Math. Comp. **18,** 233—244 (1964).

— (8) The effective order of Runge-Kutta methods. Conf. on the numer. sol. of diff. eqns., Lecture Notes in Mathematics No. **109,** 133—139, Springer 1969.

— (9) A modified multistep method for the numerical integration of ordinary differential equations. J. Assoc. Comput. Mach. **12,** 124—135 (1965).

— (10) A multistep generalization of Runge-Kutta methods with 4 or 5 stages. J. Assoc. Comput. Mach. **14,** 84—99 (1967).

Dahlquist, G.: (1) Convergence and stability in the numerical integration of ordinary differential equations. Math. Scand. **4**, 33—53 (1956).

— (2) Stability and error bounds in the numerical integration of ordinary differential equations. Trans. Roy. Inst. Tech. Stockholm No. 130, 1959.

— (3) A special stability problem for linear multistep methods. BIT **3,** 27—43 (1963).

Dahlquist, G., et al.: (1) Survey of stiff ordinary differential equations. Roy. Inst. Tech. Stockholm, Dept. of Inf. Proc. Report NA 70.11, 1971.

Donelson, J., Hansen, E.: (1) Cyclic composite multistep predictor-corrector methods. SIAM J. Numer. Anal. **8**, 137—157 (1971).

Driver, R.D.: (1) A note on a paper of Halanay on stability for finite difference equations. Arch. Rational Mech. Anal. **18**, 241—243 (1965).

Ehle, B.L.: (1) On Padé approximations to the exponential function and *A*-stable methods for the numerical solution of initial value problems. Thesis, Univ. of Waterloo 1969.

England, R.: (1) Error estimates for Runge-Kutta type solutions to systems of ordinary differential equations. Comput. J. **12**, 166—169 (1969).

Fehlberg, E.: (1) Classical fifth, sixth, seventh, and eighth order Runge-Kutta formulas with stepsize control. NASA TR 287, 1968.

— (2) Klassische Runge-Kutta-Formeln fünfter und siebenter Ordnung mit Schrittweiten-Kontrolle. Computing **4**, 93—106 (1969).

— (3) Low order classical Runge-Kutta formulas with stepsize control and their application to some heat transfer problems. NASA TR 315, 1969.

— (4) Zur numerischen Integration von Differentialgleichungen durch Potenzreihenansätze, dargestellt an Hand physikalischer Beispiele. Z. Angew. Math. Mech. **44**, 83—88 (1964).

— (5) New high-order Runge-Kutta formulas with stepsize control for systems of first and second order differential equations. Z. Angew. Math. Mech. **44**, T17—T29 (1964).

— (6) New high-order Runge-Kutta formulas with an arbitrarily small truncation error. Z. Angew. Math. Mech. **46**, 1—16 (1966).

Fox, L.: (1) The numerical solution of two-point boundary value problems in ordinary differential equations. Oxford: University Press 1957.

Gautschi, W.: (1) On inverses of Vandermonde and confluent Vandermonde matrices. Numer. Math. **4**, 117—123 (1962).

— (2) Numerical integration of ordinary differential equations based on trigonometric polynomials. Numer. Math. **3**, 381—397 (1961).

Gear, C. W.: (1) The automatic integration of stiff ordinary differential equations. Information Processing 68, Amsterdam: North Holland 1969, 187—193.

— (2) Hybrid methods for initial value problems in ordinary differential equations. SIAM J. Numer. Anal. **2**, 69—86 (1964).

— (3) The numerical integration of ordinary differential equations. Math. Comp. **21**, 146—156 (1967).

— (4) Simultaneous numerical solution of differential-algebraic equations. IEEE Trans. Circuit Theory, CT-**18**, 85—95 (1971).

— (5) The automatic integration of ordinary differential equations. Comm. Assoc. Comput. Mach. **14**, 176—179 (1971).

— (6) Numerical initial value problems in ordinary differential equations. Englewood Cliffs: Prentice-Hall 1971.

Gragg, W. B.: (1) Repeated extrapolation to the limit in the numerical solution of ordinary differential equations. Thesis, UCLA 1963.

— (2) On extrapolation algorithms for ordinary initial value problems. SIAM J. Numer. Anal. **2**, 384—403 (1965).

Gragg, W. B., Stetter, H. J.: (1) Generalized multistep predictor-corrector methods. J. Assoc. Comput. Mach. **11**, 188—209 (1964).

Groebner, W., Knapp, H., eds.: (1) Contributions to the method of Lie-series. Mannheim/Zürich: BI 1967.

Hahn, W.: (1) Theorie und Anwendung der direkten Methode von Lyapunov, Erg. Ber. No. 22, Berlin-Göttingen-Heidelberg: Springer 1959.

— (2) Stability of motion. Berlin-Heidelberg-New York: Springer 1967.

Henrici, P.: (1) Discrete variable methods in ordinary differential equations. New York-London: J. Wiley & Sons 1962.

— (2) Error propagation for difference methods. New York-London: J. Wiley & Sons 1963.

Hull, T. E., Luxemburg, W. A. J.: (1) Numerical methods and existence theorems for ordinary differential equations. Numer. Math. **2**, 30—41 (1960).

Hull, T. E., et al.: (1) Comparing numerical methods for ordinary differential equations. University of Toronto, Dept. of Comp. Science, TR No. 29, 1971.

Iri, M.: (1) A stabilizing device for unstable numerical solutions of ordinary differential equations—design principle and application of a "filter". Information Processing in Japan **4**, 65—73 (1964).

Knapp, H., Wanner, G.: (1) Numerical solution of ordinary differential equations by Gröbner's method of Lie Series. MRC Tech. Summary Rep. No. 880, 1968.

— (2) Liese, a program for ordinary differential equations using Lie-series. MRC Tech. Summary Rep. No. 881, 1968.

Kohfeld, J. J., Thompson, G. T.: (1) A modification of Nordsieck's method using an "off-step" point. J. Assoc. Comput. Mach. **15**, 390—401 (1968).

Krylov, V. I.: (1) Approximate calculation of integrals. New York: Macmillan 1962.

Lambert, J. D.: (1) Linear multistep methods with mildly varying coefficients. Math. Comp. **24**, 81—94 (1970).

— (2) Predictor-corrector methods with identical regions of stability. SIAM J. Numer. Anal. **8**, 337—344 (1971).

— (3) Computational methods in ordinary differential equations. London-New York-Sydney-Toronto: J. Wiley & Sons 1973.

Lambert, J. D., Shaw, B.: (1) A generalization of multistep methods for ordinary differential equations. Numer. Math. **8**, 250—263 (1966).

Leavitt, J. A.: (1) Methods and applications of power series. Math. Comp. **20**, 46—52 (1966).

Liniger, W.: (1) Zur Stabilität der numerischen Integrationsmethoden für Differentialgleichungen. Thesis, Univ. of Lausanne 1957.

Metzger, C.: (1) Méthodes de Runge Kutta de rang superieur à l'ordre. Thesis, Univ. of Grenoble 1967.

Mischak, R.: (1) Lineare zyklische Multischrittverfahren hoher Ordnung. Thesis, Tech. Univ. Vienna 1972.

Moore, R. E.: (1) Interval Analysis. Englewood Cliffs: Prentice Hall 1966.

Nickel, K., Rieder, P.: (1) Ein neues Runge-Kutta-ähnliches Verfahren. Int. Ser. Numer. Math. **9**, Basel: Birkhäuser 1968, 83—96.

Nørsett, S. P.: (1) A criterion for $A(\alpha)$-stability of linear multistep methods. BIT **9**, 259—263 (1969).

Nordsieck, A.: (1) On numerical integration of ordinary differential equations. Math. Comp. **16**, 22—49 (1962).

— (2) Automatic numerical integration of ordinary differential equations. AMS Proc. Symp. Appl. Math. **15**, 241—250 (1963).

Osborne, M. R.: (1) On Nordsieck's method for the numerical solution of ordinary differential equations. BIT **6**, 52—57 (1966).

Pereyra, V. L.: (1) On improving an approximate solution of a functional equation by deferred corrections. Numer. Math. **8**, 376—391 (1966).

— (2) Iterated deferred corrections for nonlinear operator equations. Numer. Math. **10**, 316—323 (1967).

Richardson, L. F.: (1) The approximate arithmetical solution by finite differences of physical problems involving differential equations. Philos. Trans. Roy. Soc. London Ser. A **210**, 307—357 (1910).

– (2) The deferred approach to the limit. Philos. Trans. Roy. Soc. London Ser. A **226**, 299—361 (1927).

Richtmyer, R. D.: (1) Difference methods for initial-value problems. New York: Interscience 1957 (2nd ed. 1967).

Riha, W.: (1) Optimal stability polynomials. Computing **9,** 37—43 (1972).

Riordan, J.: (1) An introduction to combinatorial analysis. New York: Wiley 1958 (2nd ed. 1964).

Shanks, E. B.: (1) Solution of differential equations by evaluation of functions. Math. Comp. **20**, 21—38 (1966).

Spijker, M. N.: (1) Stability and convergence of finite-difference methods. Thesis, University of Leiden 1968.

Stetter, H. J.: (1) Asymptotic expansions for the error of discretization algorithms for nonlinear functional equations. Numer. Math. **7**, 18—31 (1965).

— (2) Richardson-extrapolation and optimal estimation. Apl. Mat. **13**, 187—190 (1968).

— (3) Stability of nonlinear discretization algorithms. In: Numerical solution of partial differential equations. New York: Academic Press 1966, 111—123.

— (4) Maximum bounds for the solutions of initial value problems for partial difference equations. Numer. Math. **5**, 399—424 (1963).

— (5) Local estimation of the global discretization error. SIAM J. Numer. Anal. **8**, 512—523 (1971).

— (6) Improved absolute stability of predictor-corrector schemes. Computing **3**, 286—296 (1968).

— (7) Symmetric two-step algorithms for ordinary differential equations. Computing **5**, 267—280 (1970).

– (8) Stability properties of the extrapolation method. Conference on the numerical solution of differential equations, Lecture Notes in Math. **109**, 255—260, Springer 1969.

Timlake, W. P.: (1) A stable, k-step method of order greater than $k+2$ for the solution of ordinary differential equations. Unpublished manuscript.

Törnig, W.: (1) Über Differenzenverfahren in Rechtecksgittern zur numerischen Lösung quasilinearer hyperbolischer Differentialgleichungen. Numer. Math. **5**, 353—370 (1963).

Urabe, M.: (1) Theory of errors in numerical integration of ordinary differential equations. MRC Tech. Summary Rep. No. 183, 1960.

Varga, R.: (1) Matrix iterative analysis. Englewood Cliffs: Prentice Hall 1962.

Wachspress, E. L.: (1) Numerical solution of initial value problems. Unpublished Report, 1970.

Wanner, G.: (1) Integration gewöhnlicher Differentialgleichungen. Mannheim-Zürich: BI 1969.

Watt, J. M.: (1) Convergence and stability of discretization methods for functional equations. Comput. J. **11**, 77—82 (1968).

Widlund, O. B.: (1) A note on unconditionally stable linear multistep methods. BIT **7**, 65—70 (1967).

Subject Index*

* The page where the term has been defined is printed in bold face.

Springer Tracts in Natural Philosophy

Vol. 1 Gundersen: Linearized Analysis of One-Dimensional Magnetohydrodynamic Flows. With 10 figures. X, 119 pages. 1964. Cloth DM 22,—; US $ 7.00

Vol. 2 Walter: Differential- und Integral-Ungleichungen und ihre Anwendung bei Abschätzungs- und Eindeutigkeitsproblemen
Mit 18 Abbildungen. XIV, 269 Seiten. 1964. Gebunden DM 59—; US $ 18.70

Vol. 3 Gaier: Kontruktive Methoden der konformen Abbildung
Mit 20 Abbildungen und 28 Tabellen. XIV, 294 Seiten. 1964.
Gebunden DM 68,—; US $ 21.60

Vol. 4 Meinardus: Approximation von Funktionen und ihre numerische Behandlung
Mit 21 Abbildungen. VIII, 180 Seiten. 1964. Gebunden DM 49,—; US $ 15.60

Vol. 5 Coleman, Markovitz, Noll: Viscometric Flows of Non-Newtonian Fluids. Theory and Experiment. With 37 figures. XII, 130 pages. 1966. Cloth DM 22,—; US $ 7.00

Vol. 6 Eckhaus: Studies in Non-Linear Stability Theory
With 12 figures. VIII, 117 pages. 1965. Cloth DM 22,—; US $ 7.00

Vol. 7 Leimanis: The General Problem of the Motion of Coupled Rigid Bodies About a Fixed Point. With 66 figures. XVI, 337 pages. 1965. Cloth DM 48,—; US $ 15.30

Vol. 8 Roseau: Vibrations non linéaires et théorie de la stabilité
Avec 7 figures. XII, 254 pages. 1966. Relié DM 39,—; US $ 12.40

Vol. 9 Brown: Magnetoelastic Interactions
With 13 figures. VIII, 155 pages. 1966. Cloth DM 38,—; US $ 12.10

Vol. 10 Bunge: Foundations of Physics
With 5 figures. XII, 312 pages. 1967. Cloth DM 56,—; US $ 17.80

Vol. 11 Lavrentiev: Some Improperly Posed Problems of Mathematical Physics
With 1 figure. VIII, 72 pages. 1967. Cloth DM 20,—; US $ 6.40

Vol. 12 Kronmüller: Nachwirkung in Ferromagnetika
Mit 92 Abbildungen. XIV, 329 Seiten. 1968. Gebunden DM 62,—; US $ 19.70

Vol. 13 Meinardus: Approximation of Functions: Theory and Numerical Methods
With 21 figures. VIII, 198 pages. 1967. Cloth DM 54,—; US $ 17.20

Vol. 14 Bell: The Physics of Large Deformation of Crystalline Solids
With 166 figures. X, 253 pages. 1968. Cloth DM 48,—; US $ 15.30

Vol. 15 Buchholz: The Confluent Hypergeometric Function with Special Emphasis on its Applications. XVIII, 238 pages. 1969. Cloth DM 64,—; US $ 20.30

Vol. 16 Slepian: Mathematical Foundations of Network Analysis
XI, 195 pages. 1968. Cloth DM 44,—; US $ 14.00

Vol. 17 Gavalas: Nonlinear Differential Equations of Chemically Reacting Systems
With 10 figures. IX, 107 pages. 1968. Cloth DM 34,—; US $ 10.80

Vol. 18 Marti: Introduction to the Theory of Bases
XII, 149 pages. 1969. Cloth DM 32,—; US $ 10.20

Vol. 19 Knops, Payne: Uniqueness Theorems in Linear Elasticity
IX, 130 pages. 1971. Cloth DM 36,—; US $ 11.50

Vol. 20 Edelen, Wilson: Relativity and the Question of Discretization in Astronomy With 34 figures. XII, 186 pages. 1970. Cloth DM 38,—; US $ 12.10

Vol. 21 McBride: Obtaining Generating Functions XIII, 100 pages. 1971. Cloth DM 44,—; US $ 14.00

Vol. 22 Day: The Thermodynamics of Simple Materials with Fading Memory With 8 figures. X, 134 pages. 1972. Cloth DM 44,—; US $ 14.00

Vol. 23 Stetter: Analysis of Discretization Methods for Ordinary Differential Equations. With 12 figures. XVI, 388 pages. 1973. Cloth DM 120,—; US $ 38.10

Prices are subject to change without notice